Chitooligosaccharides

Se-Kwon Kim
Editor

Chitooligosaccharides

Prevention and Control of Diseases

Springer

Editor
Se-Kwon Kim
Department of Marine Science
and Convergence Engineering
College of Science and Technology
Hanyang University ERICA
Seoul, Korea (Republic of)

ISBN 978-3-030-92805-6 ISBN 978-3-030-92806-3 (eBook)
https://doi.org/10.1007/978-3-030-92806-3

This Springer imprint is published by the registered company Springer Nature Switzerland AG
The registered company address is: Gewerbestrasse 11, 6330 Cham, Switzerland

Preface

Natural biomaterials have gained considerable attention in recent years for the development of medical products. For example, chitooligosaccharide (COS) is a natural biopolymer and is extensively studied in medical applications. COS is a chitosan derivative and can be produced using chemical and enzyme hydrolysis techniques. The essential advantages of COSs are their solubility in the neutral condition. This solubility property of COS opens up a way to research various applications, including cosmeceutical and pharmaceutical applications.

This book aims to provide COS synthesis and its biological and biomedical applications. This book is mainly focused on medical product development using COS and its derivatives. In the initial part of the book, the chapters are focused on the chemical and enzymatic preparation of COS from marine crustacean shell wastes. In some of the book chapters, systematic and step-by-step approaches were included for the production of COS. Different enzymes involved in the development of COS production were discussed. In addition, the isolation of chitinolytic enzymes and the development of COS were also discussed. Furthermore, the influence of isolation conditions on the physicochemical COS from crustacean shell wastes was presented. Finally, some of the chapters are focused on the synthesis of COS derivatives for biological applications.

In the middle of the book, chapters are covered on the biological application of COS. These biological applications are antimicrobial, antioxidant, anti-inflammatory, anti-allergy, and anticancer activities. Chapters reveal the current status, and future trends in disease-preventing bioactivities of COS are discussed. In addition to this biological activity, the mechanistic pathways for the various biological applications, including effects of matrix metalloproteinases, inhibitory effects for cancer studies, and immunomodulatory effects of COS and their derivatives, were included in the book. The development of single COS and their roles in various bioactivities are also reviewed.

The final part of the book includes the biomedical applications of COS. Nanoparticles, hydrogels, and biocomposites made up of COS are also discussed for medical applications. In addition, wound healing materials, tissue engineering, drug delivery, and food, and nutraceutical applications of COS are included in the book. The book provides a basic understanding of COS for medical product development.

Seoul, South Korea Prof. Se-Kwon Kim

Acknowledgements

Dr. **Jae-Chul Kim** is the chairman, president, and founder of the Dongwon group. The company was established in 1969 to explore and utilize the oceans and marine resources. Chairman Kim is the pioneer in the deep-sea fishing boats in South Korea. He started tuna fishing in 1958 as the first mate of Korea's first deep-sea fishing vessel.

After graduating from National Fisheries University of Busan, he jumped into tuna fishing, and the university was renamed as Pukyong National University. Further, he became the captain of a fleet of deep-sea fishing boats and a new tuna fishing method. He developed and succeeded in catching large amounts of tuna in the South Pacific and Indian Oceans.

He founded Dongwon Industries in 1969 at the age of 35 and started tuna fishing as the first president. After that, he built a tuna processing plant and started to produce canned food. He immersed himself in business management in earnest based on the experience he learned in the sea from his youth.

Dongwon Group has successfully expanded its business from the fishery industry as the primary industry to the manufacturing industry as the secondary industry and financial services as the tertiary industry. Currently, Mr. Kim runs 30 affiliated companies.

To contribute to social welfare, he also established the Dongwon Educational Foundation. He devoted himself, his heart, and his soul to nurturing competent people who are the backbone of our society. He has provided scholarships to numerous college students and grants R&D funds for researchers. He recognized the great value and potential of underutilized marine resources from his youth. He dedicated himself to publishing technical books emphasizing the scientific importance of marine life and related research.

With his help, this book has been published, providing readers with how the scientific values of marine life can enhance human health and wellbeing.

I want to thank him sincerely for his support in publishing this book.

Seoul, South Korea

Prof. Se-Kwon Kim
Distinguished Professor

Contents

About the Editor

Prof. Se-Kwon Kim Ph.D., is presently working as a Distinguished Professor in Hanyang University and Kolmar Korea Company. He worked as a Distinguished Professor at the Department of Marine Bio Convergence Science and Technology and Director of Marine Bioprocess Research Center (MBPRC) at Pukyong National University, Busan, South Korea.

He received his M.Sc. and Ph.D. degrees from Pukyong National University and conducted his post-doctoral studies at the Laboratory of Biochemical Engineering, University of Illinois, Urbana-Champaign, Illinois, USA. Later, he became a visiting scientist at the Memorial University of Newfoundland and the University of British Colombia in Canada.

Dr. Kim served as president of the 'Korean Society of Chitin and Chitosan' in 1986–1990, and the 'Korean Society of Marine Biotechnology' in 2006–2007. To the credit for his research, he won the best paper award from the American Oil Chemists' Society in 2002. Dr. Kim was also the chairman for '7th Asia-pacific Chitin and Chitosan Symposium', which was held in South Korea in 2006. He was the chief editor in the 'Korean Society of Fisheries and Aquatic Science' during 2008-2009. In addition, he is the board member of International Society of Marine Biotechnology Associations (IMBA) and International Society of Nutraceuticals and Functional Food (ISNFF).

His major research interests are investigation and development of bioactive substances from marine resources. His immense experience of marine bioprocessing and mass-production technologies for marine bio-industry is the key asset of holding majorly funded Marine Bio projects in Korea. Furthermore, he expanded his research fields up to the development of bioactive materials from marine organisms for their applications in oriental medicine, cosmeceuticals, and nutraceuticals. To this date, he has authored around 750 research papers, 70 books, and 120 patents.

Introduction to Chitooligosaccharides

Se-Kwon Kim

1 Introduction

Chitosan is natural crustaceans-derived polysaccharides material having a different kind of molecular weight. Chitosan is not digested in our body, and the backbone structure is a resemblance with cellulose materials. No enzyme exists in the gastrointestinal tract that can degrade chitin and chitosan. Therefore, Greater 22 kDa or more molecular weight chitosan is rarely taken up by the human body (Ngo et al. 2015; Aam et al. 2010). To effectively use chitosan as active material in medical applications, it is necessary to synthesize water-soluble chitosan (Chae et al. 2005). Chitooligosaccharides are water-soluble chitosan due to their lower molecular weight. Chitooligosaccharides have lower viscosity and higher water solubility than chitosan. Chitooligosaccharides have been widely used in many potential biological, biomedical, and industrial applications. Several biological activities of chitooligosaccharides depend on their molecular weight; for example, antitumor activity depends on the molecular weight of chitooligosaccharide (Zhai et al. 2021; Xu et al. 2012; Laokuldilok et al. 2017), antioxidant activity (Laokuldilok et al. 2017; Eom et al. 2012), anti-dementia (Kim and Shin 2020; Lee et al. 2007), immune-strengthening activity (Lin et al. 2012, 2017; Das et al. 2015), anti-allergy (Vo et al. 2011), antidiabetic activity (Ju et al. 2010; Liu et al. 2007), antihypertensive (Ngo et al. 2008b; Je and Kim 2012), antihyperlipidemic (Katiyar et al. 2011) and calcium absorption promoting action (Jung et al. 2006).

Chitosan can be used for therapeutic purposes in several ways, however it is limited due to its poor solubility in neutral pH. Chitosan can be functional and as well as healthy food (Xia et al. 2011). Natural-derived materials are often considered

S.-K. Kim (✉)
Department of Marine Science and Convergence Engineering, College of Science and Technology, Hanyang University ERICA, Gyeonggi-do, Ansan 11558, Korea
e-mail: sknkim@pknu.ac.kr

S.-K. Kim (ed.), *Chitooligosaccharides*,
https://doi.org/10.1007/978-3-030-92806-3_1

biologically safe to use as medicine and functional food materials. Chitooligosaccharide is made from chitosan, a naturally occurring substance, but its safety in vivo needs to be determined (Kim 2010, 2018; Xu et al. 2008, 2010; Shen et al. 2009; Ngo et al. 2008a, b; Qian and Chen 2018; Yang et al. 2019; Vo et al. 2012; Zhao et al. 2017; Qian et al. 2010; Cho et al. 2008; Jung et al. 2006).

Safety and efficacy of the newly developed drugs or pharmaceutical substances tested in animal and clinical trials before entering into therapeutic use in the market. For example, intake of chitooligosaccharides results in specific physiological changes in the body. This reaction infers that chitooligosaccharides potentially provide a favorable response in the body. Taking chitooligosaccharide allows the body to achieve a new state of balance.

2 Biocompatibility of Chitin and Chitosan

Among natural polymers, chitin and chitosan are the most versatile biopolymers. In fisheries processing plants, about 6 to 8 million metric tons of crustacean shells are discarded every year. Despite its low solubility, it is not widely used in the pharmaceutical industry. Due to this, its use as a medical material has been severely restricted. On the other hand, deacetylated chitosan has a wide range of applications. For example, this polymer was widely used as an adsorbent for heavy metals and an enzyme carrier and chromatography resin in the early days of its development (Wang et al. 2020; Bakshi et al. 2020; Kou et al. 2020). Alternatively, chitooligosaccharides are biologically active biomaterials.

3 Safety of Chitosan Oligosaccharide

The safety of chitooligosaccharides are revealed based on the molecular weight size. Kim et al. conducted acute and subacute toxicity tests on chitooligosaccharides using white rats. As part of the compulsory toxicology test, chitosan and chitooligosaccharide were administered orally to each sex. chitosan and its oligosaccharides were administered orally to white rats for one day, and no deaths, changes in body weight, or lesions were observed. Hepatopathologically, no abnormalities were found in the liver or kidneys. In the subacute toxicity test, up to different weights of chitooligosaccharides (max 2000 mg/kg/day) orally to each male and female for 28 days. It was concluded that chitooligosaccharide is a safe substance that does not cause toxicity in vivo because death, weight change, and lesions were not observed. In addition, no abnormal symptoms were found in histopathological observation (Kim et al. 2001).

4 Organization of Book

Chitin and chitosan are the essential structural polysaccharide found in various organisms and are widely distributed in nature. These texts are taken from the different chapters of the present book. Chitooligosaccharides are the partially depolymerized low molecular weight (<4.0 kDa) derivatives of chitin or chitosan. In general, COS has a molecular weight (MW) varied between 3 and 10 kDa with a degree of depolymerization (DP) ranging from 2 to 20. However, both DP and MW fluctuate, depending on sources as well as the method of preparation. These chitooligosaccharides have received much attention due to their high solubility and exceptional biological properties than their parent polymers. Hence, mechanical, chemical, and enzymatic methods have been explored to produce chitooligosaccharides from different chitin and chitosan substrates. Among these methods, the enzymatic processes are attracting increasing interest for chitin or chitosan depolymerization applications owing to their advantages over the other methods. Hence, different specific and non-specific enzymes have been explored for the production of particular chitooligosaccharides. Enzymatic production of chitooligosaccharides may shortly expand into commercial to traditional degradation techniques. Chitosan is also prepared with alkaline deacetylation of chitin in the presence of alkali at a high temperature. To produce oligomers of chitin and chitosan (chitin oligosaccharides, NA-COS; chitooligosaccharides, COSs) and some chemical methods for synthesis their derivatives by grafting functional groups as inorganic or organic to improve biological activities. Some methods synthesize and isolate chitooligosaccharides such as carboxylated chitooligosaccharides, amino derived chitooligosaccharides, sulfated chitooligosaccharides, quaternized chitooligosaccharides, aryl chitooligosaccharides, phenolic acid conjugated chitooligosaccharides.

Chitooligosaccharides (COS) exhibit an enormously wide range of biomedical activities, including antioxidant, anti-inflammatory, antihypertensive, antidiabetic, anti-allergic, antimicrobial, anti-cancer, and anti-Alzheimer effects; one of the most vital of which is the antioxidant activity. Synthesis of COS and MMP inhibitory effects of COS to recommend these natural oligosaccharides a potential MMPIs. Overview of innate and adaptive immunity and immunomodulation effects of chitooligosaccharide is explored.

COS can easily permeate the cancer cells. COS derivatives greatly limit the formation of new blood vessels in tumors, thereby blocking tumor progression and metastasis. It has the ability to decrease free radicals in normal cells, which are implicated in cellular damage.

The nanoparticle-based drug delivery systems such as polymeric nanoparticles have attracted considerable attention as a new medical technology since these systems can alter drug pharmacokinetic profile and control drug release, resulting in a significant therapeutic drug effect. Chitooligosaccharides, a degraded product of chitosan, has been recently used as the main component of polymeric nanoparticles for drug and gene delivery due to its biocompatibility, biodegradability, water solubility, and unique physiochemical properties. In addition, possessing an active amino group in each monomer, chitooligosaccharides are facile for chemical modifications and forming various types of polymeric nano-carriers. The recent designs and developments of chitooligosaccharide-based polymeric nanoparticles for controlled drug release in the treatment of cancer and other diseases are discussed. Gene delivery based on the formation of polyion complex with chitooligosaccharide was also discussed. Chitooligosaccharide has been used in studying cardiac, nerve, and skin tissue engineering applications. Due to the cationic nature of chitooligosaccharide, the utilization of gene and drug delivery are discussed. In addition, chitooligosaccharide can be modified and tuned concerning the applications. The biocompatible, biodegradable, and non-toxic nature of chitooligosaccharide can be potentially explored for further applications. Furthermore, an attempt has been made to report chitooligosaccharide-based materials available for wound healing. Chitooligosaccharide showed positive results in controlling glucose levels, serum lipid levels, or weight gain as confirmed in both in vivo and in vitro studies. Overall, chitooligosaccharide has tremendous potential to act as a nutraceutical, which can provide health-related benefits for the consumers and can improve the quality as well as extend shelf-life of various foods. Chitooligosaccharide-based drug delivery systems (DDS) can absorb more readily, have faster release characteristics, in addition to exhibiting better biodegradability and bioaccumulation potential. These properties are central to the success of chitooligosaccharide in the delivery of a wide variety of drugs, ranging from tumor-targeting drugs to genes.

It is hard to identify which chitooligosaccharide molecules are responsible for the observed biological effects. In this scenario, access to single chitooligosaccharide molecule with accurate structure is essential, which would greatly facilitate to understand the structure–function relationships. Some chapter focuses on the preparation techniques of single chitooligosaccharide. The roles of single chitooligosaccharide in various bioactivities were presented, including antioxidant, anti-inflammatory, antitumor, biostimulant, and elicitor activity.

5 Conclusion

Chitosan is an excellent biomaterial and proved to have several biological and biomedical applications. However, due to its poor solubility, the extended application of chitosan is limited. The derivative of chitosan, chitooligosaccharides, has been studied for several application including in medical field. Recently, chitooligosaccharides are being studied in biomedical applications including tissue engineering, drug delivery, and biosensor applications. To translate into the commercial product of chitooligosaccharides in medical field, it has to pass through several clinical trials which need to be performed. Very few direct studies have been conducted on the safety of chitooligosaccharides at preclinical level. Due to biocompatible, biodegradable, easy to secrete from the body, no side effect of chitooligosaccharides can be potential materials for medial product development.

References

Aam BB, Heggset EB, Norberg AL, Sørlie M, Vårum KM, Eijsink VG (2010) Production of chitooligosaccharides and their potential applications in medicine. Mar Drugs 8:1482–1517
Bakshi PS, Selvakumar D, Kadirvelu K, Kumar N (2020) Chitosan as an environment friendly biomaterial–a review on recent modifications and applications. Int J Biol Macromol 150:1072–1083
Chae SY, Jang M-K, Nah J-W (2005) Influence of molecular weight on oral absorption of water soluble chitosans. J Control Release 102:383–394
Cho E-J, Rahman A, Kim S-W, Baek Y-M, Hwang H-J, Oh J-Y, Hwang H-S, Lee S-H, Yun J-W (2008) Chitosan oligosaccharides inhibit adipogenesis in 3T3-L1 adipocytes. J Microbiol Biotechnol 18:80–87
Das SN, Madhuprakash J, Sarma P, Purushotham P, Suma K, Manjeet K, Rambabu S, Gueddari NEE, Moerschbacher BM, Podile AR (2015) Biotechnological approaches for field applications of chitooligosaccharides (COS) to induce innate immunity in plants. Crit Rev Biotechnol 35:29–43
Eom T-K, Senevirathne M, Kim S-K (2012) Synthesis of phenolic acid conjugated chitooligosaccharides and evaluation of their antioxidant activity. Environ Toxicol Pharmacol 34:519–527
Je J-Y, Kim S-K (2012) Chitooligosaccharides as potential nutraceuticals: production and bioactivities. Adv Food Nutr Res 65:321–336
Ju C, Yue W, Yang Z, Zhang Q, Yang X, Liu Z, Zhang F (2010) Antidiabetic effect and mechanism of chitooligosaccharides. Biol Pharm Bull 33:1511–1516
Jung W-K, Moon S-H, Kim S-K (2006) Effect of chitooligosaccharides on calcium bioavailability and bone strength in ovariectomized rats. Life Sci 78:970–976
Katiyar D, Singh B, Lall AM, Haldar C (2011) Efficacy of chitooligosaccharides for the management of diabetes in alloxan induced mice: a correlative study with antihyperlipidemic and antioxidative activity. Eur J Pharm Sci 44:534–543
Kim S-K (2010) Chitin, chitosan, oligosaccharides and their derivatives: biological activities and applications. CRC Press
Kim S-K (2018) Healthcare using marine organisms. CRC Press
Kim S-K, Park P-J, Yang H-P, Han S-S (2001) Subacute toxicity of chitosan oligosaccharide in Sprague-Dawley rats. Arzneimittelforschung 51:769–774
Kim S-K, Shin K-H (2020) Large scale enzymatic production of chitooligosaccharides and their biological activities. Food Sci Ind 53:2–32
Kou S, Peters L, Mucalo M (2020) Chitosan: a review of sources and preparation methods. Int J Biol Macromol
Laokuldilok T, Potivas T, Kanha N, Surawang S, Seesuriyachan P, Wangtueai S, Phimolsiripol Y, Regenstein JM (2017) Physicochemical, antioxidant, and antimicrobial properties of chitooligosaccharides produced using three different enzyme treatments. Food Biosci 18:28–33
Lee D-H, Lee D-H, Lee J-S (2007) Characterization of a new antidementia β-secretase inhibitory peptide from Saccharomyces cerevisiae. Enzyme Microb Technol 42:83–88
Lin S-M, Jiang Y, Chen Y-J, Luo L, Doolgindachbaporn S, Yuangsoi B (2017) Effects of Astragalus polysaccharides (APS) and chitooligosaccharides (COS) on growth, immune response and disease resistance of juvenile largemouth bass, Micropterus salmoides. Fish Shellfish Immunol 70:40–47
Lin S, Mao S, Guan Y, Lin X, Luo L (2012) Dietary administration of chitooligosaccharides to enhance growth, innate immune response and disease resistance of Trachinotus ovatus. Fish Shellfish Immunol 32:909–913
Liu B, Liu W-S, Han B-Q, Sun Y-Y (2007) Antidiabetic effects of chitooligosaccharides on pancreatic islet cells in streptozotocin-induced diabetic rats. World J Gastroentero 13:725
Ngo D-H, Vo T-S, Ngo D-N, Kang K-H, Je J-Y, Pham HN-D, Byun H-G, Kim S-K (2015) Biological effects of chitosan and its derivatives. Food Hydrocoll 51:200–216
Ngo D-N, Kim M-M, Kim S-K (2008) Chitin oligosaccharides inhibit oxidative stress in live cells. Carbohydr Polym 74:228–234

Ngo D-N, Qian Z-J, Je J-Y, Kim M-M, Kim S-K (2008) Aminoethyl chitooligosaccharides inhibit the activity of angiotensin converting enzyme. Process Biochem 43:119–123
Qian L, Chen L (2018) Immune protective effects of chitooligosaccharides on mice genital tract infected by Chlamydia trachomatis. Am J Reprod Immunol 79:e12815
Qian Z-J, Eom T-K, Ryu B-M, Kim S-K (2010) Angiotensin I-converting enzyme inhibitory activity of sulfated chitooligosaccharides with different molecular weights. J Chitin Chitosan 15:75–79
Shen K-T, Chen M-H, Chan H-Y, Jeng J-H, Wang Y-J (2009) Inhibitory effects of chitooligosaccharides on tumor growth and metastasis. Food Chem Toxicol 47:1864–1871
Vo T-S, Kim J-A, Ngo D-H, Kong C-S, Kim S-K (2012) Protective effect of chitosan oligosaccharides against FcεRI-mediated RBL-2H3 mast cell activation. Process Biochem 47:327–330
Vo T-S, Kong C-S, Kim S-K (2011) Inhibitory effects of chitooligosaccharides on degranulation and cytokine generation in rat basophilic leukemia RBL-2H3 cells. Carbohydr Polym 84:649–655
Wang W, Meng Q, Li Q, Liu J, Zhou M, Jin Z, Zhao K (2020) Chitosan derivatives and their application in biomedicine. Int J Mol Sci 21:487
Xia W, Liu P, Zhang J, Chen J (2011) Biological activities of chitosan and chitooligosaccharides. Food Hydrocoll 25:170–179
Xu Q, Dou J, Wei P, Tan C, Yun X, Wu Y, Bai X, Ma X, Du Y (2008) Chitooligosaccharides induce apoptosis of human hepatocellular carcinoma cells via up-regulation of Bax. Carbohydr Polym 71:509–514
Xu W, Huang H-c, Lin C-j, Jiang Z-F (2010) Chitooligosaccharides protect rat cortical neurons against copper induced damage by attenuating intracellular level of reactive oxygen species. Bioorg Med Chem Lett 20:3084–3088
Xu W, Jiang C, Kong X, Liang Y, Rong M, Liu W (2012) Chitooligosaccharides and N-acetyl-D-glucosamine stimulate peripheral blood mononuclear cell-mediated antitumor immune responses. Mol Med Report 6:385–390
Yang Y, Xing R, Liu S, Qin Y, Li K, Yu H, Li P (2019) Immunostimulatory effects of Chitooligosaccharides on RAW 264.7 mouse macrophages via regulation of the MAPK and PI3K/Akt signaling pathways. Mar Drugs 17:36
Zhai X, Li C, Ren D, Wang J, Ma C, Abd El-Aty A (2021) The impact of chitooligosaccharides and their derivatives on the in vitro and in vivo antitumor activity: a comprehensive review. Carbohydr Polym 118132
Zhao L, Sun T, Wang L (2017) Chitosan oligosaccharide improves the therapeutic efficacy of sitagliptin for the therapy of Chinese elderly patients with type 2 diabetes mellitus. Ther Clin Risk Manag 13:739

Chemical Preparation of Chitooligosaccharides

Shashikant Joshi, Divya Nataraj, and Narendra Reddy

1 Introduction

Chitosan oligosaccharides commonly known as chitooligosaccharides or CHOS are the degraded forms of chitosan or chitin. CHOS are obtained by degradation of chitosan by acid, enzymes or a combination of both and are distinguishable by their low degree of polymerization (DP < 20), low molecular weight (<3900 Da) and >90% degree of deacetylation. Unlike chitin or chitosan, CHOS is soluble in water and is also proven to be biocompatible, non-toxic and non-allergic. CHOS have also shown antimicrobial, anticancer, anti-diabetic properties and also possesses ability to treat Alzheimer's, hypertension, hypocholesterolemic and other diseases including activity against human immunodeficiency virus (HIV). Hence, considerable attempts have been made to obtain CHOS using different approaches and with different properties for biomedical, pharmaceutical, nutraceuticals, cosmetics and many other uses (Naveed et al. 2019; Phil et al 2018). The structure of CHOS and the simple process of deriving CHOS from chitin/chitosan is shown in Fig. 1 along with the presence of reactive functional groups mainly hydroxyl, amino/acetamido (Fig. 2) which enables CHOS to have good chemical and biological activities. The properties and functionality of CHOS obtained are dependent on the preparation methods used.

CHOS with considerably different properties have been obtained when chemical, enzymatic and irradiation approaches were used (Fig. 3). Recent studies using electrospray ionization time-of-flight mass spectrometry have also shown that the formation of CHOS is dependent on the ratio of GlcN and GlcNAc in chitin/chitosan.

S. Joshi (✉)
Swakit Biotech Private Limited, Bangalore, India
e-mail: info@swakitbiotech.com

D. Nataraj · N. Reddy
Center for Incubation, Innovation, Research and Consultancy, Bangalore, India
e-mail: narendra.r@ciirc.jyothyit.ac.in

S.-K. Kim (ed.), *Chitooligosaccharides*,
https://doi.org/10.1007/978-3-030-92806-3_2

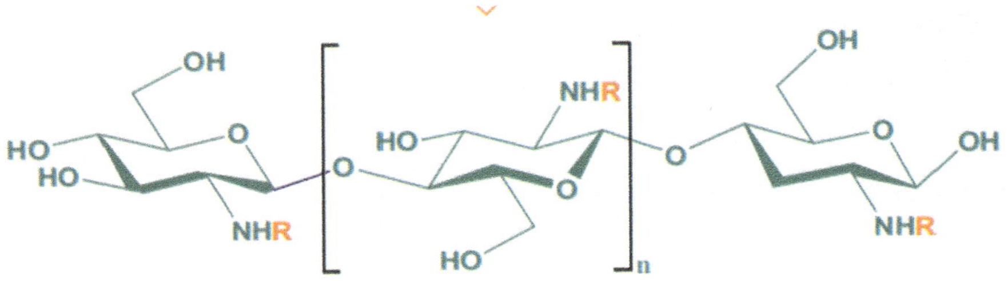

Fig. 1 Chemical structure of CHOS (Naveed et al. 2019). Reproduced with permission from Elsevier

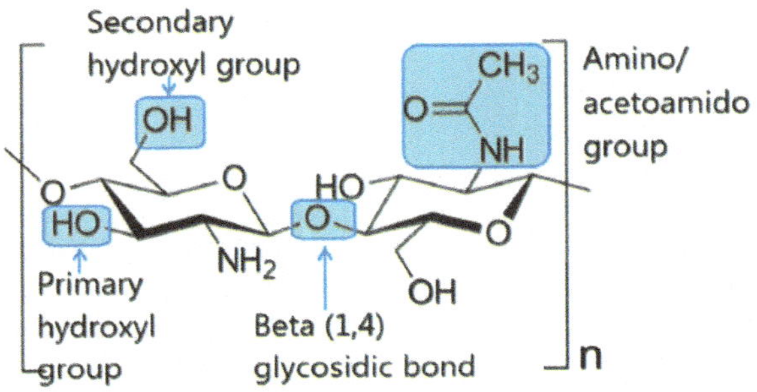

Fig. 2 The reactive groups in CHOS responsible for the chemical and biological activities (Naveed et al. 2019). Reproduced with permission from Elsevier

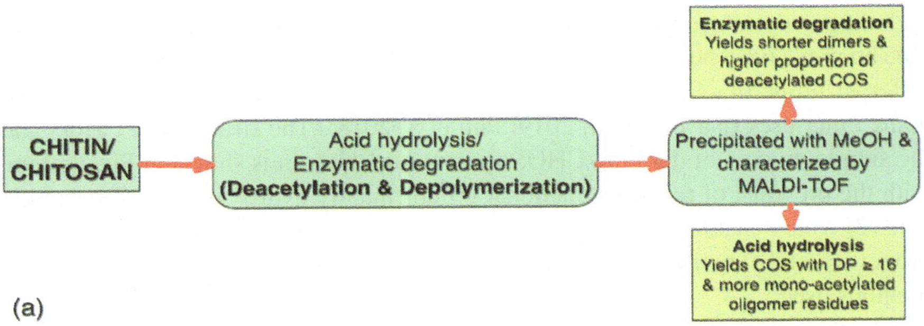

Fig. 3 Process of producing chitin/chitosan (Naveed et al. 2019). Reproduced with permission from Elsevier

Considerably distinct products are obtained depending on the ratio and the treatment protocol followed and properties of the initial polymers used. Degradation by peroxide occurs due to cleavage of glycosidic bond scission whereas nitric acid causes nitrosation and rearrangement of terminal GlCn residues leading to the formation of 2,5 -AM monomers (Allison et al. 2020). In this chapter, we present the various chemical approaches used for the synthesis and the properties of the CHOS obtained.

1.1 *Chemical Preparation by Oxidative Reactions*

1.1.1 Hydrolysis Using Hydrogen Peroxide

Oxidation reaction using strong oxidizing agents, mainly H_2O_2, is a common approach to convert chitosan into CHOS. The mechanism of degradation of chitosan and formation of CHOS by hydrogen peroxide in the presence of acid is shown in Fig. 4 (Purwanto et al. 2019).

In one study, chitosan having molecular weight of 124 kDa and degree of deacetylation of 64.9% was treated with hydrogen peroxide and acetic acid (AA) in different molar ratios at 50 °C for 8 h. CHOS were separated after precipitation and centrifugation. CHOS with MW between 0.5 and 2.1 kDa and yield of 14–36% were obtained depending on the ratio of H_2O_2/AA (Tishchenko et al. 2011). Oxidative depolymerization to obtain CHOS was considered to be more viable and economical compared to microwave irradiation or enzymatic hydrolysis. However, the properties of CHOS obtained were considerably affected by the concentration of peroxide and treatment conditions used. Peroxide reduced degree of polymerization even under ambient conditions and higher concentration of temperatures caused larger decrease. Chemical changes to the structure of chitosan including formation of carboxyl groups and deamination were noticed. Degradation of chitosan was dependent on the molecular weight with lower molecular weights leading to considerably larger decrease in molecular weights with up to 40% loss of amino groups when molecular weight was 1.2×10^3 (Qin et al. 2002). The effects of molecular weight of chitosan on the depolymerization by H_2O_2 were also confirmed in another study. Optimum conditions for depolymerization were determined to be peroxide concentration of 4% at 30 °C for 4 h which provided a yield of 57% and molecular weight of 4.2×10^3 (Tanasale et al. 2019). Treating chitosan with NaOH was found to increase swelling and ability of peroxide to hydrolyze chitosan (Minh et al. 2017). Chitosan with a degree of deacetylation of 84% and molecular weight of 1.1×10^3 kDa was treated

$$R\text{-}NH_2 + H^+ \leftrightarrow R\text{-}NH_3^+$$
$$H_2O_2 \leftrightarrow H^+ + HOO^-$$
$$H_2O_2 + R\text{-}NH_2 + H^+ \leftrightarrow R\text{-}NH_3^+ + HOO^- + H^+$$

The hydroxyl radical (HO*) is easily produced from the decomposition of hydroperoxide anion as it is not stable, and the reactions in equation (11) and (12) take place.

$$HOO^- \rightarrow OH^- + O^*$$
$$H_2O_2 + HOO^- \rightarrow HO^* + O_2^*. + H_2O$$

Eventually, the hydroxyl radical (OH*) attacks the glycosidic linkage in chitosan to produce oligomer chitosan according to the reactions in equation (13) and (14).

$$(GlcN)_m - (GlcN)_n + OH^* \rightarrow (GlcN^*)_m - (GlcN)_n + H_2O$$
$$(GlcN^*)_m - (GlcN)_n + H_2O \rightarrow (GlcN)_m + (GlcN)_n$$

Fig. 4 Reactions involved in the formation of chitosan oligosaccharides due to degradation of chitosan by hydrogen peroxide in the presence of an acid (Purwanto et al. 2019). Reproduced with permission through open access publishing

with NaOH solution for 2–14 h at room temperature. Later, H_2O_2 solution of 0.1 to 0.6% was added and the solution maintained at room temperature for 3–18 h. Products formed were filtered, washed and collected. Up to 660% higher swelling of chitosan was possible after treating with NaOH. Considerable decrease in the viscosity and molecular weight was observed with increase in NaOH concentration or treatment time (Fig. 5). Low molecular weight chitosan (1.4×10^2 kDa) and viscosity of 100 cps could be obtained after treating with 0.3% peroxide solution for 12 h at room temperature (Minh et al. 2017).

Chitin or chitosan was heated in 30% peroxide at 90 °C for different times and samples were collected to determine the types of monomers and CHOS formed. H_2O_2 at low concentrations caused cleavage of 1, 4-β-D glycosidic bonds in chitin/chitosan resulting in the formation of oligosaccharides. It was further noticed that GlcN-GlcN bonds were more susceptible to hydrolysis than GlcNAc-GlcN or GlcNAc-GlcNAc residues leading to different monomers (Allison et al. 2020). The hydrolysis of the glycosidic linkages in GlcNAc4 had rate constants about 50 times higher compared to those in GlcN4 due to the N-acetyl groups catalyzing the reaction and presence of positively charged amino groups (Einbu 2007a, b). Hydrolysis of chitosan into CHOS using H_2O_2 could be enhanced using phosphotungstic acid as a catalyst. Chitosan of 41×10^4 Da and 93.5% deacetylation was dissolved in 1% acetic acid into which 30% H_2O_2 and the catalyst (1 g) was added and heated at 65 °C for 40 min. Dissolved CHOS were precipitated using NaOH and later washed and collected (Xia et al. 2013). A yield of 92.3% with CHOS having DP of 7 was obtained depending on the reaction conditions used. Various CHOS could be obtained from chitosan after treating it with 0.5–5% peroxide at various temperatures and times. Yield of CHOS increased with increasing peroxide. In another study, it was reported that optimum reaction conditions for CHOS production using H_2O_2-phosphotungstic acid as a catalyst was when 30% peroxide, 0.02 g of catalyst and reaction temperature of 70 °C and reaction time of 30 min were used. It was suggested that free radicals formed by the reaction between the catalyst and peroxide caused rupture of glycosidic linkages and formation of carboxyl groups (Huang et al. 2007). This mechanism has been

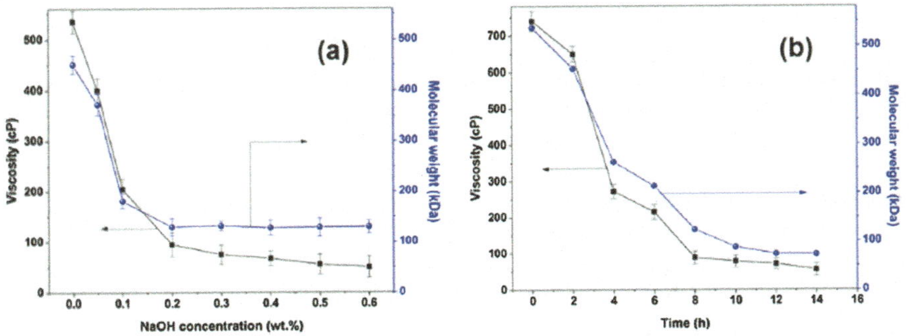

Fig. 5 Changes in the molecular weight and viscosity of chitosan with respect to changes in H_2O_2 + NaOH concentration (Minh et al. 2017). Reproduced with permission from Springer Nature

confirmed using infra red spectroscopy and ^{1}H NMR studies (Tian et al. 2004). It was proposed that depolymerization decreases degree of crystallinity and the action of peroxide varies in the crystalline and amorphous regions. In the crystalline region, depolymerization occurred due to peeling off where as it was in a penetrating pattern in amorphous regions (Fig. 6). As reported in other studies, decrease in degree of deacetylation and/or increase in concentration of peroxide caused higher degree of depolymerization (Tian et al. 2004). Increasing peroxide concentration decreased water insoluble fraction from 97 to 24% whereas increased water soluble chitosan from 13 to 67%. However, increase in peroxide concentration above 1.5 M showed negative influence and decreased the yield and molecular weight.

The influence of molecular weight of chitosan on the structure and properties of CHOS obtained was studied by Tian et al. About 20 g of chitosan was dissolved in 2% acetic acid at 20 °C and later heated with 30 ml of 30% H_2O_2 at 70 °C for 30 min. The solution was neutralized, CHOS were precipitated, washed with ethanol to remove peroxide and vacuum dried to collect the oligomers. Some of the properties of the products obtained after treating chitosan with peroxide at different conditions are given in Table 1. Considerable decrease in molecular weight and crystallinity can be seen with increasing reaction time. Solubility of CHOS was found to be dependent on molecular weights, distribution of acetyl groups along the main chain and also

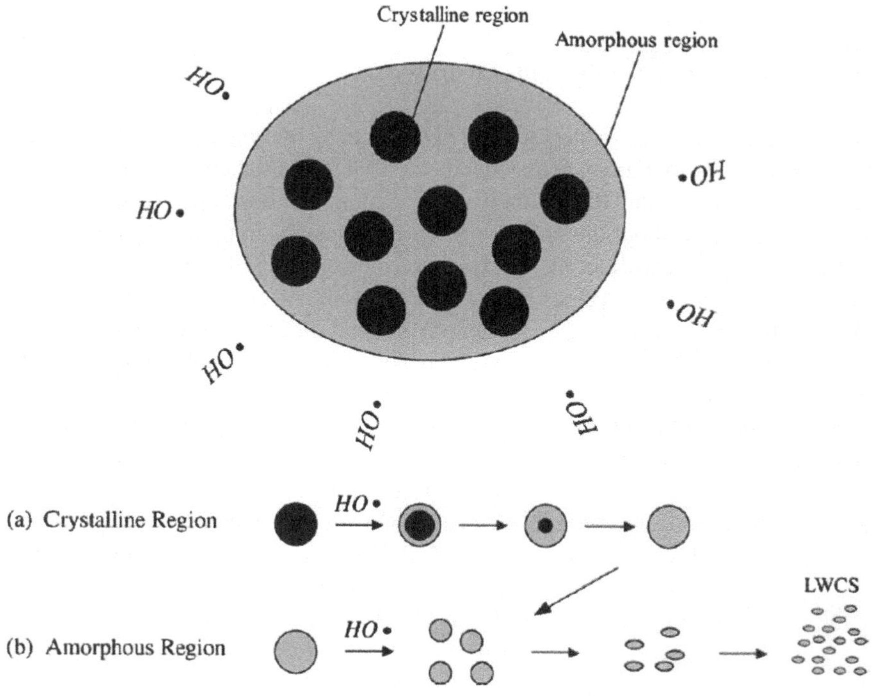

Fig. 6 Schematic representation of the mechanism of depolymerization of chitosan in the crystalline and amorphous regions (Tian et al. 2004). Reproduced with permission from Elsevier

Table 1 Effect of reaction conditions on the properties of hydrolyzed products obtained after treating chitosan with hydrogen peroxide (Tian et al. 2015)

Reaction time, h	Ethanol, %	MW x (10^3)	Polydispersity index (PDI)	Degree of acetylation, %	Crystallinity, %	Water solubility	Yield, %
0.5	–	5.52	1.8	3.2	50.9	No	71.5
1	–	5.18	1.8	3.6	51.5	No	23.3
1	50	4.67	1.5	4.3	49.9	No	25.3
1	75	2.43	1.4	10.1	23.5	Yes	10.5
2	50	3.82	1.4	5.8	35	No	18.4
2	75	2.75	1.5	7.5	11.9	Yes	30
2	87.5	1.72	1.4	15.7	6.2	Yes	8.4

on the degree of acetylation and pH. It was suggested that MW between 3.82 and 4.67×10^3 probably belong to the insoluble-soluble transition region. The pH of the solution determines the degree of charged units and hence strongly affected solubility. Changes in the configurations of the helix structure of chitosan with changing pH and consequently the changes in the amount of solubility are shown in Fig. 7 (Tian et al. 2015).

1.1.2 Catalyst Assisted Peroxide Hydrolysis

A new pathway of oxidative degradation of chitosan by hydrogen peroxide under the action of catalysts containing active metal oxide MO-(O_2) sites was proposed by Ma et al. (2013). It was found that oxidative degradation of chitosan was possible by the MO (O_2) species without the need for peroxide. Changes in the degradation ratios of chitosan with and without the catalysts are shown in Fig. 8. It was hypothesized that the active MO (O_2) species reacts with chitosan and peroxide and recreates a new degradation cycle leading to efficient hydrolysis. The catalyst could be recycled 4 times without any loss of reactivity and hence considered suitable for generating CHOS. This mechanism was further confirmed and it was shown that the "peroxo" species was more efficient in degrading chitosan than hydrogen peroxide. The W(O_2) sites could cleave the glycosidic linkages resulting in low molecular weights and dissolution in water. A new catalyst $(TBA)_3\{PO_4[WO(O_2)_2]_4\}$ was proposed to be highly efficient for degradation of chitosan of up to 96% even after recycling 7 times (Ma et al. 2014). The peroxo species $K_2[W_2O_3(O_2)_4]$ was found to be more efficient than hydrogen peroxide in hydrolyzing chitosan. Degradation rates were dependent on the conditions used for the hydrolysis (Fig. 9).

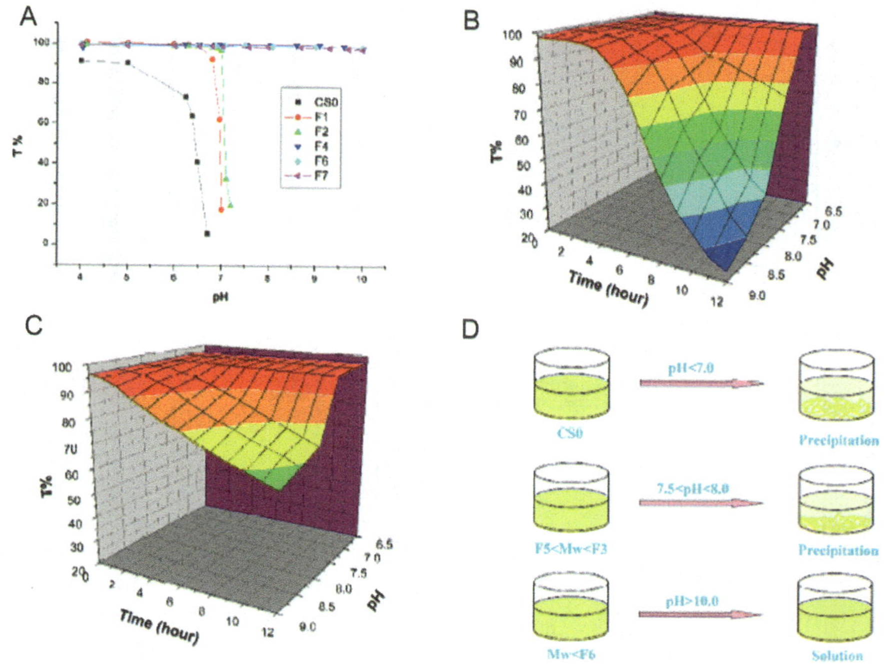

Fig. 7 Influence of pH on the solubility of chitosan treated at different conditions (**A**); pH and time effect on solubility (**B**) and (**C**) and representation of the solubility and MW relationship (**D**) (Tian et al. 2015). Reproduced with permission from Elsevier

Fig. 8 Changes in the degradation ratio of chitosan with two different Mo-(O2) generating catalysts in comparison to H_2O_2 (Ma et al. 2014). Reproduced with permission through open access publication

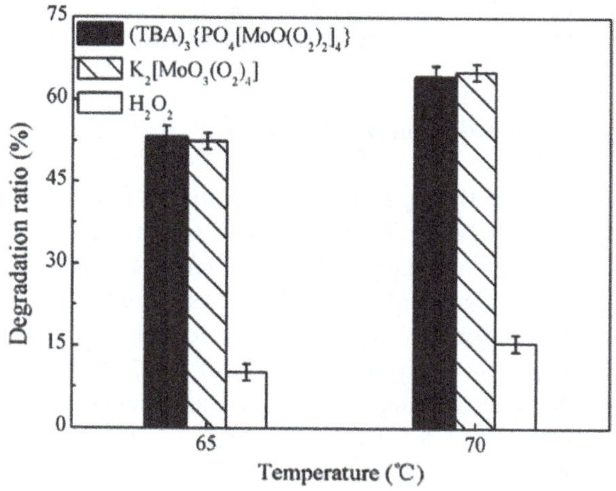

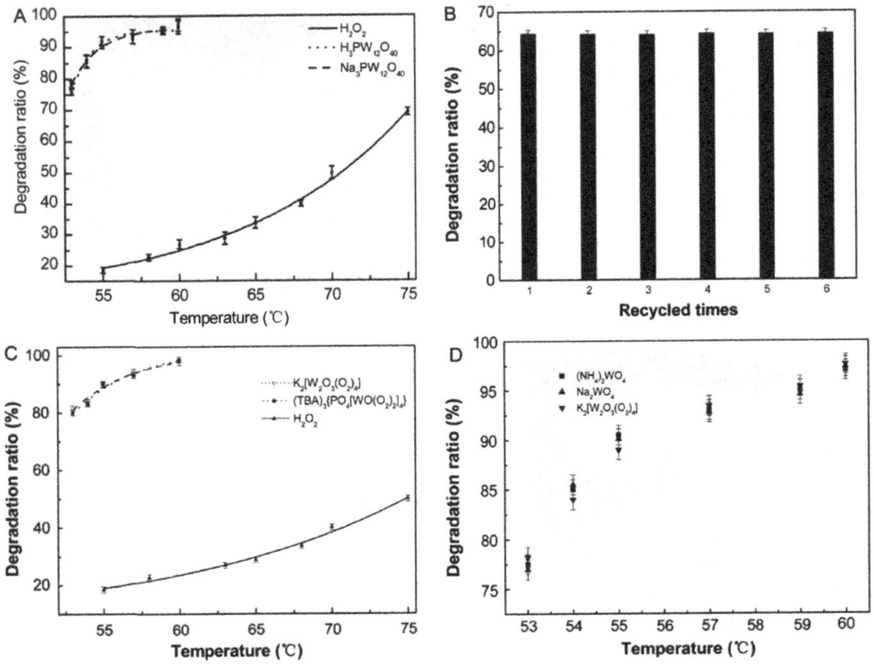

Fig. 9 The oxidative degradation of chitosan, (**A**) $H_3PW_{12}O_{40}$ (0.35 μmol), $Na_3PW_{12}O_{40}$ (0.35 μmol) and H_2O_2 without any additives. (**C**) $(TBA)_3\{PO_4[WO(O_2)_2]_4\}$ (1.05 μmol), $K_2[W_2O_3(O_2)_4]$ (2.1 μmol) and H2O2 without any additives. (**D**) Na_2WO_4 (4.2 mmol), $(NH_4)_2WO_4$ (4.2 mmol) and $K_2[W_2O_3(O_2)_4]$ (2.1 μmol). Under the same reaction conditions: H_2O_2 (1 mL), H_2O_2 (5 mL), 20 min. (**B**) Recycling of the filtrate at 65 °C, 1 mL H_2O_2. (Ma et al. 2014). Reproduced with permission through open access publishing

1.1.3 Irradiation Assisted Hydrolysis Using Hydrogen Peroxide

Combining hydrogen peroxide treatment with Y-irradiation could provide CHOS with Mw between 5000 and 10,000 which was considered feasible for commercial scale production (Duy et al. 2011; Hien et al. 2012). Chitosan with 70% deacetylation and molecular weight of $90.5X10^3$was treated with 1% H_2O_2 and gamma irradiated at 4–16 kGy. Irradiation promoted conversion of chitosan into CHOS without affecting the degree of acetylation (Hien et al. 2012). Not only gamma but even microwave irradiation was found to have substantial benefits in converting chitosan into CHOS. The process of degradation time was decreased along with molecular weights when microwave irradiation was used. Lower molecular weight CHOS obtained through microwave irradiation also had better antimicrobial properties (Sun et al. 2007). In addition to γ or microwave, even UV irradiation was reported to assist in depolymerization of chitosan and production of CHOS using H_2O_2. Chitosan purified with acetic acid was dissolved in water to which peroxide was added. The solution was subject to UV irradiation at 60 °C and pH 5 for 1 to 20 min. UV lamp used produced a UV light in the range of 254 nm and intensity of 0.256 J/cm^2/min. Combining UV irradiation with peroxide substantially decreased the MW of chitosan due to reduction in crystallinity and crystal structure of the hydrolysates. 1, 4-β-d-glucoside linkages of chitosan were degraded without affecting the overall structure (Najafabadi et al. 2018).

1.1.4 Hydrolysis by Combination of Hydrogen Peroxide & Formic Acid

In a similar study, it was found that oxidative degradation of chitosan can be done using formic acid but with the presence of hydrogen peroxide as an oxidizing agent. Molecular weight, reaction time and temperature were found to influence the degradation reaction. Degradation was faster in the initial 15 min of reaction with an activation energy of 99.8 kJ/mol. The β-glycosidic linkages were cleaved resulting in formation of lower molecular weight chitosan but the structure of chitosan was retained (Purwanto et al. 2019). CHOS could also be obtained at low temperatures from high molecular weight chitosan using a combination of peroxide and ascorbic acid complex. CHOS with DP between 10 and 40 could be obtained with optimum ratio of 1:1 of ascorbic acid and peroxide. Degradation of chitosan was found to be due to generation of free radicals and was dependent on temperature with an activation energy of 5 kJ (Fedoseeva et al. 2006).

2 Acid Hydrolysis of Chitosan

2.1 Degradation by Hydrochloric Acid

Degradation of chitosan to CHOS has been extensively and preferably done using various acids specifically hydrochloric acid or nitrous acid (Table 2) (Jeon et al. 2000). In a simple approach, chitin/chitosan was treated with concentrated (36.5%) HCl at 50 or 70 °C and the hydrolysates obtained were precipitated, filtered and collected (Novikov 2004). Cabrera et al. dissolved chitosan in acetic acid and later treated with concentrated HCl at 72 °C for 30 min. Residue obtained was dissolved in water and later precipitated (pH 6.5) using 10 M NaOH. In addition to water, CHOS was also precipitated using different concentrations of methanol (Cabrera and Cutsem 2005). It was reported that CHOS having DP up to 11 was soluble in 70 and 80% methanol whereas 90% methanol was able to dissolve CHOS with DP of up to 8 and 17% of the initial weight of chitosan could be extracted. In the CHOS obtained, oligomers with DP of up to 16 were found. Optimum conditions for degradation of chitosan and generation of oligosaccharides were found to be after treating with 35% hydrochloric acid for 2 h at 80 °C and precipitating with or without dilution. It was reported that the concentration of HCl did not affect the hydrolysis but time of hydrolysis caused substantial changes in yield and the composition of the oligomers (Table 3). A yield

Table 2 Yield (ppm) of N-Acetylglucosamine, Glucosamine and various CHOS obtained from shrimp chitosan (Chang et al. 2001) after treating for various times. (GlcN)2 to (GlcN)9 indicate chitobioase to chitonanose. Reproduced with permission from American Chemical Society

Product	60 °C	70 °C	80 °C	90 °C	100 °C
GlcNAc	0	17	67	135	176
GlcN	17	59	214	444	585
(GlcN)2	31	110	363	603	811
(GlcN)3	45	123	384	482	605
(GlcN)4	30	81	281	212	227
(GlcN)5	52	98	391	262	155
(GlcN)6	89	169	220	67	36
(GlcN)7	95	64	298	97	0
(GlcN)8	0	113	53	0	0
(GlcN)9	0	105	0	0	0
Total	360	938	2271	2301	2595

Table 3 Various types of acids and reaction conditions used for hydrolysis of chitosan to obtain various oligosaccharides (Lee et al. 1999)

Type of acid	Reaction conditions	DP
Hydrochloric acid	35%, 53 °C, 48 h	2–5
Nitrous acid	11%, 100 °C, 34 h	2–6
	35%, 80 °C, 2 h	2–6
	35%, 53 °C, 13 h	2–6
	1%, 100 °C, 5 h	15 and 30
	35%, 72 °C, 1 h	2–15 and 20–40
	0.4 mM, RT, 15 h	9–17
Phosphoric acid	70 mM, 0 °C, 9 h	14 and 37
	60%, 70 °C, 4 h	8
	85%, RT, 4 weeks	7 and 17
Hydrogen fluoride	100%, 20 °C, 19 h	5–10

of 70% could be obtained when methanol was used for coagulation and higher MW oligomers could be obtained with pentamers and hexamers being the major oligomers obtained (Lee et al. 1999). In one of the earliest studies on acid hydrolysis of chitosan, chitosan was treated with HCl of various concentrations at 100 °C. After treatment, the hydrolyzed solution was diluted with cold water and pH adjusted to 10 for the precipitate to form. The precipitates were washed with water followed by methanol, acetone and ether (Rogozhin et al. 1988). CHOS with molecular weights ranging from 2.5 to 17.7 were obtained. It was suggested that acid hydrolysis occurs due to protonation of the glycoside oxygen and subsequent decomposition of the oxonium ion. Electrostatic repulsion of hydroxonium ions due to protonation of the ring oxygen was also considered to lead to hydrolysis. The hydrolysis of chitosan by

HCl is reported to take place in three distinct steps that include depolymerization, production of N-acetyl-glucosamine and conversion of the N-acetyl-glucosamine into glucosamine and acetic acid. Many variations occur due to the changes in the chemical concentrations and reaction conditions and the properties of chitosan. For instance, even during acid hydrolysis, the glycosidic linkage at the non-reducing end is hydrolyzed more rapidly than the other glycosidic linkage in the chain. Similarly, the rate of hydrolysis of the glycosidic linkages was ten times higher than the rate of N-deacetylation when high concentration (12 M) of HCl are used (Mourya et al. 2011).

In another study, it has been reported that the activation energy for the hydrolysis of the N-acetyl linkage of GlCNAc was between 102 and 116 kJ/mol when the HCl concentration was between 3 and 12 M. Comparatively, the activation energy for the hydrolysis of the glycosidic linkage of GlcNAc-GlcNAc was between 110 and 112 kJ/mol at 3 to 12 M HCl (Einbu 2007). However, the activation energies for de-N-acetylation and depolymerization were similar at all acid concentrations. It was therefore suggested that temperature and not acid concentration should be used to control the formation of various oligomers when chitosan is hydrolyzed using HCl (Einbu 2007). In another study, CHOS was prepared by treating chitosan with 6 mol/L of HCl at 70 °C for 2 h. Hydrolyzate obtained was separated and precipitated using ethanol at 4 °C and later lyophilizing to form a powder. Further separation of CHOS was done by gel filtration using ion-exchange column (CM-Sephadex C-25). Chitooligomers with DP between 2 and 12 were obtained. Five high purity fractions were separated to obtain three high purity components glucosamine trimer, pentamer and hexamer having single DP were also obtained (Li et al. 2012).

Conventional acid hydrolysis of chitosan uses high concentrations (12 M) of HCl and a series of purification steps to generate the oligosaccharides. In a different approach, the potential of converting chitosan into CHOS using dilute HCl and obtaining higher yields was examined. In this study, chitosan was treated with 1.8 M HCl and stirred under reflux at 100 °C for 2 h as shown in Fig. 10. The treated chitosan was centrifuged and the solution was passed through zeolite absorbents having different porosities and pore sizes. The CHOS formed were completely absorbed by the zeolite and later collected by sonication in demineralized (DM) water (Ibrahim et al. 2016). CHOS with highest DP of 9 and molecular weights between 179 and 1468 were obtained with yields considerably higher than those previously reported. CHOS with considerably higher DP could be obtained when hydrochloride salt was added into 12 M HCl solution and heated at 72 °C. The solution was evaporated and later re-dissolved in water and neutralized to pH3. Separation of various oligomers was done using an Amicon ultrafiltration cell where smaller oligomers were separated by washing with sodium salt. Pure oligomers with DP up to 15 and fractions with DP up to 37 could be obtained using this approach (Domard and Cartier 1989).

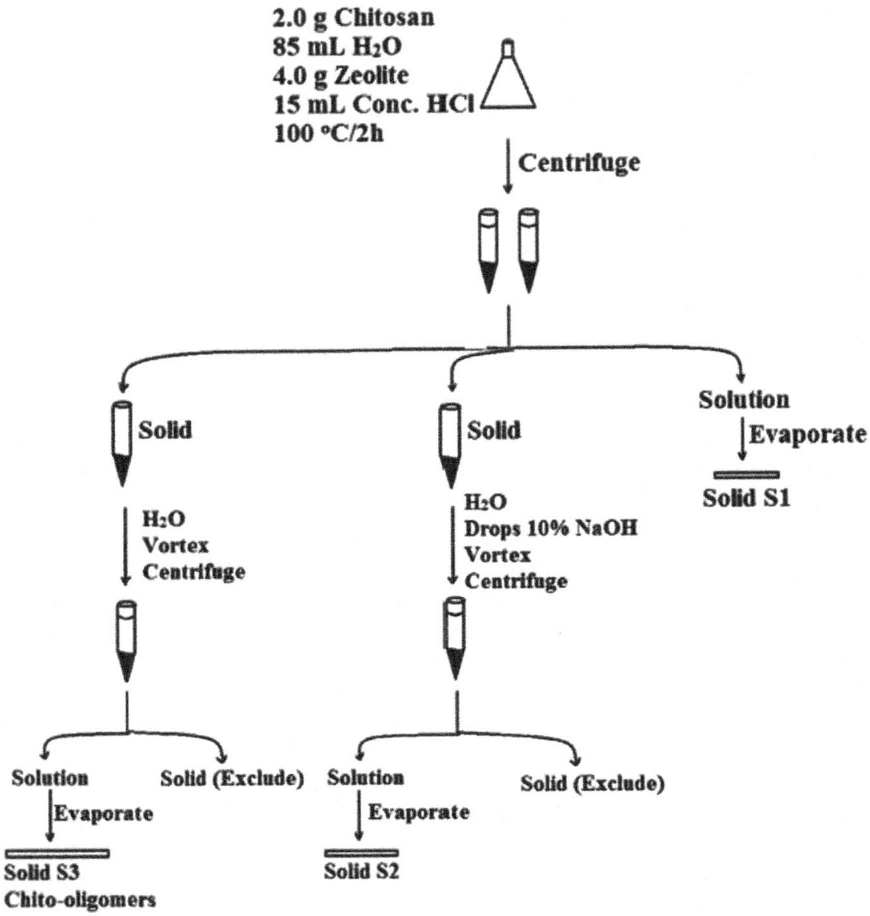

Fig.10 Process used for the production of CHOS using HCl hydrolysis and zeolite as the absorbent (Ibrahim et al. 2016). Reproduced with permission from MDPI through open access publication

2.2 *Degradation by Acetic Acid*

Instead of using HCl, it has been demonstrated that treating chitosan with 5% acetic acid at 40 °C for 30 min results in water soluble CHOS. Solubilized components had a molecular weight of 166 kDa, viscosity of 0.85cP and deacetylation degree of 93% (Santoso et al. 2020). Molecular weight of the CHOS was dependent on the temperature and time of treatment (Fig. 11). CHOS having DP between 2 and 12 but with varying degrees of N-acetylation (0–90%) were developed using a two-step approach. In the first step, various glucosamine (GlcN) oligomers were obtained by acid hydrolysis and later, the oligomers obtained were subject to partial acetylation using hydro-alcoholic solution of acetic anhydride. CHOS with DPs between 3 and

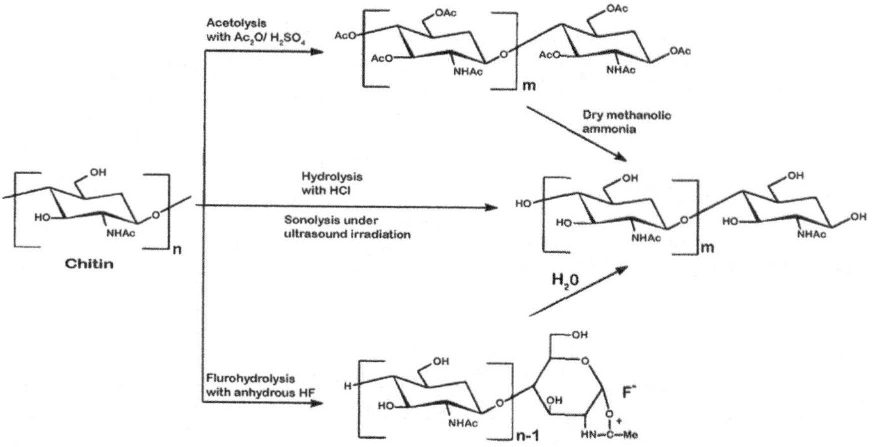

Fig. 11 Effect of temperature and time on the molecular weight of CHOS obtained after treating with 5% acetic acid (Santoso et al. 2020). Reproduced with permission through open access publishing

7 were obtained and the DP was directly related to the degree of N-acetylation (Trombotto et al. 2008).

Low concentration (0.2 and 1% V/V) of acetic acid was able to hydrolyze and produce CHOS but with the assistance of sonication (Savitri et al. 2014). Chitosan was dissolved in acetic acid and sonication was done at 500 W, 20 kHz and 50% amplitude at 40 or 60 °C for 30 min. After sonication, the hydrolyzed substances were precipitated and the dissolved substances were separated using high pressure liquid chromatography (HPLC) and characterized using LC/Q-TOF MS. Molecular weight of the hydrolysates obtained was dependent on the time, temperature, concentration of acetic acid and sonication (Table 4). Yield of soluble (32–50%) and insoluble (46–59%) chitosan and % crystallinity also varied and morphology

Table 4 Influence of time on the type and amount of CHOS/oligomers obtained by hydrolysis of chitosan with concentrated HCl (Lee et al. 1999). Reproduced with permission from Elsevier

Reaction time, h	Yield, %						
	(GlcN)	$(GlcN)_2$	$(GlcN)_3$	$(GlcN)_4$	$(GlcN)_5$	$(GlcN)_6$	$(GlcN)_7$
1	12.8	13.9	12.4	2.6	2.5	2.3	–
2	20.8	16.6	13	3.2	2.5	1.4	–
4	24.3	15.3	11.3	3	2.3	–	–
2	7	11.9	19.3	13	2.1	0.5	–
2	0.1	0.04	0.03	0.05	3.3	3.5	0.3
13	9.5	12.1	15.3	9.5	9.1	5.9	–
24	11.6	28.8	28.1	8.4	8	4.7	–
38	26.5	12.1	12.5	8.5	3.9	3.4	–

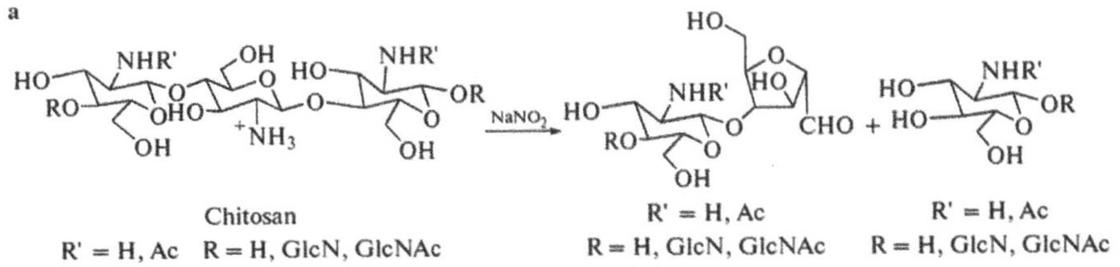

Fig. 12 Mechanism of hydrolysis of chitosan into CHOS by nitrous acid (Mourya et al. 2011). Reproduced with permission from Springer Nature

of the hydrolyzed chitosan was also different. Low concentration of acetic acid produced byproducts including 5-HMF whereas relatively higher concentrations (1%) produced glucosamine as the main product (Savitri et al. 2014).

2.3 *Degradation by Phosphoric Acid*

Phosphoric acid has also been effective to hydrolyze chitosan and produce CHOS as shown in the mechanism in Fig. 12 (Mourya et al. 2011). Chitosan with 75% deacetylation was treated with 85% phosphoric acid at room temperature for 1–2 days to form a brown solution which was maintained at room temperature for a further 1–6 weeks. Precipitates were removed and water soluble fraction (CHOS) were separated by extraction with water for 2 days. CHOS were treated with 1% trimethylamine and dried and collected (Hasegawa et al. 1993). CHOS with degree of polymerization between 5.4 and 7.3 were obtained. Effects of reaction temperature and reaction time on the formation of chitooligomers when hydrolyzed using phosphoric acid were determined by Jia and Shen (2002). In this study, chitosan was treated with 85% phosphoric acid after treating at particular time and temperature, ethanol was added to precipitate the chitosan. Free phosphoric acid and salts of amine groups were removed by treating with 1% trimethylamine in ethanol. The changes in the molecular weights of chitosan with hydrolysis time are given in Table 5. Chitosan with molecular weights between 20×10^4 and 1.9×10^4 could be obtained by varying the reaction time. Water solubility increased with decreasing molecular weight and the oligomers obtained were considered to be suitable as bioactive compounds.

2.4 *Degradation by Nitrous Acid*

In another possibility to extract CHOS, a diminutive depolymerization approach was used to obtain chitosan oligomers using nitrous acid and hydrochloric acid. Conditions used for hydrolysis and the properties of oligomers obtained are given in Table 6.

Table 5 Effect of reaction condition on the molecular weights (Mn and Mw) and polydispersity index (PDI) of chitosan obtained (Savitri et al. 2014)

Reaction conditions				Mn	Mw	PDI
Temp, °C	Time, min	Acetic acid, %	Sonication			
60	30	0.8	Yes	59,312	137,896	2.32
60	30	0.5	Yes	67,607	221,789	3.28
60	30	0.3	Yes	75,373	241,845	3.21
60	30	0.2	Yes	80,353	302,544	3.77
40	30	1	Yes	67,313	222,409	3.3
40	30	0.6	Yes	67,558	255,030	3.77
40	30	0.5	Yes	68,072	251,904	3.7
40	30	0.3	Yes	70,485	237,503	3.37
60	30	1	No	79,580	345,529	4.34
60	30	0.6	No	83,864	360,203	4.3
60	30	0.2	No	102,790	420,968	4.1
40	30	0.8	No	86,486	345,529	3.99
40	30	0.6	No	90,667	365,893	4.04
40	30	0.5	No	92,634	376,930	4.07
40	30	0.2	No	95,553	407,962	4.27

Table 6 Changes in the properties of chitosan after treating with 85% phosphoric acid at 60 °C at different times (Jia and shen 2002)

Reaction time, h	[η] (ml/g)	Ms ($\times 10^{-4}$)	Yield, %
1	154	16.4	89.3
2	137	15.5	87.9
3	132	15	83.6
4	127	7.4	79.2
5	120	6.9	77.2
6	109	5.1	68.4
7	107	5	71.2
8	66.4	3.5	71.4
9	59.6	2.4	66.7
12	52.6	2.1	58.2
15	49.3	1.9	49.2

The oligomers obtained showed excellent resistance against fungi and were considered suitable for protecting wood (Hussain et al. 2012). Effect of chitosans with two considerably different degrees of deacetylation (0.59 and 0.01) on the hydrolysis into CHOS using nitrous acid was studied (Tømmeraas 2001). For conversion into oligomers, the chitosan was dissolved in acetic acid and cooled to 4 °C. A 1.5 mmol solution of nitrous acid ($NaNO_2$) was added into the solution and reaction was carried out at 4 °C for 12 h under dark. Products obtained were centrifuged at 5000 rpm for 10 min and filtered to remove insoluble fractions. The soluble fractions were lyophilized to obtain the CHOS. A relatively lower concentration (1.13 mmol of $NaNO_2$) was used to convert the N-deacetylated chitosan into oligomers. Fully N-acetylated oligomers were stable whereas the N-deacetylated oligomers reacted intermolecularly between 2-amino groups and the M-units leading to cleavage of glycosidic bonds and the formation of 5-hydroxymethylfurfural as shown in Fig. 13 (Tømmeraas et al. 2001) (Table 7).

2.5 *Depolymerization Using Potassium Per-Sulfate*

In a different approach, the possibility of obtaining CHOS using potassium persulfate (KPS) was explored (Hsu et al. 2002). Chitosan dissolved in 2% acetic acid at 70 °C to which about 1 g of KPS was added. Addition of KPS immediately initiated the degradation by the creation of free radicals. KPS initially dissociates into anionic radicals which are attracted to the cationic amino groups in the chitosan ring. The presence of free radicals at C-4 carbon leads to be cleavage of the glycosidic C–O–C bond and subsequent hydrolysis. Molecular weight of chitosan decreased from 615,900 to 17,400 depending on the concentration of KPS and degradation time used. Similarly, intrinsic viscosity decreased from 1150 to 55.6 ml/g when the degradation time was 60 min (Hsu et al. 2002).

3 Conclusions

Chemical methods of CHOS preparation in lab scale or industrial scale manufacturing can be chosen from wide range of alternative processes to a desired polymeric length [DP]. One can produce CHOS with a fairly precise DP with small tolerance for specific applications. On the other hand, for industrial scale production one has to consider weight loss in the process of CHOS production & economics. Reproducibility needs to be ascertained pertaining to the quality of chitin/chitosan. Excessive variation in DP of CHOS is a challenge for processing into products, despite major progress in research & production of CHOS. The outcome of CHOS conversion processes can be controlled by controlling the starting chitosan and the extent to which the degradation reaction is allowed to occur.

Techniques for further purification of CHOS as well as for sequence determination are now available, but are still quite challenging to exploit. Scaling up purification methods at an economically acceptable cost is another challenge, meaning that, from an economical point of view, it is probably cheaper to produce CHOS mixtures that are enriched for a bioactivity, rather than producing truly pure compounds. It is conceivable that the further development of CHOS as bioactive molecules may include further functionalization by chemical methods, for example by coupling

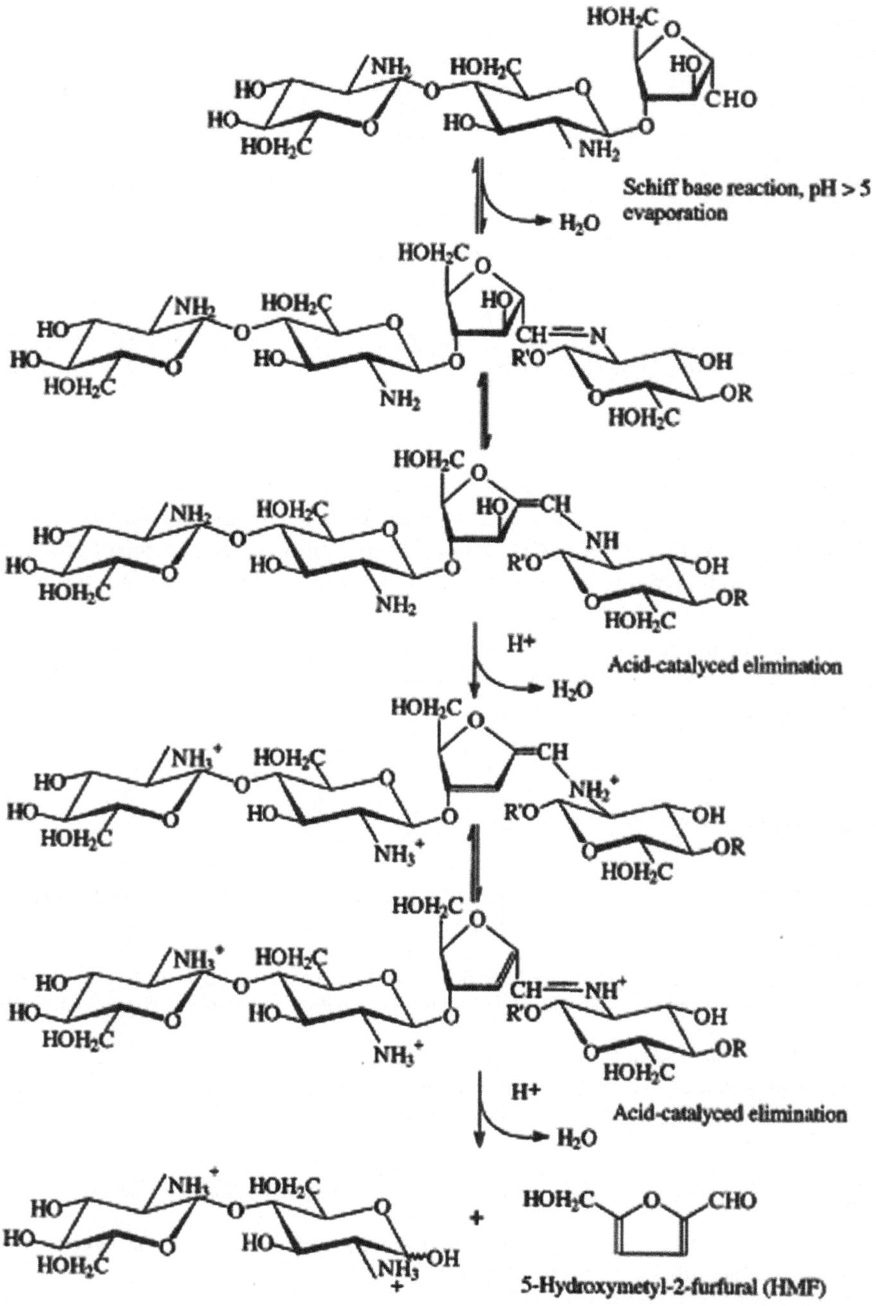

Fig. 13 Schematic representation of the possible mechanism for hydrolysis of chitosan by nitrous acid, chain cleavage and formation of 5-hydroxymethylfurfural (Tømmeraas et al. 2001). Reproduced with permission from Elsevier

Table 7 Treatment conditions, oligomers formed and their properties (Hussain et al. 2012)

Treatment conditions	Hydrolyzed product	Properties		
		Yield, g	Mn	DP
6.9 g of nitrous acid in 20 ml water added to 1 l of 0.1 M HCl chitosan I at 50 °C	Oligomer 1	23.7	2527	14
6.9 g of nitrous acid in 20 ml water added to 1 l of 0.1 M HCl chitosan II at 50 °C	Oligomer 2	19.2	2780	16
5.17 g of nitrous acid in 20 ml water added to 250 ml of 0.1 M HCl chitosan I at 50 °C	Oligomer 3	6	742	4
5.17 g of nitrous acid in 20 ml water added to 250 m3 of 0.1 M HCl chitosan II at 50 °C	Oligomer 4	6.6	724	4

groups to the reducing end. These improved methods for producing (almost) pure, well-characterized CHOS will help to create a knowledge base for understanding how CHOS exerts bioactivities. CHOS have a remarkably wide spectrum of possible bioactivities. While highly promising, there is no doubt that these bioactivities need to be substantiated by further studies with well-defined CHOS preparations, as well as by fundamental research on the molecular mechanism behind the activity. Only then the great promise of converting an abundant bio-resource, chitin to CHOS for various applications will be feasible.

References:

Allison CL, Lutzke A, Reynolds MM (2020) Identification of low molecular weight degradation products from chitin and chitosan by electrospray ionization time-of-flight mass spectrometry. Carb Res 493:108046–108059

Cabrera JC, Van Cutsem P (2005) Preparation of chitooligosaccharides with degree of polymerization higher than 6 by acid or enzymatic degradation of chitosan. Biochem Engg J 25(2):165–172

Chang KL, Tai MC, Cheng FH (2001) Kinetics and products of the degradation of chitosan by hydrogen peroxide. J Agric Food Chem 49(10):4845–4851

Domard A, Cartier N (1989) Glucosamine oligomers: 1. Preparation and characterization. Intl J Biol Macromol 11(5):297–302

Duy NN, Van Phu D, Anh NT, Hien NQ (2011) Synergistic degradation to prepare oligochitosan by γ-irradiation of chitosan solution in the presence of hydrogen peroxide. Rad Phy Chem 80(7):848–853

Einbu A, Vårum KM (2007a) Depolymerization and de-N-acetylation of chitin oligomers in hydrochloric acid. Biomacromol 8(1):309–314

Einbu A, Grasdalen H, Vårum KM (2007b) Kinetics of hydrolysis of chitin/chitosan oligomers in concentrated hydrochloric acid. Carb Res 342(8):1055–1062

Fedoseeva EN, Smirnova LA, Sorokina MA, Pastukhov MO (2006) Radical degradation of chitosan under the action of a redox system. Russian J Appl Chem 79(5):845–849

Hasegawa M, Isogai A, Onabe F (1993) Preparation of low-molecular-weight chitosan using phosphoric acid. Carb Polym 20(4):279–283
Hien NQ, Van Phu D, Duy NN, Lan NT (2012) Degradation of chitosan in solution by gamma irradiation in the presence of hydrogen peroxide. Carb Polym 87(1):935–938
Hsu SC, Don TM, Chiu WY (2002) Free radical degradation of chitosan with potassium persulfate. Polym Deg Stab 75(1):73–83
Huang QZ, Wang SM, Huang JF, Zhuo LH, Guo YC (2007) Study on the heterogeneous degradation of chitosan with hydrogen peroxide under the catalysis of phosphotungstic acid. Carb Polym 68(4):761–765
Hussain I, Singh T, Chittenden C (2012) Preparation of chitosan oligomers and characterization: their antifungal activities and decay resistance. Holzforschung 66(1):119–125
Ibrahim KA, El-Eswed BI, Abu-Sbeih KA, Arafat TA, Al Omari MM, Darras FH, Badwan AA (2016) Preparation of chito-oligomers by hydrolysis of chitosan in the presence of zeolite as adsorbent. Mar Drugs 14(8):43–51
Jeon YJ, Shahidi F, Kim SK (2000) Preparation of chitin and chitosan oligomers and their applications in physiological functional foods. Food Rev Intl 16(2):159–176
Jia Z, Shen D (2002) Effect of reaction temperature and reaction time on the preparation of low-molecular-weight chitosan using phosphoric acid. Carb Polym 49(4):393–406
Lee MY, Var F, Shin-ya Y, Kajiuchi T, Yang JW (1999) Optimum conditions for the precipitation of chitosan oligomers with DP 5–7 in concentrated hydrochloric acid at low temperature. Process Biochem 34(5):493–500
Li K, Xing R, Liu S, Li R, Qin Y, Meng X, Li P (2012) Separation of chito-oligomers with several degrees of polymerization and study of their antioxidant activity. Carb Polym 88(3):896–903
Ma Z, Wu Y, He Y, Wu T (2013) A novel protocol for the oxidative degradation of chitosan with hydrogen peroxide catalyzed by peroxomolybdate in aqueous solution. Rsc Adv 3(30):12049–12051
Ma Z, Wang W, Wu Y, He Y, Wu T (2014) Oxidative degradation of chitosan to the low molecular water-soluble chitosan over peroxotungstate as chemical scissors. PLoS ONE 9(6):e100743–e100751
Minh NC, Cuong HN, Phuong PT, Schwarz S, Stevens WF, Van Hoa N, Trung TS (2017) Swelling-assisted reduction of chitosan molecular weight in the solid state using hydrogen peroxide. Polym Bul 74(8):3077–3087
Mourya VK, Inamdar NN, Choudhari YM (2011) Chitooligosaccharides: synthesis, characterization and applications. Polym Sci A 53(7):583–612
Najafabadi SA, Honarkar H, Moghadam M, Mirkhani V, Tahriri M, Tayebi L (2018) UV irradiation-H_2O_2 system as an effective combined depolymerization technique to produce oligosaccharides from chitosan. Bio-Design Manuf 1(1):62–68
Naveed M, Phil L, Sohail M, Hasnat M, Baig MM, Ihsan AU, Shumzaid M, Kakar MU, Khan TM, Akabar MD, Hussain MI (2019) Chitosan oligosaccharide (COS): an overview. Intl J Biol Macromol 129:827–843
Novikov VY (2004) Acid hydrolysis of chitin and chitosan. Russian J Appl Chem 77(3):484–487
Phil L, Naveed M, Mohammad IS, Bo L, Bin D (2018) Chitooligosaccharide: an evaluation of physicochemical and biological properties with the proposition for determination of thermal degradation products. Biomed Pharmacotherapy 102:438–451
Purwanto E, Connor J, Ngothai Y (2019) The kinetics oxidative degradation of chitosan in formic acid with the presence of hydrogen peroxide. IOP Conf Ser Mater Sci Eng 703(1):012041–012049
Qin CQ, Du YM, Xiao L (2002) Effect of hydrogen peroxide treatment on the molecular weight and structure of chitosan. Polym Deg Stab 76(2):211–218
Rogozhin SV, Gamzazade AI, Chlenov MA, Leonova YY, Sklyar AM, Dotdayev SK (1988) The partial acidic hydrolysis of chitosan. Polymer Science USSR 30(3):607–614
Santoso J, Adiputra KC, Soerdirga LC, Tarman K (2020) Effect of acetic acid hydrolysis on the characteristics of water soluble chitosan. IOP Conf Ser Earth Environ Sci 414(1):012021–012033

Savitri E, Juliastuti SR, Handaratri A, Roesyadi A (2014) Degradation of chitosan by sonication in very-low-concentration acetic acid. Polym Deg Stab 110:344–352
Sun T, Zhou D, Xie J, Mao F (2007) Preparation of chitosan oligomers and their antioxidant activity. Eur Food Res Technol 225(3):451–456
Tanasale MF, Bijang CM, Rumpakwara E (2019) Preparation of chitosan with various molecular weight and its effect on depolymerization of chitosan with hydrogen peroxide using conventional technique. Intl J Chem Tech Res 12(1):112–120
Tian F, Liu Y, Hu K, Zhao B (2004) Study of the depolymerization behavior of chitosan by hydrogen peroxide. Carb Polym 57(1):31–37
Tian M, Tan H, Li H, You C (2015) Molecular weight dependence of structure and properties of chitosan oligomers. RSC Adv 5(85):69445–69452
Tishchenko G, Šimůnek J, Brus J, Netopilík M, Pekárek M, Walterová Z, Koppová I, Lenfeld J (2011) Low-molecular-weight chitosans: preparation and characterization. Carb Polym 86(2):1077–1081
Tømmeraas K, Vårum KM, Christensen BE, Smidsrød O (2001) Preparation and characterisation of oligosaccharides produced by nitrous acid depolymerisation of chitosans. Carb Res 333(2):137–144
Trombotto S, Ladavière C, Delolme F, Domard A (2008) Chemical preparation and structural characterization of a homogeneous series of chitin/chitosan oligomers. Biomacromolecules 9(7):1731–1738
Xia Z, Wu S, Chen J (2013) Preparation of water soluble chitosan by hydrolysis using hydrogen peroxide. Intl J Biol Macromol 59:242–245

Enzymatic Production of Different Types of Chitooligosaccharides

P. V. Suresh

1 Introduction

Enzymes are biocatalysts that play a significant role in biochemical and metabolic activities in all living cells. Enzymes from different source has been widely investigated and increasingly explored in the commercial production of various commodities such as food, chemicals, pharmaceuticals, etc. (Chandrasekharan 2015; Liu and Kokare 2017; Gurpilhares et al. 2019). Over the time, enzymes and enzymatic processing have developed as a crucial and central part of the technologies used by different industries to manufacture a wide range of products for numerous applications due to their stereospecific catalytic nature and requisite of mild catalytic conditions (Jung and Park 2014; Suresh 2019; Trincone 2019). Further benefits of enzymes and green enzymatic processing are their low energy requirements, high efficiency, safety, low cost and sustainable nature compared to chemical catalysts (Chandrasekharan 2015; Suresh et al. 2015; Fernandes 2016; Venugopal 2021). Enzymatic digestion/depolymerization of organic compounds, particularly proteins and carbohydrates, provides novel products with promising applications in various fields, including foods, nutrition, agriculture, biomedicine, pharmaceutical, chemical and other industries. In addition, enzymatic biotransformation of organic molecules can offer novel compounds with enhanced properties for specific applications in different industries.

Exploration of polysaccharides, particularly chitin and chitosan from renewable biomass, is receiving much interest in recent years due to their unique functional and biological properties. These amino polysaccharides and their hydrolytic derivatives are playing a progressively vital role in the advancement of nutrition, functional

P. V. Suresh (✉)
Department of Meat and Marine Sciences, CSIR-Central Food Technological Research Institute, Mysore, India
e-mail: drsureshpv@cftri.res.in

Academy of Scientific and Innovative Research, Ghaziabad, India

S.-K. Kim (ed.), *Chitooligosaccharides*,
https://doi.org/10.1007/978-3-030-92806-3_3

foods, cosmetics, medicines, pharmaceuticals, agriculture, etc. owing to their diverse biofunctional capacities, biocompatibility, non-toxicity, biodegradability, etc. and these promising uses have vast commercial and societal benefits (Kim 2010; Zargar et al. 2015; Ahsan et al. 2018; Shang et al. 2018; Sun et al. 2020; Joseph et al. 2021). Specific, commercially viable and simple enzymatic approaches using specific glycoside hydrolases and other enzymes are evolving as green chemistry/technologies for chitin and chitosan processing (Suresh et al. 2015; Suresh 2019; Sun et al. 2020).

Chitin is an amino polysaccharide that extensively occurs in Nature. It is a linear homopolymer of β-(1 → 4)-linked *N*-Acetyl-D-glucosamine (*N*-acetyl-2-amino-2-deoxy-D-glucose, GlcNAc) units (Figs. 1 and 2). Chitin is distributed ubiquitously in Nature as a structural polymer in the exoskeletons of insects and crustaceans and the cell walls of most fungi and some algae (Synowiecki and Al-Khateeb 2003; Feisal and Montarop 2010; Muzzarelli et al. 2012; Suresh 2019). It is the second most abundant and renewable biopolymer/polysaccharide on Earth, next to cellulose (Kurita 2006; Muzzarelli et al. 2012). Despite the widespread occurrence of chitin, currently, the principal sources of commercial chitin constitute marine crustaceans (e.g. shrimp, crab, lobster, etc.) processing discards/by-products (Rinaudo 2006; Harish and Tharanathan 2007; Aranaz et al. 2009; Nidheesh and Suresh 2015a; Younes and Rinaudo 2015; Suresh et al. 2018). Production of chitin from fungal biomass was reported (Kasongo et al. 2020). Chitin is a semi-crystalline/amorphous polysaccharide. Chitin can occur in three different semi-crystalline allomorphic forms in Nature: (1) α form, consists of antiparallel (↑↓↑) GlcNAc chains, (2) β form, contains parallel (↑↑↑) chains of GlcNAc and (3) γ form consists a set of two parallel chains alternating with a single antiparallel (↑↑↓) chain. Generally, all forms of chitin are comparatively resistant to depolymerization. Among the three forms of chitin, α and β forms are the most widely distributed and abundant in Nature (Rinaudo 2006).

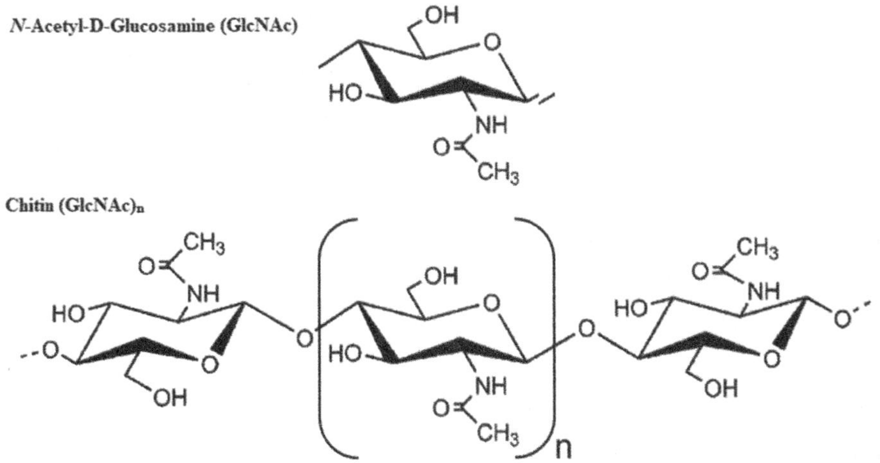

Fig. 1 Structure of *N*-Acetyl-D-glucosamine and chitin

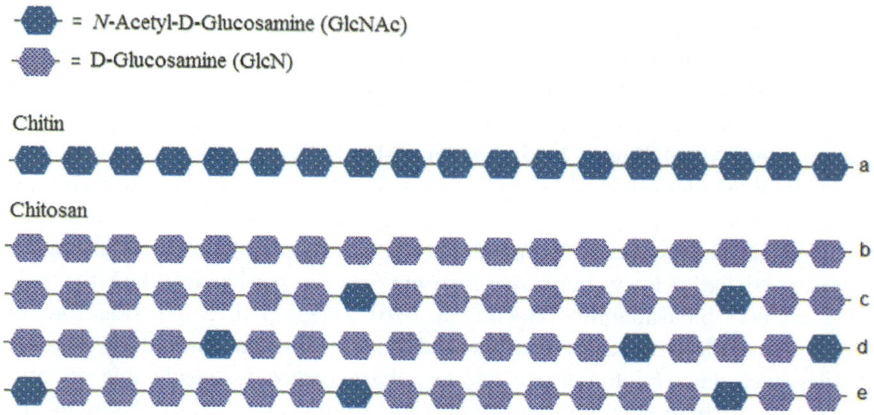

Fig. 2 Schematic presentation of chitin (**a**), chitosan with full D-glucosamine residues (**b**), partially acetylated chitosan (**c**), partially acetylated chitosan with *N*-Acetyl-D-glucosamine residue at reducing end (**d**) and partially acetylated chitosan with *N*-Acetyl-D-glucosamine residue at non-reducing end (**e**)

Chitosan is also a natural amino polysaccharide. But in Nature, its distribution is restricted to few organisms. It is a copolymer of β-(1 → 4)-GlcNAc and β-(1 → 4)-D-glucosamine (2-amino-D-glucose, GlcN) (Figs. 2 and 3). Usually, chitosan represents a collective name for a family of heterogeneous compounds composed of GlcNAc and GlcN units, with the latter must exceed 70% (Rinaudo 2006; Harish and Tharanathan 2007). Further, the GlcNAc and GlcN units are arbitrarily dispersed in chitosan and do not stand jointly in the polymer molecules (Thadathil and Velappan 2014). Hence, chitosan is generally considered as the derivative of chitin. In contrast to chitin, native chitosan occurs in a meagre quantity in nature. It is mainly found as a structural element in the cell wall of Zygomycetes fungi and also appears in smaller amounts in the cell wall of green algae (*Chlorella* sp.) and some insect exoskeleton in Nature (Muzzarelli et al. 2012; Thadathil and Velappan 2014). However, chitosan can be prepared by de-*N*-acetylation of chitin by chemical or enzymatic methods.

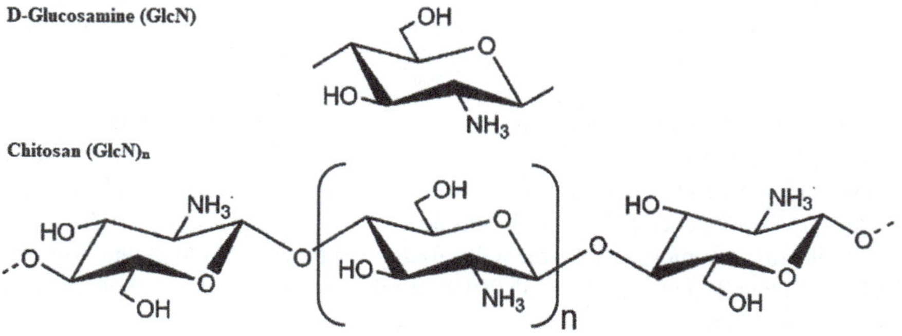

Fig. 3 Structure of D-glucosamine and chitosan

Therefore, most factories produce chitosan, despite chitin from marine crustaceans processing discards due to its extensive commercial and biological importance higher than chitin (Rinaudo 2006; Harish and Tharanathan 2007; Aranaz et al. 2009; Suresh et al. 2018, Morin-Crini et al. 2019a, b). Production of chitosan from fungal biomass was reported (Abdel-Gawad 2017; Ahmad et al. 2020).

All the above-described allomorphic forms of chitin are insoluble in most common solvents, while chitosan moderately dissolves in aqueous acid solutions. Albeit chitin and chitosan possess many biofunctional characteristics, their potential application is restricted in many fields due to some reasons, particularly their low solubility and poor bioavailability (Hoell et al. 2010; Zargar et al. 2015). Therefore, the depolymerized/hydrolyzed derivatives of chitin and chitosan, particularly oligosaccharides/oligomers (chitooligosaccharides), have received considerable attention owing to their high solubility and remarkable biofunctional properties than their parent polymer (Harish and Tharanathan 2007; Shahidi and Abuzaytoun 2005; Kim and Rajapakse 2005; Rinaudo 2006; Jung and Park 2014; Naidheesh et al. 2015a, b; Nidheesh and Suresh 2015a, b; Shang et al. 2018; Suresh 2019; Kaczmarek et al. 2019; Sun et al. 2020). Source of different chitin and chitosan substrate and chitooligosaccharides is illustrated in Fig. 4.

Chitooligosaccharides are the partially depolymerized derivatives of chitin $[(GlcNAc)_{2-20}]$ or chitosan $[(GlcN)_{2-20}]$. The depolymerization of chitin or chitosan and production of chitooligosaccharides can be achieved by mechanical, chemical and enzymatic (biocatalytic) methods or a combination of them. The mechanical methods for chitin or chitosan hydrolysis include ultrasonication (Baxter et al. 2005), hydrothermal treatment, hydrodynamic cavitation (Wu et al. 2014) and electromagnetic, gamma rays or microwave irradiation (Zainol et al. 2009). However, these mechanical methods have been scarcely investigated and are not well-suited for producing well-defined chitooligosaccharides (Kaczmarek et al. 2019; Noa Miguez et al. 2021). Chemical procedures for chitooligosaccharides preparation comprise degradative methods, such as chitin or chitosan depolymerization by strong acid hydrolysis or oxidation, and chemical synthesis strategies (Noa Miguez et al. 2021). Chemical methods for the synthesis of chitooligosaccharides have progressed substantially in the last decades. Still, they usually result in low yields. They may require protection/deprotection steps, the use of hazardous chemicals, the formation of hazardous waste, high production cost, formation of a large amount of monomers, etc. (Je and Kim 2012; Suresh and Kumar 2012; Thadathil et al. 2014; Kaczmarek et al. 2019).

As compared to mechanical and chemical methods, enzymatic production of chitooligosaccharides from chitin or chitosan substrate is very attractive since it requires milder reaction conditions (relatively low temperatures and slightly acidic pH) and does not generate harmful by-products or wastes (Sinha et al. 2016), which makes it safer and more environmentally friendly approach (Sheldon and Woodley 2018; Kaczmarek et al. 2019; Suresh 2019). In addition, strategies based on biocatalysis present a higher efficiency and allow easier control of the product composition due to inherent enzyme stereospecificity (Kim and Rajapakse 2005; Ilankovan et al. 2006; Wee et al. 2009; Jung and Park 2014; Naidheesh et al. 2015a, b). However, the

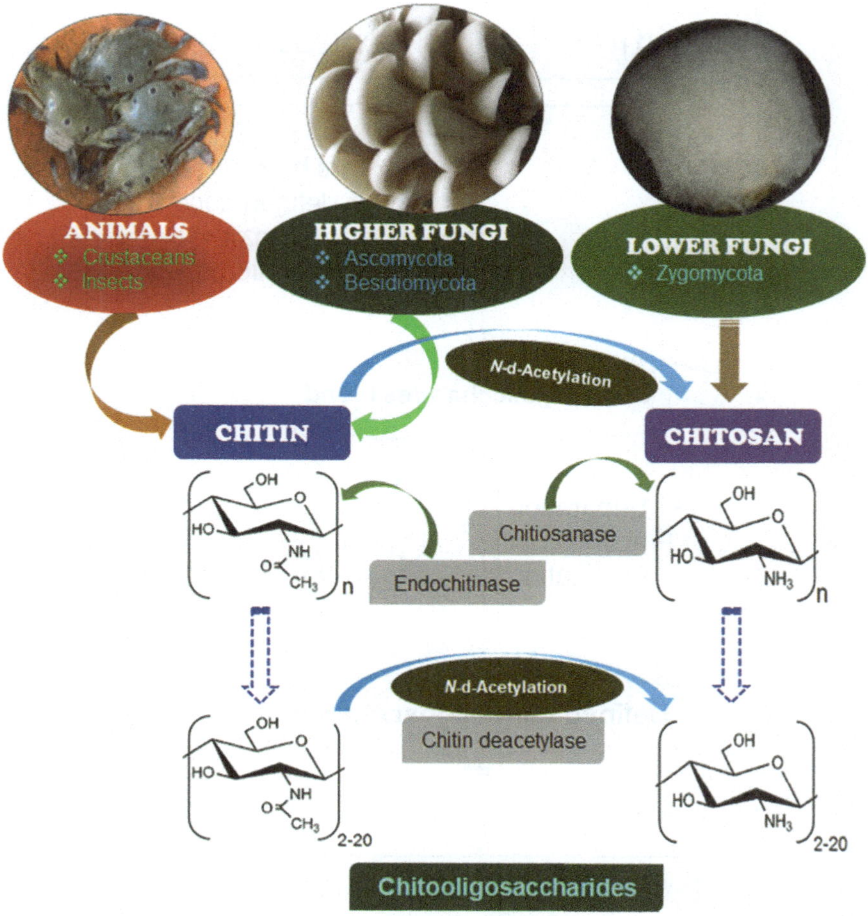

Fig. 4 Source of chitin, chitosan and chitooligosaccharides

physicochemical properties of the chitin or chitosan substrate substantially affect the chemical composition of the final products (Noa Miguez et al. 2021). In principle, it should be possible to develop enzymes and enzymatic processes that enable the production of partially depolymerized derivatives (chitooligosaccharides) of chitin or chitosan with consistent quality and biofunctional properties (Hoell et al. 2010). Though, synthesis of desired chitooligosaccharides in terms of the degree of polymerization (DP), degree of de-*N*-acetylation (DD), the pattern of acetylation (PA) by enzymatic depolymerization approaches is not straightforward (Jung and Park 2014).

Source of different substrates and enzymatic production of chitooligosaccharides are illustrated schematically in Figs. 4 and 5.

The water-soluble low molecular mass chitooligosaccharides have high commercial value due to their versatile biofunctional capacity and wide applications in various

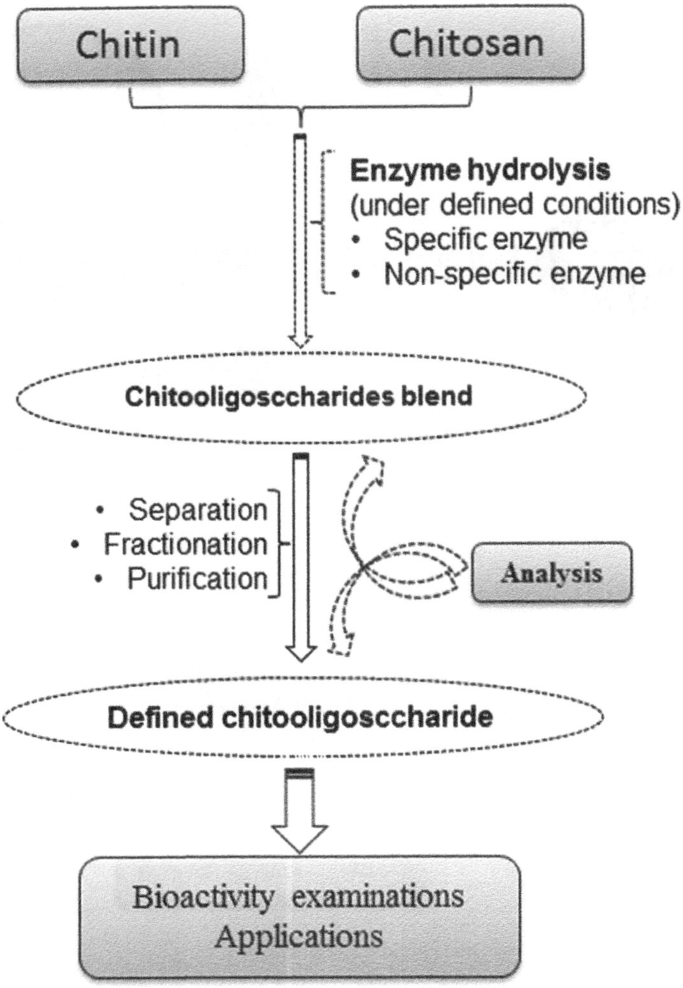

Fig. 5 Schematic presentation of enzymatic production defined chitooligosaccharides

fields (Muzzarelli et al. 2012; Thadathil and Velappan 2014; Sun et al. 2020; Noa Miguez et al. 2021). Consequently, in the recent decade, there has been an increasing demand for the production of potential chitooligosaccharides for specific applications (Wee et al. 2009; Xia et al. 2011; Nguyen et al. 2014; Pechsrichuang et al. 2013; Thadathil and Velappan 2014; Naidheesh et al. 2015a; Nidheesh and Suresh 2015b). Along this line of discussion, few reviews have been published that report different aspects of chitinase, chitosanase and enzymatic synthesis of chitooligosaccharides and applications of chitooligosaccharides (Jung and Park 2014; Thadathil and Velappan 2014; Suresh 2019; Sun et al. 2020; Noa Miguez et al. 2021). This chapter is focused mainly on the various technological approaches for the enzymatic depolymerization of chitin and chitosan and the prospective production of different types of chitooligosaccharides. The different bioactivities and various applications of chitin, chitosan chitooligosaccharides are not subjects of discussion in this chapter.

2 Enzymes Acting on Chitin and Chitosan

Enzymatic hydrolysis of chitin and chitosan can be catalyzed by several glycoside hydrolases (GHs) (Jung and Park 2014; Suresh 2019), consisting of specific enzymes, like chitinases (Oyeleye and Normi 2018; Singh et al. 2021) and chitosanases (Thadathil and Velappan 2014; Li et al. 2015; Suresh 2019). In addition, non-specific enzymes such as other glycosidases (e.g. cellulases, glycanases, pectinases, muramidase, etc.), proteases (e.g. pepsin) and lipases have also been applied in chitin and chitosan depolymerization (Kittur et al. 2005; Xia et al. 2008; Montilla et al. 2013). A large number of factors can influence the outcome of an enzymatic depolymerization reaction of any polysaccharides. In chitooligosaccharides production from chitin or chitosan, both the choice of enzyme and the selection of substrate apply an extensive control on the outcome (Noa Miguez et al. 2021). Various specific and non-specific enzymes involved in chitin or chitosan depolymerizations for chitooligosaccharides formation are briefly discussed below.

2.1 *Specific Enzymes in the Formation of Chitooligosaccharides*

2.1.1 Chitinases

Chitinases are a group of enzymes, can catalyze the hydrolysis of β-(1 → 4) glycosidic bond of GlcNAc-GlcNAc linkages of chitin polymer (Stoykov et al. 2015; Oyeleye and Normi 2018). According to the recommendations of the Nomenclature Committee of the International Union of Biochemistry and Molecular Biology (IUBMB) in 1992, the chitinases are grouped into two classes (Class 1 and Class 2) based on the substrate specificity and mode of action (Reetarani et al. 2000; Beygmoradi et al. 2018) (https://www.qmul.ac.uk/sbcs/iubmb/enzy me/). Class 1: Endochitinases [EC 3.2.1.14; (1 → 4)-2-acetamido-2-deoxy-β-D-glucan glycanohydrolase], and Class 2: β-*N*-acetylhexosaminidases (EC 3.2.1.52). The current IUBMB enzyme nomenclature has included former EC 3.2.1.29 and EC 3.2.1.30 exo-chitinases enzyme classes under a single enzyme class, known as β-N-acetylhexosaminidases (Reetarani et al. 2000; Hoell et al. 2010; Suresh 2019). Nevertheless, the existing terminology for chitinolytic enzymes is confusing to some authors (Sun et al. 2020), owing in part to the point that the previous classification system has not been fully rejected (Palanivelu and Vaishna 2013; Vincenzi et al.

2014). The endochitinases are the specific enzymes acting on chitin polymer at the internal sites and randomly cleave the GlcNAc-GlcNAc linkages, thus exclusively release low molecular mass oligosaccharides of GlcNAc with different DP as the end product (Fig. 6). Whereas the β-*N*-acetylhexosaminidases act on chitin oligosaccharides [$(GlcNAc)_{2–20}$] from the reducing end and produce monomeric GlcNAc as the sole end product (Reetarani et al. 2000; Oyeleye and Normi 2018; Suresh 2019; Sun et al. 2020). In this chapter, we follow the terminology suggested by IUBMB. In addition, we discuss only the endochitinases that catalyze the formation of oligosaccharides/oligomers from chitin molecules/biomass.

Chitinases activities are widely distributed in Nature in various groups of living organisms. Chitinases are reported from various organisms, including those that lack chitin like plants, bacteria, viruses and vertebrates, similarly in chitin containing organisms like fungi, insects and crustaceans (Oyeleye and Normi 2018). These organisms synthesize them for different physiological roles. Along this subject line, many reviews have been published by various research teams (Kasprzewska 2003; Guan et al. 2009; Hoell et al. 2010; Stoykov et al. 2015; Oyeleye and Normi 2018; Malik and Preety 2019; Sun et al. 2020). The structure and action mechanism of endochitinases have been reported (Cantarel et al. 2009; Hoell et al. 2010; Liu et al. 2015; Kidibule et al. 2018; Noa Miguez et al. 2021). Based on the similarity of amino acid sequence, the endochitinases belongs to two-glycoside hydrolase (GH) families, GH-18 and GH-19 (Cantarel et al. 2009; Liu et al. 2015), which displayed different catalytic mechanisms such as retaining and inverting processes. Family GH18 comprises bacterial and fungal chitinases (Horn et al. 2006; Huang et al. 2012), while family GH19 mainly contains plant chitinases (Malik and Preety 2019; Suresh 2019). Family GH18 enzymes use the substrate-assisted retaining double displacement mechanism, whereas family GH19 enzymes employ the inverting

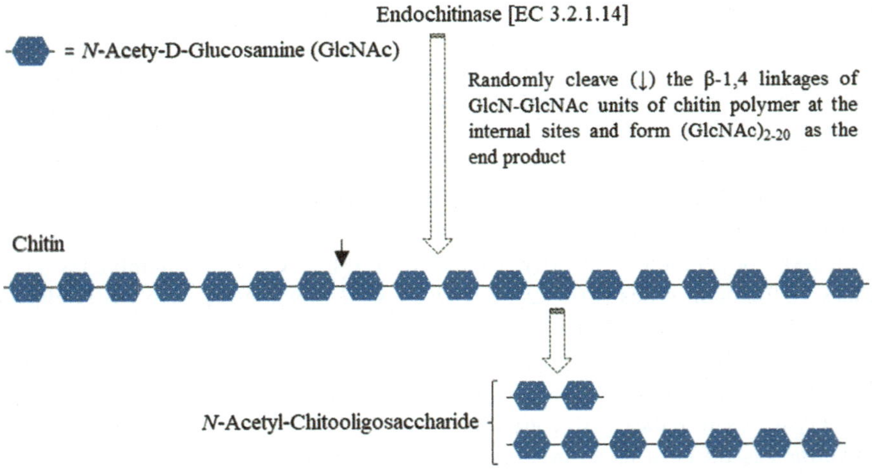

Fig. 6 Schematic presentation of endochitinase activity on chitin polymer and formation of *N*-acetyl-oligosaccharides [$(GlcNAc)_{2–20}$]

direct displacement mechanism (Hoell et al. 2010). Chitinases are also competent for hydrolyzing β-(1 → 4) glycosidic linkage of GlcNAc-GlcN and GlcN-GlcNAc units, allowing degradation of partly acetylated chitosan, but their degree of activity is significantly lower than chitosanases (Noa Miguez et al. 2021). Further, chitinases never cleave the β-1,4-glycoside GlcN-GlcN bonds of chitosan chains, differentiating these enzymes from chitosanases (Jung and Park 2014). Chitinases belonging to family GH-18 need a GlcNAc moiety at subsite-1 (Kidibule et al. 2018), while family GH-19 chitinases do not require GlcNAc moiety at this subsite. Due to this cause, family GH-18 chitinases produce oligosaccharides with GlcNAc residues at their reducing end (Stoykov et al. 2015; Lienemann et al. 2009). Conversely, oligosaccharides formed by family GH-19 chitinases have *N*-acetylated or de-*N*-acetylated reducing ends (Noa Miguez et al. 2021). Both types of chitinases have been explored to synthesize *N*-acetyl chitooligosaccharides [(GlcNAc) 2–20] from chitin bioresource.

2.1.2 Chitosanases

Chitosanases (EC 3.2.1.132, chitosan *N*-acetylglucosaminohydrolase) catalyze the hydrolysis of β-(1 → 4)-glycosidic linkages of chitosan polymer. Chitosanases are generally endo-acting enzymes, which randomly cleave the chitosan polymer at the internal sites and release a mixture of oligosaccharides, exclusively as the end product (Thadathil and Velappan 2014; Suresh 2019) (Figs. 7, 8 and 9). Chitosanase are widely distributed in Nature and have been produced, purified and characterized by several groups of organisms, including viruses (Sun et al. 1999), bacteria (Oh et al. 2011), fungi (Nidheesh et al. 2015a) and plants (Osswald et al. 1994). The biochemical characteristics, structure functional relationship, etc., of chitosanases produced from various sources, have been reviewed (Thadathil and Velappan 2014; Singh et al. 2019; Sun et al. 2020; Noa Miguez et al. 2021). Shinya and Fukamizo (2017) have studied the interaction between chitosan and its related enzymes.

According to the comparison of amino acid sequence, chitosanases belongs to five GH families, viz: GH-5, GH-8, GH-46, GH-75 and GH-80 (Wang et al. 2008; Hoell et al. 2010; Shinya and Fukamizo 2017) and are capable of hydrolyzing different types of β-(1 → 4)-glycosidic linkages in the chitosan molecules. Most bacterial chitosanases belong to the GH-46 family (Saito et al. 1999), and chitosanases from fungi include in the GH-75 family (Zhu et al. 2012). Chitosanases from GH-46, GH-75, and GH-80 families are thought to be a substrate (chitosan) specific, though GH-5, GH-7 and GH-8 chitosanases also showed other enzyme actions, like cellulase and transglycosylation activities (Sun et al. 2020). Chitosanases are grouped into three subclasses based on the specificity of cleavage sites, viz: (1) subclass I chitosanases, which cleave both GlcN-GlcN and GlcNAc-GlcN linkages (Fig. 7), (2) subclass II chitosanases, which cleave only GlcN-GlcN linkages (Fig. 8) and (3) subclass III chitosanases, which cleave both GlcN-GlcN and GlcN-GlcNAc linkages (Fig. 9) in the chitosan molecules (Jung and Park 2014; Thadathil and Velappan 2014; Hoell et al. 2010; Weikert et al. 2017) The hydrolysis of GlcN-GlcN linkages is the

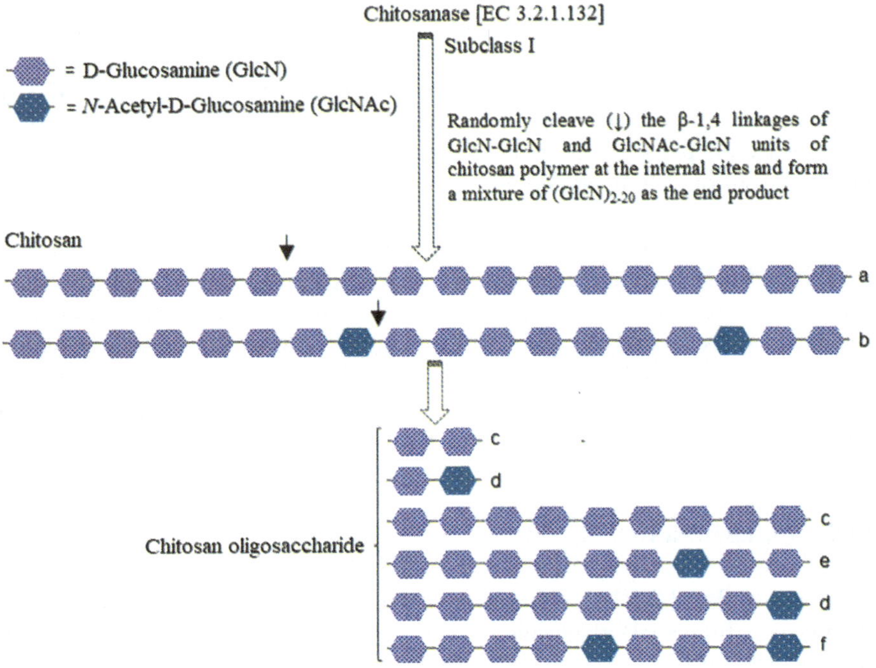

Fig. 7 Schematic presentation of chitosanase subclass I activity on chitosan polymers and formation of chitosan oligosaccharides [$(GlcN)_{2-20}$]. Chitosan polymer with full GlcN units (**a**), partially acetylated chitosan (**b**), [$(GlcN)_{2-20}$] with full GlcN units (**c**), partly acetylated [$(GlcN)_{2-20}$] (**e**), partly acetylated [$(GlcN)_{2-20}$] with GlcNAc units at reducing end (**d, f**)

common property of all these chitosanases (Figs. 7, 8 and 9). While, chitosanases are unable to hydrolyze GlcNAc-GlcNAc linkages in the partly acetylated chitosan, and this property distinguishes chitosanases from chitinases (Hoell et al. 2010). A new classification system for chitosanases employing their uniqueness and inclinations for subsites (−2) to (+2) have been proposed (Weikert et al. 2017).

The above-described specific enzymes acting on chitin and chitosan are synthesized by various organisms, including microorganisms, plants and animals and found abundantly in Nature (Hoell et al. 2010; Suresh 2019; Sun et al. 2020). The molecular structure and function as well as other aspects of chitinase (Oyeleye and Normi 2018; Sun et al. 2020; Noa Miguez et al. 2021), human chitinases (Guan et al. 2009), plant chitinases (Kasprzewska 2003) and microbial chitinases (Hoell et al. 2010; Stoykov et al. 2015) have been reviewed. The chitosanases, particularly of microbial chitosanases, have been reviewed by the various research team (Fukamizo et al. 2000; Lukas et al. 2012; Thadathil and Velappan 2014; Weikert et al. 2017; Langner and Göhre 2016; Shinya and Fukamizo 2017; Singh et al. 2019; Sun et al. 2020; Tzelepis and Karlsson 2021). Among the different groups of organisms, which produce specific enzymes acting on chitin and chitosan, the microorganisms, mainly bacteria and fungi, have commercial significance concerning the production of these

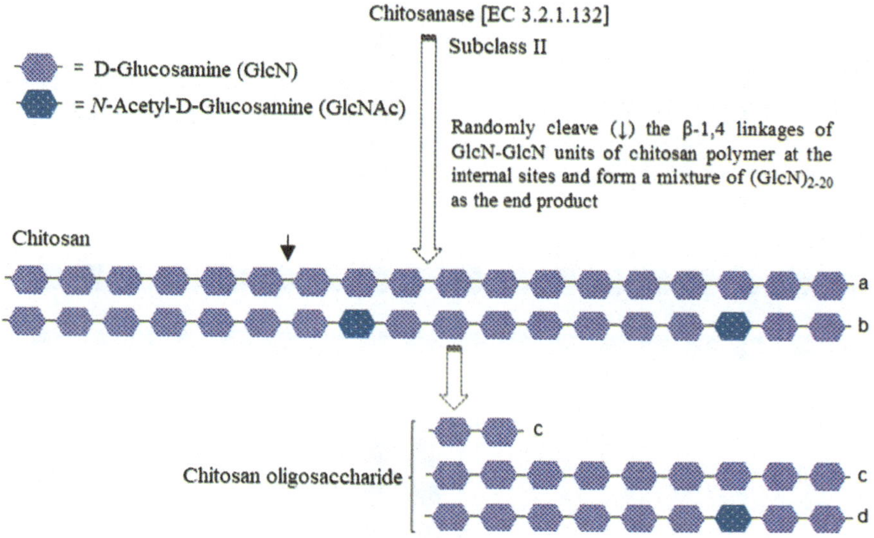

Fig. 8 Schematic presentation of chitosanase subclass II activity on chitosan polymers and formation of chitosan oligosaccharides [$(GlcN)_{2-20}$]. Chitosan polymer with full GlcN units (**a**), partially acetylated chitosan (**b**), [$(GlcN)_{2-20}$] with full GlcN units (**c**), partly acetylated [$(GlcN)_{2-20}$] (**d**)

enzymes as well as their application in chitooligosaccharides synthesis from different chitin or chitosan substrate/biomass (Thadathil and Velappan 2014; Naidheesh et al. 2015a, b; Santos-Moriano et al. 2018).

2.2 *Other Enzymes Involved in the Synthesis of Chitooligosaccharides*

2.2.1 Non-specific Enzymes Acting on Chitin and Chitosan

As well as the specific enzymes such as chitinases and chitosanases, few non-specific enzymes have exhibited some potential to identify the β-(1 → 4)-glycosidic linkages of chitin and chitosan to depolymerize them (Muzzarelli 1997; Pantaleone et al. 1992; Jung and Park 2014). The non-specific enzymes, which exhibited chitin or chitosan depolymerization activity was reported to few glycosidases (e.g. cellulases, lysozyme, pectinases, amylases, etc.) (Amano and Ito 1978; Kittur et al. 2005; Santos-Moriano et al. 2016), proteases (e.g. papain, pronase) (Pan et al. 2016), lipases (Lee et al. 2008) and chitin-active Lytic Polysaccharide Monooxygenases (LPMOs) (Vaaje-Kolstad et al. 2010; Mutahir et al. 2018). However, the rate of chitin or chitosan depolymerization by these non-specific enzymes can be far below the rate exhibited by the specific enzymes (Aktuganov et al. 2019; Noa Miguez et al. 2021).

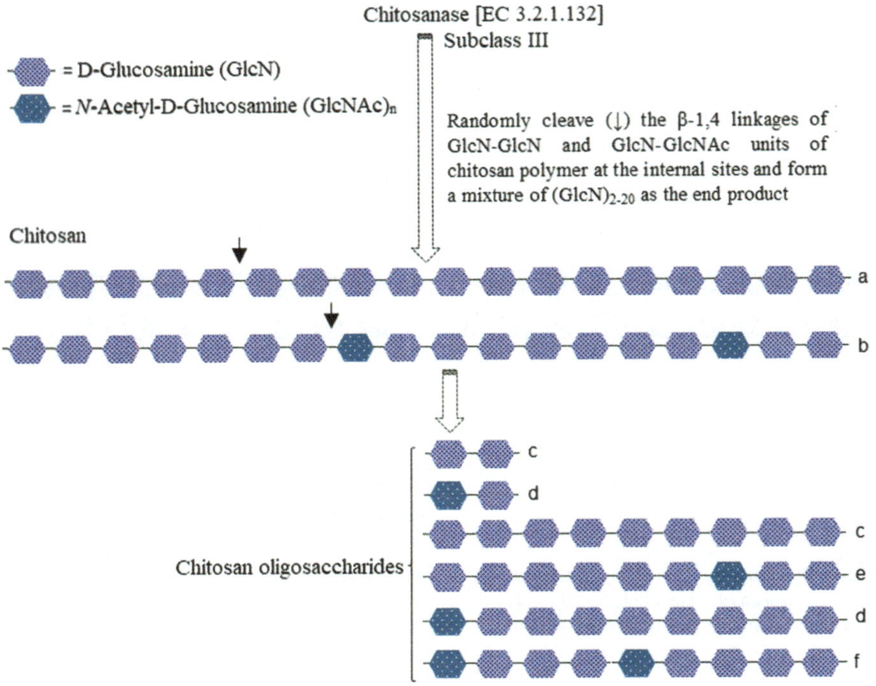

Fig. 9 Schematic presentation of chitosanase subclass III activity on chitosan polymers and formation of chitosan oligosaccharides [$GlcN)_{2-20}$]. Chitosan polymer with full GlcN units (**a**), partially acetylated chitosan (**b**), [$(GlcN)_{2-20}$] with full GlcN units (**c**), partly acetylated [$(GlcN)_{2-20}$] (**e**), partly acetylated [$(GlcN)_{2-20}$] with GlcNAc units at non-reducing end (**d**, **f**)

2.2.2 Chitin Deacetylase in the Synthesis of Chitooligosaccharides

Chitin deacetylases (EC 3.5.1.41; Chitin amidohydrolase) catalyze the hydrolysis of *N*-acetamido bonds of chitin, converting it to chitosan (Tsigos et al. 2000; Zhao et al. 2010; Jung and Park 2014; Suresh 2019). Chitin deacetylases activities have been reported in various microorganisms, particularly fungi (Aranaz et al. 2009, 2014; Suresh et al. 2011a). However, the activity of these chitin deacetylases is severely restricted by the insolubility of the chitin polymer. Chitin deacetylases have exhibited potential in synthesizing specific chitooligosaccharides (Tokuyasu et al. 1997, 2000). However, presently, there is no commercially viable method existing to produce chitin deacetylases and their uses in the synthesis of specific chitooligosaccharides (Suresh et al. 2011a; Zhao et al. 2010; Suresh 2019).

2.2.3 Chitinases and Chitosanases with Transglycosylation Activity

The endoglycosidase-catalyzed transglycosylation has emerged recently as a competent technique for creating novel oligosaccharides with potential bio-functionalities (Mangas-Sánchez and Adlercreutz 2015; Li and Wang 2016). Endochitinases belonging to the GH-18 family and a few chitosanases have exhibited some amount of transglycosylation capacity (Sirimontree et al. 2014; Noa Miguez et al. 2021). These specific enzymes act on chitin or chitosan molecules with transglycosylation capacity that can move the cleaved oligosaccharide molecules to a proper acceptor forming new glycosidic bonds and generating precise chitooligosaccharides (Tanabe et al. 2003; Purushotham and Podile 2012; Mallakuntla et al. 2020). It was reported that chitinases with transglycosylation activity are more significant than chitinases with hydrolytic activity for application in food and biomedical fields owing to their capacity to change crystalline chitin to soluble long-chain oligosaccharide (Oyeleye and Normi 2018).

2.2.4 Chitin Active Lytic Polysaccharide Monooxygenases (LPMOs)

LPMOs (EC 1.14.99.53–56) are enzymes capable of breaking glycolic linkages in the polysaccharides with crystalline nature via oxidizing either C1 or C4 of the glucopyranose moiety (Vaaje-Kolstad et al. 2017; Mutahir et al. 2018; Kaczmarek et al. 2019 LPMOs are comparatively new enzymes that have been discovered in 2010 and an extensively studied enzyme in current years (Gaber et al. 2020). LPMOs have been allotted to auxiliary activity (AA) families AA9, AA10, AA11, AA13, AA14 and AA15 in the Carbohydrate-Active Enzymes (CAZy) database (Tandrup et al. 2018). These enzymes act a leading role in the degradation of sugar-based biopolymers (such as cellulose, hemicellulose, chitin and starch) and have an encouraging implication for biomass transformation (Wang et al. 2021). LPMOs are potent tools in the degradation of various types of biomass, including marine biomasses such as chitins (Vaaje-Kolstad et al. 2010; Mekasha et al. 2017 Nakagawa et al. 2020). In distinction to GHs, such as chitinases and chitosanases, LPMOs are potent for directly breaking glycolic linkages in extremely crystallized chitin, enhancing its accessibility for succeeding enzymatic processing (Mutahir et al. 2018). Consequently, considerable attention has been received in chitin active LPMOs research. So far, LPMOs with chitinolytic activities have been recognized in carbohydrate AA families of AA10, AA11 and AA15 (Hemsworth et al. 2015). The chitin-active LPMO was first reported in *Serratia marcescens* AA10 by Vaaje-Kolstad et al. (2010). Currently, the chitinase active LPMO has been reported in various groups of organisms, including bacteria, fungi, viruses and arthropods (Chiu et al. 2015; Sabbadin et al. 2018). Among these organisms' fungi and bacteria are the most important producer of chitinolytic LPMOs (Kaczmarek et al. 2019). Nakagawa et al. (2013) investigated the enzymatic hydrolysis of α-chitin with varying particle size and crystallinity produced by mechanical pre-treatment using a cockatiel of chitinase and LPMO (from *S. marcescens*).

Mekasha et al. (2017) studied the percentage of chitinases, an LMPO and a β-*N*-acetylhexosaminidase from *S. marcescens* to depolymerize crustacean (crab and shrimp) chitins.

3 Chitooligosaccharides and Types of Chitooligosaccharides

Chitooligosaccharides can be defined as the partially hydrolyzed/depolymerized products/derivatives of chitin or chitosan. It can be β-(1 → 4) linked homo or hetero-oligomers/oligosaccharides of GlcNAc and/or GlcN (Yang and Yu 2014; Sun et al. 2020). Usually, chitooligosaccharides have a DP up to 20 and an average molecular mass of <4.0 kDa. Chitooligosaccharides are highly soluble in water. They can exhibit different DP, DD and PA depending on the type of chitin or chitosan substrates used in their production and the extent of their depolymerization. These characteristics of chitooligosaccharides significantly affect their different biofunctional properties (Hamer et al. 2015). The chitooligosaccharides can be classified into three distinct types based on their composition (Santos-Moriano et al. 2019; Noa Miguez et al. 2021) viz: (1) fully acetylated chitin oligosaccharides (*FA*-COS or *N*-acetyl-COS) (composed exclusively of GlcNAc residues (Fig. 6), (2) partly acetylated chitosan oligosaccharides (*PA*-COS) (formed mainly by GlcN residues with few GlcNAc residues) (Figs. 7, 8 and 9) and (3) fully deacetylated chitosan oligosaccharides (*FD*-COS) (composed of GlcN residues only) (Figs. 7, 8 and 9). These chitooligosaccharides are very high-value natural products with numerous promising bioactivities and potential applications in various fields, such as nutrition, functional foods, agriculture, cosmetics, pharmaceutical and biomedicine and so on (Moon et al. 2017; Marmouzi et al. 2019; Aktuganov et al. 2019; Ismail et al. 2020). The reader can refer to other chapters of this book for more information on these aspects of chitooligosaccharides. In general, these chitooligosaccharides synthesized by depolymerization of chitin or chitosan are heterogeneous blends of different types of oligosaccharides, still with extended hydrolysis time (Jung and Park 2014) (Figs. 7, 8 and 9), which describes the low reproducibility rate and conflicting results when their biofunctional capacities are checked. Consequently, several methods have been studied for the separation and purification of chitooligosaccharides, such as metal affinity (Le Dévédec et al. 2008); gel filtration (Sørbotten et al. 2005) and ion exchange (Haebel et al. 2007), chromatography and ultrafiltration (Lopatin et al. 2009). Preparative scale separation of chitooligosaccharides is usually based on their mass/size by gel filtration chromatography (Jung and Park 2014). The following sections briefly describe the development of enzymatic processes towards the synthesizing of specific type of chitooligosaccharides. As detailed above, the three types of chitooligosaccharides (Figs. 6, 7, 8 and 9) can be produced by enzymatic approaches using different enzymes.

Generally, high-performance liquid chromatography (HPLC) alone or in combination with matrix-assisted laser desorption ionization-time of flight (MALDI-TOF) mass spectrometer was used to analyze the oligosaccharides obtained from the hydrolysis of chitin or chitosan substrate. In addition to HPLC and MALDI-TOF, silica gel thin layer (TLC) was also widely used in the separation and identification of oligosaccharides liberated by the hydrolysis of chitin or chitosan substrate.

4 Enzymatic Production of Chitooligosaccharides

4.1 Enzymatic Production of Chitooligosaccharides by Specific Enzymes

4.1.1 Enzymatic Production of *N*-Acetyl-COS/*FA*-COS from Chitin

As described above, *N*-Acetyl-COS/*FA*-COS is the partly depolymerized derivative of chitin (Fig. 6). *N*-Acetyl-COS/*FA*-COS and its derivatives prepared from chitin have exhibited good solubility in water and various biological properties and can be explored in a wide variety of chemical and biological industries (Synowiecki and Al-Khateeb 2003; Oyeleye and Normi 2018). Endochitinases are the specific enzymes investigated in chitin depolymerization and synthesis of *N*-acetyl-COS/*FA*-COS (Fig. 6). As noted previously, the specific enzymes acting on chitin molecules are usually produced as a chitinolytic enzyme complex. Hence, to develop a high yield of *N*-acetyl-COS/*FA*-COS from chitin biomass, the enzyme preparation must comprise higher endochitinase activity and relatively low or nil β-*N*-acetylhexosaminidases activity (Aloise et al. 1996; Suresh 2019). Chitinases preparations from various groups of microorganisms, mainly from bacteria and fungi, have been explored in the manufacture of *N*-acetyl-COS/*FA*-COS (GlcNAc)2–20 from diverse chitin substrates (Takiguchi and Shimahara 1988, 2003; Woo and Park 2003; Tanaka et al. 1999; Kuk et al. 2006; Laura et al. 2006; Jung et al. 2007; Suresh et al. 2011b; Wang et al. 2012; Thadathil et al. 2014). Because of the strong crystalline nature and insolubility in most common solvents, in general, chitin is subjected to pre-treatment with strong acids (such as HCl/H2SO4 for colloidal chitin or H2PO4 for swollen chitin) before enzymatic digestion to enhance the accessibility of the substrate to the enzyme (Jung and Park 2014).

Preparation of *N*-acetyl-COS/*FA*-COS (GlcNAc)2–20 using bacterial chitinases from *Paenibacillus illiosensis* KJA424 (Jung et al. 2007), *Bacillus* sp. WY22 (Woo and Park 2003), *Pyrococcus kodakaraensis* KOD1 (Tanaka et al. 1999), *Aeromonas* sp. GJ-18 (Kuk et al. 2006), *Bacillus cereus* TKU027 (Wang et al. 2012), *Vibrio anguillarum* E-383a (Takiguchi and Shimahara 1988), and hyperthermophilic archaeon *Thermococcus kodakaraensis* KOD1 (Tanaka et al. 2003) have reported. Jung et al. (2007) was reported the highest *N*-acetyl-COS/*FA*-COS formation of 2.21 ± 0.13 mg/l from swollen chitin after 24 h of hydrolysis with a chitinase preparation of *P. illiosensis* KJA-424. Production of *N*-acetyl chitotriose as a primary product from a colloidal and glycol chitin employing the chitinase prepared from *Bacillus* sp. WY22 has been reported (Woo and Park 2003). *N*-acetyl chitobiose [$(GlcNAc)_2$] was prepared by a chitinase from a marine *V. anguillarum* E-383a from colloidal chitin (Takiguchi and Shimahara 1988). Production of *N*-acetyl chitobiose [$(GlcNAc)_2$] from colloidal chitin by using a chitinase from the hyperthermophilic archaeon *T. kodakaraensis* KOD1 was reported (Tanaka et al. 2003). Kuk et al. (2006) reported the production of *N*-acetyl chitobiose [$(GlcNAc)_2$] after 7 days of hydrolysis of swollen α-chitin and powdered β-chitin with a yield of 78.9% and 56.6%, respectively, using a crude chitinase preparation of *Aeromonas* sp. GJ-18. *N*-acetyl-COS [$(GlcNAc)_{2-5}$] was produced from partially deacetylated chitin (60% DD) using the culture supernatant prepared by a chitinolytic *B. cereus* TKU027 (Wang et al. 2012).

Preparation of *N*-acetyl-COS/*FA*-COS [$(GlcNAc)_{2-20}$] using fungal chitinases from *Lecanicillium fungicola* (Laura et al. 2006); *Penicillium monoverticillium* CFR 2 (Suresh et al. 2011b) and *Aspergillus flavus* CFR 10 and *Fusarrium oxysporum* CFR 8 (Thadathil et al. 2014) have been reported. Recently, a chemo-enzymatic synthesis of *N*-acetyl-COS using the fungal chitinase from *Thermomyces lanuginosus* ITCC 8895 have been reported (Kumar et al. 2021). The *N*-acetyl-CO/*FA*-COS was produced from different chitin substrates employing the chitinases obtained from *L. fungicola* (Laura et al. 2006). The highest yield (mmol/l) of *N*-acetyl COS obtained was 0.17 from a crystalline α-chitin, 2.77L from a deacetylated α-chitin and 4.44 from a deacetylated β-chitin (Laura et al. 2006). In a study, Thadathil et al. (2014) reported the production of *N*-acetyl-COS using the chitinases preparation from A. *flavus* CFR 10 and *F. oxysporum* CFR 8 with the highest yield (mmol/l) of 10.4 ± 0.28 after 6 h hydrolysis and 10.2 ± 0.01 after 6 h hydrolysis 30 h, respectively. Suresh et al. (2011b) reported a maximum *N*-acetyl-COS production (mmol/l) of 6.81 ± 0.03 from colloidal chitin, 0.36 from a deproteinized shrimp shell and 0.60 ± 0.02 from a demineralized crab shell using the crude chitinase obtained from *P. monoverticillium* CFR 2. These authors observed that *N*-acetyl chitohexose (~62.43%) was the primary product of the hydrolysate from colloidal chitin by HPLC analysis. Production of *N*-acetyl-COS by the chitinase from a novel *Streptomyces chilikensis* strain RC1830 and its various bioactivities was reported (Behera et al. 2020).

Recently, a chemo-enzymatic method for synthesizing short-chain *N*-acetyl-COS using ionic liquids and a chitinase from *T. lanuginosus* ITCC 8895 has been reported (Kumar et al. 2021). The authors produced 1.10 ± 0.89 mg/ml of *N*-acetyl chitobiose [$(GlcNAc)_2$] and 1.07 ± 0.92 mg/mL of *N*-acetyl chitotriose [$(GlcNAc)_3$] from the Trihexyltetradecylphosphonium bis(2,4,4-trimethylpentyl) phosphinate treated chitin substrate.

4.1.2 Enzymatic Depolymerization of Chitosan and Production of Chitosan Oligosaccharides

Chitosan oligosaccharides [$(GlcN)_{2\text{–}20}$] are the partly degraded derivative of chitosan, and the only positively charged oligosaccharides occur in Nature (Thadathil and Velappan, 2014; Sun et al. 2020). They possess exceptional biological functions, such as antioxidant, antimicrobial, antitumor, immunoenhancing, etc., activities for a wide range of potential applications in various fields (Sun et al. 2020). Consequently, there has been increasing attention in converting chitosan to specific chitosan oligosaccharides (Thadathil and Velappan 2014; Suresh 2019). As described above (Figs. 7, 8 and 9), the partial depolymerization of chitosan can develop two types of chitosan oligosaccharides, such as (1) *FD*-COS and (2) *PD*-COS, depending on the type of chitosan substrates employed. The enzymatic production of *FD*-COS or *PD*-COS from chitosan substrates can be carried out effectively using specific chitosanases (Kim and Rajapakse 2005; Naidheesh et al. 2015a, b; Nidheesh 2016; Suresh 2019). It was well reported that the viscosity of chitosan declines quickly when added with chitosanase, in accord with depolymerization and production of the oligosaccharides (Jung and Park 2014). Chitosanase prepared from a different group of fungi and bacteria has been explored in producing oligosaccharides [$(GlcN)_{2\text{–}20}$] from various chitosan substrates.

Enzymatic production of oligosaccharides [$(GlcN)_{2\text{–}20}$] from different chitosan substrates employing microbial chitosanase has been reported by many research teams (Shimosaka et al. 1995; Kim et al. 1998; Kurakake et al. 2000; Cheng and Li 2000; Choi et al. 2004; Gao et al. 2009; Pechsrichuang et al. 2013; Nguyen et al. 2014; Naidheesh et al. 2015a, b; Nidheesh 2016). Shimosaka et al. (1995) reported the production of chitobiose [$(GlcN)_2$] and chitotriose [$(GlcN)_3$] from chitopentose [$(GlcN)_5$] and chitosan using crude chitosanase from *Acinetobacter* sp. CHB10. Production of chitobiose [$(GlcN)_2$] was reported from chitosan mixture with 75%–85% DD using a chitosanase obtained from *B. cereus* S1 (Kurakake et al. 2000). A purified chitosanase (22.5 kDa) from *A. fumigatus* KH-94 was used for the production of chitobiose [$(GlcN)_2$], chitotriose [$(GlcN)_3$] and chitotetraose [$(GlcN)_4$] from chitohexaose [$(GlcN)_6$] (Kim et al. 1998). In addition to the production of chitosan oligosaccharides with DP 3–6 (50% yields), the 22.5 kDa chitosanase from *A. fumigatus* KH-94 has produced oligosaccharides with higher DP (> DP 7, 50% yield) from chitosan (Kim et al. 1998). Choi et al. (2004) reported the reaction pattern and production of different oligosaccharides from chitohexose [$(GlcN)_6$] using the chitosanase produced from *Bacillus* sp. KCTC 0377BP. A purified chitosanase produced by *B. cereus* D-11 hydrolyzed chito-oligomers [$(GlcN)_{5\text{–}7}$] into chitobiose [$(GlcN)_2$], chitotriose [$(GlcN)_3$] and chitotetraose [$(GlcN)_4$] as the end products (Gao et al. 2009). Production of oligosaccharides [$(GlcN)_{2\text{–}5}$] from chitosan using a chitosanase obtained from *Aspergillus* sp. Y2K has been reported (Cheng and Li 2000). In this study, a chitosan-degrading soil fungus (*Aspergillus* sp. Y2K) was grown in a minimal medium containing chitosan as the sole carbon source. Further, the purified chitosanase were applied for the hydrolysis of chitosan substrate and production of chitosan oligosaccharides. They reported that chitotriose

[$(GlcN)_3$], chitotetraose [$(GlcN)_4$] and chitopentaose [$(GlcN)_5$], were the main products in the chitosan hydrolysate (Cheng and Li 2000). Naidheesh et al. (2015a; b) reported chitosan oligosaccharides production employing crude chitosanase from a soil fungi *Purpureocillium lilacinum* CFRNT12. The chitosanase was prepared from *P. lilacinum* CFRNT12 utilizing solid-state fermentation (SSF) using shrimp (*Penaeus* sp.) by-products as the substrate. Later the crude chitosanase was concentrated using the dialysis method employing polyethylene glycol (mol. wt. 20,000). The chitosanase (169.92 unit/ml) was tested on two different chitosan substrates [crystalline (>22 mesh size) and colloidal chitosan]. The highest yield of oligosaccharides (4.43 mmol/l from colloidal chitosan and 1.7 mmol/l from crystalline chitosan) was observed after 24 h of digestion. Chitotriose [$(GlcN)_3$] and chitotetraose [$(GlcN)_4$] were the main products of hydrolysis (Naidheesh et al. 2015a; b; Nidheesh 2016).

A purified chitosanase (46 kDa) was obtained from *Bacillus* sp. HW-002 produced chitobiose [$(GlcN)_2$] as a main end product from chitosan (Lee et al. 1996). Park et al. (1999) have reported the digestion of chitopentose [$(GlcN)_5$] and chitohexose [$(GlcN)_6$] into chitobiose [$(GlcN)_2$] and chitotriose [$(GlcN)_3$] by a purified chitosanase (34 kDa) produced from *Matsuebacter chitosanotabidus* 3001. Sáncheza et al. (2017) investigated the chitooligosaccharides production by two different enzymatic processes. The first process used chitosanase enzymatic hydrolysis using a commercial chitosanase from *Streptomyces griseus,* and the second consisted of a two-step procedure based on chemical hydrolysis followed by chitosanase hydrolysis. It was observed that chitooligosaccharides produced in the second process were composed of 63% of fully deacetylated sequences (Sáncheza et al. 2017).

In addition, the potential of recombinant chitosanase in the depolymerization of different chitosan substrates and production of oligosaccharides have been reported (Chen et al. 2012; Pechsrichuang et al. 2013; Cheng et al. 2021). Preparation of chitosan oligosaccharides using a thermostable chitosanase produced by recombination of a gene responsible for chitosanase synthesis from *A. fumigatus* in *Pichia pastoris* (Chen et al. 2012). The authors reported that 3 g of enzyme preparation hydrolyzed 200 kg of chitosan into oligosaccharides after 24 h at 60 °C. Pechsrichuang et al. (2013) reported the preparation of oligosaccharides [$(GlcN)_{2-6}$] from different chitosan substrates using a recombinant *B. subtilis* chitosanase. In a recent study, Cheng et al. (2021) has reported a pilot-scale production of chitosan oligosaccharides using a recombinant chitosanase. In this work, a Vector pET20b-csn containing the *B. circulans* MH-K1 chitosanase gene was transformed into *Escherichia coli* strain BL21 (DE3) for recombinant protein (chitosanase) expression. Further, the recombinant chitosanase was prepared by growing recombinant *E. coli* strain BL21 (DE3) in a Luria–Bertani (LB) medium. The intracellular recombinant chitosanase was extracted from the cell pellet and purified. Further, the semi-purified recombinant chitosanase was used in the pilot-scale production of chitosan oligosaccharides by mixing ~2500 units to a 10 litter chitosan solution (5%) for 72 h, at ambient temperature. The author reported the formation of a digestion product having molar masses of 833 ± 222 g/mol with multiple DP [chitotriose $(GlcN)_3$ to chitotetrose $(GlcN)_4$] (Cheng et al. 2021).

All the above-described studies revealed that application of specific enzymes (chitosanases) for the partial depolymerization of chitosan are desirable enzymatic processes for the preparation of oligosaccharides [$(GlcN)_{2-20}$]. Though, the application of chitosanase in chitosan depolymerization is restricted due to various reasons, including the high cost of commercially available chitosanase, lack of availability in bulk volume and so on (Thadathil and Velappan 2014; Suresh 2019; Cheng et al. 2021). Hence, some alternative approaches have been reported on the production of chitosan oligosaccharides from different chitosan substrates, such as enzyme immobilization recycling methods (Kim and Rajapakse 2005; Ming et al. 2006; Kuroiwa et al. 2008), increasing the enzymatic concentration by stepwise substrate mixing (Ming et al. 2006) and continual production in a dual-reactor system (Jeon and Kim 2000; Santos-Moriano et al. 2016). Ming et al. (2006) have used a reactor with an immobilized chitosanase on agar gel-coated multidisc impeller for the production of chitosan oligosaccharides and attained yields of 20–45% for the addition of chitopentaose [$(GlcN)_5$] and chitohexaose [$(GlcNAc)_6$], dependent on the surface activity of the enzyme. Preparation of chitosan oligosaccharides from chitosan using different immobilized enzyme reactors like batch and column reactor and ultrafiltration membrane reactor and continuous production of oligosaccharides by a dual-reactor system have been reported (Kim and Rajapakse 2005). Kuroiwa et al. (2008) studied the development of a chitosanase immobilized on amylose-coated magnetic nanoparticles for the production of chitosan oligosaccharides. The authors have reported that in the digestion reaction employing the immobilized chitosanases, pentamers [$(GlcN)_5$] and hexamers [$(GlcNAc)_6$], of chitosan oligosaccharide were formed at a high yield (40%), which is higher than that of conventional methods for producing these oligosaccharides (Kuroiwa et al. 2008). Jeon and Kim (2000) developed a dual-reactor system with a packed-bed reactor (PBR) to minimize the viscosity of chitosan, followed by an ultrafiltration membrane reactor to form chitosan oligosaccharide of the preferred DP.

Fungal biomasses are an excellent source for chitosan extraction and oligosaccharides production (Mahata et al. 2014; Abdel-Gawad et al. 2017; Ahmad et al. 2020). In a study, Mahata and co-workers (2014) were hydrolyzed intact mycelium of *Rhizopus oligosporus* NRRL2710 with a chitosanase preparation from *Streptomyces* sp. N174 and produced a mixture of functionally active chitooligosaccharides. This study shows the potential of bioconversion of abundant fungal chitin or chitosan biomass to high-value biomolecules with no ecological issues.

4.2 Chitooligosaccharides Production by Non-specific Enzymatic Depolymerization

Because of the high cost of commercial chitinase and chitosanase, the use of cheap non-specific enzymes such as glucanase, lipase, lysozyme, cellulase, amylase, papain and pectinase (derived from bacteria, fungi, mammal and plant sources) have been explored for chitin or chitosan depolymerization and synthesis of their oligosaccharides (Pantaleone et al. 1992; Terbojevich et al. 1996; Muzzarelli 1997; Kittur et al. 2005; Ilankovan et al. 2006; Lee et al. 2008; Pan et al. 2016; Santos-Moriano et al. 2016). A pectinase preparation from *A. niger* has been explored for the hydrolysis of chitosan and production of a mixture of low molecular weight chitosan (LMWC) (86%) and oligosaccharides (4.8%) (Kittur et al. 2005). Production of *N*-acetyl COS (~81% yield) from de-crystallized chitin using the commercial bovine pepsin has been reported (Ilankovan et al. 2006). Chitosan with 80% DD was digested using the cellulase prepared from *A. niger* to produce chitooligosaccharides with DP 3–11 (Xie et al. 2009). Zhang et al. (1999) effectively applied a blend of cellulase, α-amylase and proteinase to produce chitooligosaccharides with DP 5–17. Wu (2012) has reported the production of chitosan oligosaccharides (DP 2–8) using the commercial α-amylase from a chitosan substrate prepared from the skin of *Clanis bilineata* larvae. Enzymatic preparation of chitosan oligosaccharides (DP 1–6) by a commercial lipase yield 93.8% after 24 h digestion at 37 °C was reported (Lee et al. 2008). Among these non-specific enzymes, papain (a cysteine protease) was very attractive because of its low cost and wide acceptance in the food industry and the ability to depolymerize chitosan efficiently (Pan et al. 2016).

In addition to free non-specific enzyme processes, immobilized non-specific enzymes such as papain (Lin et al. 2002), lysozyme (Aiba 1994), pronase (Kumar et al. 2004) and α-amylase (Santos-Moriano et al. 2016) have been tested for chitooligosaccharides production. In a work, chitosanlytic activity present in a commercial α-amylase (from *B. amylolyquefaciens*, BAN) was covalently immobilized onto glyoxal agarose beads and used for the continuous production of chitooligosaccharides in a dual-reactor system (Santos-Moriano et al. 2016). All the studies confer its potential in the commercial preparation of chitooligosaccharides since these enzymes are commercially available in bulk quantity, inexpensive than chitinase and chitosanase and very simple to manage the digestion processes (Jung and Park 2014; Suresh 2019).

4.3 Enzymatic Synthesis of Chitooligosaccharides by the Transglycosylation Activity of Chitinase and Chitosanase

As detailed above, chitin and chitosan oligosaccharides can be prepared using various techniques, including mechanical, chemical and enzymatic methods. However, all these methods, including enzymatic hydrolysis (either specific or nonspecific) of chitin and chitosan, generally produce a complex oligosaccharides mixture (*N*-acetyl COS/*FA*-COS, *FD*-COS, *PA*-COS), which has been grouped according to the DA or DD, DP or molecular weight, and the molecular weight distribution (poly-disparity,

PD) (Li et al. 2016). Therefore, the bioactivities of these oligosaccharides are habitually analyzed using heterogeneous and/or relatively poorly characterized oligosaccharides mixtures. Hence, it is difficult to determine which molecules are responsible for the detected biofunctional properties. To overcome these problems, the enzymatic synthesis of specific chitooligosaccharides has been investigated by transglycosylation (Akiyama et al. 1995; Tokuyasu et al. 1997, 2000; Kobayashi et al. 1996, 2006; Li et al. 2016). The transglycosylation methods allow regioselective synthesis under mild conditions and without complicated protection (Li et al. 2016). GlcNAc-(1 → 4)-GlcN and β-GlcNAc-(1 → 4)-β-GlcNAc(1 → 4)-β-GlcNAc-(1 → 4)-GlcN was synthesized using the chitin deacetylase from *Colletotrichum lindemuthianum* (Tokuyasu et al. 1997, 2000).

The synthesis of chitin or chitosan oligosaccharides (*N*-acetyl COS or *FD*-COS) using a transglycosylation reaction catalyzed by lysozyme and an *N*-acetyl COS or *FD*-COS with DP4-12 have reported utilizing *N, N′, N″*-tri(monochloro)acetylchitotriose and *N*-acetyl chitotriose [$(GlcNAc)_3$] as glycosyl donors followed by *N*-demonochloroacetylation (Akiyama et al. 1995). The transglycosylation reaction activity of a chitosanase from *B. cereus* was used to produce high-DP chitosan oligosaccharides (DP2-7) (Hsiao et al. 2008).

The synthesis of single oligosaccharides with higher DP such as *N*-acetyl chitopentaose [$(GlcNAc)_5$] using a recombinant *E. coli* harbouring the nodC gene (encoding chitooligosaccharide synthase) has been reported (Zhang et al. 2007). A one-pot chemo-enzymatic technique for regio- and stereospecific production of *N*-acetyl chitoheptaose [$(GlcNAc)_7$] utilizing 1, 2-oxazoline derivative of [$(GlcNAc)_5$] and [$(GlcNAc)_2$] as starting materials by defined transglycosylation reaction by a mutant chitinase has also been described (Yoshida et al. 2012).

4.4 *Synthesis of Chitooligosaccharides by Whole-Cell Biocatalytic Approaches*

Most chitin and chitosanase synthesizing microorganisms require chitin or chitosan or their derivates as source nutrients (for growth) and an inducer for production of respective enzymes (chitinase or chitosanase) (Jung and Park 2014; Suresh 2019). In a study, Liang et al. (2012) have reported the direct production of *N*-acetyl COS [$(GlcNAc)_2$, $(GlcNAc)_4$, $(GlcNAc)_5$ and $(GlcNAc)_6$] using the fermentation approaches by growing a chitosanase producing *B. cereus* TKU022 in a medium containing 1.5% shrimp head powder after 48 h of incubation. Chitooligosaccharides production was reported in the culture supernatant of *Acinetobacter calcoaceticus* TKU024 from shellfish by-products along with chitosanases after 96 h fermentation (Wang et al. 2011). Even though the quantity of end products of this approach is meagre, these types of research have significance to the development processes for the direct production of chitooligosaccharides from the abundant chitin bio-resources.

5 Conclusion and Future Prospects

Chitooligosaccharides, the partially degraded derivative of polymer chitin or chitosan, possess numerous unique biological activities that describe these biomolecules' significance in various fields. However, the biological properties of these chitooligosaccharides have been significantly influenced by many factors, including their chemical composition, molecular mass, the extent of deacetylation and acetylation pattern. Therefore, the production of well-defined selective chitooligosaccharides has immense value to understand their bioactive effect more profound and, therefore, their upcoming applications in nutrition and biomedicine. The enzymatic processes of chitin and chitosan depolymerization, employing specific GH enzymes such as endochitinase and chitosanase, has achieved considerable progress in chitooligosaccharides production. Further, current advances in endochitinases and chitosanases with transglycosylation activity and chitin active LPMOs research have improved the enzymatic production of defined chitooligosaccharides. Today, all these enzymes are progressively emerged as powerful tools for the enzymatic preparation of chitooligosaccharides, specifically when a precise and distinct method is essential. Regardless of much progress in the past decade, currently, commercial-scale productions of pure chitooligosaccharides with defined characteristics by enzymatic approaches is still a task. The main limitations of the commercial-scale enzymatic processes recline on the high cost of commercially available specific enzymes, non-availability in bulk volume and so on, even though these specific enzymes are copious in Nature (Jung and Park 2014; Thadathil and Velappan 2014; Suresh 2019; Cheng et al. 2021). Nevertheless, numerous recent investigations report the significance of the enzymatic methodologies for chitooligosaccharides production and might soon advance it into economical to conventional chemical methods. Hence, future research must focus on developing highly efficient and appropriate technologies by properly selecting the chitin or chitosan substrate and the enzyme source towards making highly defined chitooligosaccharides for wide application in biomedicine and other fields.

References

Abdel-Gawad KM, Hifney AF, Fawzy MA, Gomaa M (2017) Technology optimization of chitosan production from *Aspergillus niger* biomass and its functional activities. Food Hydrocoll 63:593–601

Ahmad FB, Akmal MMH, Amran A, Hasn MH (2020) Characterization of chitosan from extracted fungal biomass for piezoelectric application. IOP Conf Ser Mater Sci Eng 778:012034. https://doi.org/10.1088/1757-899X/778/1/012034

Ahsan SM, Thomas M, Reddy KK, Sooraparaju SG, Asthana A, Bhatnagar I (2018) Chitosan as biomaterial in drug delivery and tissue engineering. Int J Biol Macromol 110:97–109

Aiba SI (1994) Preparation of *N*-acetylchitooligosaccharides from lysozymatic hydroxylate of partially *N*-acetylated chitosans. Carbohydr Res 261:297–306

Akiyama K, Kawazu K, Kobayashi A (1995) A novel method for chemo-enzymatic synthesis of elicitor-active chitosan oligomers and partially *N*-deacetylated chitin oligomers using *N*-acylated chitotrioses as substrates in a lysozyme-catalyzed transglycosylation reaction system. Carbohydr Res 279:151–160

Aktuganov GE, Melentiev AI, Varlamov VP (2019) Biotechnological aspects of the enzymatic preparation of bioactive chitooligosaccharides. Appl Biochem Microbiol 55:323–343

Aloise PA, Lumme M, Haynes CA (1996) *N*-Acetyl-D-glucosamine production from chitin waste using chitinase from *Serratia marcescens*. In: Muzzarelli RAA (ed) Chitin enzymology, vol 2. European Chitin Society, Grottammare, pp 581–594

Amano K, Ito E (1978) The action of lysozyme on partially deacetylated chitin. Eur J Biochem 78:97–104

Aranaz I, Mengíbar M, Harris R, Paños I, Miralles B, Acosta N, Heras Á (2009) Functional characterization of chitin and chitosan. Curr Chem Biol 3(2):203–230

Baxter S, Zivanovic S, Weiss J (2005) Molecular weight and degree of acetylation of high-intensity ultrasonicated chitosan. Food Hydrocoll 19:821–830

Behera HT, Mojumdar A, Das SR, Jema S, Ray L. (2020) Production of N-acetyl chitooligosaccharide by novel *Streptomyces chilikensis* strain RC1830 and its evaluation for anti-radical, anti-inflammatory, antiproliferative and cell migration potential. Bioresour Technol Rep 11:100428. https://doi.org/10.1016/j.biteb.2020.100428

Beygmoradi A, Homaei A, Hemmati R, Santos-Moriano P, Hormigo D, Fernández-Lucas J (2018) Marine chitinolytic enzymes, a biotechnological treasure hidden in the ocean? Appl Microbiol Biotechnol 102:9937–9948

Cantarel BL, Coutinho PM, Rancurel C, Bernard T, Lombard V, Henrissat B (2009) The carbohydrate-active EnZymes database (CAZy): an expert resource for glycogenomics. Nucleic Acids Res 37:D233–D238

Chandrasekaran M (ed) (2015) Enzyme in food and beverage processing. CRC Press, Tyler and Francis, Boca Raton, FL

Chen X, Zhai C, Kang L, Li C, Yan H, Zhou Y, Yu X, Ma L (2012) High-level expression and characterization of a highly thermostable chitosanase from *Aspergillus fumigatus* in *Pichia pastoris*. Biotechnol Lett 34(4):689–694

Cheng L, Li YK (2000) An *Aspergillus* chitosanase with potential for large-scale preparation of chitosan oligosaccharides. Biotechnol Appl Biochem 32:197–203

Cheng CY, Tsai CH, Liou PJ, Wang CH (2021) Pilot-scaleproduction of chito-oligosaccharides using an innovative recombinant chitosanase preparation approach. Polymers 13:290. https://doi.org/10.3390/polym13020290

Chiu E, Hijnen M, Bunker RD, Boudes M, Rajendran C, Aizel K et al (2015) Structural basis for the enhancement of virulence by viral spindles and their in vivo crystallization. Proc Natl Acad Sci 112:3973–3978

Choi YJ, Kim EJ, Piao Z, Yun YC, Shin YC (2004) Purification and characterization of chitosanase from *Bacillus* sp. strain KCTC 0377BP and its application for the production of chitosan oligosaccharides. Appl Environ Microbiol 70:4522–4531

Feisal K, Montarop Y (2010) Chitin research revisited. Mar Drugs 8:1988–2012

Fernandes P (2016) Enzymes in fish and seafood processing. Front Bioeng Biotech 4:59. https://doi.org/10.3389/fbioe.2016.00059

Fukamizo T, Juffe AH, Vogel JH, Honda Y, Tremblay H, Boucher I, Neugebaufr WA, Brzezinski R (2000) Theoretical calculation of pKa reveals an important role of Arg205 in the activity and stability of *Streptomyces* sp N174 chitosanase. J Biol Chem 275:25633–25640

Gaber Y, Rashad B, Hussein R, Abdelgawad M, Ali NS, Dishisha T, Várnai A (2020) Heterologous expression of lytic polysaccharide monooxygenases (LPMOs). Biotechnol Adv 43:107583. https://doi.org/10.1016/j.biotechadv.2020.107583

Gao XA, Jung WJ, Kuk JH, Park RD (2009) Reaction pattern of *Bacillus cereus* D-11 chitosanase on chito-oligosaccharide alcohols. J Microbiol Biotechnol 19:358–361

Guan SP, Mok YK, Koo KN, Chu KL, Wong W (2009) Chitinases: biomarkers for human diseases. Protein Pept Lett 16:490–498
Gurpilhares DB, Sette, LD,Pessoa Jr A (2019) Microbial enzymes and potential use in algal polysaccharide modifications. In: Trincone A (ed), Enzymatic technologies for marine polysaccharides. CRC Press, Taylor & Francis Group, Boca Raton, FL, pp 375–394
Haebel S, Bahrke S, Peter MG (2007) Quantitative sequencing of complex mixtures of heterochitooligosaccharides by MALDI-linear ion trap mass spectrometry. Anal Chem 79:5557–5566
Hamer SN, Cord-Landweh S, Biarnés X, Planas A, Waegeman H, Moerschbacher BM, Kolkenbrock S (2015) Enzymatic production of defined chitosan oligomers with a specific pattern of acetylation using a combination of chitin oligosaccharide deacetylases. Sci Rep 5:8716. https://doi.org/10.1038/srep08716
Harish PKV, Tharanathan RN (2007) Chitin/chitosan: modifications and their unlimited application potential-an overview. Trends Food Sci Tech 18(3):117–131
Hemsworth GR, Johnston EM, Davies GJ, Walton PH (2015) Lytic polysaccharide monooxygenases in biomass conversion. Trends Biotechnol 33(12):747–761
Hoell IA, Vaaje-Kolstad G, Eijsink VGH (2010) Structure and function of enzymes acting on chitin and chitosan. Biotechnol Genet Eng Rev 27:331–366
Horn SJ, Sorbotten A, Synstad B, Sikorski P, Sorlie M, Varum KM, Eijsink VGH (2006) Endo/exo mechanism and processivity of family 18 chitinases produced by *Serratia marcescens*. FEBS J 273:491–503
Hsiao YC, Lin YW, Su CK, Chiang BH (2008) High degree polymerized chitooligosaccharides synthesis by chitosanase in the bulk aqueous system and reversed micellar micro reactors. Process Biochem 43(1):76–82
Huang QS, Xie XL, Liang G, Gong F, Wang Y, Wei XQ, Wang Q, Ji ZL, Chen QX (2012) The GH18 family of chitinases: their domain architectures, functions and evolutions. Glycobiology 22:23–34
Ilankovan P, Hein S, Ng CH, Trung TS, Stevens WF (2006) Production of *N*-acetyl chitobiose from various chitin substrates using commercial enzymes. Carbohydr Polym 63:245–250
Ismail SA, El-Sayed HS, Fayed B (2020) Production of prebiotic chitooligosaccharide and its nano/microencapsulation for the production of functional yoghurt. Carbohydr Polym 234:115941. https://doi.org/10.1016/j.carbpol.2020.115941
Je JY, Kim SK (2012) Chitooligosaccharides as potential nutraceuticals: production and bioactivities. Adv Food Nutr Res 65:321–336
Jeon YJ, Kim SK (2000) Production of chitooligosaccharides using an ultrafiltration membrane reactor and their antibacterial activity. Carbohydr Polym 41:133–141
Joseph SM, Krishnamoorthy S, Paranthaman R, Moses JA, Anandharamakrishnan C (2021) A review on source-specific chemistry, functionality, and applications of chitin and chitosan. Carbohydr Polym Technol Appl 2:100036. https://doi.org/10.1016/j.carpta.2021.100036
Jung WJ, Park RD (2014) Bioproduction of chitooligosaccharides: present and perspectives. Mar Drugs 12(11):5328–5356
Jung WJ, Souleimanov A, Park RD, Smith DL (2007) Enzymatic production of N-acetyl chitooligosaccharides by crude enzyme derived from *Paenibacillus illinosensis* KJA-424. Carbohydr Polym 67:256–259
Kaczmarek MB, Struszczyk-Swita K, Li X, Szczesna-Antczak M, Daroch M (2019) Enzymatic modifications of chitin, chitosan, and chitooligosaccharides. Front Bioeng Biotechnol 7:243. https://doi.org/10.3389/fbioe.2019.00243
Kasongo KJ, Tubadi DJ, Bampole LD, Kaniki TA, Kanda NJM, Lukumu ME (2020) Extraction and characterization of chitin and chitosan from *Termitomyces titanicus*. SN Appl Sci 2:406. https://doi.org/10.1007/s42452-020-2186-5
Kasprzewska A (2003) Plant chitinases-regulation and function. Cell Mol Biol Lett 8(3):809–824
Kidibule PE, Santos-Moriano P, Jiménez-Ortega E, Ramírez-Escudero M, Limón MC, Remacha M, Plou FJ, Sanz-Aparicio J, Fernández-Lobato M (2018) Use of chitin and chitosan to produce

new chitooligosaccharides by chitinase Chit42: Enzymatic activity and structural basis of protein specificity. Microb Cell Fact 17:47. https://doi.org/10.1186/s12934-018-0895-x

Kim SK, Rajapakse N (2005) Enzymatic production and biological activities of chitosan oligosaccharides (COS): a review. Carbohydre Polym 62(4):357–368

Kim SY, Shon DH, Lee KH (1998) Purification and characteristics of two types of chitosanases from *Aspergillus fumigatus*. J Microbiol Biotechnol 8:568–574

Kim SK (ed) (2010) Chitin, chitosan, oligosaccharides and their derivatives: Biological activities and applications. CRC Press, Taylor and Francis Group, Boca Raton, FL. https://doi.org/10.1201/EBK1439816035

Kittur FS, Vishu Kumar AB, Varadaraj MC, Tharanathan RN (2005) Chitooligosaccharides-preparation with the aid of pectinase isozyme from *Aspergillus niger* and their antibacterial activity. Carbohydr Res 340(6):1239–1245

Kobayashi S, Kiyosada T, Shoda S (1996) Synthesis of artificial chitin: irreversible catalytic behavior of a glycosyl hydrolase through a transition state analogue substrate. J Am Chem Soc 118:13113–13114

Kobayashi S, Makino A, Matsumoto H, Kunii S, Ohmae M, Kiyosada T, Makiguchi K, Matsumoto A, Horie M, Shoda SI (2006) Enzymatic polymerization to novel polysaccharides having a glucose-N-acetylglucosamine repeating unit, a cellulose-chitin hybrid polysaccharide. Biomacromol 7:1644–1656

Kuk JH, Jung WJ, Jo GH, Kim KY, Park RD (2006) Production of N, N'-diacetylchitobiose from chitin using temperature-sensitive chitinolytic enzyme preparations of *Aeromonas* sp. GJ-18. World J Microbiol Biotechnol 22:135–139

Kumar ABV, Gowda LR, Tharanathan RN (2004) Non-specific depolymerization of chitosan by pronase and characterization of the resultant products. Eur J Biochem 271:713–723

Kumar M, Madhuprakash J, Balan V, Singh AK, Vivekanand V, Pareek N (2021) Chemoenzymatic production of chitooligosaccharides employing ionic liquids and Thermomyces lanuginosus chitinase. Bioresour Technol 337:125399. https://doi.org/10.1016/j.biortech.2021.125399

Kurakake M, You S, Nakagawa K, Sugihara M, Komaki T (2000) Properties of chitosanase from *Bacillus cereus* S1. Curr Microbiol 40:6–9

Kurita K (2006) Chitin and chitosan: functional biopolymers from marine crustaceans. Mare Biotechnol 8(3):203–226

Kuroiwa T, Noguchi Y, Nakajima M, Sato S, Mukataka S, Ichikawa S (2008) Production of chitosan oligosaccharides using chitosanase immobilized on amylose-coated magnetic nanoparticles. Process Biochem 43:62–69

Langner T, Göhre V (2016) Fungal chitinases: function, regulation, and potential roles in plant/pathogen interactions. Curr Genet 62:243–254

Laura RC, del Marıa CM, Sergio H, Sergio R, Keiko S (2006) Enzymatic hydrolysis of chitin in the production of oligosaccharides using *Lecanicillium fungicola* chitinases. Process Biochem 41:1106–1110

Le Dévédec F, Bazinet L, Furtos A, Venne K, Brunet S, Mateescu MA (2008) Separation of chitosan oligomers by immobilized metal affinity chromatography. J Chromatogr A 1194:165–171

Lee HW, Choi JW, Han DP, Park MJ, Lee NW, Yi DH (1996) Purification and characteristics of chitosanase from *Bacillus* sp. HW-002. J Microbiol Biotechnol 6:19–25

Lee DX, Xia WS, Zhang JL (2008) Enzymatic preparation of chitooligosaccharides by commercial lipase. Food Chem 111:291–295

Li H, Fei Z, Gong J, Yang T, Xu Z, Shi J (2015) Screening and characterization of a highly active chitosanase based on metagenomic technology. J Mol Catal B Enzym 111:29–35

Li K, Ronge X, Song L, Pengcheng L (2016) Advances in preparation, analysis and biological activities of single chitooligosaccharides. Carbohydr Polym 139:178–190

Li C, Wang L-X (2016) Endoglycosidases for the synthesis of polysaccharides and glycoconjugates. Adv Carbohydr Chem Biochem 73. http://dx.doi.org/https://doi.org/10.1016/bs.accb.2016.07.001

Liang TW, Hsieh JL, Wang SL (2012) Production and purification of a protease, a chitosanase, and chitin oligosaccharides by *Bacillus cereus* TKU022 fermentation. Carbohydr Res 362:38–46
Lienemann M, Boer H, Paananen A, Cottaz S, Koivula A (2009) Toward understanding of carbohydrate binding and substrate specificity of a glycosyl hydrolase 18 family (GH-18) chitinase from *Trichoderma harzianum*. Glycobiology 19:1116–1126
Lin H, Wang H, Xue C, Ye M (2002) Preparation of chitosan oligomers by immobilized papain. Enzyme Microb Technol 31:588–592
Liu X, Kokare C (2017) Microbial enzymes of use in industry. Biotechnol Microb Enzyme 267–298
Lopatin SA, Derbeneva MS, Kulikov SN, Varlamov VP, Shpigun OA (2009) Fractionation of chitosan by ultrafiltration. J Anal Chem 64:648–651
Lukas H, Simone Z, Verena SS (2012) Fungal chitinases: diversity, mechanistic properties and biotechnological potential. Appl Microbiol Biotechnol 93:533–543
Mahata M, Shinya S, Masaki E, Yamamoto T, Ohnuma T, Brzezinski R, Mazumder TK, Yamashita K, Narihiro K, Fukamizo T (2014) Production of chitooligosaccharides from *Rhizopus oligosporus* NRRL2710 cells by chitosanase digestion. Carbohydr Res 383:27–33
Malik A, Pretty (2019) Purification and properties of plant chitinases: A review. J Food Biochem e12762. https://doi.org/10.1111/jfbc.12762
Mallakuntla MK, Vaikuntapu PR, Bhuvanachandra B, Podile AR (2020) Selection and mutational analyses of the substrate interacting residues of a chitinase from *Enterobacter cloacae* subsp. cloacae (EcChi2) to improve transglycosylation. Int J Biol Macromol 165:2432–2441
Mangas-Sánchez J, Adlercreutz P (2015) Enzymatic preparation of oligosaccharides by transglycosylation: a comparative study of glucosidases. J Mol Catal B Enzym 122:51–55
Marmouzi I, Ezzat SM, Salama MM, Merghany RM, El-Desoky AAM, AM, Mohamed SO (2019) Recent updates in pharmacological properties of chitooligosaccharides. Biomed Res Int 2019:4568039. https://doi.org/10.1155/2019/4568039
Mekasha S, Byman IR, Lynch C, Toupalová H, Anděra L, Næs T et al (2017) Development of enzyme cocktails for complete saccharification of chitin using monocomponent enzymes from *Serratia marcescens*. Process Biochem 56:132–138
Miguez N, Kidibule P, Santos-Moriano P, Ballesteros AO, Fernandez-Lobato M, Plou FJ (2021) Enzymatic synthesis and characterization of different families of chitooligosaccharides and their bioactive properties. Appl Sci 11:3212. https://doi.org/10.3390/app11073212
Ming M, Kuroiwa T, Ichikawa S, Sato S, Mukataka S (2006) Production of chitosan oligosaccharides by chitosanase directly immobilized on an agar gel coated multi disk impeller. Biochem Eng J 28(3):289–294
Montilla A, Ruiz-Matute AI, Corzo N, Giacomini C, Irazoqui G (2013) Enzymatic generation of chitooligosaccharides from chitosan using soluble and immobilized glycosyltransferase (Branchzyme). J Agric Food Chem 61:10360–10367
Moon C, Seo DJ, Song YS, Hong SH, Choi SH, JungWJ(2017) Antifungal activity and patterns of *N*-acetyl-chitooligosaccharide degradation via chitinase produced from Serratia marcescens PRNK-1. Microb Pathog 113:218–224
Morin-Crini N, Lichtfouse É, Torri G, Crini G (2019a) Fundamentals and applications of chitosan. In: Crini G, Lichtfouse É (eds) Chitin and chitosan-history, fundamentals and innovations. Sustainable agriculture reviews 35. Springer, Switzerland, p 49–123
Morin-Crini N,· Lichtfouse É, Torri G, Crini G (2019b). Applications of chitosan in food, pharmaceuticals, medicine, cosmetics, agriculture, textiles, pulp and paper, biotechnology, and environmental chemistry. Environ Chem Lett 17:1667–1692
Mutahir Z, Mekasha S, Loose JSM, Abbas F, Vaaje-Kolstad G, Eijsink VGH et al (2018) Characterization and synergistic action of a tetra-modular lytic polysaccharide monooxygenase from *Bacillus cereus*. FEBS Lett 592:2562–2571
Muzzarelli RAA (1997) Depolymerization of chitins and chitosans with hemicellulase, lysozyme, papain and lipase. In: Muzzarelli RAA, Peter MG (eds) Chitin hand book. European Chitin Society, Grottammare, Italy, pp 153–165

Muzzarelli RAA, Boudrant J, Meyer D, Manno N, DeMarchis M, Paoletti MG (2012) Current views on fungal chitin/chitosan, human chitinases, food preservation, glucans, pectins and inulin: a tribute to Henri Braconnot, precursor of the carbohydrate polymers science, on the chitin bicentennial. Carbohydr Polym 87:995–1012

Nakagawa YS, Eijsink VGH, Totani K, Vaaje-Kolstad G (2013) Conversion of α-chitin substrates with varying particle size and crystallinity reveals substrate preferences of the chitinases and lytic polysaccharide monooxygenase of *Serratia marcescens*. J Agric Food Chem 61:11061–11066

Nakagawa YS, Kudo M, Onodera R, Ang LZP, Watanabe T, Totani K, Eijsink VGH, Vaaje-Kolstad G (2020) Analysis of four chitin-active lytic polysaccharide monooxygenases from streptomyces griseus reveals functional variation. J Agric Food Chem 68(47):13641–13650

Nguyen AD, Huang CC, Liang TW, Nguyen VB, Pan PS, Wang SL (2014) Production and purification of a fungal chitosanase and chitooligomers from *Penicilium janthinellum* D4 and discovery of the enzyme activators. Carbohydr Polym 108:331–337

Nidheesh T, Pal GK, Suresh PV (2015a) Chitooligomers preparation by chitosanase produced under solid state fermentation using shrimp by-products as substrate. Carbohydr Polym 121:1–9

Nidheesh T, Pal GK, Suresh PV (2015b) Enzymatic degradation of chitosan and production of D-glucosamine by solid substrate fermentation of exo-β-D-glucosaminidase (exo-chitosanase) by *Penicillium decumbens* CFRNT15. Int Biodeterior Biodegrad 97:97–106

Nidheesh T (2016) Microbial production of chitosan oligosaccharides and its potential applications. Ph.D. Thesis, The Academy of Scientific and Innovative Research (AcSIR), Ghaziabad, India

Nidheesh T, Suresh PV (2015a) Optimization of conditions for isolation of high quality chitin from shrimp processing raw byproducts using response surface methodology and its characterization. J Food Sci Technol 52(6):3812–3823

Nidheesh T, Suresh PV (2015b) Functional polysaccharides from shellfish by-products. In: Sachindra NM, Mahendraker SM (eds) Fish processing by-products: quality assessment and applications. Stadium Press LLC, Houston, TX. p 299–345

Oh CH, De Zoysa M, Kang DH, Lee YD, Whang IS, Nikapitiya C, Heo SJ, Yoon KT, Affan A, Lee JH (2011) Isolation, Purification, and enzymatic characterization of extracellular chitosanase from marine bacterium *Bacillus subtilis* CH2. J Microbiol Biotechnol 21:1021–1025

Osswald WF, Shapiro JP, Doostdar H, McDonald RE, Niedz RP, Nairn CJ, Hearn CJ, Mayer RT (1994) Identification and characterization of acidic hydrolases with chitinase and chitosanase activities from sweet orange callus tissue. Plant Cell Physiol 35:811–820

Oyeleye A, Normi YM (2018) Chitinase: Diversity, limitations, and trends in Engineering for suitable applications. Biosci Rep 38:BSR2018032300. https://doi.org/10.1042/BSR20180323

Palanivelu P, Vaishna L (2013) Molecular cloning and characterization of a chitinase gene from the thermophilic fungus, *Thermomyces lanuginosus*. J Pharm Biol Sci 5:99–107

Pan AD, Zeng HY, Foua GB, Alain C, Li YQ (2016) Enzymolysis of chitosan by papain and its kinetics. Carbohydr Polym 135:199–206

Pantaleone D, Yalpani M, Scollar M (1992) Unusual susceptibility of chitosan to enzymic hydrolysis. Carbohydr Res 237:325–332

Park JK, Shimono K, Ochiai N, Shigeru K, Kurita M, Ohta Y, Tanaka K, Matsuda H, Kawamukai M (1999) Purification, characterization, and gene analysis of a chitosanase (ChoA) from *Matsuebacter chitosanotabidus* 3001. J Bactriol 181:6642–6649

Pechsrichuang P, Yoohat K, Yamabhai M (2013) Production of recombinant *Bacillus subtilis* chitosanase, suitable for biosynthesis of chitosan-oligosaccharides. Bioresour Technol 127:407–414

Purushotham P, Podile AR (2012) Synthesis of long-chain chitooligosaccharides by a hypertransglycosylating processive endochitinase of *Serratia proteamaculans* 568. J Bacteriol 194:4260–4271

Reetarani SP, Vandana G, Deshpande MV (2000) Chitinolytic enzymes: an exploration. Enzyme and Microb Technol 26:473–483

Rinaudo M (2006) Chitin and chitosan: properties and applications. Prog Polym Sci 31(7):603–632

Sabbadin F, Hemsworth GR, Ciano L, Henrissat B, Dupree P, Tryfona T et al. (2018) An ancient family of lytic polysaccharide monooxygenases with roles in arthropod development and biomass digestion. Nat Commun 9. https://doi.org/10.1038/s41467-018-03142-x

Saito A, Fujii T, Yoneyama T, Redenbach M, Ohno T, Watanabe T et al (1999) High-multiplicity of chitinase genes in *Streptomyces coelicolor* A3. Biosci Biotechnol Biochem 63:710–718

Sáncheza Á, Mengíbara M, Rivera-Rodríguez G, Moerchbacher B, Acosta N, Heras A (2017) The effect of preparation processes on the physicochemical characteristics and antibacterial activity of chitooligosaccharides. Carbohydr Polym 157:251–257

Santos-Moriano P, Woodley JM, Plou FJ (2016) Continuous production of chitooligosaccharides by an immobilized enzyme in a dual-reactor system. J Mol Catal B Enzym 113:211–217. https://doi.org/10.1016/j.molcatb.2016.09.001

Santos-Moriano P, Kidibule PE, Alleyne E, Ballesteros AO, Heras A, Fernandez-Lobato M, Plou FJ (2018) Efficient conversion of chitosan into chitooligosaccharides by a chitosanolytic activity from *Bacillus thuringiensis*. Process Biochem 73:102–108

Santos-Moriano P, Kidibule P, Míguez N, Fernández-Arrojo L, Ballesteros AO, Fernández-Lobato M, Plou FJ (2019) Tailored enzymatic synthesis of chitooligosaccharides with different deacetylation degrees and their anti-inflammatory activity. Catalysts 9:405. https://doi.org/10.3390/catal9050405

Shahidi F, Abuzaytoun R (2005) Chitin, chitosan, and co-products: chemistry, production, applications, and health effects. Adv Food Nutr Res 49:93–135

Shang Q, Jiang H, Cai C, Hao J, Li G, Yu G (2018) Gut microbiota fermentation of marine polysaccharides and its effects on intestinal ecology: an overview. Carbohydr Polym 179:173–185

Sheldon RA, Woodley JM (2018) Role of biocatalysis in sustainable chemistry. Chem Rev 118:801–838

Shimosaka M, Nogawa M, Wang XY, Kumehara M, Okazaki M (1995) Production of two chitosanases from a chitosan-assimilating bacterium, *Acinetobacter* sp. strain CHB101. Appl Environ Microbiol 61:438–442

Shinya S, Fukamizo T (2017) Interaction between chitosan and its related enzymes: a review. Int J Biol Macromol 104:1422–1435

Singh R, Weikert T, Basa S, Moerschbache BM (2019) Structural and biochemical insight into mode of action and subsite specificity of a chitosan degrading enzyme from *Bacillus spec.* MN Sci Rep 9:1132. https://doi.org/10.1038/s41598-018-36213-6

Singh RV, Sambyal K, Negi A, Sonwani S, Mahajan R (2021) Chitinases production: a robust enzyme and its industrial applications. Biocatal Biotransform 39:161–189

Sinha S, Chand S, Tripathi P (2016) Recent progress in chitosanase production of monomer-free chitooligosaccharides: bioprocess strategies and future applications. Appl Biochem Biotechnol 180:883–899

Sirimontree P, Suginta W, Sritho N, Kanda Y, Shinya S, Ohnuma T, Fukamizo T (2014) Mutation strategies for obtaining chitooligosaccharides with longer chains by transglycosylation reaction of family GH18 Chitinase. Biosci Biotechnol Biochem 78:2014–2021

Sørbotten A, Horn SJ, Eijsink VG, Vårum KM (2005) Degradation of chitosans with chitinase B from *Serratia marcescens*. Production of chito-oligosaccharides and insight into enzyme processivity. FEBS J 272:538–549

Stoykov YM, Pavlov AI, Krastanov AI (2015) Chitinase biotechnology: production, purification, and application. Eng Life Sci 15:30–38

Sun L, Adams BJ, Gurnon JR, Ye Y, Van Etten JL (1999) Characterization of two chitinase genes and one chitosanase gene encoded by *Chlorella* Virus PBCV-1. Virology 263:376–387

Sun H, Gao L, Xue C, Mao X (2020) Marine-polysaccharide degrading enzymes: status and prospects. Compr Rev Food Sci Food Saf 19:2767–2796

Suresh PV (2019) Enzymatic technologies of chitin and chitosan. In: Trincone A (ed) Enzymatic technologies for marine polysaccharides. CRC Press, Taylor and Francis Group, Boca Raton, FL, pp 449–491

Suresh PV, Kumar PA (2012) Enhanced degradation of α-chitin materials prepared from shrimp processing byproduct and production of *N*-acetyl-D-glucosamine by thermoactive chitinases from soil mesophilic fungi. Biodegradation 23(4):597–607

Suresh PV, Kumar PA, Sachindra NM (2011a) Thermo active β-*N*-acetylhexosaminidase production by a soil isolate of *Penicillium monoverticillium* CFR 2 under solid state fermentation: parameter optimization and application for *N*-acetyl chitooligosaccharides preparation from chitin. World J Microbiol Biotechnol 27(6):1435–1447

Suresh PV, Sachindra NM, Bhaskar N (2011b) Solid state fermentation production of chitin deacetylase by *Colletotrichum lindemuthianum* ATCC 56676 using different Substrates. J Food Sci Technol 48(3):349–356

Suresh PV, Sakhare PZ, Sachindra NM, Halami PM (2014) Extracellular chitin deacetylase production in solid state fermentation by native soil isolates of *Penicillium monoverticillium* and *Fusarium oxysporum.* J Food Sci Technol 51(8):1594–1599

Suresh PV, Nidheesh T, Pal GK (2015) Enzymes in seafood processing. In: Chandrasekaran M (ed) Enzymes in food and beverage processing. CRC Press, Tylor and Francis, Boca Raton, FL, pp 353–377

Suresh PV, Tanaji GK, Lidiya CJ (2018) Sustainable valorization of seafood processing by-products/discards. In: Singhania RR, Agarwal RA, Kumar RP, Sukumaran RK (eds) Waste to Wealth. Springer, Singapore, pp 11–139

Synowiecki J, Al-Khateeb NA (2003) Production, properties, and some new applications of chitin and its derivatives. Crit Rev Food Sci Nutr 43:145–171

Takiguchi Y, Shimahara K (1988) *N*, N-Diacetylchitobiose production from chitin by *Vibrio anguillarum* strain E-383a. Lett Appl Microbiol 6:129–131

Tanabe T, Morinaga K, Fukamizo T, Mitsutomi M (2003) Novel chitosanase from streptomyces griseus HUT 6037 with transglycosylation activity. Biosci Biotechnol Biochem 67:354–364

Tanaka T, Fujiwara S, Nishikori S, Fukui T, Takagi M, Imanka T (1999) A unique chitinase with dual active sites and triple substrate binding sites from the hyperthermophilic *Pyrococcus kodakaraensis* KOD1. Appl and Environ Microbiol 65:5338–5344

Tanaka T, Fukui T, Atomi H, Imanaka T (2003) Characterization of an exo-β-D-glucosaminidase involved in a novel chitinolytic pathway from the hyperthermophilic archaeon *Thermococcus kodakaraensis* KOD1. J Bact 185:5175–5181

Tandrup T, Frandsen KEH, Johansen KS, Berrin JG, Lo Leggio L (2018) Recent insights into lytic polysaccharide monooxygenases (LPMOs). Biochem Soc Trans 46:1431–1447

Terbojevich M, Cosani A, Muzzarelli RAA (1996) Molecular parameters of chitosans depolymerized with the aid of papain. Carbohydr Polym 29:63–68

Thadathil N, Velappan SP (2014) Recent developments in chitosanase research and its biotechnological applications: a review. Food Chem 150:392–399

Thadathil N, Kuttappan AP, Vallabaipatel E, Kandasamy M, Velappan SP (2014) Statistical optimization of solid state fermentation conditions for the enhanced production of chitinases by mesophilic soil fungi using response surface methodology and their application in the reclamation of shrimp processing by-products. Ann Microbiol 64:671–681

Tokuyasu K, Ono H, Kameyama MO, Hayashi K, Mori Y (1997) Deacetylation of chitin oligosaccharides of DP 2–4 by chitin deacetylase from *Colletotrichum lindemuthianum.* Carbohydr Res 303(3):353–358

Tokuyasu K, Ono H, Mitsutomi M, Hayashi K, Mori Y (2000) Synthesis of a chitosan tetramer derivative, beta-D-GlcNAc-(1→4)-beta-D-GlcNAc-(1→4)-beta-D-GlcNAc-(1→4)-D-GlcN through a partial *N*-acetylation reaction by chitin deacetylase. Carbohydr Res 325(3):211–215

Trincone A (ed) (2019) Enzymatic technologies for marine polysaccharides. CRC Press, Taylor and Francis Group, Boca Raton, FL

Tsigos I, Martinou A, Kafetzopoulos D, Bouriotis V (2000) Chitin deacetylase: new, versatile tools in biotechnology. Trends Biotechnol 18:305–311

Tzelepis G, Karlsson M (2021) The fungal chitinases. In: Zaragoza O, Casadevall A (eds) Encyclopedia of mycology. Elsevier, pp 23–31

Vaaje-Kolstad G, Forsberg Z, Loose JS, Bissaro B, Eijsink VG (2017) Structural diversity of lytic polysaccharide monooxygenases. Curr Opin Struc Biol 44:67–76

Vaaje-Kolstad G, Westereng B, Horn SJ, Liu Z, Zhai H, Sørlie M et al. (2010) An oxidative enzyme boosting the enzymatic conversion of recalcitrant polysaccharides. Science 80(330):219–222

Venugopal V (2021) Valorization of seafood processing discards: bioconversion and bio-refinery approaches. Front Sustain Food Syst 5:611835. https://doi.org/10.3389/fsufs.2021.611835

Vincenzi S, Bierma J, Wickramasekara SI, Curioni A, Gazzola D, Bakalinsky AT (2014) Characterization of a grape class IV chitinase. J Agric Food Chem 62(24):5660–5668

Wang SL, Tseng WN, Liang TW (2011) Biodegradation of shellfish wastes and production of chitosanases by a squid pen-assimilating bacterium, *Acinetobacter calcoaceticus* TKU024. Biodegradation 22:939–948

Wang SL, Liu CP, Liang TW (2012) Fermented and enzymatic production of chitin/chitosan oligosaccharides by extracellular chitinases from *Bacillus cereus* TKU027. Carbohydr Polym 90:1305–1313

Wang SL, Chen SJ, Wang CL (2008) Purification and characterization of chitinases and chitosanases from a new species strain *Pseudomonas* sp. TKU015 using shrimp shells as a substrate. Carbohydr Res 343(7):1171–1179

Wang D, Li Y, Zheng Y, Hsieh YSY (2021) Recent advances in screening methods for the functional investigation of lytic polysaccharide monooxygenases. Front Chem 9:653754. https://doi.org/10.3389/fchem.2021.653754

Wee YJ, Reddy LVA, Chung KC, Ryu HW (2009) Optimization of chitosanase production from *Bacillus* sp. RKY3 using statistical experimental designs. J Chem Technol Biotechnol 84:1356–1363

Weikert T, Niehues A, Cord-Landwehr S, Hellmann MJ, Moerschbacher BM (2017) Reassessment of chitosanase substrate specificities and classification. Nat Commun 8(1):1698. https://doi.org/10.1038/s41467-017-01667-1

Woo CJ, Park HD (2003) An extracellular *Bacillus* sp. chitinase for the production of chitotriose as a major chitinolytic product. Biotechnol Lett 25:409–512

Wu S (2012) Preparation of chitooligosaccharides from *Clanis bilineata* larvae skin and their antibacterial activity. Int J Biol Macromol 51(5):1147–1150

Wu Y, Huang Y, Zhou Y, Ren X, Yang F (2014) Degradation of chitosan by swirling cavitation. Innov Food Sci Emerg Technol 23:188–193

Xia W, Liu P, Liu J (2008) Advance in chitosan hydrolysis by non-specific cellulases. Bioresour Technol 99:6751–6762

Xia W, Liu P, Zhang J, Chen J (2011) Biological activities of chitosan and chitooligosaccharides. Food Hydrocoll 25(2):170–179

Xie Y, Hu J, Wei Y, Hong X (2009) Preparation of chitooligosaccharides by the enzymatic hydrolysis of chitosan. Polym Degrad Stab 94:1895–1899

Yang Y, Yu B (2014) Recent advances in the synthesis of chitooligosaccharides and congeners. Tetrahedron 70:1023–1046

Yoshida N, Tanaka T, Noguchi M, Kobayashi A, Ishikura K, Ikenuma T et al (2012) One-pot chemoenzymatic route to chitoheptaose via specific transglycosylation of chitopentaose-oxazoline on chitinase-template. Chem Lett 41(7):689–690

Younes I, Rinaudo M (2015) Chitin and chitosan preparation from marine sources. Structure, properties and applications. Mar Drugs 13:1133–1174

Zainol I, Akil HM, Mastor A (2009) Effect of irradiation on the physical and mechanical properties of chitosan powder. Mater Sci Eng C 29:292–297

Zargar V, Asghari M, Dashti A (2015) A review on chitin and chitosan polymers: structure, chemistry, solubility, derivatives, and applications. Chem Bio Eng Rev 2(3):204–226

Zhang H, Du Y, Yu X, Mitsutomi M, Aiba S (1999) Preparation of chitooligosaccharides from chitosan by acomplex enzyme. Carbohydr Res 320:257–260

Zhang D, Wang PG, Qi Q (2007) A two-step fermentation process for efficient production of penta-*N*-acetyl-chitopentaose in recombinant *Escherichia coli*. Biotechnol Lett 29(11):1729–1733

Zhao Y, Park RD, Muzzarelli RAA (2010) Chitin deacetylases: properties and applications. Mar Drugs 8:24–46

Zhu XF, Tan HQ, Zhu C, Liao L, Zhang XQ, Wu M (2012) Cloning and overexpression of a new chitosanase gene from *Penicillium* sp. D-1. AMB Express 2(1):13. https://doi.org/10.1186/2191-0855-2-13

Synthesis of Chitooligosaccharides Derivatives

Dai-Nghiep Ngo and Se-Kwon Kim

1 Introduction

Chitosan is prepared with alkaline deacetylation of chitin in the presence of alkali at high temperature. It consists of β-(1 → 4)-2-amino-D-glucose residues and water soluble by making salt with various acids on the amino groups of D-glucosamine units. Furthermore, partially acetylated chitosan having about 50% D-glucosamine units is only able to dissolve in water (Rinaudo 2006).

Chitin can be obtained from crustacean shells, mollusks, insects, and fungal cell walls (Kim et al. 2006; Yen et al. 2007). However, chitin and chitosan are currently manufactured from the major wastes in industrial sea-food. Chitin and chitosan are commercially prepared by a chemical method the following: Crustacean shells are treated with an aqueous 35% sodium hydroxide solution at room temperature overnight or at 80–90 °C to remove protein. Products, deproteinized shells are then neutralized and minerals are removed by treating with an aqueous 3–5% hydrochloric acid solution at room temperature to afford a white or slightly pink precipitate of chitin. Then the chitin is *N*-deacetylated with an aqueous 40–45% sodium hydroxide solution at 90–120 °C for 4–5 h to form chitosan. Another method friendly with environment is enzymatic method that would gradually replace the chemical method using protease under acidic conditions for the removal of both proteins and minerals at the same time to afford chitin, and the successive treatment with chitin-N-deacetylase

D.-N. Ngo (✉)
Department of Biochemistry, Faculty of Biology-Biotechnology, University of Science, Ho Chi Minh City 700000, Vietnam
e-mail: ndnghiep@hcmus.edu.vn

Vietnam National University, Ho Chi Minh City 700000, Vietnam

S.-K. Kim
Department of Marine Science and Convergence Engineering, College of Science and Technology, Hanyang University, Gyeonggi-do 11558, Korea

S.-K. Kim (ed.), *Chitooligosaccharides*,
https://doi.org/10.1007/978-3-030-92806-3_4

(EC 3.5.1.41) that catalyzes the *N*-deacetylation reaction of chitin into gives chitosan (Rodde et al. 2008; Rinaudo 2006).

Chitin and chitosan were applied in broad range of lot of fields such as agriculture, water-waste treatment and food processing. Furthermore, their new applications are concentrated on producing higher and higher valuable products by means of cosmetics, drug carriers, gen carriers food additives, semi-permeable membranes, and biopharmaceuticals. Chitin and its derivatives have a number of biological functions to serve as useful and safe bioactive substances in various applications because of specific bioactivities such as antimicrobial activity, immuno-enhancing effects, hypocholesterolemic activity (Ngo et al. 2008), antitumor, antioxidant (Qin et al. 2004; Liang et al. 2007; Ngo et al 2012; Trinh et al. 2014a), anticoagulant activity (Huang et al. 2003; Vongchan et al 2003), inhibitor of enzyme (e.g. β- secretase cause Alzheimer's disease, angiotensin I converting enzyme cause hypertension, prolyl endopeptidase, matrixmetalloproteinase-2, -9) (Je et al. 2007; Ngo et al. 2008, 2007; Yoon et al. 2009; Kim and Kim 2006), anti-inflammation (Trinh et al. 2014b), stimulation of wound healing (Okamura et al. 2005); Especially, biocompatibility, biodegradability, polyelectrolyte properties, presence of reactive functional groups and ability to modify the structure via chemical modifications, have attracted their applicability in biomedical and pharmaceutical industries (Sashiwa and Aiba 2004; Ahmed and Aljaeid 2016).

Even though chitin and chitosan were known to have very interesting functional properties in many areas, their high molecular weights, high viscosity and water insolubility, are the major limiting factors for its utilization in vivo. Furthermore, in order to improve absorption in animal intestines, chitin and chitosan were converted to oligomers. In the recent researches, the aim increases biological activities of them, oligomers were grafted functional groups to produce their derivatives. It is expected that oligomers are absorbed and possess bioactivity in the in vivo system, using oligomers and their derivatives are more efficient than that of chitin and chitosan. This chapter was focused on the production of oligomers of chitin (chitin oligosaccharides, NA-COS), chitosan (chitosan oligosaccharides, COS), and synthesis of their derivatives.

2 The Manufacture of Chitin Oligosaccharides and Chitooligosaccharides

In the manufacture of oligomers of chitin and chitosan (chitin oligosaccharides, NA-COSs and chitooligosaccharides, COSs, respectively), like other polysaccharides, chitin and chitosan can also be cleaved by hydrolyzing agents due to the presence of rather unstable glycosidic bonds. Degradation of their O-glycosidic linkages by different methods leads to the production of NA-COSs and COSs varying in the degree of polymerization as well as number and sequence of glucosamine (GlcN) and *N*-acetyl glucosamine (GlcNAc) units.

Some of these methods include acid hydrolysis (Einbu et al. 2007; Einbu and Vårum 2007; Ngo et al. 2008; 2009), enzymatic hydrolysis (Jeon and Kim 2000a, 2000b; Jung et al. 2007; Wu et al. 2008; Xie et al. 2012; Bui et al 2017), ultrasonic degradation (Yue et al. 2008), oxidative degradation (Shao et al. 2003), combination of gamma irradiation and enzymatic method (Nguyen et al 2007) and combination of gamma irradiation and hydrogen peroxide (Nguyen et al. 2011). Moreover, electromagnetic radiation, sonication, and mechanical energy generated by microfiltration can also be used for the production of their oligomers without employing chemical agents. Absorption of energy by chitosan molecules in any of the above methods results in scission of chemical bonds, and if the broken bond belongs to the backbone of the polymer (O-glycosidic bond), their molecular weight will be decreased (Einbu and Vårum 2007; Xing et al. 2005a, b).

Especially, Jeon and Kim (2000a) prepared chitooligosaccharides with various molecular weights using enzymatic method combined with a UF (ultrafiltration) membrane reactor system and improved continuously by connecting to an immobilized enzyme column reactor in which chitosanase from *Bacillus sp.* was absorbed on chitin as a carrier for immobilization. A 1% chitosan solution was prepared by dispersing chitosan in ratio 1L of water, dissolving it and stirring by adding 400 ml of 1 M lactic acid and making up to 15 L with water; the pH was adjusted to be 5.5 with a saturated sodium bicarbonate ($NaHCO_3$) solution. The chitosan was hydrolyzed by enzymatic reaction in the reactor system and fractioned by passing through UF membrane of molecular weight cut-off (MWCO) 10 kDa, 5 kDa, 3 kDa, and 1 kDa (Jeon and Kim 2000b). However, chitin oligosaccharides are currently prepared using primary acid hydrolysis and chitooligosaccharides were prepared by the major enzymatic hydrolysis. In addition, there are a lot of reports focusing on manufacturing chitooligosaccharides more than chitin oligosaccharides. (Kim and Rajapakse 2005).

Chitin (degree of deacetylation < 10%) which was prepared from crab shells was ground to a powder and placed into a reactor, adding 12 M hydrochloride with 10% content, stirring at 40 °C for various times which depended on the desired molecular weight of NA-COSs. Reaction was stopped by adding distilled water, centrifuged at 10,000 g to remove insoluble residues after hydrolytic chitin solution was neutralized using a 25% sodium hydroxide solution. Afterwards the hydrolysates were separated using Ultrafiltration (UF) membranes, which were molecular weight cut off (MWCO) 3 and 1 kDa, respectively. Using UF membrane to isolate two kinds of NA-COSs, NA-COS passed out through MWCO 3 kDa membrane but did not pass out through 1 kDa membrane, namely NA-COS 1–3 kDa and NA-COS passed out through MWCO 1 kDa membrane, NA-COS below 1 kDa. Supernatant was desalted and purified using a Micro Acilyzer G3 and then the solution was subjected to a spray drying. Finally, a light yellow powder of NA-COSs was obtained with different molecular weights. (Ngo et al. 2009) Fig. 1.

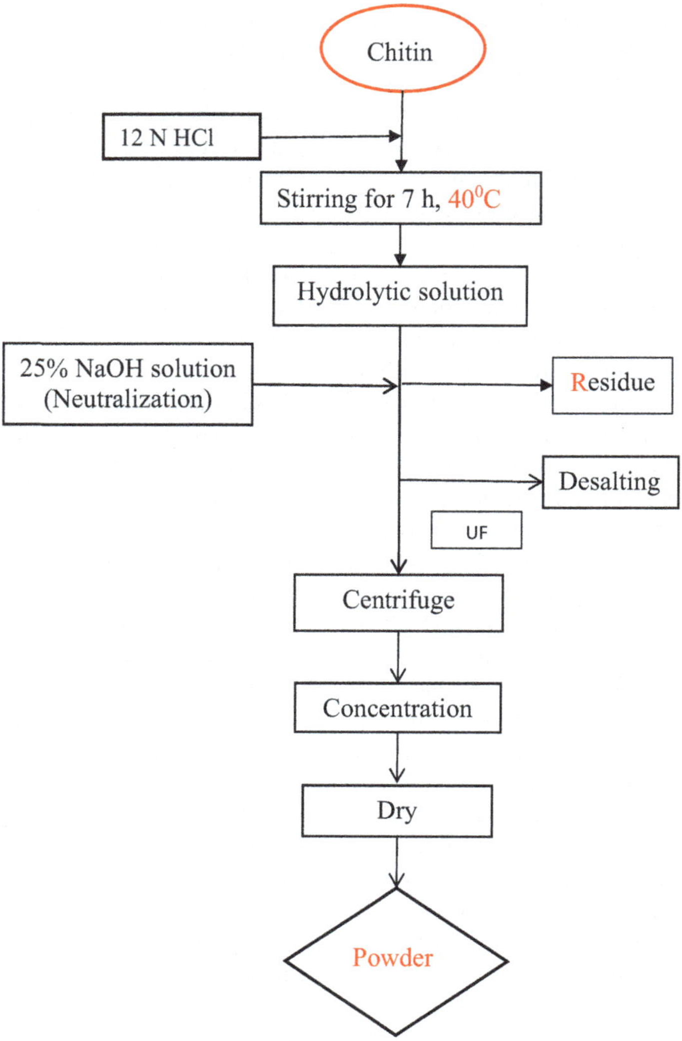

Fig. 1 Manufacturing process of chitin oligosaccharides (NA-COS) by HCl

3 Synthesis of Chitooligosaccharides Derivatives

In researches of the past 20 years, the biological activities of chitin and its derivatives attracted a greater attention such as antioxidant, anti-microorganism, anti-inflammation, and enzymatic inhibition. Most of the reports showed that biological activities of them depend on different molecular weights (MW), low MW has antioxidant activity higher than high MW (Kim and Thomas 2007). They also depend on degree of deacetylation, the higher deacetylation is higher antioxidant activity, based on the results obtained from studies carried out using electron spin trapping technique, chitooligosaccharides (COSs) with molecular weight range (1–3 kDa) and highly deacetylated (90%) COSs have been identified to have a higher potential to scavenge different radicals such as DPPH, hydroxyl, superoxide and alkyl radicals (Park et al. 2004). Furthermore, researches aim to improve biological activities, chitosan, a biopolymer, is the possibility of undergoing a wide range of modifications, both chemical and physical. Chitosan has three categories of reactive functional groups on its backbone which can be chemically modified with either inorganic or organic compounds with the amino group on the C-2 position and the primary and secondary hydroxyl groups on the C-3 and C-6 positions, respectively to enhance its activities (Alishahi and Aïder 2012; Lizardi-Mendoza et al. 2016). The derivatives of chitosan such as N, O-carboxymethyl chitosan (NOCC), hydroxypropylated chitosan (HPC), N-benzoylhexanoyl chitosan (NBHC), sulfated chitosan, low molecular weight carboxymethyl chitosan (LMWCMC), chitosan gallic acid, amino derivatized chitosan, and quaternary chitosan were prepared by chemical modification, and have been assayed their antioxidant capacity (Xie et al. 2001; Xing et al. 2005b; Sun et al. 2007; Pasanphan and Chirachanchai 2008; Je and Kim 2006; Guo et al. 2006). Xie et al. (2001) prepared hydroxypropyl chitosan maleic acid sodium and carboxymethyl chitosan sodium maleic acid sodium by graft copolymerization, and investigated hydroxyl radical scavenging effects. Different molecular weights of sulfated chitosan were synthesized and tested for antioxidant activity, the results suggested that their antioxidant activity depended on molecular weight and low molecular weight effected stronger than the other (Xing et al. 2005a; b). LMWCMC was synthesized to have superoxide anion scavenging activity (Sun et al. 2007). A novel synthetic chitosan derivative, chitosan-gallic acid has a wide range of antioxidant activity including alkyl and hydroxyl radicals (Pasanphan and Chirachanchai 2008). Je and Kim (2006) also synthesized novel amino derivatized chitosan (aminoethyl-chitosan, AEC, dimethylaminoethyl-chitosan, DMAEC, diethylaminoethyl-chitosan, DEAEC) with different DD (50 and 90%). AEC with 90% DD showed the highest scavenging effects against hydroxyl and superoxide anion radical. The results also showed that their antioxidant effect depends on substituted group. Furthermore, based on methods used for synthesizing chitosan derivatives, chitooligosaccharides (COSs) derivatives are also synthesized by grafting functional groups into C-2, C-3 and C-6 positions of the pyranose ring in COS structure. When comparing scavenging effect of quaternized COS with carboxylated COS, Huang et al. (2006a; b) found that their scavenging effect against hydroxyl and alkyl radicals depends on not only substituted group but also the special structure of group. Rajapakse et al. (2007) and Mendis et al. (2007) have investigated carboxylated COS which are non-toxic and exhibited inhibition of free radical mediated oxidation of cellular biomolecules in live cells.

Until now, some chemical methods to improve biological activities of chitooligosaccharides (COSs) using grafting functional groups on their back-bones are as follows:

3.1 Synthesis of Carboxylated Chitooligosaccharides

The carboxylated chitooligosaccharides (CCOS) were synthesized by grafting carboxyl group ($COCH_2CH_2COO$-) to the amino position of pyranose unit of COS. COS in 10% acetic acid with adding methanol was mixed to form a solution at room temperature and dropwise different amounts of succinic anhydride dissolved in acetone for 1 h to make CCOS with different substitution degrees. CCOS is easily soluble in water/media, therefore it can be used for various applications. (Huang et al. 2005a, b Esfandiarpour-Boroujeni et al. 2016).

3.2 Synthesis of Amino Derived Chitooligosaccharides

Amino derived COS in accordance with our previous method (Ngo et al. 2008, 2019; Yoon et al. 2009; Hong et al 2016). Aqueous 3.0 M (20 mL) 2-chlorethylamino hydrochloride for synthesizing aminoethyl chitooligosaccharide (AE-COS) or 2-dimethylamino-ethylchloride hydrochloride for dimethylaminoethyl chitooligosaccharide (DMAE-COS) or 2-diethylamino-ethylchloride hydrochloride for diethylaminoethyl chitooligosaccharide (DEAE-COS) was added to COS (0.40 g) while stirring at 40 °C. Sodium hydroxide (3.0 M, 20 mL) was added to the reaction mixture dropwise, and continuously stirred for 48 h. After reaction, the solution was filtered. Subsequently, the reaction mixture was acidified with 0.1 N HCl, and dialyzed against water for 2 days. The amino derived COSs were freeze dried to give AE-COS (0.334 g), DMAE-COS (0.342 g) and DEAE-COS (0.345 g), respectively. COSs, hydrolytic products of chitosan, have amino groups or acetamide groups at C-2, they were modified to make amino COS derivatives when the hydroxyl group at C-6 position was by aminoethyl or dimethylaminoethyl or diethylaminoethyl group because the C-6 hydroxyl groups had the highest reactivity for aminoethylation, while the structure of COS was maintained; the products formed was completely soluble in water. (Scheme 1).

3.3 Synthesis of Sulfated Chitooligosaccharides

The sulfated chitooligosaccharides (SCOS) were synthesized by the introduction of sulfonated groups at the 3 and 6 positions of COS. The mixture which contains dimethylformamide (DMF) solution (60 ml) is added to 1.5 g COS, followed by the addition of dichloroacetic acid (1.2 ml) is reacted for 24 h (A solution). The chlorosulfonic acid (10 ml) was slowly added to DMF solution (60 ml) at 0–4 °C to prepare sulfonated reagent (B solution). B solution is heated to 60 °C and slowly added to A solution to react for 2 h. After that, reacted mixture is adjusted to pH 7.0 with sodium hydroxide solution (20%). The supernatant after centrifugation to

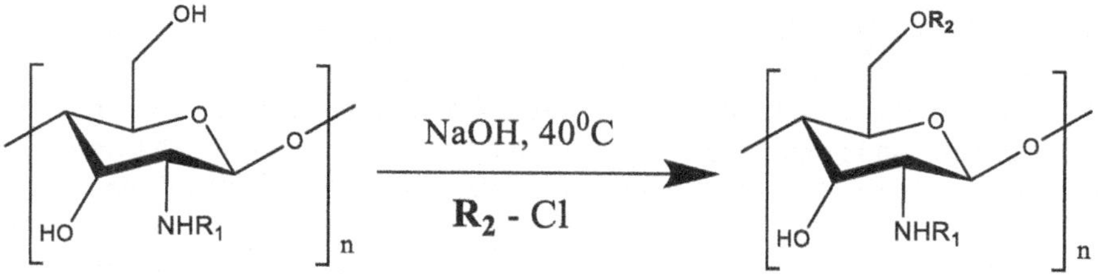

AE-COS : R_1=H, $COCH_3$; R_2=$(CH_2)_2NH_2$
DMAE-COS : R_1=H, $COCH_3$; R_2=$(CH_2)_2N(CH_3)_2$
DEAE-COS : R_1=H, $COCH_3$; R_2=$(CH_2)_2N(CH_2CH_3)_2$

Scheme 1 Synthesis of amino chitooligosaccharide derivatives

remove residues is washed with ethanol to obtain the precipitate. Then, the precipitate is dissolved in water and dialyzed for desalting. The solution is concentrated and freeze dried to isolate SCOS powder with 12.8% sulfur content and a degree of sulfation (DS) value of 2.1. (Liu et al 2019; Huang et al 2003, 2005a).

3.4 *Synthesis of Quaternized Chitooligosaccharides*

The quaternized chitooligosaccharides (QCOS) were synthesized by grafting the quaternary ammonium substituents to the amino position of pyranose unit of COS. A certain amount of COS was dissolved in distilled water, and then glycidyltrimethylammonium chloride (GTMA) (ratio 1 COS: 3 GTMA) was added with stirring. The mixture was reacted at 50 °C for 12 h under pH 5.5 conditions using acetic acid. After the reaction, solution was mixed in acetone, QCOS was precipitated and filtered. Finally precipitated QCOS was dried with 90% degree of substitution comparing with COS. (Feng et al. 2015; Huang et al. 2006a, b; Guo et al 2006).

3.5 *Synthesis of N-aryl Chitooligosaccharides*

Synthesis of N-aryl chitooligosaccharides (aryl-COS) was a two-step reaction procedure. The first step of aryl-COS synthesis involved Schiff base formation as an intermediate. In the second step, the reduction of the Schiff base provided aryl-COS. The COS (1 g) were dissolved in 100 ml of 69% ethanol (v/v), 1% acetic acid (v/v) and adjusted the pH to 5. Thereafter, benzaldehyde, 4-hydroxybenzaldehyde or methyl benzaldehyde (2 g) was added to the mixture, and continuously stirred at room temperature for 12 h. After Schiff base was formed, 0.1 g Sodium borohydride (NaBH4) was added to the reaction mixture and stirred overnight (12–14 h). After that, the reaction mixture was neutralized with 15% (w/v) sodium hydroxide and centrifuged at 6,000 rpm for 30 min. Subsequently, the solution was filtered to obtain a precipitate. The precipitate was dialyzed against distilled water using dialysis membranes molecular weight cut off 1 kDa in order to remove salt. After dialysis, solution in dialysis bag was concentrated and precipitated in acetone. The precipitate was washed with diethyl ether several times to remove the un-reacted aldehyde. Finally, the product was dried at room temperature to give benzyl-COS, 4-hydroxybenzyl-COS, methyl benzyl-COS. (Trinh et al. 2014a, b; Guinesi and Cavalheiro 2006).

3.6 Synthesis of Phenol Acid Conjugated Chitooligosaccharides

The phenolic acid conjugated chitooligosaccharides (PA-c-COSs) were synthesized by N-conjugation of various phenolic acids to C-2 position of glucosamine monomer in COSs according to the method described by Eom et al. (2012, 2013). One gram of COS was dissolved in 100 ml of methanol solution. One of eight kinds of phenolic acids such as caffeic, ferulic, 4-hydroxybenzoic, protocatechuic, *p*-coumaric, syringic, sinapinic, and vanillic acids were mixed with one equivalent of dicyclohexylcarbodiimide (DCC), 1-hydroxybenzotriazole and triethylamine (TEA) (1:3) and added to the COS mixture. Then, the mixture was stirred at room temperature for 24 h. The resultant solution was precipitated in acetone and washed with acetone repeatedly, following freeze drying to obtain phenolic acid conjugated COSs.

Another synthetic process of chitooligosaccharides derivative for specific phenolic acid is gallic acid with COSs is described by Trinh et al. 2014a, b COS was dissolved in distilled water, methanol and then adjusted to pH 6.8 with triethylamine to obtain a solution A. Gallic acid was dissolved in methanol and dicyclohexylcarbodiimide (DCC dissolved in methanol) reacted with gallic acid to obtain a solution B. The solution B was slowly added to solution A and stirred in a water bath at 30 °C, 150 rpm for 5 h and then filtered using filter paper. The solution was added to diethyl ether at 2 °C, filtered to obtain a precipitate. The precipitate was dissolved in 20 ml distilled water and then freeze dried to obtain gallyl-COS with 43.5% yield. (Scheme 2) (Ngo et al. 2011a, b; Vo et al 2017). It should also be noted that some synthetic method can replace the DCC by carbodiimide (EDC) and N-Hydroxysuccinimide (NHS).

In addition, myself and my research groups have been studying phenolic acid conjugated COSs based on the method of Liu's group (2013, 2015) to graft gallic acid or syringic acid on pyranose ring of COSs using reaction catalysts environmentally friendly with reaction mechanism as shown in Scheme 3, it has been suggested that: the first solution ascorbic acid is present as a di-acid and can further react with H_2O_2 to generate hydroxyl radical (HO•). The second, the HO• attracts hydrogen from amino and hydroxyl group of COSs resulting in COSs macro radicals. Finally the corresponding phenolic acid molecules become acceptors of COSs macroradicals, thus forming COSs-phenolic acid conjugates.

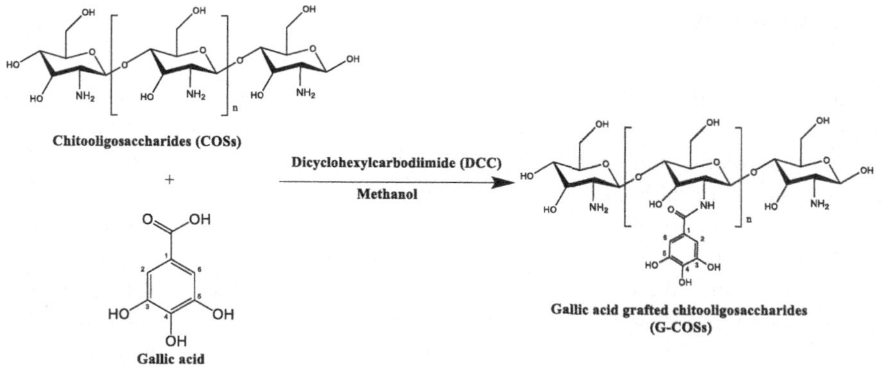

Scheme 2 Synthesis of gallic acid conjugated chitooligosaccharides

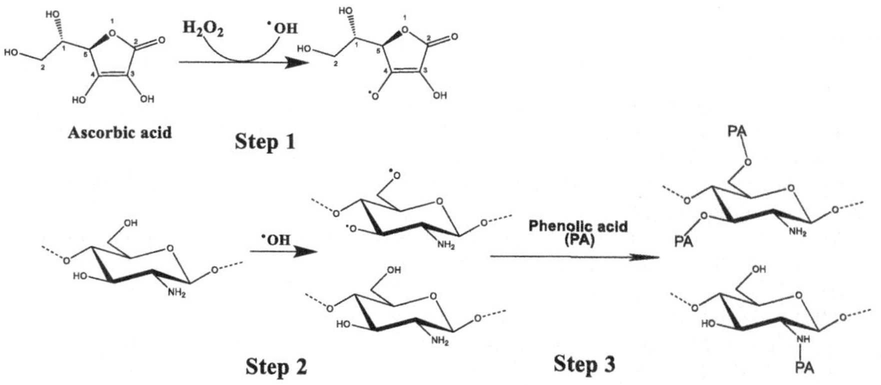

Scheme 3 The mechanism of phenolic acid conjugated with COSs

We successfully synthesized gallic acid and syringic acid conjugated COSs. The COSs (0.5 g) was dissolved in 25 ml distilled water and then 1 mL of 1.0 M H_2O_2 containing 0.054 g of ascorbic acid was added. After 30 min, the phenolic acid was added to the mixture at different amounts with the different mass ratios of phenolic acid conjugated onto COSs chain ranged. The pH value of the mixture changed to 5 and was stirred at room temperature for 6 h. After reaction, the mixture was filtered with filter membrane 1 kDa to remove unreacted phenolic acid and catalyst. Finally, the resulting solution was lyophilized using freeze dry system to obtain PA-c-COSs powder. (Bui and Ngo 2019; Bui et al. 2021).

All of chitooligosaccharides derivatives have been analyzed by thin layer chromatography with silica gel plate, and UV–Vis, Infrared spectra were recorded as clear signals on KBr plates with a Spectrum 2000 FT-IR spectrophotometer and proton and carbon NMR (^{1}H and ^{13}C NMR) spectra were recorded in D_2O on spectrometer. Molecular mass of COSs derivatives was determined by MALDI-TOF- MS spectrometer.

4 Conclusion

Chitooligosaccharides (COSs) which were made from hydrolytic chitosan using acid or enzyme were easily dissolved in water. They were modified COSs structure by introducing functional groups at C-2, C-3 or C-6 positions on pyranose ring using some chemical methods for synthesis and their derivatives by grafting functional groups as inorganic or organic to improve biological activities. Some methods synthesize and isolate chitooligosaccharides such as carboxylated chitooligosaccharides, amino derived chitooligosaccharides, sulfated chitooligosaccharides, quaternized chitooligosaccharides, aryl chitooligosaccharides, phenolic acid conjugated chitooligosaccharides. All COSs enhance biological activities and can be applied to various fields.

References

Ahmed TA, Aljaeid BM (2016) Preparation, characterization, and potential application of chitosan, chitosan derivatives, and chitosan metal nanoparticles in pharmaceutical drug delivery. Drug Des Devel Ther 10:483–507. https://doi.org/10.2147/DDDT.S99651

Alishahi A, Aïder M (2012) Applications of chitosan in the seafood industry and aquaculture: a review. Food Bioproc Tech 5:817–830

Bui VH, Dao QA, Ngo DN (2017) Optimization of chitosan hydrolysis by Cellulase Enzyme to produce chitooligosaccharide. Sci Tech Dev J 20(K3):74–82. https://doi.org/10.32508/stdj.v20iK3.1095

Bui VH, Vo NHT, Ngo DN (2021) Antioxidant effect of syringic acid grafted chitooligosaccharides in RAW264.7 Cells. Molecular and cellular biology in medicine. Part of IFMBE proceeding book series (IFMBE, V 85), MME 2020. In: 8th International conference on the development of biomedical engineering in Vietnam. Online 26 August 2021. Springer, pp 501–516

Bui VH, Ngo DN (2019) The research determines appropriate parameters in the synthesis process of syringic acid grafted chitooligosaccharides. Sci Tech Dev J 22(3):317–323. https://doi.org/10.32508/stdj.v22i3.1289

Einbu A, Vårum KM (2007) Depolymerization and de-N-acetylation of chitin oligomers in hydrochloric acid. Biomacromol 8:309–314

Einbu A, Grasdalen H, Vårum KM (2007) Kinetics of hydrolysis of chitin/chitosan oligomers in concentrated hydrochloric acid. Carbohydr Res 342:1055–1062

Eom TK, Senevirathne M, Kim SK (2012) Synthesis of phenolic acid conjugated chitooligosaccharides and evaluation of their antioxidant activity environ. Toxicol Pharmacol 34(2):519–527. https://doi.org/10.1016/j.etap.2012.05.004

Eom TK, Ryu BM, Lee JK et al (2013) β-secretase inhibitory activity of phenolic acid conjugated chitooligosaccharides J Enzyme Inhib Med Chem 28(1):214–217

Esfandiarpour-Boroujeni S, Bagheri-Khoulenjani S, Mirzadeh H (2016) Modeling and optimization of degree of folate grafted on chitosan and carboxymethyl-chitosan. Prog Biomater 5:1–8

Feng H, Xia W, Shan C et al (2015) Quaternized chitosan oligomers as novel elicitors inducing protection against *B. cinerea* in Arabidopsis. Int J Biol Macromol 72:364–369

Guinesi LS, Cavalheiro ETG (2006) Influence of some reactional parameters on the substitution degree of biopolymeric Schiff bases prepared from chitosan and salicylaldehyde. Carbohydr Polym 65:557–561

Guo Z, Liu H, Chen X et al (2006) Hydroxyl radicals scavenging activity of N-substituted chitosan and quaternized chitosan. Bioorganic Med Chem Lett 16:6348–6350

Hong S, Ngo DN, Kim MM (2016) Inhibitory effect of aminoethyl-chitooligosaccharides on invasion of human fibrosarcoma cells. Environ Toxicol Pharmacol 45:309–314
Huang R, Du Y, Yang J (2003) Preparation and anticoagulant activity of carboxybutyrylated hydroxyethyl chitosan sulfates. Carbohydr Polym 51:431–438
Huang R, Mendis E, Kim SK (2005) Factor affecting the free radical scavenging behavior of chitosan sulfate. Int J Biol Macromol 36:120–127
Huang R, Rajapakse N, Kim SK (2006) Structural factors affecting radial scavenging activity of chitooligosaccharides (COS) and its derivatives. Carbohydr Polym 63:122–129
Huang R, Mendis E, Rajapakse N et al (2006) Strong electronic charge as an important factor for anticancer activity of chitooligosaccharides (COS). Life Sci 78:2399–2408
Huang R, Mendis E, and Kim SK (2005a) Improvement of ACE inhibitory activity of chitooligosaccharides (COS) by carboxyl modification. Biorg Med Chem 13(11):3649–3655. https://doi.org/10.1016/j.bmc.2005.03.034
Je JY, Kim SK (2006) Reactive oxygen species scavenging activity of aminoderivatized chitosan with different degree of deacetylation. Bioorg Med Chem 14:5989–5994
Je JY, Kim EK, Ahn CB et al (2007) Sulfated chitooligosaccharides as prolyl endopeptidase inhibitor. J Nutr Biochem 18:31–38
Jeon YJ, Kim SK (2000) Production of chitooligosaccharides using an ultrfiltration membrane reactor and their antibacterial activity. Carbohydr Polym 41:133–141
Jeon YJ, Kim SK (2000) Continuous production of chitooligosaccharides using a dual reactor system. Process Biochem 35:623–632
Jung WJ, Souleimanov A, Park RD et al (2007) Enzymatic production of N-acetyl chitooligosaccharides by crude enzyme derived from *Paenibacillus illioisensis* KJA-424. Carbohydr Polym 67:256–259
Kim MM, Kim SK (2006) Chitooligosaccharides inhibit activation and expression of matrix metalloproteinase-2 in human dermal fibroblasts. FEBS Lett 580:2661–2666
Kim SK, Rajapakse N (2005) Enzymatic production and biological activities of chitosan oligosaccharides (COS): a review. Carbohydr Polym 62:357–368
Kim KW, Thomas RL (2007) Antioxidative activity of chitosans with varying molecular weights. Food Chem 101:308–313
Kim SK, Ngo DN, Rajapakse N (2006) Therapeutic prospectives of Chitin, Chitosan and their derivatives. J Chitin Chitosan 11:1–10
Liang TW, Chen YJ, Yen YH et al (2007) The antitumor activity of the hydrolysates of chitinous materials hydrolyzed by crude enzyme from *Bacillus amyloliquefaciens* V656. Process Biochem 42:527–534
Liu J, Lu JF, Kan J et al (2013) Synthesis of chitosan-gallic acid conjugate: structure characterization and in vitro anti-diabetic potential. Int J Biol Macromol 62:321–329. https://doi.org/10.1016/j.ijbiomac.2013.09.032
Liu J, Wu HT, Lu JF et al (2015) Preparation and characterization of novel phenolic acid (hydroxybenzoic and hydroxycinnamic acid derivatives) grafted chitosan microspheres with enhanced adsorption properties for Fe(II). Chem Eng J 262:803–812. https://doi.org/10.1016/j.cej.2014.10.041
Liu Y, Yu F, Zhang B et al. (2019) Improving the protective effects of aFGF for peripheral nerve injury repair using sulfated chitooligosaccharides
Mendis E, Kim MM, Rajapakse N et al (2007) An in vitro cellular analysis of the radical scavenging efficacy of chitooligosaccharides. Life Sci 80:2118–2127
Lizardi-Mendoza J, Monal WMA, Valencia FMG (2016) Chitosan in the preservation of agricultural commodities: chapter 1: chemical characteristics and functional properties of chitosan. Elsevier Inc., pp 3–31. https://doi.org/10.1016/B978-0-12-802735-6.00001-X
Ngo DN, Qian ZJ, Je JY et al (2008) Aminoethyl chitooligosaccharides inhibit the activity of angiotensin converting enzyme. Process Biochem 43:119–123
Ngo DH, Qian ZJ, Ngo DN et al (2011) Gally chitooligosaccharides inhibit intracellular free radical-mediated oxidation. Food Chem 128:974–981. https://doi.org/10.1016/j.foodchem.2011.03.128

Ngo DN, Kim MM, Kim SK (2012) Protective effects of aminoethyl-chitooligosaccharides against oxidative stress in mouse macrophage RAW 264.7 cells. Int J Biol Macromol 50:624–631. https://doi.org/10.1016/j.ijbiomac.2012.01.036

Ngo DH, Ngo DN, Kim SK et al (2019) Antiproliferative effect of aminoethyl-chitooligosaccharide on human lung A549 cancer cells. Biomol 9:195. https://doi.org/10.3390/biom9050195

Ngo DN, Kim MM, Kim SK (2007) Effects of chitin oligosaccharides on production of reactive oxygen species and matrix metalloproteinases in live cells. Advances in chitin science. Ankara, pp 355–359

Ngo DN, Lee SH, Kim MM et al (2009) Production of chitin oligosaccharides with different molecular weights and their antioxidant effect in RAW 264.7 cells. J Funct Foods 1(2):188–198

Ngo DH, Qian ZJ, Vo TS et al (2011) Antioxidant activity of gallate-chitooligosaccharides in mouse macrophage RAW264.7 cells. Carbohydr Polym 84(4):1282–1288. https://doi.org/10.1016/j.carbpol.2011.01.022

Nguyen AD, Vo TPK, Pham QA et al (2007) Study on hydrolysis of chitosan by cellulase combined with gamma irradiation. Advances in chitin science, vol 10. Antalya. 96–100

Nguyen ND, Dang VP, Nguyen TA et al (2011) Synergistic degradation to prepare oligochitosan by γ-irradiation of chitosan solution in the presence of hydrogen peroxide Radiat. Phys Chem 80(7):848–853

Okamura Y, Nomura A, Minami S et al (2005) Effects of chitin/ chitosan and their oligomers/monomers on release of type I collagenase from fibroblasts. Biol Macromol 6:2382–2384

Park PJ, Je JY, Kim SK (2004) Free radical scavenging activities of differently deacetylated chitosans using an ESR spectrometer. Carbohydr Polym 55:17–22

Pasanphan W, Chirachanchai S (2008) Conjugation of gallic acid onto chitosan: an approach for green and water-based antioxidant. Carbohydr Polym 72:169–177

Qin C, Zhou B, Zeng L et al (2004) The physicochemical properties and antitumor activity of cellulase-treated chitosan. Food Chem 84:107–115

Rajapakse N, Kim MM, Mendis E et al (2007) Inhibition of free radical-mediated oxidation of cellular biomolecules by carboxylated chitooligosaccharides. Bioorg Med Chem 15:997–1003

Rinaudo M (2006) Chitin and chitosan: properties and applications. Prog Polym Sci 31:603–632

Rødde RH, Einbu A, Vårum KM (2008) Seasonal study of the chemical composition and chitin quality of shrimp shells obtained from northern shrimp (*Pandalus borealis*) Carbohydr. Polym 71:388–393

Sashiwa H, Aiba SI (2004) Chemically modified chitin and chitosan as biometerials. Prog Polym Sci 29:887–908

Shao J, Yang Y, Zhong Q (2003) Studies on preparation of oligoglucosamine by oxidative degradation under microwave irradiation. Polym Degrad Stab 82(3):395–398. https://doi.org/10.1016/S0141-3910(03)00177-0

Sun T, Zhou D, Mao F et al (2007) Preparation of low-molecular-weight carboxymethyl chitosan and their superoxide anion scavenging activity. Eur Polym J 43:652–656

Trinh MDL, Ngo DH, Tran DK et al (2014) Prevention of H2O2-induced oxidative stress in chang liver cells by 4-hydroxybenzyl-chitooligomers carbohydr. Polym 103:502–509

Trinh MDL, Dinh MH, Ngo DH et al (2014) Protection of 4-hydroxybenzyl-chitooligomers against inflammatory responses in chang liver cells. Int J Biol Macromol 66:1–6. https://doi.org/10.1016/j.ijbiomac.2014.01.064

Vo TS, Ngo DH, Bach LG et al (2017) The free radical scavenging and anti-inflammatory activities of gallate-chitooligosaccharides in human lung epithelial A549 cells. Process Biochem 54:188–194. https://doi.org/10.1016/j.procbio.2017.01.001

Vongchan P, Sajomsang W, Kasinrerk W (2003) Anticoagulant activities of the chitosan polysulfate synthesized from marine crab shell by semi-heterogeneous conditions. Sci Asia. 29:115–120

Wu H, Yao Z, Bai X (2008) Anti-angiogenic activities of chitooligosaccharides. Carbohydr Polym 73:105–110

Xie W, Xu P, Liu Q (2001) Antioxidant activity of water-soluble chitosan derivatives. Bioorg Med Chem Lett 11:1699–1701

Xie H, Jia Z, Huang J et al (2012) Preparation of low molecular weight chitosan by complex enzymes hydrolysis. Chem Reag 3:180–186

Xing R, Liu S, Yu H et al (2005) Salt-assisted acid hydrolysis of chitosan to oligomers under microwave irradiation. Carbohydr Res 340:2150–2153

Xing R, Liu S, Guo Z et al (2005) Relevance of molecular weight of chitosan and its derivatives and their antioxidant activities in vitro. Bioorg Med Chem 13:1573–1577

Yen MT, Tseng YH, Li RC et al (2007) Antioxidant properties of fungal chitosan from shiitake stipes. LWT 40:255–261

Yoon NY, Ngo DN, Kim SK (2009) Acetylcholinesterase inhibitory activity of novel chitooligosaccharide derivatives. Carbohydr Polym 78:869–872

Yue W, Yao P, Wei Y et al (2008) Synergetic effect of ozone and ultrasonic radiation on degradation of chitosan. Polym Degrad Stab 93:1814–1821

Biological Preparation of Chitooligossacarides-Based Hydrogel Systems for Drug Delivery Systems

Safrina Dyah Hardiningtyas, Rizfi Fariz Pari, and Kustiariyah Tarman

1 Introduction

Chitin is the second most important natural polymer in the world and mainly extracted from shrimp and crab shells. The application of chitin in form of its derivative product chitosan has been widely used in many fields. However, chitosan which is soluble in acidic aqueous media caused its application still limited. To extend the application of chitosan as well to increase its activity, chitosan was further hydrolyzed to obtain chitooligosaccharides (COS) which is water soluble. Chitooligosaccharides can be prepared biologically using microbial enzymes. Most reported bacteria in degrading chitin include *Bacillus, Pseudomonas, Flovobacterium*, and *Cytophaga*; and for fungi were *Mucor, Aspergillus, and Mortierella.* Recombinant DNA technology has been applied by altering genetic material outside an organism as their products to increase the production of chitinase.

Chitooligosaccharides differ in the degree of deacetylation and degree of polymerization. Chitooligosaccharides have been applied increasingly in pharmaceutical and biomedical industry. The characteristics of COS which are water soluble, solubility in physiological pH, biocompatible, and biodegradable widen its application. In addition, the biological activity of COS was higher compared to that of native chitosan polymer. Chitooligosaccharides likewise increased the richness and diversity of the gut microbial community. In pharmaceutical industry, the application of COS includes drug delivery systems, hydrogel, and as a carrier for delivery of metal or microbial.

S. D. Hardiningtyas · R. F. Pari · K. Tarman (✉)
Department of Aquatic Product Technology, Fisheries and Marine Sciences, IPB University, Bogor, Indonesia
e-mail: kustiaz@apps.ipb.ac.id

K. Tarman
Division of Marine Biotechnology, Center for Coastal and Marine Resources Studies, IPB University, Bogor, Indonesia

S.-K. Kim (ed.), *Chitooligosaccharides*,
https://doi.org/10.1007/978-3-030-92806-3_5

2 Biological Preparation of Chitooligosaccharides (COS)

Chitooligosaccharides (COS) are a product of degraded chitin or chitosan that can be prepared by the enzymatic hydrolysis process. Glycoside hydrolases are types of enzymes in COS production. Substrate for COS production might be in a form of chitin or chitosan in varied properties where the degree of deacetylation, degree of polymerization, and molecular weight determine the process and the property of the product. A combination of enzyme and substrate properties possibly work in a direct process where the enzyme is used directly to produce COS, and in indirect where a combination of the non-enzymatic process with the enzymatic process (Miguez et al. 2021). According to product composition, COS is classified into fully acetylated chitooligosaccharides (faCOS) that composed only N-Acetyl-D-Glucosamine, partially acetylated (paCOS) that composed a combination of D-glucosamine and N-Acetyl-D-Glucosamine and fully deacetylated chitooligosaccharides (fdCOS) that composed only D-glucosamine. Chitooligosaccharides possess biological activity which depends on their structure, degree of polymerization, molecular weight, degree of acetylation, fraction of acetylation, and pattern of acetylation (Li et al. 2016). Enzyme involved in chitin and chitosan conversion are chitinases (EC 3.2.1.14), chitosanases (EC 3.2.1.132), chitooligosaccharides deacetylases (EC 3.5.1.105), and lytic polysaccharide monooxygenases (LPMOs) (EC 1.14.99.53–56) (Kaczmarek et al. 2019). As the enzyme determines product properties, various sources of enzymes to produce COS from various organisms have been observed.

Enzymes to produce COS are produced by microorganisms found in soil where shrimp or crab shell waste accumulates. Chitinases are generally found in organisms that need to reshape chitin on their own structure, or dissolve and digest the chitin from fungi or animals. In contrary to the expectation, soil and water sediment have higher chitinolytic bacteria and fungi than shrimp waste, only 4% of the total heterotrophic bacteria and 25–60% of the total fungi, but their number is lower than the bacteria (Brzezinka et al. 2008; Suresh 2012). The highest number of chitinolytic bacteria from shrimp waste was found in their head. Chitinolytic microorganisms were highly identified in soil and lake sediment. Soil and the rhizosphere are containing abundant chitinolytic microorganisms, and actinobacteria are the most abundant. The chitinolytic bacteria isolated from the soil are more active than the chitinolytic bacteria isolated from water and sediments. Genus *Streptomyces* existed in almost all Actinobacteria found in soil and sediment. Bacteria that have the ability to degrade chitin include *Bacillus, Pseudomonas, Flovobacterium*, and *Cytophaga*. Fungi with the ability to degrade chitin include *Mucor, Aspergillus, and Mortierella* (Schlegel 2006). Although chitinolytic microorganisms are abundance, chitinase expression was low for industrial production, therefore recombinant technology for chitinase was applied.

To enhance the production of chitinase, recombinant DNA technology is being used by altering genetic material outside an organism as their products. This technology involves the insertion of chitinase DNA fragments from chitinolytic microorganisms using an appropriate vector (Khan et al. 2016). *E. coli* BL21, *Saccharomyces cereviseae, Lactobacillus plantarum, Pichia pastoris* are microorganisms that have been used for chitinase, chitosanase, or other enzyme production. The expression vector selection for COS production is mostly based on the wanted expression system, application of the enzyme, application of the chitooligosacharides, and vector size.

Some microorganisms produce many chitinases, for example, *Trichoderma* sp. which have 4 types or more enzymes. *Trichoderma harzianum* is a powerful biocontrol agent against plant pathogenic fungi from soil that produced four types of endo-chitinase (52, 42, 33, and 31 kDa) (Carsolio 1999). The hydrolysis of colloidal chitin with chitinase Chit42 for 24 h mainly yielded N-acetyl-chitobiose, N-acetyl-D-glucosamine, and N-acetyl-chitotriose. The DD of the synthesized COS was lower than 5%. In addition, trace amounts of N-acetyl-chitotetraose were found. Over-express Chit42 into *Pichia pastoris* increased chitinase production 100 times to 2.9 mg/mL (15 U/mL; 5.2 U/mg) (Kidibule et al. 2018). *Trichoderma harzianum* also produces another endo-chitinase Chit33. Expression of Chit33 in *Pichia pastoris* produced a mixture of fully acetylated COS and partially acetylated COS from chitinolytic materials, majority yielded as D-glucosamine-(N-Acetyl-D-Glucosamine)$_2$ (Boer et al. 2007). Another overexpressing experiment was done from *Trichoderma* sp., such as Enc1 from *T. harzianum* in expressed 3 mg/L of chitinase, T25-1 from *T. harzianum* in *S. cereviseae* expressed 200 mg/L of chitinase (Draborg et al. 1996), Ech42 from *T. atroviride* P1 in *P. pastoris* expressed 6.2 g/L of chitinase (Pérez-Martínez et al. 2007), and Tachi1 from *T. asperellum* in *P. pastoris* expressed 6.2 g/L (Tang et al. 2012).

The abundant of metagenome data was also utilized to obtain highly potential chitinases. Chitin deacetylase CDA20 expressed in *E. coli* was able to hydrolyze N-acetyl-D-glucosamine and chitobiose simultaneously into high-purity D-Glucosamine. This ability simplifies the industrial production process and saves production costs (Guan et al. 2020). A marine bacterium *Shewanella baltica* produces chitin oligosaccharide deacetylase, sbCOD. The sequence identity between chitinase in *Shewanella* sp. and *Vibrio* sp was very high (up to 99.0%). As the production is naturally low, a recombinant protein showed specific activity 10.2 U/mg of protein against (N-acetyl-D-glucosamine) (Hirano et al. 2017). The COS deacetylases VcCOD from *Vibrio cholerae* acts on the second acetyl group at the non-reducing end. The VcCOD enzyme showed better deacetylation effect on COS with low DP (Li et al. 2007).

Plant also have chitinase enzyme which works to produce COS. The COS deacetylases NodB from *Rhizobium* sp. acts in symbiosis between legume plants and rhizobial bacteria. It was expressed in *E. coli* BL21 (DE3) using pET-32a-nodB expression plasmid that encodes Trx-NodB fusion protein. This enzyme converts a single-chain deacetylases from COS at the non-reducing end. (Chambon et al. 2017). The COS deacetylases MrCDA from fungus *Mucor rouxii* attack any sequence in the chitosan molecule without preference. It could deacetylate COS with DP1-7 but not the acetyl group on the reducing end when the DP is 3,5,7, and the final product will be a monoacetylated product (Tsigos et al. 1999).

Chitosanases are found in both bacteria and fungi, and act to catalyze the breakdown of chitosan for completely or partially deacetylated of chitin. Chitosanase could produce partially acetylated COS and high fully deacetylated COS. The abundant of metagenome data was utilized to obtain highly potential chitosanases. The enzyme from the expression of Chitosanases Chis5 from *Sphaerisporangium album*, a Gram-positive and aerobic strain, in *E. coli* produce short chain chitooligosaccharides with different degree of polymerization DP2 (341.1484), DP3 (502.2134), DP4 (663.2793), and DP5 (824.3433). The highest content was in DP3. (Guan et al. 2020).

Chitosanase enzymatic hydrolysis and the second consisted of a two-step procedure based on chemical hydrolysis followed by chitosanase hydrolysis. Directly use of chitosanase give better antibacterial activity with 27% of fully deacetylated sequences. Using two-step procedure, the COS product compose of 63% of fully deacetylated sequences. The CPS have the ability to inhibit Escherichia coli and Listeria monocytogenes (Sánchez et al. 2017).

Chitosanolytic is also found in *Bacillus amyloliqufaciens* which can produce fully deacetylated products. Some protease from the bacteria demonstrate the ability to hydrolyze N-acetyl residues in chitosan (Kumar and Tharanathan 2004; Pan et al. 2016). Chit600 from *B. amyloliqufaciens* was chosen to hydrolyze chitosan for 24 h. The products were identified as chitobiose, chitotriose, glucosamine, chitrotetraose and chitopentaose. The degree of deacetylation in the reaction mixture was always between 95 and 100% (Santos-Moriano et al. 2018).

Another method to produce foodgrade COS was by expressing the enzyme in lactic acid bacteria. *Bacillus subtilis* ATCC23857 have chitosanase, CsnA, which was fused to different anchoring motifs of *L. plantarum.* The gene makes covalent attachment to the surface of the cell, either on N-terminal lipoprotein anchor or C-terinal cell wall anchor. The fusion proteins were expressed in *L. plantarum.* The highest chitosanase activities obtained in *L. plantarum* was 1360 U g dry cell weight. The enzyme has ability to break down chitosan and produce COS which support growth of *L. plantarum* (Nguyen et al. 2016). Chitosan degradation with degree of deacetylation between 70 and 85%) (Low DD) by chitosanases probably produce high amount of *pa*COS. However, the combination of this type of chitosans with chitinases could be more effective because these enzymes require at least one GlcNAc moiety at the -1 position in the active site, thus assuring the presence in the synthesized products of a GlcNAc residue on their reducing ends.

Metagenomic analysis is applied to determine chitosanase with a high specific activity. Chitosanase expressed in *E. coli* was reached as much 2373 U/mg specific activity from CsnA gene. Another source of gene expressing chitosanase enzyme was *Bacillus* sp. A gene from *Bacillus subtilis* 168 that expressed in *E. coli,* produced a chitosanase enzyme with specific activity 904.7 U/mg (Pechsrichuang et al. 2013). Chitosanase enzyme from *Bacillus* sp. MET 1299 expressed in *E. coli* have specific activity 476 U/mg, respectively (Kim et al. 2004). Chitosanase that expressed in *E.*

coli performed higher specific activity than the wild-type chitosanase. The wild-type chitosanase have specific activity 217 U/mg and 59.1 U/mg produced by *Streptomyces roseolus* and *Pseudomonas* sp. TKU015 (Jiang et al. 2012).

3 Application of Chitooligosaccharide in Drug Delivery Systems

Chitooligosaccharide (COS) is a low molecular weight depolymerization product of chitosan that differs in the degree of deacetylation (DD) and degree of polymerization (DP). Since the COS properties are water soluble, solubility in physiological pH, biocompatible, biodegradable (Tabassum et al. 2021), it has attracted increasing attention in pharmaceutical and biomedical applications. COS also demonstrates the greater biological activities to native chitosan polymer, including antitumor, immune-enhancing, and antibacterial activities (Kim and Rajapakse 2005). In addition, COS increased the richness and diversity of the gut microbial community (Guo-Ji et al. 2021). The application of COS in the pharmaceutical field includes drug delivery systems, hydrogel, and as a carrier for delivery of metal or microbial.

a. Drugs delivery system

COS exhibits polycationic behavior because it has free amino groups with shorter chains. Therefore, COS has the ability to bind to polyanionic complexing agents (eg, DNA, calcium alginate, hyaluronic acid) and negatively charged cell and tissue membranes (Gåserød, et al. 1998). In addition, COS has a low molecular weight so that it can increase its hydrophilic properties. Because of these two behaviors, COS can be more suitable for use as a drug delivery system (Vårum and Smidsrd 2004). Yang and Hon (2009) reported that the loading of drugs can be improved by the use of low molecular weight chitosan. The rate of drug release can be increased if the hydrophilic fraction is higher in the polymer matrix (Zupancic et al. 2016).

In drug delivery systems applications, COS is used as a carrier for drug, gene, or biomolecules delivery, such as DNA, heparin, insoluble anticancer drugs, curcumin, etc., into target cells or organs. COS-based drug delivery systems were prepared by several methods, such as absorption, ionic gelation, chemical conjugation, modified with polymer conjugates, emulsification, Layer-by-layer (LbL) technique, etc.

COS can be used for developing a protective and non-toxic gene delivery system. The negatively charged plasmid DNA can bind to polycationic COS effectively, which the stable DNA complex was successfully achieved using COS with a DD of > 65%. The transgene expression in vitro for COS with a DD of 85% was 36–70-times higher than a DD of 65% and a DD of 51%, respectively. The rapid onset of action and the highest expression of the luciferase gene was achieved on COS-based DNA polyplexes (DP 15–21) with 260-fold greater than for chitosan (Koping-Hoggård et al. 2001). Zhang and Mi reported that COS also can bind to an anionic polysaccharide drug, low molecular weight heparin (LMWH). LMWH has poor membrane permeability as the negatively charged mucous membrane repels the molecule. The addition of LMWH to COS (DP = 6, pH = 7.4) enhanced the relative absorption to the maximum degree into the intestinal epithelia; were 9.2 and 13.2 times greater than when LMWH was used alone in the small and large intestine, respectively. None of the chitosan oligomers at any concentration induced any necrosis on intestinal epithelial cells (Zhang et al. 2014).

COS can interact with negatively charged biomolecules, i.e., glycated whey protein isolate (gWPI) to form nanoparticles. Yu et al. produced the COS-glycated whey protein isolate (gWPI) nanoparticles for stabilization of pickering emulsion on curcumin delivery system. The result showed the viscoelasticity and stability of the emulsions stabilized with gWPI–COS were higher in the curcumin delivery system than in the emulsions formed individually by WPI, gWPI, or WPI–COS. gWPI–COS nanoparticles also could decrease free energy, thereby preventing the flocculation of the formed droplet and remain stable during storage for curcumin delivery (Yu et al. 2021). In addition, COS also can bind with pectin to develop hydrogel microcapsule carriers mimicking red blood cells. This hydrogel was designed to carry hemoglobin (Hb) as an oxygen therapeutics (Crouse et al. 2015; Cherwin et al. 2019). Cherwin et al. observed that pectin-COS-based microcapsules with a diameter less than 10 microns exhibited oxygen encapsulation efficiencies of around 99.99%, which were produced by an electro-spinning technique using COS with a DD of 95% (Cherwin et al. 2019). The shelf life of these therapeutic drugs extended by several months (Kheir et al. 2013).

The amphiphilic COS can be modified chemically conjugate using hydrophobic molecules or other polymers for fabricating self-assembled polymeric nanoparticles, therefore facilitating delivery of insoluble drugs. The modification conjugates of COS is possible due to its availability for coupling with the primary amino groups and the hydroxyl groups of each polymer subunit and the cationic nature, which allows ionic cross-linking (Agnihotri et al. 2004). Hydrophobically-modified amphiphilic COS derivatives can facilitate drug delivery in cancer therapy by improving the solubility of insoluble drugs, drug targeting, and enhancement of absorption (Du et al. 2009; Wei et al. 2010). Zhang et al. (2012) reported that the amphiphilic oleic acid–chitooligosaccharide conjugate proved to improve the therapeutic efficacy of loading drugs with translocation into cells efficiently. The same author also synthesized an amphiphilic all-trans-retinoic acid (ATRA)–chitooligosaccharide (RCOS) conjugate to form self-assembled polymeric nanoparticles to facilitate the co-delivery of ATRA and paclitaxel (PTX). The cytotoxicity of PTX-loaded RCOS nanoparticles to HepG2 cells was higher than the PTX plus ATRA solution. This is due to the RCOS nanoparticles could be rapidly and continuously taken up by HepG2 cells via endocytosis and transported into the nucleus, and the uptake rates increased with particle concentration (Zhanga et al. 2015).

COS can be used for modifying the surface charge of polymer nanoparticles or liposomes. Hu et al. (2020) developed the positively charged AZD9291 (Osimertinib)-poly (lactic-co-glycolic acid) (PLGA) nanoparticles with chitooligosaccharides (COS). AZD-PLGA-COS NPs have good biocompatibility and biological affinity with smaller particle sizes and higher EE. The positive charge on the surface of the AZD-PLGA-COS NPs promotes cellular uptake by increasing the affinity with negatively charged cell membranes. COS was found to inhibit the expression of immune checkpoint PD-L1. Seong et al. (2018) prepared multi-layered liposomes with chitooligosaccharide (COS) and N-succinyl-chitosan using layer-by-layer (LbL) technique. They were used to coat a conventional liposome for enhanced stability and the permeation of quercetin through the skin. In addition, COS can be used for developing lycopene delivery systems by sonication, freeze-drying, and kneading methods. The complexing lycopene with chitooligosaccharides enhanced solubility stability under different storage conditions and its antioxidant activity (Bockuviene et al. 2021).

b. Hydrogel

On the designing of hydrogels, COS might be better suited as a hydrogel precursor than high molecular weight chitosan because it has low of degree polymerization (DP), molecular weight, and high degree deacetylation (DD). These properties could improve the structural and functional properties of hydrogels, which is increasing the percentage of swelling, increasing the solubility in water (Ahmed 2015; Li et al. 2015), to ensure there are no adverse reactions to animal cells whose surfaces are maintained at pH 7–8 (Shorrock and Rees 1988). Chitosan can be dissolved in physiological pH if it has a DD of 40 and 60% (Sorlier et al. 2001; Vårum et al. 1994). The oxygen encapsulation efficiency in hydrogel microcapsules was increased more than 99% when used with chitosan of 95% DD (Cherwin et al. 2019). In addition, the better proliferation of olfactory envelope cells on the chitosan scaffold is possible by increasing the DD from 72 to 85%, whereas DD below 75% induces cellular stress and has reduced tensile strength (Foster et al. 2015).

COS can function better as the hydrophilic polymer matrix in hydrogels. COS-based hydrogel has potential as biomaterial for several treatments, including acne treatment, wound healing, and as an antibacterial and antioxidant agent. Park et al. (2018) reported that the chitooligosaccharide (COS)-gallic acid (GA) conjugated showed an excellent antibacterial activity against *Propionibacterium acnes*. COS-GA was loaded on a polyvinyl alcohol (PVA) hydrogel for acne treatment. Kim et al. (2021) were used a cationic chitooligosaccharide (COS) to prepare hydrogel with the anionic ϒ-PGA/DA conjugated by the dynamic physical cross-linking. The COS was induced by gelation through intermolecular π-cation and ionic interactions. the ϒ-PGA/DA/COS hydrogels have great prospects as a biomaterial with self-healing and tissue adhesive properties. Chandika et al. (2021) developed a poly (vinyl alcohol) (PVA)/methacrylate kappa-carrageenan (κ-CaMA) composite hydrogel encapsulated with a chitooligosaccharide (COS) as a wound dressing, The encapsulation was prepared in a cassette via repeated freeze/thaw cycles, photo-cross-linking, and chemical cross-linking. The results showed that the PVA/κ-CaMA/COS (Pκ-CaC) hydrogel had a uniformly thick, highly porous three-dimensional architecture with uniformly distributed pores, a high fluid absorption, and retention capacity without disturbing its mechanical stability, and good in vitro biocompatibility. The Pκ-CaC hydrogel could enhance the rate of wound closure and reach complete healing by day 14. The COS-salicylic acid (SA) conjugates were loaded to oxidized alginate-gelatin hydrogel to improve its wound healing behavior. The hydrogel with the COS-SA conjugate induced poli resulted in increased proliferation of collagen at the wound sites, and maximized the wound closure rate. In addition, COS-SA conjugates did not exhibit cytotoxicity in the mouse model (Oh et al. 2021). COS-based dressings can protect against bacterial infections more effectively because of its enhanced antibacterial activity (Park et al. 2018) COS has also been determined to support the growth of embryonic stem cells (Willenberg et al. 2006). Stiffer hydrogels resulting from the high DD of COS can allow for better proliferation and dispersion of the cells on the scaffold (Sacco et al. 2018).

Besides being used for drug delivery, COS can also be used for the delivery of nanomaterials for cancer therapy or wound healing. Bae et al. (2012) developed chitosan-DOPA-stabilized ferrimagnetic iron oxide nanocubes (Chito-FIONs) as an effective heat nano mediator for cancer thermotherapy. A bioinspired surface modification approach using chitosan-DOPA conjugates produced highly stable and water-dispersible ferrimagnetic iron oxide nanocubes. Chito-FIONs were prepared using oil-in-water emulsification method with ultrasonication-assisted. Chito-FIONs were composed of multiple 30-nm-sized FIONs encapsulated by a chitosan polymer shell. The magnetically guided Chito-FIONs successfully eradicated target cancer cells through caspase-mediated apoptosis. In addition, cerium-oxide nanoparticles, an antioxidant capable of scavenging reactive oxygen species and free radicals (that lead to the apoptosis of retinal pigment epithelium cells associated with age-related macular degeneration), can be combined with COS to develop a hydrogel system that achieves sustained release of the drug otherwise insoluble in aqueous solutions (Cai and McGinnis 2016; Karakoti et al. 2008; Kong et al. 2011).

References

Agnihotri SA, Mallikarjuna NN, Aminabhavi TM (2004) Recent advances onchitosan-based micro- and nanoparticles in drug delivery. J Control Release 100:5–28

Ahmed EM (2015) Hydrogel: Preparation, characterization, and applications: A review. J Adv Res 6(2):105–121. https://doi.org/10.1016/j.jare.2013.07.006

Bae K, Park M, Do MJ, Lee N, Ryu JH, Kim GW, Kim CG, Park TG, Hyeon T (2012) Chitosan oligosaccharide-stabilized ferrimagnetic iron oxide nanocubes for magnetically modulated cancer hyperthermia. ACS Nano 6(6):5266–5273. https://doi.org/10.1021/nn301046w

Bockuviene A, Zalneravicius R, Sereikaite J (2021) Preparation, characterization and stability investigation of lycopene-chitooligosaccharides complexes. Food Biosci 40:100854

Boer H, Simolin H, Cottaz S, Soderlund H, Koivula A (2007) Heterologous expression and site-directed mutagenesis studies of two *Trichoderma harzianum* chitinases, Chit33 and Chit42 *Escherichia coil*. Protein Expr Purif 51(2):216–226. https://doi.org/10.1016/j.pep.2006.07.020

Brzezinska MS, Lalke-Porczyk E, Donderski W (2008) Occurrence and activity of microorganisms in shrimp waste. Curr Microbiol 57:580–587

Cai X, McGinnis JF (2016) Nanoceria: A potential therapeutic for dry AMD. Ad Exp Med Biol 854:111–118. https://doi.org/10.1007/978-3-319-17121-0_16

Carsolio C, Benhamou N, Haran S, Cortés C, Gutiérrez A, Chet I, Herrera-Estrella A (1999) Role of the Trichoderma harzianum endochitinase gene, ech42, in mycoparasitism. Appl Environ Microbiol 65(3):929–935. https://doi.org/10.1128/AEM.65.3.929-935.1999

Chambon R, Pradeau S, Fort S, Cottaz S, Armand S (2017) High yield production of Rhizobium NodB chitin deacetylase and its use for in vitro synthesis of lipo-chitinoligosaccharide precursors. Carbohyd Res 442:25–30

Chandika P, Kim MS, Khan F, Kim YM, Heo SY, Oh GW, Kim NG, Jung WK (2021) Wound healing properties of triple cross-linked poly (vinyl alcohol)/methacrylate kappa-carrageenan/chitooligosaccharide hydrogel. Carbohydr Polym 269:118272

Cherwin A, Namen S, Rapacz J, Kusik G, Anderson A, Wang Y et al (2019) Design of a novel oxygen therapeutic using polymeric hydrogel microcapsules mimicking red blood cells. Pharmaceutics 11(11). https://doi.org/10.3390/pharmaceutics11110583

Crouse JZ, Mahuta KM, Mikulski BA, Harvestine JN, Guo X, Lee JC, Kaltchev MG, Midelfort KS, Tritt CS, Chen J et al (2015) Development of a microscale red blood cell-shaped pectin-oligochitosan hydrogel system using an electrospray-vibration method: preparation and characterization. J Appl Biomater Funct Mater 13:326–331

Draborg H, Christgau S, Halkier T, Rasmussen G, Dalboge H, Kauppinen S (1996) Secretion of an enzymatically active *Trichoderma harzianum* endochitinase by *Saccharomyces cerevisiae.* Curr Genet 29(404–9):34

Du YZ, Wang L, Yuan H, Wei XH, Hu FQ (2009) Preparation and character-istics of linoleic acid-grafted chitosan oligosaccharide micelles as a carrier fordoxorubicin. Colloids Surf, B 69:257–263

Foster LJR, Ho S, Hook J, Basuki M, Marçal H (2015) Chitosan as a biomaterial: influence of degree of deacetylation on its physiochemical, material and biological properties. PloS One 10(8)

Gåserød O, Smidsrød O, Skjåk-Bræk G (1998) Microcapsules of alginate-chitosan-I: a quantitative study of the interaction between alginate and chitosan. Biomaterials 19(20):1815–1825. https://doi.org/10.1016/S0142-9612(98)00073-8

Guo Ji X, Lin Chang K, Chen M, Liang ZL, Osman A, Yin H, Ming ZL (2021) In vitro fermentation of chitooligosaccharides and their effects on human fecal microbial community structure and metabolites. LWT 144:111224. https://doi.org/10.1016/j.lwt.2021.111224

Guan F, Han Y, Yan K, Zhang Y, Zhang Z, Wu N, Tian J (2020) Highly efficient production of chitooligosaccharides by enzymes mined directly from the marine metagenome. Carbohydr Polym 234:115909

Hirano T, Shiraishi H, Ikejima M, Uehara R, Hakamata W, Nishio T (2017) Chitin oligosaccharide deacetylase from *Shewanella baltica* ATCC BAA-1091. Biosci Biotechnol Biochem 81(3):547–550

Hu X, Chen S, Yin H et al (2020) Chitooligosaccharides-modified PLGA nanoparticles enhance the antitumor efficacy of AZD9291 (Osimertinib) by promoting apoptosis. Int J Biol Macromol. https://doi.org/10.1016/j.ijbiomac.2020.06.154

Jiang X, Chen D, Chen L, Yang G, Zou S (2012) Purification, characterization, and action mode of a chitosanase from *Streptomyces roseolus* induced by chitin. Carbohydr Res 355:40–44

Kaczmarek MB, Struszczyk-Swita K, Li X, Szczę sna-Antczak M, Daroch M (2019) Enzymatic modifications of Chitin, Chitosan, and Chitooligosaccharides. Front Bioeng Biotechnol 7:243.

Karakoti AS, Monteiro-Riviere NA, Aggarwal R, Davis JP, Narayan RJ, Self WT et al. (2008) Nanoceria as antioxidant: synthesis and biomedical applications. JOM (Warrendale, Pa.: 1989) 60(3):33–37

Khan S, Ullah MW, Siddique R, Nabi G, Manan S, Yousaf M, Hou H (2016) Role of recombinant DNA technology to improve life. Int J Genomics 2405954. https://doi.org/10.1155/2016/2405954

Kheir JN, Polizzotti BD, Thomson LM, O'Connell DW, Black KJ, Lee RW et al (2013) Bulk manufacture of concentrated oxygen gas-filled microparticles for intravenous oxygen delivery. Adv Healthcare Mater 2(8):1131–1141. https://doi.org/10.1002/adhm.201200350

Kidibule PE, Santos-Moriano P, Jimenez-Ortega E, Ramirez-Escudero M, Limón MC, Remacha M, Plou FJ, Sanz-Aparicio J, Fernandez-Lobato M (2018) Use of chitin and chitosan to produce new chitoolig, osaccharides by chitinase chit42: enzymatic activity and structural basis of protein specificity. Microb Cell Fact 17:47

Kim SK, Rajapakse N (2005) Enzymatic production and biological activities of Chitosan Oligosaccharides (COS): a review. Carbohydr Polym 62:357–368. https://doi.org/10.1016/j.carbpol.2005.08.012

Kim HC, Lee JN, Kim E, Kim MH, Park WH (2021) Self-healable poly(γ-glutamic acid)/chitooligosaccharide hydrogels via ionic and π-interactions. Mater Lett 297:129987. https://doi.org/10.1016/j.matlet.2021.129987

Kim PI, Kang TH, Chung KJ, Kim IS, Chung K (2004) Purification of a constitutive chitosanase produced by *Bacillus* sp. MET 1299 with cloning and expression of the gene. FEMS Microbiol Lett 240(1):31–39. https://doi.org/10.1016/j.femsle.2004.09.006

Kong L, Cai X, Zhou X, Wong LL, Karakoti AS, Seal S, et al (2011) Nanoceria extend photoreceptor cell lifespan in tubby mice by modulation of apoptosis/survival signaling pathways. Neurobiol Dis 42(3):514–523. https://doi.org/10.1016/j.nbd.2011.03.004.

Koping-H oggård M, Tubulekas I, Guan H, Edwards K, Nilsson M, Vårum KM et al (2001) Chitosan as a nonviral gene delivery system. Structure–property relationships and characteristics compared with polyethylenimine in vitro and after lung administration in vivo. Gene Therapy 8(14):1108–1121. https://doi.org/10.1038/sj.gt.3301492

Kumar ABV, Tharanathan RN (2004) A comparative study on depolymerization of chitosan by proteolytic enzymes. Carbohydr Polym 58:275–283

Li B, Wang L, Xu F, Gang X, Demirci U, Wei D, *et al* (2015) Hydrosoluble, UV-crosslinkable and injectable chitosan for patterned cell-laden microgel and rapid transdermal curing hydrogel in vivo. Acta Biomaterialia 22:59–69. https://doi.org/10.1016/j.actbio.2015.04.026

Li X, Wang LX, Wang X, Roseman S (2007) The chitin catabolic cascade in the marine bacterium *Vibrio cholerae*: characterization of a unique chitin oligosaccharide deacetylase. Glycobiology 17(12):1377–1387

Li K, Xing R, Liu S, Li P (2016) Advances in preparation, analysis and biological activities of single chitooligosaccharides. Carbohydr Polym 139:178–910. https://doi.org/10.1016/j.carbpol.2015.12.016

Miguez N, Kidibule P, Santos-Moriano P, Ballesteros AO, Fernandez-Lobato M, Plou FJ (2021) Enzymatic synthesis and characterization of different families of chitooligosaccharides and their bioactive properties. Appl Sci 11:3212. https://doi.org/10.3390/app11073212

Nguyen HM, Mathlesen G, Stelzer EM, Pham ML, Kuczkowska K, Mackenzie A, Agger JW, Eijsink VGH, Yamabhai M, Peterbauer CK, Haltrich D, Nguyen TH (2016) Display of a β-mannanase and a chitosanase on the cell surface of *Lactobacillus plantarum* towards the development of whole-cell biocatalysts. Nguyen et al. Microb Cell Fact 15:169

Oh GW, Kim SC, Kim TH, Jung WK (2021) Characterization of an oxidized alginate-gelatin hydrogel incorporating a COS-salicylic acid conjugate for wound healing. Carbohydr Polym 252:Article 117145. https://doi.org/10.1016/jcarbpol.2020.117145

Pan AD, Zeng H-Y, Foua GB, Alain C, Li Y-Q (2016) Enzymolysis of chitosan by papain and its kinetics. Carbohydr Polym 135:199–206

Park HH, Ko SC, Oh GW, Jang YM, Kim YM, Park WS et al (2018) Characterization and biological activity of PVA hydrogel containing chitooligosaccharides conjugated with gallic acid. Carbohyd Polym 198:197–205. https://doi.org/10.1016/j.carbpol.2018.06.070

Park WS, Choi I-Whan, Jung W-Kyo, (2018) Characterization and biological activity of PVA hydrogel containing chitooligosaccharides conjugated with gallic acid. Carbohyd Polym. https://doi.org/10.1016/j.carbpol.2018.06.070

Pechsrichuang P, Yoohat K, Yamabhai M (2013) Production of recombinant Bacillus subtilis chitosanase, suitable for biosynthesis of chitosan-oligosaccharides. Bioresour Technol 127:407–414. https://doi.org/10.1016/j.biortech.2012.09.130

Pérez-Martínez AS, De León-Rodríguez A, Harris LJ, Herrera-Estrella A, Barba de la Rosa AP (2007) Overexpression, purification and characterization of the *Trichoderma atroviride* endochitinase, Ech42, in *Pichia pastoris*. Protein Expr Purif 55(1):183–188. https://doi.org/10.1016/j.pep.2007

Sacco P, Furlani F, de Marzo G, Marsich E, Paoletti S, Donati I (2018) Concepts for developing physical gels of chitosan and of chitosan derivatives. Gels 4(3):67. https://doi.org/10.3390/gels4030067

Santos-Moriano P, Fernandez-Arrojo L, Mengibar M, Belmonte-Reche E, Peñalver P, Acosta FN, Ballesteros AO, Morales JC, Kidibule P, Fernandez-Lobato M et al (2018) Enzymatic production of fully deacetylated chitooligosaccharides and their neuroprotective and anti-inflammatory properties. Biocatal Biotransform 36:57–67

Sánchez Á, Mengíbar M, Rivera-Rodríguez G, Moerchbacher B, Acosta N, Heras A (2017) The effect of preparation processes on the physicochemical characteristics and antibacterial activity of chitooligosaccharides. Carbohyd Polym 157:251–257

Schlegel HG (2006) General microbiology. Polish Scientific Publishers, Warsaw, Poland (in polish)

Seong JS, Yun ME, Park SN (2018) Surfactant-stable and pH-sensitive liposomes coated with N-succinyl-chitosan and chitooligosaccharide for delivery of quercetin. Carbohyd Polym 181:659–667. https://doi.org/10.1016/j.carbpol.2017.11.098

Shorrock CJ, Rees WD (1988) Overview of gastroduodenal mucosal protection. Am J Med 84(2A):25–34. https://doi.org/10.1016/0002-9343(88)90251-3

Sorlier P, Denuzi'ere A, Viton C, Domard A (2001) Relation between the degree of acetylation and the electrostatic properties of chitin and chitosan. Biomacromolecules 2(3):765–772. https://doi.org/10.1021/bm015531

Suresh PV (2012) Biodegradation of shrimp processing bio-waste and concomitant production of chitinase enzyme and N-acetyl-D-glucosamine by marine bacteria: production and process optimization. World J Microbiol Biotechnol 28:2945–2962. https://doi.org/10.1007/s11274-012-1106-2

Tabassum N, Ahmed S, Ali MA (2021)Review: Chitooligosaccharides and their structural-functional effect on hydrogels. Carbohydr Polym 261:117882

Tang W, Li Y, Liu L, Zhang J, Xian H (2012) Characterization and production optimization of a chitinase (Tachi1) from *Trichoderma asperellum* in recombinant *Pichia pastoris* expression system. Wei Sheng Wu Xue Bao 52(3):345–352

Tsigos I, Zydowicz N, Martinou A, Domard A, Bouriotis V (1999) Mode of action of chitin deacetylase from *Mucor rouxii* on N-acetylchitooligosaccharides. Eur J Biochem 261(3):698–705

Vårum K, Smidsrød O (2004) Structure–Property relationship in chitosans. In S. Dumitriu (ed.) Polysaccharides. CRC Press

Vårum KM, Ottøy MH, Smidsrød O (1994) Water-solubility of partially N- acetylated chitosans as a function of pH: effect of chemical composition and depolymerisation. Carbohyd Polym 25(2):65–70. https://doi.org/10.1016/0144-8617(94)90140-6

Wei XH, Niu YP, Xu YY, Du YZ, Hu FQ, Yuan H (2010) Salicylic acid-grafted chitosan oligosaccharide nanoparticle for paclitaxel delivery. J Bioact Compat Polym 25:319–335

Willenberg BJ, Hamazaki T, Meng F-W, Terada N, Batich C (2006) Self-assembled copper-capillary alginate gel scaffolds with oligochitosan support embryonic stem cell growth. J Biomed Mater Res Part A 79(2):440–450. https://doi.org/10.1002/jbm.a.30942

Yang H-C, Hon M-H (2009) The effect of the molecular weight of chitosan nanoparticles and its application on drug delivery. Microchem J 92(1):87–91. https://doi.org/10.1016/j.microc.2009.02.001

Yu J, Wang Q, Zhang H, Qin X, Chen H, Corke H, Hu Z, Liu G (2021) Increased stability of curcumin-loaded pickering emulsions based on glycated proteins and chitooligosaccharides for functional food application. LWT-Food Sci Technol 148:111742

Zhang J, Chen XG, Huang L, Han JT, Zhang XF (2012) Self-assembledpolymeric nanoparticles based on oleic acid-grafted chitosan oligosaccharide:Biocompatibility, protein adsorption and cellular uptake. J Mater Sci Mater Med 23:1775–1783

Zhanga J, Hana J, Zhanga X, Jianga J, Xua M, Zhanga D, Jing H (2015) Polymeric nanoparticles based on chitooligosaccharide as drugcarriers for co-delivery of all-trans-retinoic acid and paclitaxel. Carbohydr Polym 129:25–34

Zhang H, Mi J, Huo Y, Huang X, Xing J, Yamamoto A et al (2014) Absorption enhancing effects of chitosan oligomers on the intestinal absorption of low molecular weight heparin in rats. Int J Pharm 466(1):156–162. https://doi.org/10.1016/j.ijpharm.2014.03.010

Zupancic S, Sinha-Ray S, Kristl J, Yarin AL (2016) Long-term sustained ciprofloxacin release from PMMA and hydrophilic polymer blended nanofibers. Mol Pharm 13(1):295–305. https://doi.org/10.1021/acs.molpharmaceut.5b00804

Matrix Metalloproteinases Inhibitory Effects of Chitooligosaccharides

Noel Vinay Thomas, A. Salomy Monica Diyya, Dlzar Dlshad Ghafoor, and Se-Kwon Kim

1 Introduction

Metalloproteinases, a family of proteolytic enzymes called endopeptidases and are known to be involved in the tissue remodelling, organ development processes by degradation of extracellular matrix [ECM]. These endopeptidases majorly include enzymes that are calcium- and zinc-dependent, Matrix Metalloproteinases [MMPs] that are known for their involvement in several pathological conditions like inflammatory, vascular and autoimmune disorders, and carcinogenesis as well as physiological processes like wound healing, bone resorption, uterine involution and organogenesis (GROSS and LAPIERE 1962; Egeblad and Werb 2002; Lee and Murphy 2004; Page-McCaw et al. 2007). Scientific investigations by various research groups across the globe have confirmed the role of MMPs in various severe clinical conditions like arthritis (Wright et al. 2014; Pap and Korb-Pap 2015), cancer (Shay et al. 2015; Buralla and Parthasarathy 2019), chronic obstructive pulmonary disease [COPD] (McGarry Houghton 2015; Shevchenko et al. 2019). Recent scientific studies have also revealed the sepsis related mortality and its relevancy with MMPs unregulated

N. V. Thomas (✉) · D. D. Ghafoor
College of Science, Department of Biomedical Science, Komar University of Science and Technology, Sulaymaniyah, Kurdistan Region 46001, Iraq
e-mail: noel.thomas@komar.edu.iq

A. S. M. Diyya
College of Medicine, Department of Pharmacy, Komar University of Science and Technology, Sulaymaniyah, Kurdistan Region 46001, Iraq

D. D. Ghafoor
College of Science, Department of Chemistry, University of Sulaymaniyah, Kurdistan Region, Sulaymaniyah 46001, Iraq

S.-K. Kim
Department of Marine Science and Convergence Engineering, College of Science and Technology, Hanyang University, Gyeonggi-do 11558, Korea

S.-K. Kim (ed.), *Chitooligosaccharides*,
https://doi.org/10.1007/978-3-030-92806-3_6

Table 1 Classification of Matrix metallo proteases

Classification of Matrix metallo proteases [Reference: Thomas and Kim (2014)]	
Type	Enzyme group
Collagenases	MMP-1, -8, -13, -18
Gelatinases	MMP-2, -9
Elastases	MMP-12
Stromelysins	MMP-3, -7, -1, -11, -26, -27
Membrane type specific MMPs	MMP-14, -15, -16, -17, -24, -25
Other MMPs	MMP-19, -20, -28, -21, -22, -23

expressions (Vandenbroucke et al. 2011; Martin et al. 2014; Wessels and Cousins 2015) In addition, the MMPs have been extensively studied and have been reported about their involvement in tumour invasion, angiogenesis, metastasis, transformation of cancer cells, signal transduction, and apoptosis and thus considered the best targets for the cancer related disorders (Overall and López-Otín 2002; Thomas and Kim 2011a, b). A list of types and enzyme groups of MMPs is shown in Table 1.

In general, the intercellular regulation and cell matrix adhesion phenomena are regulated in an organized fashion. But when the expression MMPs is dis-regulated such organized phenomena is thrown out of control and leads to various types of human cancers (Bourboulia and Stetler-Stevenson 2010). Natural matrix metalloproteinase inhibitors [MMPIs] which are called the Tissue inhibitors of metalloproteinase [TIMP] commonly regulate or prevent the over expressions of MMPs. Factors like solubility and interaction of TIMPs with Pro-MMPs does determine the inhibitory activity of TIMPs (Tilson 2000; Burduk et al. 2019). Several pathological responses associated with MMPs have made them as therapeutic targets for the better management of human cancers. Till now, many synthetic MMPIs have been reported. Synthetic drugs like Batimastat (BB-94), Marimastat (BB-516) have been successful in lowering the expression of MMPs. However, improper metabolism, low oral bioavailability, poor solubility, side effects like musculoskeletal pain and inflammation, complications and the risk of increased drug toxicity is still a big challenge (Thomas and Kim 2010, 2011a, 2013, 2014; b; Kang and Kim 2012; Thomas et al. 2020).

2 The Role of MMPs in Cancer Progression

The world's leading causes for mortality is cancer and hence majority of the world's biomedical research teams are continuously researching on molecular events that take place during carcinogenesis and the signalling pathways participating in cancer progression. In addition (Gialeli et al. 2011) studies conducted over more than 50 years have provided enormous evidence that confirms the role of MMPs as the principal mediators of the alterations observed in the microenvironment during cancer progression (Page-McCaw et al. 2007). In the process of carcinogenesis, tumour cells are involved in several interactions with the tumour microenvironment [TME] which involves extracellular matrix [ECM], growth factors and cytokines associated with ECM, as well as surrounding cells (endothelial cells, fibroblasts, macrophages, mast cells, neutrophils, pericytes and adipocytes) (Deryugina and Quigley 2006; Murphy 2008). Migration, invasion, metastasis and angiogenesis are the four major hallmarks of cancer that are dependent on the surrounding microenvironment. MMPs play a major role in these processes and degrade various cell adhesion molecules, thereby modulating cell–cell and cell–ECM interactions. And key MMPs in relation to the stages of cancer progression is reviewed elsewhere (Gialeli et al. 2011).

3 Chitosan/COS Sources and Methods for the Synthesis of COS

The commercial source of chitin is majorly obtained from crustacean shells because of their ready availability and high content. Chitosan is a natural biopolymer that is obtained by the deacetylation of chitin (Gagné and Simpson 1993). Chitins and chitosans are produced from biological wastes obtained from the marine organisms for commercial use. However, terrestrial organisms like insects, terrestrial crustaceans and mushrooms are considered as alternative resources for the production of chitin and chitosan (Nwe and Stevens 2008). A comprehensive review about the Chitin and COS sources, synthesis and their biological effects has been published elsewhere (Thomas et al. 2015). A schematic representation of chitosan production is shown in Fig. 1.

4 Methods to Prepare COS

A research study is conducted on Marine *Streptomyces* sp for its chitinolytic activity on agar plates with colloidal chitin and was reported to have highest activity. This strain is identified as Streptomyces champavatii AZ-1 by thin layer chromatography. The strain AZ-1 produced a multi-chitinolytic enzyme complex on submerged and solid-state fermentation at 37 °C after 144, 168 h incubation time intervals. Chitinases secreted in the culture supernatant were partially purified. Production of N-acetyl-D-glucosamine (GlcNAc) and N-acetyl chitooligosaccharides is important to biotechnology sectors because of their potent biomedical applications. N-acetyl chitooligosaccharides which exhibit anti-tumour activity are produced from colloidal chitin by the use of crude enzyme obtained from *S. champvatii* AZ-1 which resulted in a good yield of chitopentaose $(GlcNAc)_5$ and chitohexaose $(GlcNAc)_6$ (Hosny et al. 2010).

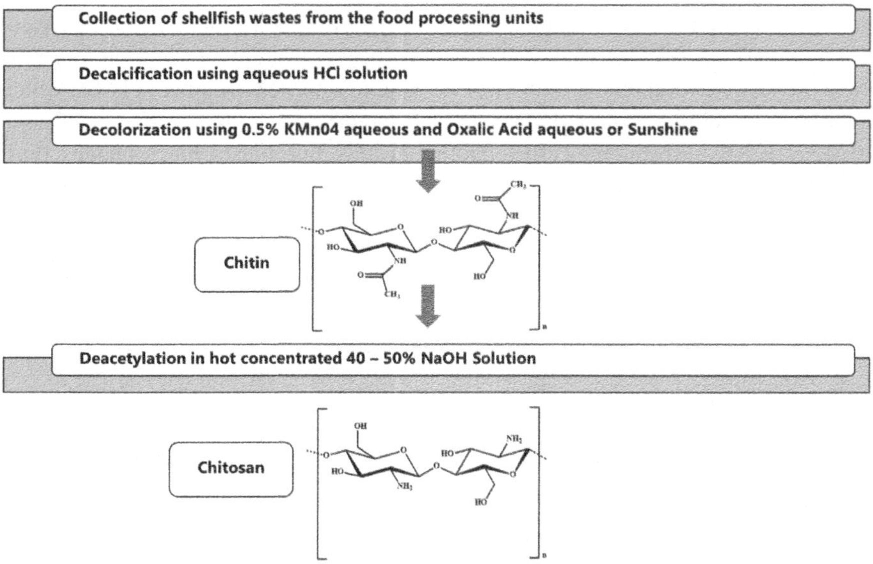

Fig. 1 Chitosan production—a schematic representation

Chitosan is to be subjected to hydrolysis which results in the cleavage of unstable O-glycosidic linkages. Thus, producing COS with varying degree of polymerization. Different hydrolytic mechanisms like acid hydrolysis (Il'ina and Varlamov 2004), enzymatic hydrolysis (Zhang et al. 1999; Kuroiwa et al. 2002) and other processes like oxidative degradation (Mao et al. 2004), ultra-sonic degradation (Chen and Chen 2000), chemo-enzymatic (Akiyama et al. 1995) and recombinant approaches (Samain et al. 1997) are followed to produce COS.

5 Preparation of Chitin Oligomers

Chitin and chitosan oligomers are composed of beta-(1 → 4) linked *N*-acetylglucosamine and glucosamine units, respectively, with approximately ten residues or less. The two hydrolytic methods which include chemical hydrolysis (acidic hydrolysis) and enzymatic hydrolysis are used to prepare the oligomers from chitin and chitosan.

6 Chemical Hydrolysis

In the recent days, the commercially available hexameric forms of chitin oligomers are prepared by the partial hydrolysis of chitin with concentrated hydrochloric acid followed by column chromatographic fractionation (Rupley 1964). The conventional procedure (Kurita 1998) to obtain the chitin oligomers is shown in Fig. 2.

In the process of extracting chitin and chitosan from the shell wastes, shrimp or crab shells are first treated with 3–5% aqueous NaOH solution to remove proteins attached to the shells and thereby prevent the contamination of chitin products with proteins. Deproteinized shells are then neutralized, and calcium is removed by treating with 3–5% aqueous HCl solution to afford a white or slightly pink precipitate of chitin. Then the chitin is *N*-deacetylated with 40–45% NaOH to form chitosan with a cationic nature. The resulting crude sample is dissolved in 2% acetic acid and the supernatant is neutralized with aqueous NaOH solution which results in purified chitosan as a white precipitate (Hirano 1996).

Most of the acidic hydrolyses procedures reported production of oligomers with a low degree of polymerization (DP), mainly from monomer to tetramer in quantitative yield or their processes are not convenient. It is disadvantageous that the yields of the oligomers with relatively higher DP, such as pentamer to heptamer, are low because physiological effects have been expressed mainly by high DP oligomers. According to Domard and Carter, the oligomers with a wide distribution of glucosamine can be separated up to DP 15 in pure form and up to DP 37 by the hydrolysis in 12 M HCl at 72 °C over 90 min followed by steric exclusion chromatography with aqueous solutions of ammonium acetate as eluents (Domard and Cartier 1989).

The chemical hydrolysis has its own loop-holes which include low yield of COS and presence of larger amounts of monomeric D-glucosamine units (Uchida et al.

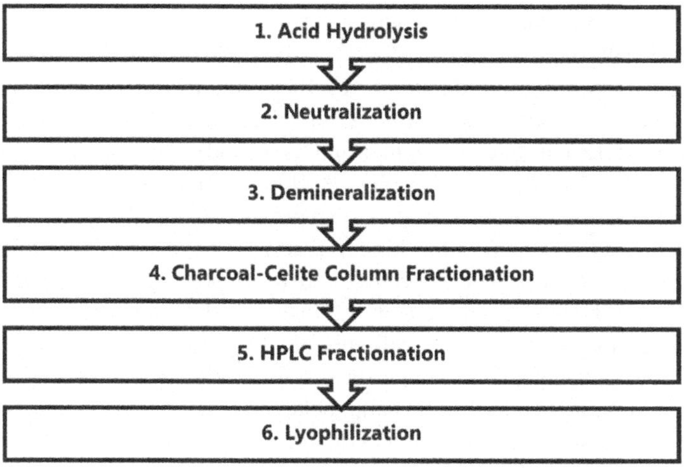

Fig. 2 Conventional procedure to obtain the chitin oligomers in chemical hydrolysis

1989), an alternative method called enzymatic hydrolysis to synthesize COS is implemented.

7 Enzymatic Hydrolysis

As the chitosan is susceptible to a number of enzymes, chitosan is said to have broad substrate specificity (Aiba 1994a). The chitin hydrolyzing enzyme, chitosanase is generally used in the production of COS. This enzyme is found in various different organisms like fungi (Muzzarelli et al. 1994; Kim et al. 1998), bacteria (Vårum et al. 1996).

A few scientific reports suggest that some of the carbohydrases and proteases can also hydrolyze chitosan to produce COS with various molecular weights (Aiba 1994b; Zhang et al. 1999). The four different types of glycosidic bonds resulted in differential deacetylations of chitosan. These include linkages between two *N*-acetylated units (A–A), acetylated and deacetylated units (A–D), deacetylated and acetylated units (D–A) and two deacetylated units (D–D). The specific cleaving ability of chitosanases on the four different glycosidic linkages in partially *N*-deacetylated chitosan is determined by the identity of the reducing and non-reducing ends and DD of chitosan.

To obtain chitosan oligomers by enzymatic reactions, chitosanase and cellulase are usually used. Microorganisms like bacteria and fungi and a few terrestrial plants are known to produce chitosanases, like chitinases (Somashekar and Joseph 1996). Usually chitosanases have a lower molecular mass when compared to chitinases (Grenier 1990). The utilization of the microbial chitosanases results in effective COS yield. However, they are exceedingly expensive to be used on an industrial scale. Hence the utilization of other enzymes that are commercially available are being considered and are used to produce COS with relatively less expenses (Izume and Ohtakara 1987).

In the earlier period of production of COS enzymatically, the batch reactors are used, where chitosan and chitosanase are mixed and are allowed to breakdown the glycosidic bonds (Zhang et al. 1999; Jeon and Kim 2000). This approach reported that the optimum conditions for hydrolyzing 1% chitosan (DD, 89%) using chitosanase from *Bacillus pumilus* BN-262 are 45 °C, 5.5 pH and for 1 h. The main disadvantage of the batch reactor system is the higher cost associated with large quantities of expensive enzymes that cannot be reused. Also, in these systems lower yields are generally resulted due to the limited ability to control DP and hence.

the final product is obtained as a mixture of COS with broader molecular weight range. Therefore, many studies are performed to develop new strategies to produce COS with desired molecular weights and make these systems economical.

Figure 3: Schematic representation of dual reactor system employed in enzymatic hydrolysis of Chitin to obtain COS

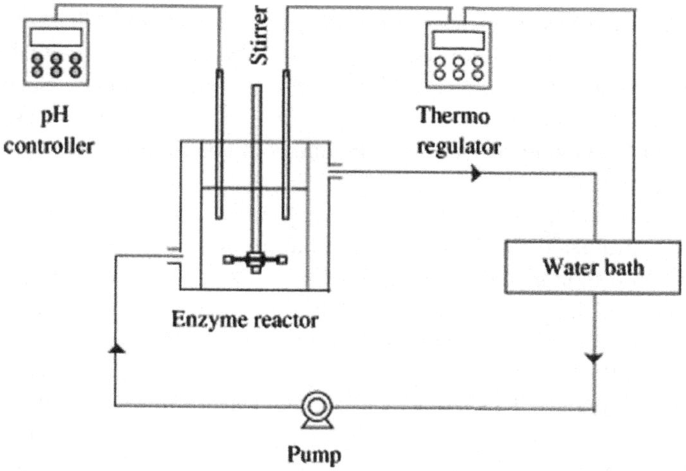

Fig. 3 Typical schematic diagram of a batch reactor for the hydrolysis of chitosan, using free enzyme

Another method of producing COS is by the usage of column reactors which have the immobilized enzyme. A new method to produce COS with higher DP was designed by using an immobilized chitosanase is reported by Jeon et al. (Jeon 1998).

8 COS as MMP Inhibitors

COS that was prepared from chitosan by depolymerization with enzyme has shown anti-angiogenic on chicken chorioallantoic membrane (CAM) angiogenesis and on proliferation, migration and tube formation of human umbilical vein endothelial cells (HUVECs) induced by human hepatoma carcinoma cells (HCC) culture fluid. This research group has suggested that the anti-angiogenic activity is possibly by COS interfering with the interaction of pro-angiogenic factors and their receptors, and thereby blocking the signal conduction or by influencing the expression and production of anti-angiogenic factors in the endothelial cells (Wu et al. 2008) They also correlated their research results that inhibition is possible by the deactivation of MMP-2 (Kim and Kim 2006) because MMPs play a major role in angiogenic process.

Another research group recently reported the anti-angiogenic effects of COS with different concentration and degree of polymerization (DP). They used the culture of human umbilical vein endothelial cells (HUVECs) and carried out the cell proliferation activity, cell morphology, cell migration and angiogenesis. Their results indicated that COS at a high concentration of 400 μg/mL (COS(400)) and DP of 6 (Chitinhexaose Hydrochloride, COS6) had inhibitory effect on angiogenic activities of HUVECs. Specifically, COS(400) and COS6 inhibited cell proliferation activity, cell migration and vascular endothelial cell growth factor (VEGF) expression of HUVECs. While COS at a concentration of <400 μg/mL with a suitable degree of polymerization, DP < 6, had exhibited less significant effects. Such findings clearly suggest the efficiency of COS in inhibiting the expressions of MMPs, hence exhibiting anti-angiogenic effects. This feature could be exploited in controlling the cancers (Huang et al. 2021).

On the other hand, Chitooligosaccharides (COS) and N-acetyl-D-glucosamine (NAG), were evaluated for their effects on the proliferation and differentiation of peripheral blood mononuclear cells (PBMCs) and sarcoma 180 transplanted mice [tumour model]. Administration of COS at the dose of 100 mg/kg and NAG at a dose of 300 mg/kg or 500 mg/kg can not only promote the differentiation of PBMCs and the secretion of IL-2 and IFN-γ, but can also inhibit the expression of VEGF mRNA in sarcoma 180 tumours. This research group has clearly reported that effects of COS and NAG are dose-dependent especially by reducing the Vascular endothelial growth factor (VEGF) expression levels and hence exhibiting the anti-angiogenesis effects (Xu et al. 2012).

The oligomers or the monomers of chitin or chitosan have profound effect on the MMP-1 activity. A few researchers stated that the MMP-1 release from adult and neonatal fibroblasts is influenced by oligomers and monomers of chitin and chitosan (Okamura et al. 2005). Chitosan inhibits the activity of MMP-2. The atomic force microscopy revealed the binding specificity of MMP to Chitosan is high and the colorimetric assay demonstrated the non-competitive inhibition of MMP-2 by chitosan (Mayssara and Abo Hassanin Supervised 2014). Various biological applications of chitosan and its derivatives include inhibitory effect on matrix metalloproteinases, hypertension, coagulation, cancer, obesity and microbial growth (Ngo et al. 2015). As the MMP-9 has a significant role in the development of malignant tumours, derivatives of chitin and chitosan can be considered as the potential therapeutic avenue in treatment of cancer. The MMP-9 expression is inhibited by COS (1–3, 3–5 and 5–10 kDa) (van Ta et al. 2006). The chitosan and chitosan oligosaccharides have wide medical applications because of their biomedical properties (Zou et al. 2016). The anti-tumour activity of COS by a research study showed the inhibition of the cell proliferation and metastasis because of their MMP inhibitory activity. As matrix metalloproteinase inhibitors are categorized as significant therapeutic agents for metastasis and many chronic inflammation conditions, COS are categorized as potent anti-cancer agents (Zhai et al. 2021). Chitosan showed anticancer activity by reducing the growth of cancer cells (Šimat et al. 2020). The antiproliferative activity against human colon cancer cells is exhibited by the chitin and its derivatives (Sayari et al. 2016). A few studies have confirmed that chitin and chitosan and their derivatives have inhibited MMP-2 and -9 in human fibrosarcoma cells (van Ta et al. 2006).

Among many biomedical properties of COS, antioxidant property is extensively studied on various derivatives of chitin and chitosan. The water-soluble derivatives of chitosan and chitin have potential antioxidant properties. Carboxymethyl-chitosan (CM-chitosan) and carboxymethyl-chitin (CM-chitin) decreased the formation of intracellular reactive oxygen species, protein, DNA and lipid peroxidation. MMP-2 and MMP-9 expression is reduced by CM-chitin and CM-chitosan without any cytotoxicity effects. CM-chitin has shown greater MMP inhibitory activity than CM-chitosan by down regulating the transcription of A-1 and NF-Potential antioxidant activity. The CM-chitin and CM-chitosan lessens the radical-induced damage and thus rendering them the potential antioxidant and MMP inhibitory activity (Kong et al. 2010). The DDA and MW of chitosan are the important factors that decides the scavenging ability of chitosan. The researchers reported that higher antioxidant activity is achieved with lower MW of chitosan (30 kDa). The results stated that high rate of DDA and low MW of chitosan has higher antioxidant activity (Kim 2018). Studies conducted on the derivatives of chitosan which includes N-furoyl chitosan (NF-CS) and N-furoyl chitosan oligosaccharide (NF-COS) showed significant antioxidant properties with their radical scavenging activities (Zhao et al. 2013).

A research group from South Korea has used carboxymethylation reaction and synthesized water-soluble derivatives of chitosan and chitin, carboxymethyl-chitosan (CM-chitosan) and carboxymethyl-chitin (CM-chitin) and were tested for the antioxidative and matrix metalloproteinase-2 and -9 (MMP-2 and -9) inhibitory effects using human fibrosarcoma cells [HT1080]. This group has reported that CM-chitosan and CM-chitin have successfully reduced the expression levels of MMP-2 and -9 in gelatin zymography, RTPCR and western blot analysis with almost zero cytotoxic influence. In their observations, it was evident that CM-chitin exhibited a higher MMP inhibitory effect via transcriptional down-regulations of activator protein-1 (AP-1) and nuclear factor kB (NF-kB). Thus, emphasizing the therapeutic effects of COS and, they recommended the nutraceutical value of CM-chitosan and CM-chitin (Kong et al. 2010).

The therapeutic potential of sulfated glucosamine (SGlc) which can alleviate the inflammation associated with arthritis has been reported by Rajapakse and et al. They have evaluated the MMP-2 and MMP-9 inhibitory effects of SGIc in human fibrosarcoma cells. They reported that post treatment of the HT1080 with SGIc has shown the downregulation of transcription factor, NF-κB along with low levels production of both NF-κB p50 and p65 proteins that directly affected activation process of MMP-2 and MMP-9 expressions. Hence suggesting the SGlc to be an effective drug in relieving the symptoms of osteoarthritis (Rajapakse et al. 2007).

Decreased expressions of MMP-3 and MMP-9 were reported by the treatment or administration of glucosamine sulfate in osteosarcoma cell lines, MG-63 and SaOS-2, in vitro. These findings clearly suggested that treating the osteosarcoma cell lines with glucosamine sulfate at the doses 10, 50 or 100 μg/ml as shown significant reduction of MMP expression in both cell lines. However, to be more precise, MMP expression in both cell lines was observed only for MMP-3, whereas MMP-9 expression was seen decreased only in SaOS-2 cells. Hence suggesting the efficacy of Glucosamine sulfate as potential anti-sarcoma therapeutic (Pohlig et al. 2016).

The administration of glucosamine sulfate combined with meloxicam has benefitted the patients with osteoarthritis by reduction of clinical symptoms, pain relief and reduction of serum C-terminal telopeptide of type I collagen (CTX-I), C-terminal telopeptide of type II collagen (CTX-II), cartilage oligomeric matrix protein (COMP) and MMP-3. They reported that a combination of 0.5 g of glucosamine sulfate capsule in addition to meloxicam has shown the reduction of serum CTX-I, CTX-II, MMP-3 and COMP along with reduction of clinical symptoms (Zhijun et al. 2019).

9 Conclusion

In the recent past, research communities across the globe have shown an enormous interest in the understanding the role of the matrix metalloproteinases in various human health related ailments that could be dangerous to survival if not treated properly. It is also evident that there is an immediate need of intense research findings for the discovery and development of MMPIs. However there are many chemically synthesized MMPIs available in the market, perhaps due to the intense side effects observed due to their usage has alarmed the researchers across the world to look for MMPIs from natural resources. One such a natural resource is chitosan. Chitosan and its derivatives have been reported to have an greater application in food, medicinal and pharmacological industries. The advancements in the research technical know-hows has enabled the researchers to prepare the chitosan derivatives, COS at different molecular weights and also work with the efficient functional group additions to enhance their biological effects. There is enormous amount of scientific data to prove the biological activities of COS including antioxidation, antiinflammation, anti microbial, anti cancer and anti-HIV effects. In this part of write-up we have focused on the MMP inhibition effects of COS and found limited scientific observations. Hence we strongly recommend that there is a huge scope of enhancing the research especially in inhibition effects of COS on all classes of MMPs. Importantly as Chitosan and COS are already recommended as safe and efficient functional ingredients, they can be used in the field of nutraceutical and functional food that might combat the over expressions of MMPs and regulate the human health.

Acknowledgements The authors are thankful to **Dr. Jayachandran Venkatesan** for redrawing the chemical structures and helping with the preparation of content for the figures shown in this article. Samples are available from the authors. Sincere thanks to **Prof. Dr. Salahalddin S Ali,** President, **Komar University of Science and Technology** [KUST], Sulaymaniyah 46001, Kurdistan Region, Iraq for his continuous support and encouragement for the successful completion of this article. The authors would like to extend their sincere thanks to **Komar Research Center** at KUST, Sulaymaniyah 46001, Kurdistan Region, Iraq for their consistent support.

References

Aiba S (1994a) Preparation of N-acetylchitooligosaccharides by hydrolysis of chitosan with chitinase followed by N-acetylation. Carbohydrate Research 265:323–328. https://doi.org/10.1016/0008-6215(94)00243-6

Aiba S, ichi, (1994) Preparation of N-acetylchitooligosaccharides from lysozymic hydrolysates of partially N-acetylated chitosans. Carbohyd Res 261:297–306. https://doi.org/10.1016/0008-6215(94)84025-3

Akiyama K, Kawazu K, Kobayashi A (1995) A novel method for chemo-enzymatic synthesis of elicitor-active chitosan oligomers and partially N-deacetylated chitin oligomers using N-acylated chitotrioses as substrates in a lysozyme-catalyzed transglycosylation reaction system. Carbohyd Res 279:151–160. https://doi.org/10.1016/0008-6215(95)00288-X

Bourboulia D, Stetler-Stevenson WG (2010) Matrix metalloproteinases (MMPs) and tissue inhibitors of metalloproteinases (TIMPs): positive and negative regulators in tumor cell adhesion. Semin Cancer Biol 20:161–168

Buralla KK, Parthasarathy V (2019) Roles of matrix metalloproteinases (Mmps) in cancer progression: a review. Int Res J Pharm 10:12–15. https://doi.org/10.7897/2230-8407.100371

Burduk PK, Sawicki P, Szylberg L et al. (2019) Expression of matrix metalloproteinase-2/9 and tissue inhibitor of metalloproteinase-1/2 as predictive factors in oropharyngeal squamous cell carcinoma. Iranian J Otorhinolaryngol 31:153–161. https://doi.org/10.22038/ijorl.2018.27737.1905

Chen RH, Chen JS (2000) Changes of polydispersity and limiting molecular weight of ultrasonic treated chitosan. Adv Chitin Sci 4(4):361–366

Deryugina EI, Quigley JP (2006) Matrix metalloproteinases and tumor metastasis. Cancer Metastasis Rev 25:9–34

Domard A, Cartier N (1989) Glucosamine oligomers: 1. Preparation and characterization. Int J Biol Macromol 11:297–302. https://doi.org/10.1016/0141-8130(89)90023-8

Egeblad M, Werb Z (2002) New functions for the matrix metalloproteinases in cancer progression. Nat Rev Cancer 2:161–174

Gagné N, Simpson BK (1993) Use of proteolytic enzymes to facilitate the recovery of chitin from shrimp wastes. Food Biotechnol 7:253–263. https://doi.org/10.1080/08905439309549861

Gialeli C, Theocharis AD, Karamanos NK (2011) Roles of matrix metalloproteinases in cancer progression and their pharmacological targeting. FEBS J 278:16–27

Grenier J (1990) Some pathogenesis-related proteins are chitosanases with lytic activity against fungal spores. Mol Plant Microbe Interact 3:401. https://doi.org/10.1094/mpmi-3-401

GROSS J, LAPIERE CM (1962) Collagenolytic activity in amphibian tissues: a tissue culture assay. Proc Natl Acad Sci U. S. A. 48:1014–1022. https://doi.org/10.1073/pnas.48.6.1014

Hirano S (1996) Chitin biotechnology applications. Biotechnol Annu Rev. Elsevier 237–258

Hosny AEDMS, El-Shayeb NA, Abood A, Abdel-Fattah AM (2010) A potent chitinolytic activity of marine Actinomycete sp. and enzymatic production of chitooligosaccharides. Aust J Basic Appl Sci 4:615–623

Huang X, Jiao Y, Zhou C (2021) Impacts of chitosan oligosaccharide (COS) on angiogenic activities. Microvasc Res 134. https://doi.org/10.1016/j.mvr.2020.104114

Il'ina AV, Varlamov VP (2004) Hydrolysis of chitosan in lactic acid. Appl Biochem Microbiol 40:300–303. https://doi.org/10.1023/B:ABIM.0000025956.98250.30

Izume M, Ohtakara A (1987) Preparation of d-glucosamine oligosaccharides by the enzymatic hydrolysis of chitosan. Agric Biol Chem 51:1189–1191. https://doi.org/10.1271/bbb1961.51.1189

Jeon YJ (1998) Production of chitosan oligosaccharides using chitin-immobilized enzyme. Korean J Biotechnol Bioeng 13:147–154

Jeon YJ, Kim SK (2000) Production of chitooligosaccharides using an ultrafiltration membrane reactor and their antibacterial activity. Carbohyd Polym 41:133–141. https://doi.org/10.1016/S0144-8617(99)00084-3

Kang KH, Kim SK (2012) Microalgae as sources of biomaterials and pharmaceuticals

Kim MM, Kim SK (2006) Chitooligosaccharides inhibit activation and expression of matrix metalloproteinase-2 in human dermal fibroblasts. FEBS Lett 580:2661–2666. https://doi.org/10.1016/j.febslet.2006.04.015

Kim SY, Shon DH, Lee KH (1998) Purification and characteristics of two types of chitosanases from Aspergillus fumigatus KH-94. J Microbiol Biotechnol 8:568–574

Kim S (2018) Competitive biological activities of chitosan and its derivatives: antimicrobial, antioxidant, anticancer, and anti-inflammatory activities. Int J Polym Sci. https://doi.org/10.1155/2018/1708172

Kong CS, Kim JA, Ahn B, et al (2010) Carboxymethylations of chitosan and chitin inhibit MMP expression and ROS scavenging in human fibrosarcoma cells. Process Biochem 45. https://doi.org/10.1016/j.procbio.2009.09.004

Kurita K (1998) Chemistry and application of chitin and chitosan. Polym Degrad Stab 59:117–120. https://doi.org/10.1016/S0141-3910(97)00160-2

Kuroiwa T, Ichikawa S, Hiruta O et al (2002) Factors affecting the composition of oligosaccharides produced in chitosan hydrolysis using immobilized chitosanases. Biotechnol Prog 18:969–974. https://doi.org/10.1021/bp020011r

Lee MH, Murphy G (2004) Matrix metalloproteinases at a glance. J Cell Sci 117:4015–4016

Mao S, Shuai X, Unger F et al (2004) The depolymerization of chitosan: effects on physicochemical and biological properties. Int J Pharm 281:45–54. https://doi.org/10.1016/j.ijpharm.2004.05.019

Martin G, Asensi V, Montes AH, et al (2014) Role of plasma matrix-metalloproteases (MMPs) and their polymorphisms (SNPs) in sepsis development and outcome in ICU patients. Sci Rep 4. https://doi.org/10.1038/srep05002

Mayssara A, Abo Hassanin Supervised A (2014) 済無No title no title no title. Pap Knowl. Toward Media Hist Doc 8:3035–3040

McGarry Houghton A (2015) Matrix metalloproteinases in destructive lung disease. Matrix Biol 44–46:167–174

Murphy G (2008) The ADAMs: signalling scissors in the tumour microenvironment. Nat Rev Cancer 8:929–941

Muzzarelli RAA, Ilari P, Tarsi R et al (1994) Chitosan from absidia coerulea. Carbohyd Polym 25:45–50. https://doi.org/10.1016/0144-8617(94)90161-9

Ngo DH, Vo TS, Ngo DN et al (2015) Biological effects of chitosan and its derivatives. Food Hydrocoll 51:200–216. https://doi.org/10.1016/j.foodhyd.2015.05.023

Nwe N, Stevens WF (2008) Production of chitin and chitosan and their applications in the medical and biological sector. Recent Res Biomed Asp Chitin Chitosan 978:161

Okamura Y, Nomura A, Minami S, Okamoto Y (2005) Effects of chitin/chitosan and their oligomers/monomers on release of type I collagenase from fibroblasts. Biomacromol 6:2382–2384. https://doi.org/10.1021/bm050092q

Overall CM, López-Otín C (2002) Strategies for MMP inhibition in cancer: innovations for the post-trial era. Nat Rev Cancer 2:657–672

Page-McCaw A, Ewald AJ, Werb Z (2007) Matrix metalloproteinases and the regulation of tissue remodelling. Nat Rev Mol Cell Biol 8:221–233

Pap T, Korb-Pap A (2015) Cartilage damage in osteoarthritis and rheumatoid arthritis—two unequal siblings. Nat Rev Rheumatol 11:606–615

Pohlig F, Ulrich J, Lenze U et al (2016) Glucosamine sulfate suppresses the expression of matrix metalloproteinase-3 in osteosarcoma cells in vitro. BMC Complement Altern Med 16. https://doi.org/10.1186/s12906-016-1315-6

Rajapakse N, Mendis E, Kim MM, Kim SK (2007) Sulfated glucosamine inhibits MMP-2 and MMP-9 expressions in human fibrosarcoma cells. Bioorg Med Chem 15:4891–4896. https://doi.org/10.1016/j.bmc.2007.04.048

Rupley JA (1964) The hydrolysis of chitin by concentrated hydrochloric acid, and the preparation of low-molecular-weight substrate for lysozyme. BBA Specialized Sect Mucoprotein Mucopolysacch 83:245–255. https://doi.org/10.1016/0926-6526(64)90001-1

Samain E, Drouillard S, Heyraud A et al (1997) Gram-scale synthesis of recombinant chitooligosaccharides in Escherichia coli. Carbohyd Res 302:35–42. https://doi.org/10.1016/S0008-6215(97)00107-9

Sayari N, Sila A, Abdelmalek BE et al (2016) Chitin and chitosan from the Norway lobster by-products: antimicrobial and anti-proliferative activities. Int J Biol Macromol 87:163–171. https://doi.org/10.1016/j.ijbiomac.2016.02.057

Shay G, Lynch CC, Fingleton B (2015) Moving targets: emerging roles for MMPs in cancer progression and metastasis. Matrix Biol 44–46:200–206

Shevchenko OS, Ovcharenko IA, Todoriko LD (2019) Pathophysiological mechanisms destruction of the lung connective tissue in tuberculosis. Infus Chemother 14–20. https://doi.org/10.32902/2663-0338-2019-2-14-20

Šimat V, Elabed N, Kulawik P et al (2020) Recent advances in marine-based nutraceuticals and their health benefits. Mar Drugs 18:627. https://doi.org/10.3390/md18120627

Somashekar D, Joseph R (1996) Chitosanases—properties and applications: a review. Biores Technol 55:35–45. https://doi.org/10.1016/0960-8524(95)00144-1

Thomas NV, Kim SK (2011) Potential pharmacological applications of polyphenolic derivatives from marine brown algae. Environ Toxicol Pharmacol 32:325–335. https://doi.org/10.1016/j.etap.2011.09.004

Thomas NV, Kim SK (2013) Beneficial effects of marine algal compounds in cosmeceuticals. Mar Drugs 11:146–164. https://doi.org/10.3390/md11010146

Thomas NV, Venkatesan J, Manivasagan P, Kim S-K (2015) Production and biological activities of chitooligosaccharides (COS)—an overview. J Chitin Chitosan Sci 3:1–10. https://doi.org/10.1166/jcc.2015.1084

Thomas NV, Kim S-K (2010) Metalloproteinase inhibitors: status and scope from marine organisms. Biochem Res Int 2010:

Thomas NV, Kim SK (2011b) Health beneficial aspects of phloroglucinol derivatives from marine brown algae

Thomas NV, Kim SK (2014) Fucoidans from marine algae as potential matrix metalloproteinase inhibitors

Thomas NV, Diyya ASM, Ghafour DD, Kim S (2020) Marine Algal Phlorotannins and their biological importance. Encycl Mar Biotechnol

Tilson D (2000) Inhibition of matrix metalloproteinases: therapeutic applications. J Vasc Surg 31:1082–1083. https://doi.org/10.1067/mva.2000.105884

Uchida Y, Lzume M, Ohtakara A (1989) Preparation of chitosan oligomers with purified chitosanase and its application. In Chitin and chitosan: sources, chemistry, biochemistry physical properties and applications. Elsevier App Sc 373–382

van Ta Q, Kim MM, Kim SK (2006) Inhibitory effect of chitooligosaccharides on matrix metalloproteinase-9 in human fibrosarcoma cells (HT1080). Mar Biotechnol 8:593–599. https://doi.org/10.1007/s10126-006-6031-7

Vandenbroucke RE, Dejager L, Libert C (2011) The first MMP in sepsis. EMBO Mol Med 3:367–369

Vårum KM, Holme HK, Izume M et al (1996) Determination of enzymatic hydrolysis specificity of partially N-acetylated chitosans. Biochim Biophys Acta Gen Subj 1291:5–15. https://doi.org/10.1016/0304-4165(96)00038-4

Wessels I, Cousins RJ (2015) Zinc dyshomeostasis during polymicrobial sepsis in mice involves zinc transporter Zip14 and can be overcome by zinc supplementation. Am J Physiol Gastrointest Liver Physiol 309. https://doi.org/10.1152/ajpgi.00179.2015

Wright HL, Moots RJ, Edwards SW (2014) The multifactorial role of neutrophils in rheumatoid arthritis. Nat Rev Rheumatol 10:593–601

Wu H, Yao Z, Bai X et al (2008) Anti-angiogenic activities of chitooligosaccharides. Carbohyd Polym 73:105–110. https://doi.org/10.1016/j.carbpol.2007.11.011

Xu W, Jiang C, Kong X et al (2012) Chitooligosaccharides and N-acetyl-D-glucosamine stimulate peripheral blood mononuclear cell-mediated antitumor immune responses. Mol Med Rep 6:385–390. https://doi.org/10.3892/mmr.2012.918

Zhai X, Li C, Ren D, et al (2021) The impact of chitooligosaccharides and their derivatives on the in vitro and in vivo antitumor activity: a comprehensive review. Carbohydr Polym 266:118132. https://doi.org/10.1016/j.carbpol.2021.118132

Zhang H, Du Y, Yu X et al (1999) Preparation of chitooligosaccharides from chitosan by a complex enzyme. Carbohyd Res 320:257–260. https://doi.org/10.1016/S0008-6215(99)00154-8

Zhao D, Wang J, Tan L et al (2013) Synthesis of N-furoyl chitosan and chito-oligosaccharides and evaluation of their antioxidant activity in vitro. Int J Biol Macromol 59:391–395. https://doi.org/10.1016/j.ijbiomac.2013.04.072

Zhijun L, Rongchun C, Feixiang L, et al (2019) Therapeutic effects of combined meloxicam and glucosamine sulfate treatment on patients with osteoarthritis, and its effect on serum CTX-I, CTX-II, COMP and MMP-3. Trop J Pharm Res 18. https://doi.org/10.4314/tjpr.v18i7.28

Zou P, Yang X, Wang J et al (2016) Advances in characterisation and biological activities of chitosan and chitosan oligosaccharides. Food Chem 190:1174–1181. https://doi.org/10.1016/j.foodchem.2015.06.076

Immunomodulatory Effects of Chitooligosaccharides

Vishnupriya Govindaraj, Keyur Raval, and Ritu Raval

Abbreviation

Aβ	Amyloid β
AD	Alzheimer's Disease
AIDS	Acquired Immunodeficiency Syndrome
AKT	Protein kinase B
AUCOS	Amino Urea Chitooligosaccharide
Bax	Bcl-2-associated X
Bcl-2	B cell lymphoma-2
BUS	Busulfan
CCL	Chemokine C-C motif Ligand
CD 4+	Cluster of Differentiation 4
CD 8+	Cluster of Differentiation 8
COS	Chitooligosaccharides
CXCL	Chemokine C-X-C Motif Ligand
Cy	Cyclophosphamide
DAMP	Danger Associated Molecular Pattern
DC	Dendritic cells
EGFR	Epidermal Growth Factor Receptor
FAK	Focal Adhesion Kinase
FIZZ1	resistin–like α

V. Govindaraj · K. Raval
Department of Chemical Engineering, National Institute of Technology Karnataka, Surathkal, Mangalore, Karnataka 575025, India
e-mail: keyurnraval@nitk.edu.in

R. Raval (✉)
Department of Biotechnology, Manipal Institute of Technology, Manipal Academy of Higher Education, Manipal, Karnataka 576104, India
e-mail: ritu.raval@manipal.edu

S.-K. Kim (ed.), *Chitooligosaccharides*,
https://doi.org/10.1007/978-3-030-92806-3_7

GTMAC	Glycidyl Tri Methyl Ammonium Chloride
HCl	Hydrochloric Acid
H_2O_2	Hydrogen Peroxide
HUVEC	Human umbilical vein endothelial cells
IFN	Interferon
Ig G	Immunoglobin G
Ig M	Immunoglobin M
IL	Interleukin
iNOS	Inducible nitric acid synthase
IRF	Interferon Regulatory Factor
JMJD3	Jumonji domain containing protein
LPS	Lipopolysaccharide
MAPK	Mitogen activated protein kinase
MHC	Major Histocompatibility Complex
MMP	Metastatic-related protein
NADPH	Nicotinamide Adenine Dinucleotide Phosphate Oxidase
NA-COS	N-acetyl Chitooligosaccharides
NF-κB	Nuclear Factor - κB
NO	Nitric Oxide
NK Cells	Natural killer cells
OGSC	Ovarian Germ Stem Cells
PAMP	Pathogen Associated Molecular Patterns
PARP	Poly ADP ribose polymerase
PBMC	Peripheral Blood Mono nuclear Cells
POD	Peroxidase
PPAR	Peroxisome Proliferator activated Receptor
PRR	Pathogen Recognition Receptors
ROS	Reactive Oxygen Species
SOD	Superoxide dismutase
STAT	Signal Transducer and Activator of Transcription
TGF	Transforming Growth Factor
Th1	Type 1 T helper cells
TLR	Toll like Receptor
TNF	Tumor Necrosis Factor
XRD	X-Ray Diffraction

1 Immunity and Immune Modulation

The term immunity is defined as the ability of the body to identify and resist pathogens in order to avert diseases and tissue damage (Kumar et al. 2012). Immune system comprises of cells, chemicals and processes with roles to protect the host from foreign antigens. Traditionally the host immune system is classified as innate and adaptive immune responses. As soon as the immune system encounters a pathogen, the first response is mounted by innate immunity and if it is not able to address the antigenic attack, the second line of response with the adaptive immune arm is built. The uniqueness of adaptive immunity lies in its ability to create immunological memory with which the host can react efficiently upon subsequent exposure to the antigen. Cells of innate immunity include neutrophils, monocytes, macrophages, eosinophils, basophils, natural killer (NK) cells, dendritic cells, mast cells as their cellular component and the complement proteins mounting the humoral arm. Adaptive immunity involve cytokines, immunoglobulins as the humoral system and the B and T lymphocytes as the cellular component (Marshall et al. 2018; McComb et al. 2013).

Innate immunity, the first line of defense exists against highly conserved structures known as pathogen associated molecular patterns (PAMP) in micro-organisms and danger associated molecular patterns (DAMPs) associated with tissue damage. The response is elicited in minutes to hours (Sun et al. 2020). The function of pattern recognition receptor includes opsonization, activation of the complement cascades, phagocytosis, activation of proinflammatory signaling pathways and induction of apoptosis (Hansson et al. 2002; Janeway and Medzhitov 2002; Martell 2017).

Pathogen recognition receptors present on the innate cells include Toll like Receptors (TLR) and scavengers. There are 5 classes of scavenger receptors (Platt and Gordon 1998). Scavenger receptors are cell surface proteins expressed in monocytes, haemocytes and macrophages (Pearson 1996). Binding of endogenous proteins or pathogen with scavenger receptors results in degradation of recognized particles by endocytosis. Ten different TLR have been recognized in humans so far. The binding of the TLR with the PAMP activates the NF- κB and MAP kinase pathway which results in induction of cytokine production, inflammation and can directly result in apoptosis (Hansson et al. 2002).

The adaptive immune system is activated when innate immunity is unable to restrict the spread of the pathogen (Janeway and Medzhitov 2002). The antigen-presenting cells process the antigen and present it on its surface along with MHC molecules so that appropriate T cells can interact with it (Cruse et al. 2004). Antigen-presenting cells such as dendritic cells express CD 80 and CD 86 on their surface which in turn activates the adaptive immune system. Expression of CD 80 and CD 86 on cell surface is regulated by NF-κB transcription factor (Janeway and Medzhitov 2002). B-cell receptor and T-cell receptor are antigen-specific receptors expressed on the B cells and T cells, respectively. These receptors evolve with each incidence of infection by the process of somatic gene recombination (Netea et al. 2019). The T lymphocytes, develop in the thymus (primary lymphoid organ). It is here that they differentiate into the helper T cells and cytotoxic T cells (Keselowsky et al. 2020). T- helper (T_h) cells express $CD4^+$ on its surface which recognizes exogeneous antigens with MHC-II. T_h cells secrete cytokines which in turn activates B and T lymphocytes and other immune cells. (Kang et al. 2019; Keselowsky et al. 2020). The cytotoxic T cells express $CD8^+$ on its surface which recognizes the endogenous antigens associated with MHC-I and kills the infected cell (Kang et al. 2019). Cytokines play an important role in this cell–cell communication. Th1 cells secrete

IL-2, IL-12, IFN-γ, TNF which promotes cell mediated immune response whereas IL-4, IL5, IL-6, IL-10 are secreted by Th2 cells which promotes humoral or allergic responses (Mei et al. 2013). A summary of the cell mediated and humoral response is presented in Fig. 1.

Hence to recapitulate, the immune response involving the innate cells solely would result in rapid recognition followed by tissue inflammation, elimination of pathogen and activation of adaptive immunity while the adaptive cells would lead to activation of antigen-specific T cells, release of cytokines and also activation of B cells for antibody production. This would eventually result in the destruction of microbes and development of immune memory.

2 Positive and Negative Modulations of the Immune Response

The term Immunomodulation describes modification of the immune response with therapeutic interventions (Gea-Banacloche 2006). Immunomodulators are substances which alter the scope, type and duration of the immune response (Lebish and Moraski 1987). Immunomodulators are classified into immunostimulants or immunosuppressants.

Immunostimulants are nonspecific in nature and enhance host resistance against infection through innate and adaptive immune system. In healthy individuals, immunostimulants serve as immunopotentiators but act as immunotherapeutic agents in individuals with impaired immune system (Kumar et al. 2012). Also, immunostimulants are administered in case of infection, tumors, immunodeficiency (AIDS). In case of auto immune diseases and organ transplantation, immune suppressive drugs are given to suppress the immune system. Hence, immune modulating drugs act on specific cells of immune system aiming to inhibit or activate specific population of immune system (Bascones-Martinez et al. 2014).

The benefit of immunomodulators lies in their ability to stimulate or correct the immune system (Patil et al. 2012). Drugs such as tilorone, cyclophosphamide, levamisole, penicillamine are used clinically to control immune function but synthetic immunomodulatory drugs exhibit toxicity and side effects on the leeward side. This limitation of the chemical drugs are addressed with emphasis on natural immune modulator research (Kang et al. 2019). The nontoxic, eco-friendly biological macromolecules have gained attention in the field of medicine. Hyaluronic acid, alginate and β-glucan are some of the natural biopolymers which possess excellent immunomodulating properties (Mallakpour et al. 2021; Reddy et al. 2021). These biopolymers modulate immune response by influencing macrophage activation (Notararigo et al. 2014). The emphasis of the chapter would be on the second most abundant β-glucan which has exhibited immunomodulatory aspects; chitin.

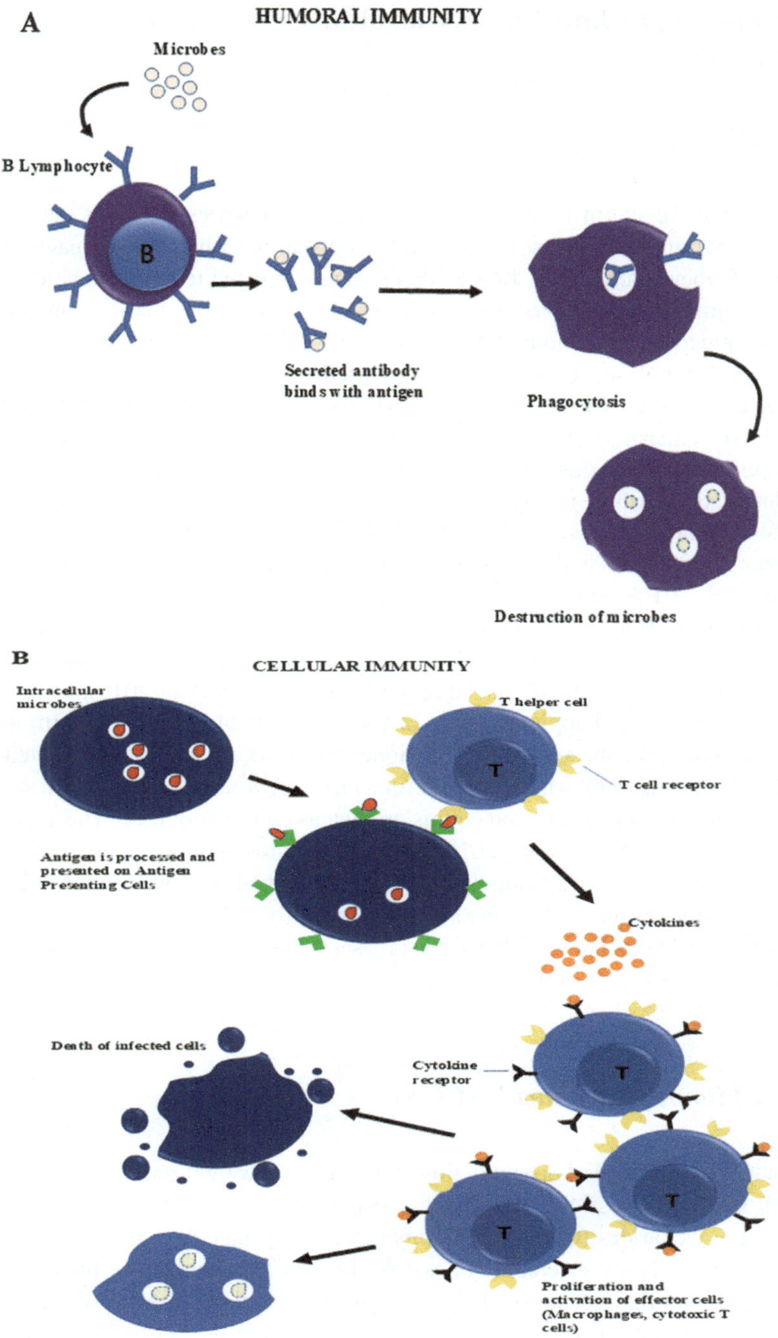

Fig. 1 Representation of the humoral (**A**) and cellular immunity (**B**)

3 Chitin and Chitooligosaccharides: The Wonder Molecules

3.1 History

Development of chitin dates back to 1799 where Charles Hatchett, an English chemist identified that a soft and elastic crystalline white color material was produced when crab, prawn, lobster shells were subjected to decalcification (Crini 2019). Chitin is a high molecular weight linear homo polysaccharide composed of N-acetyl glucosamine residues linked with β (1–4) glycosidic bonds. It is the second most abundant polysaccharide subsequent to cellulose. It is obtained from renewable feedstock and is known for its biodegradability and biocompatibility nature (Hoell et al. 2010; Khanafari et al. 2008). The main source of chitin are exoskeletons of arthropods, mollusks, diatoms and fungal cell wall (Zhao 2019).

Chitin exists in three polymorphic forms: α, β and γ forms. The α and β chitin are the more abundant forms. High stability and rigidity of α chitin is contributed by the presence of numerous hydrogen bonds which stabilizes the antiparallel arrangement (Hoell et al. 2010). This is evident from the XRD pattern of α and β chitin with crystalline reflections at 2Θ were 9.6, 19.6, 21.1, 23.7, 9.1 and 20.3°, respectively signifying the crystalline nature (Jang et al. 2004). γ chitin is composed of two parallel forms alternated with one antiparallel form (Hoell et al. 2010).

The interest toward applying chitin in various facets of life led to modifying the polymer to increase its solubility. The pioneering work in this domain started with Charles Rouget in 1859, who treated chitin with concentrated caustic potash solution (deacetylation of chitin) and named this product as chitine modifiee. This product is renamed now as chitosan (Crini 2019). Therefore, chitosan is the partial deacetylated product of chitin. It is a heteropolymer of β(1–4) linked N-acetyl glucosamine and D-glucosamine residues (Hoell et al. 2010). Generally, chitosan is insoluble in water and organic solvents but its solubility increases when pKa < 6.2. Protonated amino groups on chitosan determines its solubility and degradability (Lodhi et al. 2014).

3.2 Chitooligosaccharide (COS)

Chitin has 90% degree of deacetylation and molecular weight ranges from 30 k Da–3000 k Da whereas COS has molecular weight of 3900 Da (Mourya et al. 2011). Chitooligosaccharides (COS) are oligomers with less than 20 monomeric units per polymer chain. This conversion is possible through the chemical or the enzymatic route. The chitooligosaccharides hold greater commercial importance as compared to its parent molecule due to its low viscosity, low molecular weight and good solubility (Lodhi et al. 2014). The structure of chitooligosaccharide is represented in Fig. 2.

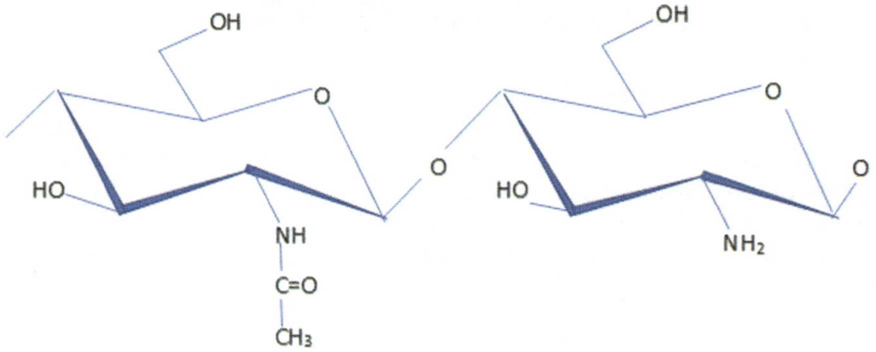

Fig. 2 The structural representation of chito oligosaccharide (COS)

3.3 Chitooligosaccharide (COS) Synthesis

The production of COS from chitin and chitosan can be performed by physical, chemical or enzymatic approach. Chitosan can be degraded by ultrasonic waves (Wu et al. 2014), microwave irradiation (Xing et al. 2017) or swirling cavitation (Wu et al. 2014). Addition of salts such as sodium chloride further accelerates the degradation of chitosan and has a major impact on reducing its molecular weight (Xing et al. 2005). Structurally simple and highly pure COS can be obtained by these physical methods. Low yield however limits the usage of physical method for large-scale production of COS (Benchamas et al. 2021). Large production of Chitooligosaccharides is achieved by chemical method at a low cost (Liang et al. 2018). Acid hydrolysis of chitin occurs in three distinct steps. Degradation of polysaccharide into small polymeric units; Production of N-acetyl glucosamine and conversion of N-acetyl glucosamine into glucosamine and acetic acid (Falk et al. 1966). Rate of acid catalysed reaction depends on fraction of N-acetylated residues (Aam et al. 2010). Mourya et al., investigated the mechanism of hydrolysis of N-acetyl linkage and glycosidic linkage. Chitosan can be hydrolyzed by HCl, nitrous acid, phosphoric acid, formic acid, acetic acid, lactic acid, hydrogen peroxide to produce COS. (Mourya et al. 2011). The downstream purification is difficult as the molecular weight of degraded COS is widely distributed (Liang et al. 2018). Also the by-products of the reaction are harmful to human and environment (Benchamas et al. 2021). Chemical method combined with enzymatic method can be explored in order to obtain low molecular weight oligomers with defined patterns (Sánchez et al. 2017). This nontoxic greener route of conversion is environmental friendly and results in a higher productivity (Benchamas et al. 2021). The enzymes used for the modification range from cellulase, pectinase, lipase, lysozyme, chitinase, pepsin, chitosanase, snailase, chitin deacetylase, neutral protease (Jung and Park 2014; Lv et al. 2016).

3.4 Physicochemical Properties of Chitooligosaccharide

Chitin due to its crystalline structure is insoluble while chitosan is soluble in dilute acidic solutions (pH < 6) (Zargar et al. 2015). These limitations are addressed with the chitooligosaccharides, which are highly soluble. Chitooligosaccharides exhibit wide biological activities such as antitumor, antioxidant, anti-inflammatory, immunomodulating, wound healing and antimicrobial effects. Unique properties of COS include excellent biocompatibility, biodegradability, easy absorption, membrane permeability and solubility (Mallakuntla et al. 2021).

4 Biological Activities of Chitooligosaccharides

4.1 Antimicrobial Activity

Degree of polymerization, molecular weight and degree of deacetylation has a strong impact on the antimicrobial activities of molecules. Chitooligosaccharides (COS) with a degree of polymerization (DP) of 32 displayed a strong inhibitory activity in a study conducted against yeast (Ganan et al. 2019). COS possesses a better antibacterial activity against gram-positive and gram-negative species (Benchamas et al. 2021). However, an increase in the number of free amino groups on chito oligomers generated a repulsive force against bacterial cell wall which resulted in decreased antibacterial activity against *E. coli* and *L. monocytogenes* (Sánchez et al. 2017). Chitooligosaccharide coupled with glycidyl trimethyl ammonium chloride (GTMAC) showed an antimicrobial activity against *S. mutans*, a gram-positive bacterium commonly seen in human dental caries (ATCC 27,351) after 2 h of exposure. (Kim et al. 2003). Bacterial cell surface consists of acidic macro molecules with which COS-GTMAC formed poly electrolyte complexes. This resulted in an impermeable layer around the bacterial cell wall thus blocking the permeation of nutrients into the cell. The reduction of the nutrient transfer resulted in a suppressed metabolic activity of the bacteria. COS showed growth inhibition of 10% against *S. mutans* which was further enhanced with COS-GTMAC to 80% (Kim et al. 2003). In another study, COS obtained from giant fresh water prawn shells displayed antibacterial activity against *Vibrio parahaemolyticus* (Chimtong 2018) and COS obtained from *Clanis bilineata* showed antibacterial activity against *Bacillus subtilis* (Wu 2012). Antifungal activity of COS against *Trichophyton rubrum* were explored using two methods such as broth dilution method and agar diffusion method. Results of microdilution method revealed that 0.25–0.5% of COS was the minimum inhibitory concentration required to suppress the growth of *T. rubrum*. Analysis of agar diffusion method further showed that 0.5 and 1% COS inhibited cell growth of *T. rubrum* after 10 days. This anti-dermatophytic activity of COS against *T. rubrum* was evaluated in guinea pig models. The study which involved an initial subjection with the infection for 7 days followed by the treatment with 5% COS for 14 days resulted in a significant improvement in the skin lesions in the model organisms (Mei et al. 2015). Amino urea chitooligosaccharide (AUCOS), a formulation of COS prepared by grafting amine carboxyl group and a hydrazine group to chitooligosaccharide backbone led to an improved antifungal activity. AUCOS showed antifungal activity against *Fusarium solani, Verticillium alboatrum and Phytophthora capsici* with inhibition rates of 60.12%, 82.95% and 85.23%, respectively (Zhu et al. 2020). Further studies illustrate the usage of COS as a food preservative with its high potency against food-borne diarrhoea causing organisms such as *Bacillus cereus, Pseudomonas aeruginosa, Candida albicans* and *Staphylococcus aureus* (El-Sayed et al. 2017; Laokuldilok et al. 2017).

4.2 Antioxidant Property

During normal metabolism, reactive oxygen species (ROS) are produced which leads to oxidative stress. To protect from ROS, the cells produce antioxidants such as catalase, glutathione peroxidase, super oxide dismutase. An uncontrolled release of ROS can have detrimental effects. This can be catered with chitin and its derivatives with their antioxidant properties (Park and Kim 2010). Studies in rat models elucidated that COS inhibited the apoptosis induced by β-Amyloid peptide (Aβ peptide) in primary hippocampal neurons by depressing Bax/Bcl-2 ratio, caspase 3 and blocked phosphorylation of c-Jun N- terminal kinase. Hence, COS can be used as a potential drug in treatment of Alzheimer's disease (Dai et al. 2013).

The antioxidant properties of COS were also evaluated in the *in*-vitro model using murine melanoma cell line B16F (Mendis et al. 2007). NF-κβ gene plays a major role in cell regulation and is activated in response to oxidative stress. In addition to that, it triggers the expression of proinflammatory genes. Hence, in disease conditions such as rheumatoid arthritis, multiple sclerosis, inflammatory bowel disease and atherosclerosis, inhibition of NF-κβ will suppress the inflammatory cytokine expression and thus severity of the disease can be reduced (T. Liu et al. 2017). Low molecular weight COS (<1 kDa) exhibited suppression of free radicals and NF-κβ gene promoter activity. In addition to that, COS exerted protective effect on oxidative damage to DNA regardless of molecular weight (Mendis et al. 2007). Superoxide dismutase (SOD) enzyme converts O_2^- to H_2O_2 and protects cells from oxidant stress (Buettner 2011). Oranges treated with 2% chitosan for 1 min reduced the disease caused by *Penicillum italicum* and *Penicillium digitatum* and also enhanced the activities of hydrogen peroxide (H_2O_2), superoxide dismutase (SOD), peroxidase (POD) and glutathione levels (Zeng et al. 2010). Qu et al., investigated the antioxidant effects of COS in high fat diet mouse model. Results revealed that 0.5% COS increased the activities of glutathione peroxidase, superoxide dismutase and catalase (Qu and Han 2016). Human umbilical vein endothelial cells (HUVEC, ECV304 cells) were employed to analyze the effect of COS against oxidative damage caused by H_2O_2. Results showed that incubation of ECV304 cells with COS (25–200 g/ml) for 24 h lowered apoptosis and induced cell cycle arrest. COS, capable of inducing intracellular antioxidants and suppressing ROS, can be employed for treating oxidative and cardiovascular diseases (Liu et al. 2009). Metals such as copper can induce neurotoxicity which leads to neurodegenerative disorders. COS elicits protective mechanism by reducing the reactive oxidation species generated by Cu(II) (Xu et al. 2010). COS inhibited the neurodegeneration induced by an insecticide, monocrotophos in *C. elegans* and also attenuated the oxidative stress (Nidheesh et al. 2016).

Antioxidant property of COS is dependent on its molecular weight (Laokuldilok et al. 2017). N-acetyl Chitooligosaccharides (NA-COS) with two different molecular weights (1–3 kDa and <1 kDa) were produced from crab chitin by acid hydrolysis in order to analyze its antioxidant properties using mouse macrophages (RAW 264.7). Results showed that glutathione levels increased with respect to time. 1–3 kDa NA-COS shows better antioxidant property than NA-COS with molecular weight < 1 kDa. Being a potential scavenger against oxidative stress in cells, NA-COS protects DNA and protein from oxidative stress (Ngo et al. 2009). Caffeic acid conjugated chitooligosaccharide showed hydroxyl radical scavenging activity and it is a potent antioxidant compound (Eom et al. 2012). Dietary supplementation of COS increased activities of superoxide dismutase, catalase, and total antioxidant capacity in pigs. Hence, COS can enhance growth performance by accelerating antioxidant and immune properties in pigs (Wan et al. 2017).

5 Use of Chitooligosaccharide in Immunomodulation

5.1 Effect of Chitooligosaccharides on Spleen and Thymus Indices

Role of COS as immunomodulator was analyzed using cyclophosphamide (cy) induced immune suppressed mice models. Cyclophosphamide is an alkylating agent widely used as an immunosuppressive agent in cancer therapy, organ transplantation and in treatment of auto immune disease. Thymus and spleen are the primary and secondary lymphoid organ respectively involved in the immune response. The study conducted revealed the counter action of COS treatment on the immunosuppression induced by cyclophosphamide (Mei et al. 2013).

The cytotoxicity studies conducted on COS showed that COS is safe even at high concentration of 1000 μg/ml. Immunomodulatory function of COS was studied using cyclophosphamide-induced mouse model. Effect of COS at three different concentrations 20, 40 and 80 mg/kg were studied. Spleen and thymus indices for normal group were 3.92 ± 0.58 and 4.98 ± 0.29, respectively. Immune system of mice was suppressed by intravenously injecting cyclophosphamide (cy). Spleen and thymus indices for cy treated model were 2.51 ± 0.37 and 2.58 ± 0.55, respectively. Spleen and thymus indices were low in cy induced model when compared with normal group. Increase in spleen and thymus indices after treatment with COS revealed that COS reversed the immunosuppression caused by cy. Spleen and thymus index were high for COS 80 mg/ml (Zhai et al. 2018). Anticancer effects of various concentration of COS on proliferation and differentiation of peripheral blood mono nuclear cells (PBMC) were investigated. COS with three different concentrations (100 μg/ml, 300 μg/ml, 500 μg/ml) were analyzed on PBMC and its respective spleen indices (0.158 ± 0.011, 0.242 ± 0.014, 0.239 ± 0.015) and respective thymus indices (0.984 ± 0.053, 1.016 ± 0.079, 1.247 ± 0.085) were obtained. COS concentration of 500 μg/ml increased the weight of spleen and thymus. Change in weight of these organs are indications of immune function and antitumor activity (Xu et al. 2012). Oral supplementation of COS at three different concentrations (100, 250 and 500 mg) per kg body weight was experimented in mice to explore its impact on spleen and thymus index (Zhang et al. 2014). Dietary supplementation of 100 mg/kg COS in broiler chicken for 21 days significantly improved spleen and thymus index from 0.67 to 0.96 and 2.49 to 3.14, respectively. Also, high concentration of IL- 1β, IL-6, TNF-α, IFN-γ were seen in birds receiving COS (0.27 pg/ml, 295.3 pg/ml, 3.8 ng/ml, 51.6 pg/ml) when compared with control (0.20 pg/ml, 186.8 pg/ml, 2.4 ng/ml, 44 pg/ml) respectively. Results proved that COS enhanced immunity (Deng et al. 2008).

5.2 *Effect of Chitooligosaccharide (COS) on Macrophages*

Upregulation and downregulation of inflammatory process are tightly regulated and any imbalance in regulation causes cellular and tissue damage. Macrophages exhibit important roles such as initiation, maintenance and resolution of inflammation process (Fujiwara and Kobayashi 2005). Macrophages originate from the circulating blood monocytes (Nagata 2009). Antigen presentation, phagocytosis and immunomodulation are the three major functions of macrophages which involve production of various cytokines and growth factors. Macrophages are activated by bacterial lipopolysaccharides, extracellular matrix proteins and cytokines such as tumor necrosis factor- α (TNF-α), interferon γ, granulocyte- monocyte colony stimulating factor. Interleukin 10 and transforming Growth factor β are anti-inflammatory cytokines which deactivate the activated macrophages (Fujiwara and Kobayashi 2005). Functional properties of macrophages determine severity of inflammation process. There exists functional heterogeneity in macrophages termed as macrophage polarization based on which macrophages are classified into activated (M1) macrophages and alternatively activated (M2) macrophages (Nagata 2009).

M1 macrophages are characterized by the presence of TLR-2, TLR-4, CD80, CD86, iNOS and MHC-II on its surface. Cytokines and chemokines such as TNF-α, IL-1β, IL-6, IL-12, CXCL9 and CXCL10 released by M1 macrophages exert positive feedback on the unpolarized macrophages. The M1 gene expression is regulated by transcription factors such as NF- κB, STAT1, STAT5, IRF3 and IRF5. M2 macrophages contain surface markers such as mannitol receptor, CD206, CD209, CD163 and FIZZ1. M2 gene expression is regulated by transcription factors such as STAT6, IRF4, JMJD3, PPAR γ and PPAR δ. The downstream signals of cytokines

(IL-4, IL-13, IL-10, IL-33 and TGF-β) results in the M2 polarization (Yao et al. 2019). This polarization has been illustrated in Fig. 3.

Since macrophages play a major role in the inflammatory process and produces various active molecules, targeting them can pave the way for controlling diseases. Common strategy to resolve inflammation is by reducing M1 polarization or by increasing M2 polarization (Wang et al. 2019).

In-vivo implantation of a biomaterial results in host reactions such as acute/chronic inflammation. Implanted site recruits monocytes from peripheral blood and monocytes mature into macrophages and secretes cytokines and other growth factors and modulate inflammatory site. Huang et al., investigated the effect of COS on macrophage polarization using RAW264.7 cells and associated angiogenic and osteogenic properties. Results showed that COS concentration of 4 mg/ml and degree of polymerization (DP) of 5 upregulated the expression of anti-inflammatory cytokines such as IL-10. Analysis of supernatant obtained from RAW264.7 culture

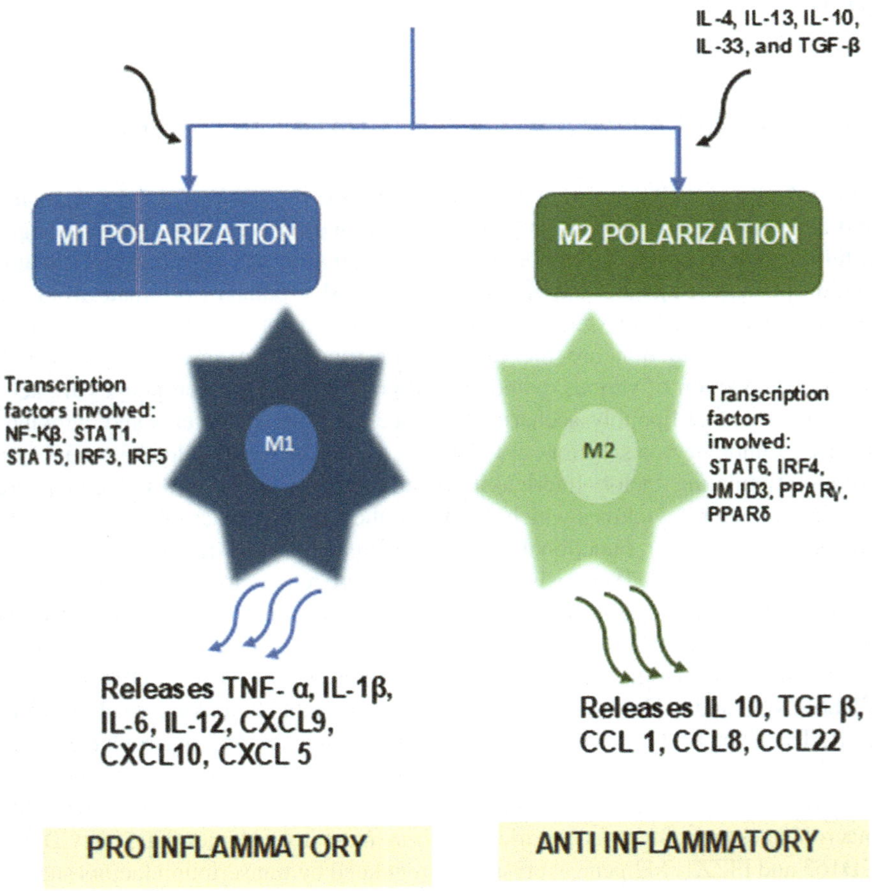

Fig. 3 M1 and M2 polarization of macrophages

containing COS showed the presence of osteogenesis and angiogenesis-related proteins. COS promoted the proliferation of macrophages. Hence, for regeneration of tissues, COS at low concentration with specific DP has the ability to modulate osteogenesis or angiogenesis process without using any inducive agent (Huang et al. 2020).

Chitosan-derived oligosaccharides resulted in effective immunomodulation in RAW264.7 cells by inducing cytokines and phagocytosis (Tsvetkov et al. 2020). Also, COS induced white blood cell proliferation, macrophagic phagocytosis thereby stimulating the secretion of NO and TNF-α (Benchamas et al. 2021). NO is a small diffusible molecule synthesized by iNOS and is responsible for eradicating the invading pathogens in macrophages. Immunostimulatory effects of COS examined in RAW 264.7 macrophages showed increase in NO expression in a dose-dependent manner. Proinflammatory cytokines play a major role in pathogen destruction and administration of COS increased TNF-α, IL-6 (Yang et al. 2019). Enzymatically prepared COS with degree of polymerization 3–7 exhibited immunostimulant properties in-vitro and in-vivo. COS administered in RAW 264.7 macrophages resulted in increased cell proliferation and phagocytic function of neutral red. Immune stimulation mechanism of COS involves initiating macrophage stimulation which is triggered by TLR-4. In human cells, TLR 4 gene is a member of Toll like receptor family. TLR are transmembrane proteins and their main purpose is to recognize ligands with PAMP. It is known for recognizing LPS and other proteins. Thus, COS recognized by a pattern recognition receptor, TLR results in activation of NF-κB signaling pathway which results in the production of various cytokines and activation of innate immune system (Phil et al. 2018). Han et al., investigated the role of mannose receptor in the uptake of chitooligosaccharide and found that it could serve as a signaling receptor in oligo chitosan induced activation of macrophage. MR is a surface C type lectin receptor expressed by macrophages and dendritic cells and it plays a key role in removing microbes bearing MR ligands (Han et al. 2005). Liu et al., utilized macrophages from Blunt snout bream to study immune stimulating effect of oligo chitosan. Results showed an increase in phagocytic activity of macrophages and genes involved in Nicotinamide Adenine Dinucleotide Phosphate oxidase (NADPH oxidase), inducible nitric acid synthase (iNOS) pathway and mRNA levels of TNF-α and IL-1β were enhanced (Liu et al. 2014). Dietary uptake of oligo chitosan increased phagocytic activity and lysosomal activity in striped Catfish (*P. hypothalamus*). Catfish fed with 100–200 mg of oligo chitosan per kg of feed showed improved growth and decreased mortality (Duy et al. 2016). Catfish fed with oligo chitosan for 45 days was challenged with pathogenic bacteria, *Edwardsiella ictaluri* in order to test antibacterial property of oligo chitosan. Lysozyme and phagocytosis play a key role in eliminating infections. Bacterial cell wall consists of peptidoglycan layer composed of N-acetyl glucosamine and N-acetyl muramic acid linked with β (1–4) glycosidic bonds. Increase in lysozyme activity showed that it cleaves β(1–4) glycosidic bonds in peptidoglycan layer and eliminates infection (Duy et al. 2016).

5.3 *Tumor Suppression*

The term cancer refers to the mass of tissue produced as a result of uncontrolled cell division. The ability of these cells to invade nearby tissues and organs is called metastasis (Benchamas et al. 2021). Programmed cell death is termed as apoptosis. Chemotherapeutic agents are known to inhibit the tumor cell proliferation by inducing apoptotic signaling pathways (Yu et al. 2018).

Induction of antitumor activity by a molecule will upregulate and downregulate various molecules in cell. Bcl-2 present in outer mitochondrial membrane binds with Bax to inhibit apoptosis as represented in Fig. 4. Bax protein present in cytosol translocates to mitochondria and triggers loss of membrane potential. Bax protein induces apoptosis. Hence, disruption of mitochondrial membrane potential, increase

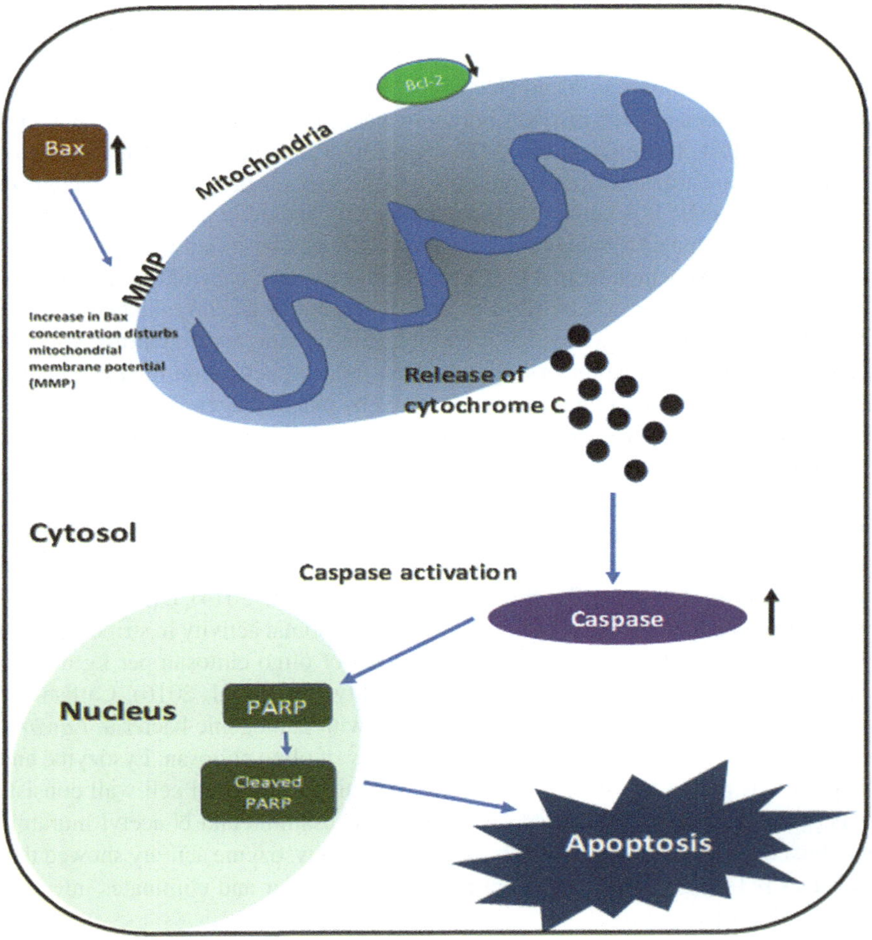

Fig. 4 Role of Bax and Bcl-2 in induction of apoptosis

in ROS and Bax levels, decrease in Bcl-2 expression results in release of cytochrome C from mitochondria into cytoplasm, activation of caspase, cleavage of poly ADP ribose polymerase (PARP) (Jiang et al. 2019; Yu et al. 2018).

Antitumor activity of COS was investigated using human breast cancer cell lines BT-474 and SUM-159. COS inhibited colony formation of breast cancer cells and induced apoptosis of tumor cells. COS concentration has direct impact on the expression of apoptotic regulators such as BAX, cleaved PARP and cleaved caspase-3 (Mallakuntla et al. 2021). Ability of chitosan oligosaccharide on suppressing lung cancer was studied in Lewis tumor-bearing mouse model. Chitosan oligosaccharide inhibited the viability and proliferation of lung carcinoma cells (A549). Increase in Bax (Bcl-2-associated X) protein expression, decrease in Bcl-2 expression were observed in immune histochemical staining. ELISA results showed increase in levels of TNF-α, IL-2, Fas, Fas-L. TNF-alpha plays a key role in enhancing host immunity and apoptosis of tumor cells. IL-2 activates effector components and Fas system acts against cancer by mediating apoptosis. Hence, supplementation of COS promotes necrosis in tumor cells (Jiang et al. 2019). Zhai et al., reported that COS induced apoptosis in human renal carcinoma cells by activating ROS-dependent ER stress signaling pathway (Zhai et al. 2019).

EGFR (Epidermal Growth Factor Receptor), FAK (Focal Adhesion Kinase), AKT (Protein Kinase B), and MAKP are upregulated in various cancer. Activation of EGFR is the major mechanism in cancer. COS inhibits the activity of signaling molecules which downregulates apoptosis. Western blot analysis of cell extracts of BT-474 cells after COS treatment showed decreased expression of EGFR and its signaling molecules such as FAK, AKT and MAKP (Mallakuntla et al. 2021). COS with DP 2–6 promoted mitosis, apoptosis and S cell cycle arrest in human colon carcinoma HCT116 cells (Zou et al. 2018). Molecular weight and the cationic property of COS has a major impact on its anticancer property. COS induced iNOS, nitric oxide synthesis and TNF- α, thereby increasing cytotoxicity in tumor cells (Huang et al. 2006). Effect of Selenium conjugated Chitooligosaccharide (Se-COS) on improving immune function and blocking gastric cancer was investigated. Se-COS inhibited metastasis and proliferation of cancer (Jiang et al. 2021). COS significantly inhibited the proliferation of human hepatocellular carcinoma (HepG2). Analysis of cell cycle-related genes revealed interesting results. Upregulation of p21 and downregulation of PCNA, Cyclin A and cdk-2 were seen. Also, COS inhibited activity of metastatic-related protein (MMP-9) in Lewis lung carcinoma cells (Shen et al. 2009). COS showed antitumor activities against PC3, HepG2 and A549 tumor cells (Park et al. 2011). Galactosylated chitooligosaccharide nanoparticles can be used as an intracellular drug delivery carrier for targeting hepatocellular carcinoma cells (Zhu et al. 2013).

COS improved ovarian and immune function of mice and increased the rate of survival of ovarian germ stem cells (OGSC) (Huang et al. 2021). Infertile mouse model was established by intraperitoneal injection of Cyclophosphamide (cy) and Busulfan (BUS) and effect of COS on OGSC was studied and results showed increased secretion of TNF-α and IL-2. TNF-α secreted by activated macrophages promotes cell proliferation and differentiation. Interleukin IL-2, IL-18 promotes the proliferation of Natural killer (NK) cells (Huang et al. 2021).

6 Conclusion

Innate and adaptive system plays a key role in eliminating infection or to control the unusual effects. If it fails to do so, disease occurs. In such cases, immunostimulants or immunosuppressive drugs are administered. COS, a natural, nontoxic, soluble and a biodegradable polysaccharide plays a major role in immunomodulation. Physiochemical properties such as molecular weight and degree of acetylation have a major impact on biological activities of COS. Spleen and thymus are organs involved in nonspecific immunity. COS treatment can reverse damage to these organs and can stimulate nonspecific immunity. Effect of COS on tumor suppression explored using mice models and cancer cell lines reveals the effectiveness of COS in tumor suppression. Mechanism of COS in diseases needs further research.

References

Aam BB, Heggset EB, Norberg AL, Sørlie M, Vårum KM, Eijsink VGH (2010) Production of chitooligosaccharides and their potential applications in medicine. Mar Drugs 8(5):1482–1517. https://doi.org/10.3390/md8051482

Bascones-Martinez A, Mattila R, Gomez-Font R, Meurman JH (2014) Immunomodulatory drugs: Oral and systemic adverse effects. Medicina Oral, Patologia Oral y Cirugia Bucal 19(1):e24–e31. https://doi.org/10.4317/medoral.19087

Benchamas G, Huang G, Huang S, Huang H (2021) Preparation and biological activities of chitosan oligosaccharides. Trends Food Sci Technol 107:38–44. https://doi.org/10.1016/j.tifs.2020.11.027

Buettner GR (2011) Superoxide dismutase in redox biology: the roles of superoxide and hydrogen peroxide. Anticancer Agent Med Chem 11(4):341–346. https://doi.org/10.2174/187152011795677544

Chimtong S (2018) Antibacterial activity of chito-oligosaccharides (COSS) from shrimp shells wastes. Adv Plant Agric Res 8(6):392–394. https://doi.org/10.15406/apar.2018.08.00356

Crini G (2019) Historical review on chitin and chitosan biopolymers. Environ Chem Lett 17:1623–1643. https://doi.org/10.1007/s10311-019-00901-0

Cruse JM, Lewis RE, Wang H (Eds.) (2004) Antigen presentation. Immunology guidebook. Acadamic Press, pp. 267–276. https://doi.org/10.1016/B978-012198382-6/50031-5

Dai X, Chang P, Zhu Q, Liu W, Sun Y, Zhu S, Jiang Z (2013) Chitosan oligosaccharides protect rat primary hippocampal neurons from oligomeric β-amyloid 1–42-induced neurotoxicity. Neurosci Lett 554:64–69. https://doi.org/10.1016/j.neulet.2013.08.046

Deng X, Li X, Liu P, Yuan S, Zang J, Li S, Piao X (2008) Effect of chito-oligosaccharide supplementation on immunity in broiler chickens. Asian Australas J Anim Sci 21(11):1651–1658. https://doi.org/10.5713/ajas.2008.80056

Duy NN, Phu DV, Quoc LA, Lan K, Hai PD, Nguyen NV, Hien NQ (2016) Effect of oligochitosan and oligo- β -glucan supplementation on growth, innate immunity and disease resistance of striped catfish (Pangasianodon hypophthalmus). Biotechnol Appl Biochem 64(4):564–571. https://doi.org/10.1002/bab.1513

El-Sayed ST, Ali AM, El-Sayed ESM, Shousha WG, Omar NI (2017) Characterization and potential antimicrobial effect of novel chitooligosaccharides against pathogenic microorganisms. J Appl Pharm Sci 7(6):006–012. https://doi.org/10.7324/JAPS.2017.70602

Eom TK, Senevirathne M, Kim SK (2012) Synthesis of phenolic acid conjugated chitooligosaccharides and evaluation of their antioxidant activity. Environ Toxicol Pharmacol 34(2):519–527. https://doi.org/10.1016/j.etap.2012.05.004

Falk M, Smith DG, McLachlan J, McInnes AG (1966) Studies on chitan (β-(1→ 4)-linked 2-acetamido-2-deoxy-d-glucan) fibers of the diatom thalassiosira fluviatilis hustedt: ii. Proton magnetic resonance, infrared, and x-ray studies. Canadian J Chem 44(19):2269–2281

Fujiwara N, Kobayashi K (2005) Macrophages in inflammation. Curr Drug Targets Inflamm Allergy 4(3):281–286. https://doi.org/10.2174/1568010054022024

Ganan M, Lorentzen SB, Agger JW, Heyward CA, Bakke O, Knutsen SH, Aam BB, Eijsink VGH, Gaustad P, Sørlie M (2019) Antifungal activity of well-defined chito-oligosaccharide preparations against medically relevant yeasts. PLoS ONE 14(1):e0210208. https://doi.org/10.1371/journal.pone.0210208

Gea-Banacloche JC (2006) Immunomodulation. Principles of molecular medicine. Humana Press, pp 893–904. https://doi.org/10.1007/978-1-4614-5338-3_17

Han Y, Zhao L, Yu Z, Feng J, Yu Q (2005) Role of mannose receptor in oligochitosan-mediated stimulation of macrophage function. Int Immunopharmacol 5(10):1533–1542. https://doi.org/10.1016/j.intimp.2005.04.015

Hansson GK, Libby P, Schönbeck U, Yan ZQ (2002) Innate and adaptive immunity in the pathogenesis of atherosclerosis. Circ Res 91(4):281–291. https://doi.org/10.1161/01.RES.0000029784.15893.10

Hoell IA, Vaaje-Kolstad G, Eijsink VGH (2010) Structure and function of enzymes acting on chitin and chitosan. Biotechnol Genet Eng Rev 27(1):331–366. https://doi.org/10.1080/02648725.2010.10648156

Huang R, Mendis E, Rajapakse N, Kim SK (2006) Strong electronic charge as an important factor for anticancer activity of chitooligosaccharides (COS). Life Sci 78(20):2399–2408. https://doi.org/10.1016/j.lfs.2005.09.039

Huang X, Chen M, Wu H, Jiao Y, Zhou C (2020) Macrophage polarization mediated by chitooligosaccharide (COS) and associated osteogenic and angiogenic activities. ACS Biomater Sci Eng 6(3):1614–1629. https://doi.org/10.1021/acsbiomaterials.9b01550

Huang Y, Ye H, Zhu F, Hu C, Zheng Y (2021) The role of chito-oligosaccharide in regulating ovarian germ stem cells function and restoring ovarian function in chemotherapy mice. Reprod Biol Endocrinol 19. https://doi.org/10.1186/s12958-021-00699-z

Janeway CA, Medzhitov R (2002) Innate immune recognition. Annu Rev Immunol 20:197–216. https://doi.org/10.1146/annurev.immunol.20.083001.084359

Jang MK, Kong BG, Jeong YI, Lee CH, Nah JW (2004) Physicochemical characterization of α-chitin, β-chitin, and γ-chitin separated from natural resources. J Polym Sci Part A Polym Chem 42(14):3423–3432. https://doi.org/10.1002/pola.20176

Jiang Z, Li H, Qiao J, Yang Y, Wang Y, Liu W (2019) Potential analysis and preparation of chitosan oligosaccharides as oral nutritional supplements of cancer adjuvant therapy. Int J Mol Sci 20(4):920. https://doi.org/10.3390/ijms20040920

Jiang Z, Chi J, Li H, Wang Y, Liu W, Han B (2021) Effect of chitosan oligosaccharide-conjugated selenium on improving immune function and blocking gastric cancer growth. Eur J Pharmacol 891:173673. https://doi.org/10.1016/j.ejphar.2020.173673

Jung WJ, Park RD (2014) Bioproduction of chitooligosaccharides: present and perspectives. Mar Drugs 12(11):5328–5356. https://doi.org/10.3390/md12115328

Kang HK, Lee HH, Seo CH, Park Y (2019) Antimicrobial and immunomodulatory properties and applications of marine-derived proteins and peptides. Mar Drugs 17(6):350. https://doi.org/10.3390/md17060350

Keselowsky BG, Acharya A, Lewis JS (2020) Innate and adaptive immunity: the immune response to foreign materials. Biomater Sci. https://doi.org/10.1016/B978-0-12-816137-1.00049-0

Khanafari A, Marandi R, Sanatei S (2008) Recovery of chitin and chitosan from shrimp waste by chemical and microbial methods. Iranian J Environ Health Sci Eng 5(1):19–24

Kim JY, Lee JK, Lee TS, Park WH (2003) Synthesis of chitooligosaccharide derivative with quaternary ammonium group and its antimicrobial activity against Streptococcus mutans. Int J Biol Macromol 32(1–2):23–27. https://doi.org/10.1016/S0141-8130(03)00021-7

Kumar D, Arya V, Kaur R, Bhat ZA, Gupta VK, Kumar V (2012) A review of immunomodulators in the Indian traditional health care system. J Microbiol Immunol Infect 45(3):165–184. https://doi.org/10.1016/j.jmii.2011.09.030

Laokuldilok T, Potivas T, Kanha N, Surawang S, Seesuriyachan P, Wangtueai S, Phimolsiripol Y, Regenstein JM (2017) Physicochemical, antioxidant, and antimicrobial properties of chitooligosaccharides produced using three different enzyme treatments. Food Biosci 18:28–33. https://doi.org/10.1016/j.fbio.2017.03.004

Lebish IJ, Moraski RM (1987) Mechanisms of immunomodulation by drugs. Toxicol Pathol 15(3):338–345. https://doi.org/10.1177/019262338701500312

Liang S, Sun Y, Dai X (2018) A review of the preparation, analysis and biological functions of chitooligosaccharide. Int J Mol Sci 19(8):2197. https://doi.org/10.3390/ijms19082197

Liu HT, Li WM, Xu G, Li XY, Bai XF, Wei P, Yu C, Du YG (2009) Chitosan oligosaccharides attenuate hydrogen peroxide-induced stress injury in human umbilical vein endothelial cells. Pharmacol Res 59(3):167–175. https://doi.org/10.1016/j.phrs.2008.12.001

Liu L, Zhou Y, Zhao X, Wang H, Wang L, Yuan G, Asim M, Wang W, Zeng L, Liu X, Lin L (2014) Oligochitosan stimulated phagocytic activity of macrophages from blunt snout bream (Megalobrama amblycephala) associated with respiratory burst coupled with nitric oxide production. Dev Comp Immunol 47(1):17–24. https://doi.org/10.1016/j.dci.2014.06.005

Liu T, Zhang L, Joo D, Sun SC (2017) NF-κB signaling in inflammation. Signal Transduct Target Ther 2:e17023. https://doi.org/10.1038/sigtrans.2017.23

Lodhi G, Kim YS, Hwang JW, Kim SK, Jeon YJ, Je JY, Ahn CB, Moon SH, Jeon BT, Park PJ (2014) Chitooligosaccharide and its derivatives: preparation and biological applications. Biomed Res Int. https://doi.org/10.1155/2014/654913

Lv M, Hu Y, Gänzle MG, Lin J, Wang C, Cai J (2016) Preparation of chitooligosaccharides from fungal waste mycelium by recombinant chitinase. Carbohyd Res 430:1–7. https://doi.org/10.1016/j.carres.2016.04.019

Mallakpour S, Azadi E, Hussain CM (2021) Chitosan, alginate, hyaluronic acid, gums, and β-glucan as potent adjuvants and vaccine delivery systems for viral threats including SARS-CoV-2: a review. Int J Biol Macromol 182:1931–1940. https://doi.org/10.1016/j.ijbiomac.2021.05.155

Mallakuntla MK, Penugurti V, Manavathi B, Podile AR (2021) Chitooligosaccharides induce apoptosis in human breast cancer cells. Carbohydr Polym Technol Appl 2:100077. https://doi.org/10.1016/j.carpta.2021.100077

Marshall JS, Warrington R, Watson W, Kim HL (2018) An introduction to immunology and immunopathology. Allergy Asthma Clin Immunol 14(s2):1–10. https://doi.org/10.1186/s13223-018-0278-1

Martell J (2017) Immunomodulation: basic concepts and clinical application. Eur Med J 5(14):22–28

McComb S, Thiriot A, Krishnan L, Stark F (2013) Chapter 1. Introduction to the immune response. Immunoproteomics. Humana Press, Totowa, NJ, pp 1–20. https://doi.org/10.1007/978-1-62703-589-7_1

Mei YX, Chen HX, Zhang J, Zhang XD, Liang YX (2013) Protective effect of chitooligosaccharides against cyclophosphamide-induced immunosuppression in mice. Int J Biol Macromol 62:330–335. https://doi.org/10.1016/j.ijbiomac.2013.09.038

Mei YX, Dai XY, Yang W, Xu XW, Liang YX (2015) Antifungal activity of chitooligosaccharides against the dermatophyte trichophyton rubrum. Int J Biol Macromol 77:330–335. https://doi.org/10.1016/j.ijbiomac.2015.03.042

Mendis E, Kim MM, Rajapakse N, Kim SK (2007) An in vitro cellular analysis of the radical scavenging efficacy of chitooligosaccharides. Life Sci 80(23):2118–2127. https://doi.org/10.1016/j.lfs.2007.03.016

Mourya VK, Inamdar NN, Choudhari YM (2011) Chitooligosaccharides: Synthesis, characterization and applications. Poly Sci Series A 53(7):583–612. https://doi.org/10.1134/S0965545X11070066

Nagata M (2009) Immune-mediated glomerular injury. Pediatr Nephrol 703–741. https://doi.org/10.1007/978-3-540-76341-3_29

Netea MG, Schlitzer A, Placek K, Joosten LAB, Schultze JL (2019) Innate and adaptive immune memory: an evolutionary continuum in the host's response to pathogens. Cell Host Microbe 25(1):13–26. https://doi.org/10.1016/j.chom.2018.12.006

Ngo DN, Lee SH, Kim MM, Kim SK (2009) Production of chitin oligosaccharides with different molecular weights and their antioxidant effect in RAW 264.7 cells. J Funct Food 1(2):188–198. https://doi.org/10.1016/j.jff.2009.01.008

Nidheesh T, Salim C, Rajini PS, Suresh PV (2016) Antioxidant and neuroprotective potential of chitooligomers in caenorhabditis elegans exposed to monocrotophos. Carbohydr Polym 135:138–144. https://doi.org/10.1016/j.carbpol.2015.08.055

Notararigo S, Casas-Engel MD, Las P, De PF, Corbí AL, López P (2014) Immunomodulation of human macrophages and myeloid cells by 2-substituted (1–3)-β-d-glucan from P. parvulus 2.6. Carbohyd Polym 112:109–113. https://doi.org/10.1016/j.carbpol.2014.05.073

Park BK, Kim MM (2010) Applications of chitin and its derivatives in biological medicine. Int J Mol Sci 11(12):5152–5164. https://doi.org/10.3390/ijms11125152

Park JK, Chung MJ, Choi HN, Park YI (2011) Effects of the molecular weight and the degree of deacetylation of chitosan oligosaccharides on antitumor activity. Int J Mol Sci 12(1):266–277. https://doi.org/10.3390/ijms12010266

Patil US, Jaydeokar AV, Bandawane DD (2012) Immunomodulators: a pharmacological review. Int J Pharm Pharm Sci 4(Suppl 1):30–36

Pearson AM (1996) Scavenger receptors in innate immunity. Curr Opin Immunol 8(1):20–28. https://doi.org/10.1016/S0952-7915(96)80100-2

Phil L, Naveed M, Mohammad IS, Bo L, Bin D (2018) Chitooligosaccharide: an evaluation of physicochemical and biological properties with the proposition for determination of thermal degradation products. Biomed Pharmacother 102:438–451. https://doi.org/10.1016/j.biopha.2018.03.108

Platt N, Gordon S (1998) Scavenger receptors: diverse activities and promiscuous binding of polyanionic ligands. Chem Biol 5(8):R193–R203. https://doi.org/10.1016/S1074-5521(98)90156-9

Qu D, Han J (2016) Investigation of the antioxidant activity of chitooligosaccharides on mice with high-fat diet. Revista Brasileira De Zootecnia 45(11):661–666. https://doi.org/10.1590/S1806-92902016001100004

Reddy MSB, Ponnamma D, Choudhary R, Sadasivuni KK (2021) A comparative review of natural and synthetic biopolymer composite scaffolds. Polymers 13(7). https://doi.org/10.3390/polym13071105

Sánchez A, Mengíbar M, Rivera-Rodríguez G, Moerchbacher B, Acosta N, Heras A (2017) The effect of preparation processes on the physicochemical characteristics and antibacterial activity of chitooligosaccharides. Carbohyd Polym 157:251–257. https://doi.org/10.1016/j.carbpol.2016.09.055

Shen KT, Chen MH, Chan HY, Jeng JH, Wang YJ (2009) Inhibitory effects of chitooligosaccharides on tumor growth and metastasis. Food Chem Toxicol 47(8):1864–1871. https://doi.org/10.1016/j.fct.2009.04.044

Sun L, Wang X, Saredy J, Yuan Z, Yang X, Wang H (2020) Innate-adaptive immunity interplay and redox regulation in immune response. Redox Biol 37:101759. https://doi.org/10.1016/j.redox.2020.101759

Tsvetkov YE, Paulovičová E, Paulovičová L, Farkaš P, Nifantiev NE (2020) Synthesis of biotin-tagged chitosan oligosaccharides and assessment of their immunomodulatory activity. Front Chem 8:1083. https://doi.org/10.3389/fchem.2020.554732

Wan J, Jiang F, Xu Q, Chen D, Yu B, Huang Z, Mao X, Yu J, He J (2017) New insights into the role of chitosan oligosaccharide in enhancing growth performance, antioxidant capacity, immunity and intestinal development of weaned pigs. RSC Adv 7(16):9669–9679. https://doi.org/10.1039/c7ra00142h

Wang Y, Smith W, Hao D, He B, Kong L (2019) M1 and M2 macrophage polarization and potentially therapeutic naturally occurring compounds. Int Immunopharmacol 70:459–466. https://doi.org/10.1016/j.intimp.2019.02.050

Wu S (2012) Preparation of chitooligosaccharides from clanis bilineata larvae skin and their antibacterial activity. Int J Biol Macromol 51(5):1147–1150. https://doi.org/10.1016/j.ijbiomac.2012.08.035

Wu Y, Huang Y, Zhou Y, Ren X, Yang F (2014) Degradation of chitosan by swirling cavitation. Innov Food Sci Emerg Technol 23:188–193. https://doi.org/10.1016/j.ifset.2014.02.001

Xing R, Liu S, Yu H, Guo Z, Wang P, Li C, Li Z, Li P (2005) Salt-assisted acid hydrolysis of chitosan to oligomers under microwave irradiation. Carbohyd Res 340(13):2150–2153. https://doi.org/10.1016/j.carres.2005.06.028

Xing R, Liu Y, Li K, Yu H, Liu S, Yang Y, Chen X, Li P (2017) Monomer composition of chitooligosaccharides obtained by different degradation methods and their effects on immunomodulatory activities. Carbohyd Polym 157:1288–1297. https://doi.org/10.1016/j.carbpol.2016.11.001

Xu W, Huang HC, Lin CJ, Jiang ZF (2010) Chitooligosaccharides protect rat cortical neurons against copper induced damage by attenuating intracellular level of reactive oxygen species. Bioorg Med Chem Lett 20(10):3084–3088. https://doi.org/10.1016/j.bmcl.2010.03.105

Xu W, Jiang C, Kong X, Liang Y, Rong M, Liu W (2012) Chitooligosaccharides and N-acetyl-D-glucosamine stimulate peripheral blood mononuclear cell-mediated antitumor immune responses. Mol Med Rep 6(2):385–390. https://doi.org/10.3892/mmr.2012.918

Yang Y, Xing R, Liu S, Qin Y, Li K, Yu H, Li P (2019) Immunostimulatory effects of chitooligosaccharides on RAW 264.7 mouse macrophages via regulation of the MAPK and PI3K/Akt signaling pathways. Mar Drug 17(1):36. https://doi.org/10.3390/md17010036

Yao Y, Xu XH, Jin L (2019) Macrophage polarization in physiological and pathological pregnancy. Front Immunol 10:792. https://doi.org/10.3389/fimmu.2019.00792

Yu J, Dong X, Wang LL, Ji H, Liu A (2018) Antitumor effects of seleno-β-lactoglobulin (Se-β-Lg) against human gastric cancer MGC-803 cells. Eur J Pharmacol 833:109–115. https://doi.org/10.1016/j.ejphar.2018.05.042

Zargar V, Asghari M, Dashti A (2015) A review on chitin and chitosan polymers: structure, chemistry, solubility, derivatives, and applications. Chem Bio Eng Rev 2(3):204–226. https://doi.org/10.1002/cben.201400025

Zeng K, Deng Y, Ming J, Deng L (2010) Induction of disease resistance and ROS metabolism in navel oranges by chitosan. Sci Hortic 126(2):223–228. https://doi.org/10.1016/j.scienta.2010.07.017

Zhai X, Yang X, Zou P, Shao Y, Yuan S, Abd El-Aty AM, Wang J (2018) Protective effect of chitosan oligosaccharides against cyclophosphamide-induced immunosuppression and irradiation injury in mice. J Food Sci 83(2):535–542. https://doi.org/10.1111/1750-3841.14048

Zhai X, Yuan S, Yang X, Zou P, Li L, Li G, Shao Y, Abd El-Aty AM, Haclmüftüoğlu A, Wang J (2019) Chitosan oligosaccharides induce apoptosis in human renal carcinoma via reactive-oxygen-species-dependent endoplasmic reticulum stress. J Agric Food Chem 67(6):1691–1701. https://doi.org/10.1021/acs.jafc.8b06941

Zhang P, Liu W, Peng Y, Han B, Yang Y (2014) Toll like receptor 4 (TLR4) mediates the stimulating activities of chitosan oligosaccharide on macrophages. Int Immunopharmacol 23(1):254–261. https://doi.org/10.1016/j.intimp.2014.09.007

Zhao L (Ed.) (2019) Oligosaccharides of chitin and chitosan. Springer Nature. https://doi.org/10.1007/978-981-13-9402-7

Zhu XL, Du YZ, Yu RS, Liu P, Shi D, Chen Y, Wang Y, Huang FF (2013) Galactosylated chitosan oligosaccharide nanoparticles for hepatocellular carcinoma cell-targeted delivery of adenosine triphosphate. Int J Mol Sci 14(8):15755–15766. https://doi.org/10.3390/ijms140815755

Zhu J, Liu S, Qin Y, Xing R, Yu H, Chen X, Li P (2020) Preparation, characterization, and antifungal evaluation of a new type of aminourea chitooligosaccharide derivatives. J Oceanol Limnol 38(3):841–850. https://doi.org/10.1007/s00343-019-9099-4

Zou P, Yuan S, Yang X, Zhai X, Wang J (2018) Chitosan oligosaccharides with degree of polymerization 2–6 induces apoptosis in human colon carcinoma HCT116 cells. Chem Biol Interact 279:129–135. https://doi.org/10.1016/j.cbi.2017.11.010

Anticancer Effects of Chitooligosaccharides

S. R. Pavan, Jayachandran Venkatesan, Se-Kwon Kim, and Ashwini Prabhu

1 Introduction

Biopolymer is a naturally generated substance that has a multitude of purposes. It is made from agricultural, seaweed (Manivasagan and Oh 2016), and bacterial cell wall resources, with enhanced biodegradability and biocompatibility (Behera et al. 2019). Synthetic polymers have several undesired effects, including poor solubility, cellular uptake, reduced biocompatibility compared to biopolymers.

Chitosan, a nontoxic biopolymer, is made from chitin. Crab, shrimp, and crawfish are the most common sources for the isolation of chitosan (Behera et al. 2019). It is degraded from chitin by deacetylation and purified by bed adsorption chromatography (Santana et al. 2015). Chitin and chitosan have a variety of biological uses, including antioxidant, anti-inflammatory, immune-stimulating, and antitumor properties (Park and Kim 2010; Kerch et al. 2011) and the prevention of metabolic illnesses such as diabetes (Kerch 2015). The modified chitosan and chitosan acts as homeostatic agents. Due to the drawbacks of poor water solubility, chitosan was further degraded using several methods such as physicochemical and enzymatic methods to get chitooligosaccharide (COS) (Zhang et al. 2010). In this study, we attempt to summarize the anticancer impacts of COS and its derivatives on various cancers.

S. R. Pavan · J. Venkatesan · A. Prabhu (✉)
Yenepoya Research Centre, Yenepoya (Deemed To Be University), Deralakatte, Mangalore, Karnataka 575018, India

S.-K. Kim
Department of Marine Science and Technology, Hanyang University, Hanyang, South Korea

S.-K. Kim (ed.), *Chitooligosaccharides*,
https://doi.org/10.1007/978-3-030-92806-3_8

2 Cancer

Different forms of cancer develop in the human body as a result of aberrant, damaged cell division and repair pathways. It also has an effect on surrounding normal tissues due to cancer cell metastasis. The increasing mortality rate in cancer-related incidents throughout the world demands the need for alternative therapy options. According to GLOBOCAN, 19.3 million new cancer cases are diagnosed and 10 million cancer-related deaths are reported globally in 2020. The primary disadvantages of conventional treatment methods are tumor environment alterations, drug resistance, and shorter half-life of the chemotherapeutic drugs (Choudhury et al. 2020). Surgery, chemotherapy, and radiation therapy are the most common conventional treatment modalities for cancer (Sudhakar 2009). These modalities have major drawbacks, such as increased growth factor generation, undesired effects on normal surrounding tissue after radiation treatment (Chinnaiyan et al. 2005), and a higher risk of secondary cancer development after surgery. Chemotherapy also has a high risk of damaging other organs in the body, including brain dissemination (Zimmermann et al. 2014). A mutation is the most important element in cancer development (Wu et al. 2016), and translocation of pro-oncogenes is the crucial factor in carcinogenesis via modulation of the signalling pathways (Mitelman et al. 2007). Targeted therapy, nanotechnology-based approaches, angiogenesis inhibitors, and immunotherapy are some of the upcoming treatment techniques used to overcome the side effects rendered by conventional treatment modalities, which have a specific anticancer effect while having fewer negative effects on normal cells (Jain 2005; El-Kenawi and El-Remessy 2013; Rosenblum et al. 2018; Farkona et al. 2016).

3 Chitooligosaccharides and Their Derivatives

Polymer-based medication delivery for cancer therapy is currently a promising approach. The most significant characteristic for anticancer drug delivery is its release in response to cues with response to the enzyme, pH, and temperature changes. Among the polymer-based drug delivery, the bio-copolymer chitosan looks encouraging, due to its low toxicity and excellent biocompatibility, while also having a higher drug encapsulation capability. Chitosan's molecular weight is determined by its production method and degree of deacetylation (DDA) (Kumara et al. 2021). COS are the derivatives of chitosan with a molecular weight of less than 3900 Da and a DDA of greater than 20. It is usually made up of N-acetylglucosamine and D-glucosamine as homo or heterooligomers. The COS is isolated using both physical and chemical techniques. Microwave (Xing et al. 2005), ultrasonication (Wu et al. 2008a; b), gamma-rays (Yoksan et al. 2004), and hydrothermal isolation (Sato et al. 2003) are the examples of physical techniques used for COS isolation. Chemical approaches are employed more than physical methods such as enzymatic degradation (Cabrera and Cutsem 2005; Aam et al. 2010), hydrogen peroxide (Lin et al.

2009), and $NaNO_2$ treatment (Morris et al. 2009). COS derivatives include amino COS, carboxylate COS, gallyl COS, sulfated COS, and phenolic acid conjugated COS. The characteristics of COS derivatives are determined by the location of the functional group on the COS, as well as molecular weight of the derived molecule, and its water solubility. COS does not exert genotoxicity or cytotoxicity on human lymphocytes and other normal cells (Fernandes et al. 2011). It has industrial uses such as food preservation and pharmaceuticals (Hamed et al. 2016). It also has anti-cancer properties due to its influence on the AKT pathway, which is crucial in cancer development and progression (Amirani et al. 2020; Zhai et al. 2021). COS has an essential characteristic in common with certain proteins in the human genome, such as YLK-39 and YLK-40 (Schimpl et al. 2012). It is gaining popularity because of its benefits in decreasing the cholesterol levels in the body, lowering high blood pressure, and providing protection against infections, arthritis (Yuan et al. 2019), and offers neuroprotection (Jiang et al. 2009).

4 Synthesis of Various COS Derivatives

4.1 Amino Derived COS

Ngo et al. produced the aminoethyl chitooligosaccharide (AE-COS) by linking the C6 position of the COS with the chloroethyl amino hydrochloride. The biological activities of the AE-COS include antioxidant effects, anti-inflammatory actions (Ngo et al. 2012), and anticancer effects (Karagozlu et al. 2010). To form amino derived COS, under stirring COS was added to the 2-chlorethylamino hydrochloride at 40 °C and then NaOH was added dropwise with concordant stirring for the next 48 h. This solution was acidified with HCl to get the AE-COS. This molecule inhibits the ACE enzyme activity, due to C6 position of the amino acetyl in the COS (Ngo et al. 2008). In BV2 murine microglial cells, it showed antioxidant activity along with anti-inflammatory effects (Ngo et al. 2012).

4.2 Carboxylate COS

Huang et al. replaced the amino position of the pyranose unit in the COS with a carboxyl group ($COCH_2CH_2COO$) to produce the carboxylase COS (CCOS). The carboxylate COS synthesis was achieved by dissolving the COS in 10% acetic acid and adding methanol to the dissolved COS solution, while stirring. Succinic acid was added dropwise to the COS solution continuously stirring for 4 h, and the pH was adjusted to 9–10 using sodium carbonate. This derivative is also known to inhibit ACE activity in vitro (Huang et al. 2005).

4.3 Phenolic Acid Conjugated COS

Eom et al. produced phenolic acid conjugated COS via an amide coupling process. The carboxylic group of several phenolic acids, such as p-coumaric acid, caffeic acid, and ferulic acid, is connected to the amino group of the COS to form the phenolic acid conjugated COS (Eom et al. 2012).

4.4 Gallyl COS

The OH group of gallic acid is connected with the amide group of COS in gallic acid-modified COS. The authors synthesized G-COS by preparing the separate two solutions, solution A and solution B. Solution A was prepared by dissolving COS in distilled water and ethanol and adjusting the pH to 6.8 using triethylamine, while solution B was prepared by dissolving the gallic acid in methanol and DCC-methanol. After constant stirring for 5 h, solution B was added to solution A at 30^0C, after which the mixture was filtered to remove the dicyclohexyl urea (Ngo et al. 2011). Jang et al. in their study, reported that gallic acid has more antioxidant capacity which makes it an attractive candidate for coupling with COS. Therefore, gallic acid grafted COS acts as a reducing agent for free radicals in the cells by increasing the intracellular GSH and SOD activity (Jang et al. 2009).

4.5 Sulfated COS

Sulfated COS was synthesized using two distinct kinds of sulfating agents. By varying the degrees of polymerization, chlorosulfonic acid and formamide bind to the NH and O positions of the COS (Lu et al. 2013). To form the sulfated COS, two different sulfate compounds were dissolved in formamide and chitosan was dissolved in formamide under constant stirring. Then the sulfated solution was added drop-wise at 50 °C for 3 h with stirring and the solution was neutralized by the addition of NaOH. Precipitation was obtained by the addition of ethanol to get sulfated COS. Sulfated COS has a property of reducing the cellular damage and DNA fragmentation induced by H_2O_2, by promoting the production of antioxidants in the cells (Lu et al. 2012).

5 Anticancer Activity of COS

Numerous studies have reported the anticancer action of COS on various cancer cells. Huang et al. reported that the charge of COS is responsible for the induction of cytotoxicity to HeLa, Hep3B, and SW480 cells in their study. Charge modified COS derivatives such as sulfated COS (SCOS0) and quaternized amino COS (QCOS) were shown to dramatically reduce the survival of cancer cells after 24 h of treatment, as measured by MTT and NR assays (Figs. 1 and 2).

The IC_{50} values of QCOS and SCOS on HeLa and SW480 were found to be 0.32, 0.40 mg/ml and 0.12, 0.50 mg/ml, respectively (Huang et al. 2006). Rajapakse et al. strengthened the suppression of angiogenesis and cancer cell invasiveness after treatment with COS. This study indicated that by inhibiting the MMP-9 transcription factor owing to AP-1 suppression, the c-FOS protein level retards the transcription factor (AP-1) and therefore reduces MMP-9 activity in human fibrosarcoma cells (Rajapakse et al. 2006). Likewise, COS with a molecular weight of 1 kDa suppressed MMP-9 expression in HT1080 cells comparatively higher than COS with a higher molecular weight of 3 kDa. The polymeric chain length of the COS determines the

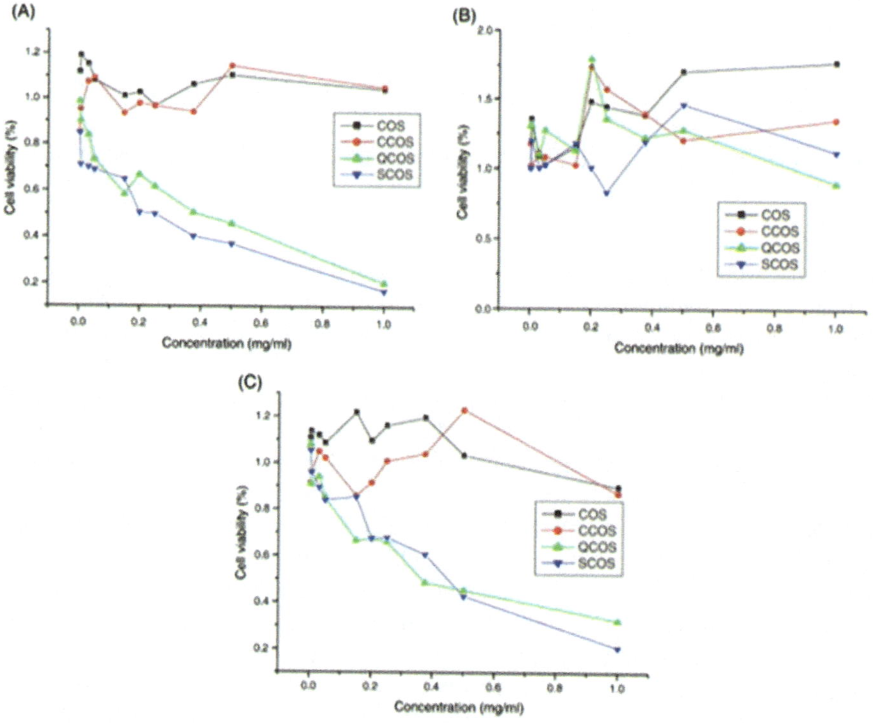

Fig. 1 Anticancer activity of COS and its derivatives on HeLa (A), Hep3B (B), and SW480 (C) cell lines assessed by MTT assay. Cells were cultured under standard conditions and incubated with different sample concentrations for 24 h prior to the assessment of viability

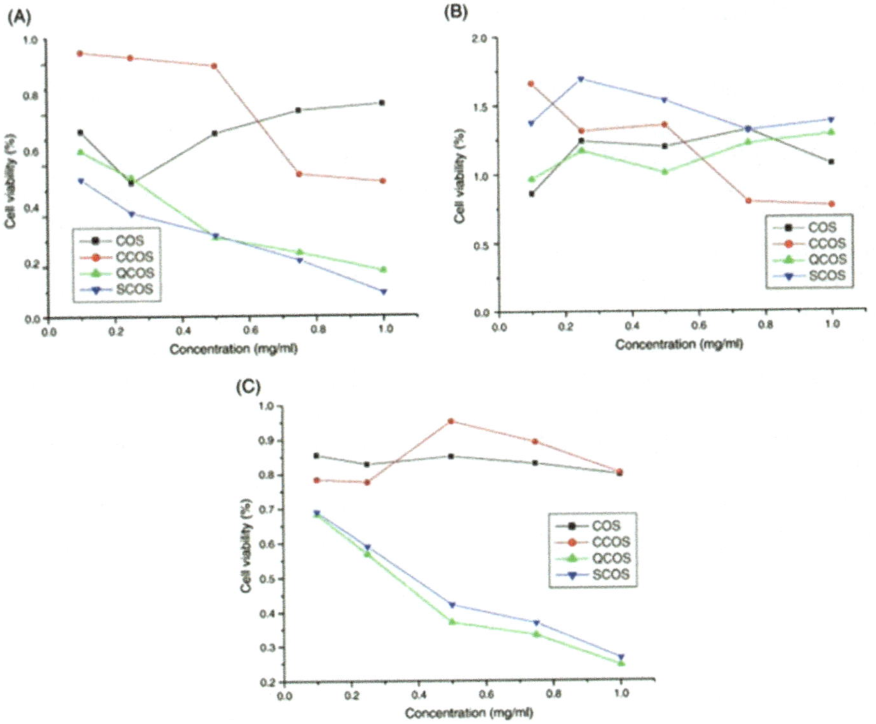

Fig. 2 Anticancer activity of COS and its derivatives on HeLa (A), Hep3B (B), and SW480 (C) cell lines as assessed by NR assay

molecular weight of the COS. Han et al. discovered that a combination of COS and tomato extract containing 2% lycopene (PLX) inhibited cancer cell growth. Using MTT assay, they were able to identify a decrease in the cell viability of prostate cancer cells (Kang et al. 2007). Furthermore, another study found that COS and N-acetyl – COS (N-acyl-CO) have antiangiogenic and cytotoxic effects under both in vitro and in vivo conditions. In vitro, the cytotoxic effects on NIH3T3 cells were considerably enhanced, which was also exerted by the antiangiogenic activity owing to endothelial cell death in N-acyl-CO treated cells compared to COS alone treatment. Anti-angiogenic effects of COS were evaluated in vivo with zebrafish embryos (Wang et al. 2007). According to Quan et al., COS restricts the activity of the enzyme Heparase, which is involved in the angiogenesis process, thereby downregulating the angiogenesis (Quan et al. 2009). COS is utilized as an adjuvant to vaccination for targeting tumor metastasis by blocking the antigen hesparase, which is elevated in the metastatic state (Yin et al. 2017). COS was extracted from chitosan hydrogel and was investigated on lung cancer cells (H460) and human lung normal cells (MRC-5) for anticancer activities. It showed the selective toxicity toward the H460 cells while the normal lung cells remained unaffected (Chokradjaroen et al. 2017a; b).

Upregulation of TIMP-1 expression by treatment with deacetylated chitohexose separated from COS by cation exchange technique using HCl was reported. This treatment also decreased uPA and VEGF expression in ECV304 cells (Xiong et al. 2009). With respect to antiangiogenic action, another study by Wu et al. found that COS reduced HUVEC cell migration caused by nitric oxide, as well as increased antioxidants like SOD in tumor-bearing mice in vivo (Wu et al. 2010). COS demonstrated antitumor efficacy in vitro against HepG2, COLO 205, and AGS cells, as indicated in Fig. 3.

COS exerted evident antitumor activity in lung cancer-bearing mice, wherein it reduced the expressions of MMP-2 and MMP-9. It was also observed that in metastatic tumors, an increase in the VEGF and COX-2 levels is responsible for increased angiogenesis (Shen et al. 2009). Another research revealed that COS with a degree of depolymerization of 2–18 has a great potential for antiangiogenic activity by reducing the migration and blood vessel development in HUVEC cells as studied

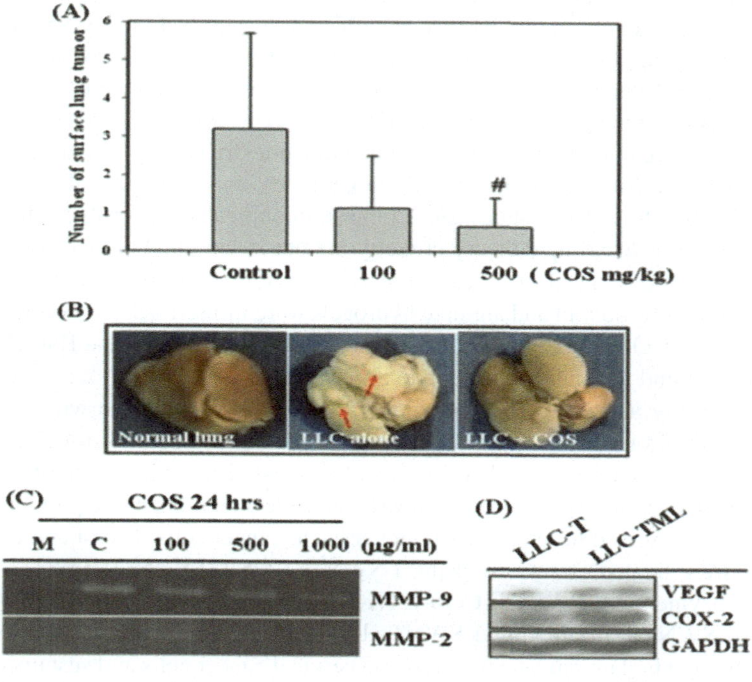

Fig. 3 Anticancer activity of COS. (A) C57BL/6 mice after s.c. injection of LLC cells were treated daily with vehicle (n = 16), COS (100 mg/kg i.p., n = 8), and COS (500 mg/kg i.p., n = 8) for 28 days after tumor implantation. Values represent the mean number of metastatic nodules in each group ± SE. # $p < 0.05$ versus control in student t test. (B) Lung morphology observation. Arrows indicated the metastatic nodules. (C) Inhibition of MMP-2 and MMP-9 proteolytic activities. LLC cells were treated with 250–1000 μg/mL COS for 24 h. The cellular lysates were processed for MMP-2 and MMP-9 activities determination by gelatin zymography assay. (D) Different expression levels of COX-2 and VEGF in the tumors metastasizing to lung and the original tumor tissues. LLC-T represents the original tumor tissues. LLC-TML represents the tumors metastasizing to lung

using CAM study (Wu et al. 2008a; b). Various acetyl percentages and degrees of polymerized COS have antiangiogenic effects in endothelial cells. Fractions of FA0.15, FA0.3, and FA0.6 with COS exhibited antiproliferative effects with no cell migration in the treated group. In the CAM assay, COS with a fraction of acetylation 0.3 (FA 0.3) treated group exhibited the highest effects in preventing the formation of new blood vessels (Wu et al. 2012). Xu et al. used mice transplanted with sarcoma cells for evaluating the anticancer effects of COS. COS (100 mg/kg) and N-acetyl-D-glucosamine (NAG) (300 mg/kg) were administered to mice. In an immunohistochemical examination, it was discovered that reductions in vasculature growth factors like VEGF, as well as upregulation of IL-2 and IFN-γ blocked angiogenesis in the experimental group. COS and NAG were shown to be nontoxic to peripheral mononuclear cells in an in vitro investigation, with less than 10% apoptosis identified using flow cytometry analysis (Xu et al. 2012).

Bromelain was utilized to produce COS by hydrolysis technique from chitin as a source. It exhibited reduction in leukemia cell proliferation. It also decreased the colorectal adenocarcinoma development in mice (Wang et al. 2008). Another study found that chitosan and its variants affected the survivability of SMMC7721 hepatocarcinoma cells. Water-soluble chitosan and sulfonated chitosan exhibited concentration-dependent anticancer action, while carboxymethyl chitosan and COS had no impact on the proliferation of SMMC7721 cells (Xie et al. 2008).

One of the significant criteria for chemotherapeutic drug delivery to tumors is the burst release of drugs from polymers, nanoparticles, and hydrogels. To overcome this problem, Cho et al. synthesized a chitosan conjugated doxorubicin hydrogel with thermosensitive drug delivery property in both in vitro and in vivo conditions. The released drug showed an anticancer activity on A549 lung adenocarcinoma cells (Cho et al. 2009). Gelatin and alginate hydrogels were infused with COS conjugated salicylic acid. COS-SA-hydrogels promoted proliferation in HDF and HaCaT cells, as well as wound healing in mice (Oh et al. 2021).

Bhatnagar et al. constructed the single-walled carbon nanotube with COS for loading with gliotoxin and other targeted molecules including p53, folic acid, and lysozyme. Targeted molecule-loaded nanotube assisted in targeted tumor therapy. P53 is involved in the DNA repair pathway, and folic acid is a surface protein that is abundantly expressed in lung, breast, and cervical cancer cells. Breast and cervical cancers were treated effectively using f-SWNTs-COS-GTX-p53, f-SWNTs-COS-GTX-lysozyme, and f-SWNTs-COS-GTX-FA. However, when compared to other modified SWNTS, the modified SWNTs loaded with p53 (f-SWNTs-COS-GTX-p53) exhibited higher anticancer activity on cervical cancer cells and sustained drug release (Bhatnagar et al. 2014).

In the context of drug delivery, one of the primary drawbacks of certain drug-loaded micelle delivery is the cellular uptake. To improve cellular absorption of the PEG-loaded medication, the micelles were charged with deoxycholic acid-modified COS (COS-DOCA). The nanoprecipitation technique demonstrated that the chemotherapeutic drug paclitaxel supplied via the modified micelles (COS-DOCA/mPEG-PDLLA) displayed 95% encapsulation activity, with an enhancement in cellular absorption. The drug-loaded COS-DOCA/mPEG-PDLLA exhibited a reduction in the tumor volume in a lung cancer xenograft model (Jiang et al. 2014). Ying et al. developed a hydrogel containing gold nanorods for the administration of doxorubicin (DOX/GNR) along with NIR radiation in another responsive drug delivery investigation. Furthermore, the drug-laden hydrogel loaded with Methoxylpoly (ethyleneglycol)-poly (-caprolactone)-acryloylchloride (PECA)/glycidyl methacrylate chitooligosaccharide (COS-GMA)/N-isopropylacrylamide (NIPAm)/acrylamide (AAm) reduced the formation of secondary tumors in breast cancer (PCNA). The tumor recurrence was reduced by 16% after treatment with DOX-PCNA-GNR hydrogels (Qu et al. 2015). In an in vivo model, dual functional nanomicelles were created for imaging and drug delivery for targeting pancreatic tumor. DTPA-Gd coupled with nanomicelles-based COSSA with an encapsulation of 99.4% of docetaxel was synthesized by Xin et al., which had an efficient anticancer activity with a higher specificity (Xin et al. 2019).

PLGA nanoparticles modified with COS were synthesized for increased cellular absorption in another work by Hu et al. Osimertinib loaded into configured PLGA nanoparticles upregulated apoptosis by increasing PARP, Bak, caspase-9, Bax, and Bcl-2 expression, with an IC_{50} value of 45.90% in H1975 non-small cell lung cancer cells (Hu et al. 2020). Liposomes containing curcumin modified using positive COS and negative phospholipids Cur@Lip-Cos-PC with enhanced drug encapsulation and hydrophilicity were used to boost the cellular absorption. Cur@Lip-Cos-PC elevated the toxicity in MCF-7 breast cancer cells. Riwang et al. injected liposomes to thiolate chitosan hydrogel for sustained drug release, which also suppressed cell proliferation in MCF-7 cells (Li et al. 2021). Chen et al. utilized glutaraldehyde as a crosslinking agent for preparing polyampholyte gels with a dual sensibility for both pH and ionic strength by combining carboxymethyl chitosan (CM-chitosan). Due to the osmotic pressure, the hydrogel's swelling characteristics aid in drug delivery (Chen et al. 2003). In breast cancer, lung cancer, and hepatoma cells, podophyllotoxin (PTT)-loaded COSSA (stearic acid) micelles demonstrated antitumor effectiveness with a higher drug loading capacity and sustained release of PTT (Huang et al. 2012). Temperature and pH-responsive PF127/HA-Ala-Chito (oligo) were cytotoxic to skin cancer cells (Chatterjee et al. 2021). De Oliveira et al. undertook a study on COS coated on Fe_3O_4 nanoparticles which improved the drug release capacity with context to the magnetic field and increased drug cellular uptake (de Oliveira et al. 2020). Elbehairi et al. used framed COS coupled Pd(II) complex of Gboxinanalog (Pd(II)COS@GbA) to suppress the proliferation of prostate cancer cells by reducing the ATP production and mitochondrial membrane breakdown (Elbehairi et al. 2020).

Another analysis reported the receptor-based liposome-loaded delivery of chemotherapeutic regimen using COS. Due to GSH responsive COS detachment from Chol-SS-COS/ES/DOX liposomes, sustained release of doxorubicin to the tumor site was achieved. The cationic liposomes are first modified with oestrogen and then treated with COS with a sulfide bond, resulting in high doxorubicin encapsulation effectiveness. This system had a greater impact on inhibiting the growth and proliferation of osteosarcoma cells (Yin et al. 2018).

According to Karagozlu et al., the treatment with water-soluble amino derivatized COS like aminoethyl-COS (AE-COS), dimethyl aminoethyl-COS (DMAE-COS), and diethyl aminoethyl aminoethyl-COS (DEAE-COS) on gastric cancer cells could result in a decrease in the percentage of live cells by upregulating the Bax and p53 biomarkers (Karagozlu et al. 2010). Sulphated COS with a molecular weight of 3-5 kDa showed an inhibition of p50 and p65 expression which are actively involved in the Nf-Kb pathway. It reduced the degradation of collagenase in the SW-1353 chondrosarcoma cells by decreasing TNF-alpha expression via the phosphorylation of IB and IKK (Ryu et al. 2012). MMP-9 activity was reduced by aminoethyl-COS (AE-COS) in human fibrosarcoma cells (Hong et al. 2016). COS also affected A549 cell growth in a concentration-dependent manner by downregulating the COX-2 expression, which is implicated in angiogenesis (Ngo et al. 2019).

In context to the anticancer action of COS on rat models, Fernandes et al. found that COS successfully prevented bladder cancer provoked by N-butyl-N-(4-hydroxybutyl) nitrosamine in an animal model (Fernandes et al. 2012). COS has also been shown to cause apoptosis and autophagy in A549 lung adenocarcinoma cells (Li et al. 2019). COS could inhibit the tumor progression by reducing the COX-2, IL-6, TNF-α, IL-1β, IL-10, and IKK-β expression in colorectal cancer in vivo. COS treatment significantly decreased the gut microbiota and mycobiota, that has been implicated in the progression of colorectal cancer (Wu et al. 2019). COS isolated from chitosan using a plasma solution and oxidizing agents (H_2O_2, $NaNO_2$, and $K_2S_2O_8$) exhibited cytotoxic action on HeLa uterine cervix carcinoma cells, however not on normal cells, indicating its cancer cell specificity (Chokradjaroen et al. 2017a; b).

For the anticancer investigation on gastric cancer, BoMi Ryu and his team designed gallic acid-linked COS. They found that G-COS dramatically increased cell death, increased Caspase-3, 9 and Bax expression, and diminished Bcl-2 expression (Ryu et al. 2017). Another study looked at how COS prevented metastasis in human stomach cancer. It inhibited metastasis in SGC 7901 gastric cancer cells by lowering the MMP-2 activity and CD147 expression (Luo et al. 2014). The biocompatible COS is employed with radiation therapy in an effort to enhance the radiosensitivity in human lung cancer. The results demonstrated an increase in cancer cell sensitivity to COS when compared to the control group, as well as a decrease in cell proliferation when treated with COS on HepG2 liver cancer cells (Han et al. 2015). The viability of colon cancer SW480 cells treated with COS was reduced with an enhanced radiosensitivity. Compared to RAY-treated cells, COS + RAY-treated cells demonstrated higher cell death percentages (Han et al. 2016).

Sun et al. reported that the ANRIL gene involved in the retardation of apoptosis and radiosensitivity of the SW480 colon cancer cells negatively affects the radiosensitive potential by the modulation of miR-181a-5p (Sun et al. 2021). COS exhibited anti-cancer efficacy in BGC823, MKN45, and SGC7901 gastric cancer cells. COS + RAY-treated cells showed decreased cell viability after 48 h of treatment by prompting apoptosis and cell cycle arrest during the G2/M phase, due to increased radiation sensitivity of the cells (Luo et al. 2016). Photothermal therapy is being studied in cancer treatment due to its substantial effects in inducing cancer cell death. The apoptotic rate of cells was enhanced when the surface of fink-blue titanium dioxide

($BTiO_2$) nanoparticles were treated with the zwitterionic COS for photothermal treatment. The COS helps to re-educate tumor-associated macrophages (TAMs) to M1, which helps to activate immunity (Zhang et al. 2019). It suppressed the tumor progression in C33A cervical squamous carcinoma cells by raising the apoptosis rate, which is accompanied by an increase in caspase-3, caspase-9, and Bcl-2 gene expression (Zhao et al. 2019). Another method to eliminate cancer cells by employing increased temperature was developed by Bae et al., who developed ferrimagnetic iron oxide nanocubes (FIONs) clubbed with COS (chitosan-FIONs) that showed anticancer ability by magnetic induced hyperthermia and regulating the caspase expression in an in vivo model (Bae et al. 2012). N-acetyl COS (N-AcCOS), which were extracted by chitin utilizing bacterial strains, exhibited antioxidant activity and impacted the survival of breast cancer cells, as well as promoted the cell migration in HEK293 normal kidney cells (Behera et al. 2020). Chol-SS-COS/DOX liposomes increased cellular uptake and death in MG63 osteosarcoma cells in response to cytoplasmic delivery, as well as improved pharmacokinetics in an animal model (Yin et al. 2017). The data suggest that COS-deacetylase homolog YDJC binds to CDC16 and is involved in SPC-induced events in A549 lung cancer cells. These findings imply that inhibiting YDJC or increasing CDC16 interaction with YDJC might be a novel approach for preventing malignant alterations in metastatic cancer cell physiological properties (Kim et al. 2018).

To investigate the mechanism of anti-metastatic effect of COS in cancer, researchers found that partly acetylated COS (paCOS) with 0.46 fractionalization of acetylation induced tumor suppression at each stage of metastasis in liver cancer (Jing et al. 2019). EMT was triggered by the ubiquitination of the protein PP2A. As a result, YDJC might be a novel target for lung cancer treatment (Kim et al. 2019). PP2-type lectin binding to COS (BGL24) was isolated and purified from a bottle of powdered extract. BGL24 was found to be more cytotoxic to triple-negative breast cancer cells (MDA-MB-231) (Mondal and Swamy 2020). COS inhibited osteosarcoma cell survival and proliferation by activating the p53/mTOR pathway through autophagy, which also increased apoptosis in both in vitro and in vivo conditions (Pan et al. 2021). COS is involved in the mitochondrial membrane disintegration, leading to the release of cytochrome-c, which causes an increase in the release of caspase-3 and caspase-9, and an increase in the apoptotic rate of HepG2 cells (Zhu et al. 2021). In an in vivo research, lower molecular weight chitosan and its N-acetyl derivatives inhibited tumor development in the sarcoma 180 (S180) with a maximal lowering capability of 33% (Qin et al. 2002). The prodrug of 5-fluorouracil linked to COS by the carbonyl group revealed anticancer efficacy in both in vitro and in vivo conditions, and it prolonged the survival time of p-388 leukemia mice with a smaller tumor size compared to the prodrug alone treated group (Ouchi et al. 1990). Using the copolymer Chitosan oligosaccharide, dual drug delivery with responsiveness to the pH changes in the tumor microenvironment was evaluated. COS was first treated with hyaluronic acid (HA) before being grafted with oleic acid (HA-CSO-g-OA). The oleic acid grafted copolymer was coupled with doxorubicin modified with HA (HA-DOX@HA-CSO-g-OA) and was shown to be cytotoxic to B16 melanoma cells with an increase in cellular absorption. It had an encapsulation efficiency of 85% (Yin et al. 2018). For melanoma therapy, the COS was treated with phenylboronic acid and loaded with PD-L1-siRNA (Liu 2020). In another research, Gboxin conjugated COS, significantly decreased the proliferation of prostate cancer cells by disrupting the mitochondrial function (Elbehairi et al. 2020).

6 Conclusion

The implementation of chitosan and its derivatives for cancer treatment has been successful in both in vitro and in vivo settings against a variety of malignant tumors. The physicochemical properties of COS promote biocompatibility toward normal cells, while simultaneously being lethal to cancer cells. COS has also shown to increase the radiosensitivity in the cells and inhibit the tumor vasculature development. COS formulation has a significant use in cancer therapy, resulting in high-water solubility and increased cellular absorption of the drugs. However, other investigations have found that the COS and chitosan crosslinkers employed in some micelles could improve the drug loading capacity and prolong the drug release in response to tumor stimuli. Many studies have shown that COS is an effective and reliable material for drug delivery and biomedical applications.

For biological and molecular research, drug delivery in cancer therapy demands the novel agents. COS, a biopolymer, is gaining popularity these days due to the lack of undesired effects. Currently, researchers are working on the characterization of established COS in terms of understanding their properties and how they may be used for biological activities. Future studies are warranted on the molecular mechanism of COS on cancer cells, as well as the effect of COS on co-signaling pathways.

The main constraint of COS and its derivatives is that the physicochemical characteristics fluctuate with pH and temperature, which causes variations in the COS attributes. This may have an impact on the functioning of COS as well as its ability to encapsulate drugs. Another important disadvantage is that the anticancer action of COS at a molecular level has not been fully investigated. In case of bulk synthesis and purification, obtaining COS with a degree of polymerization of more than 5 is extremely challenging. The COS and derivatives that are currently available in the market are quite expensive. Hence, designing the novel synthesis approaches and deriving an economical strategy constitutes a good scope for COS-based drug delivery and therapy systems for anticancer applications.

References

Aam BB, Heggset EB, Norberg AL, Sørlie M, Varum KM, Eijsink VG (2010) Production of chitooligosaccharides and their potential applications in medicine. Mar Drugs 8:1482–1517

Amirani E, Hallajzadeh J, Asemi Z, Mansournia MA, Yousefi B (2020) Effects of chitosan and oligochitosans on the phosphatidylinositol 3-kinase-AKT pathway in cancer therapy. Int J Biol Macromol 164: 456–467

Bae KH, Park M, Do MJ, Lee N, Ryu JH, Kim GW, Kim C, Park TG, Hyeon T (2012) Chitosan oligosaccharide-stabilized ferrimagnetic iron oxide nanocubes for magnetically modulated cancer hyperthermia. ACS Nano 6:5266–5273

Behera HT, Upadhyay AK, Raina V, Ray L (2019) Optimization of media components for the production of N-acetylchitooligosaccharide from chitin by *Streptomyces chilikensis* through Taguchi experimental design. J Microbiol Methods 159: 194–199

Behera HT, Mojumdar A, Das SR, Jema S, Ray L (2020) Production of N-acetyl chitooligosaccharide by novel *Streptomyces chilikensis* strain RC1830 and its evaluation for anti-radical, anti-inflammatory, anti-proliferative and cell migration potential. Bioresour Technol Rep 11:100428

Bhatnagar I, Venkatesan J, Kim SK (2014) Polymer functionalized single walled carbon nanotubes mediated drug delivery of gliotoxin in cancer cells. J Biomed Nanotech 10:120–130

Cabrera JC, Van Cutsem P (2005) Preparation of chitooligosaccharides with degree of polymerization higher than 6 by acid or enzymatic degradation of chitosan. Biochem Eng J 25:165–172

Chatterjee S, Hui PC, Siu WS, Kan CW, Leung PC, Wanxue C, Chiou JC (2021) Influence of pH-responsive compounds synthesized from chitosan and hyaluronic acid on dual-responsive (pH/temperature) hydrogel drug delivery systems of Cortex Moutan. Int J Biol Macromol 168:163–174

Chen L, Du Y, Huang R (2003) Novel pH, ion sensitive polyampholyte gels based on carboxymethyl chitosan and gelatin. Polym Int 52:56–61

Chinnaiyan P, Huang S, Vallabhaneni G, Armstrong E, Varambally S, Tomlins SA, Chinnaiyan AM, Harari PM (2005) Mechanisms of enhanced radiation response following epidermal growth factor receptor signaling inhibition by erlotinib (Tarceva). Cancer Res 65: 3328–3335

Cho YI, Park S, Jeong SY, Yoo HS (2009) In vivo and in vitro anti-cancer activity of thermo-sensitive and photo-crosslinkable doxorubicin hydrogels composed of chitosan–doxorubicin conjugates. Eur J Pharm Biopharm 73:59–65

Chokradjaroen C, Rujiravanit R, Watthanaphanit A, Theeramunkong S, Saito N, Yamashita K, Arakawa R (2017) Enhanced degradation of chitosan by applying plasma treatment in combination with oxidizing agents for potential use as an anticancer agent. Carbohydr Polym 167:1–1

Chokradjaroen C, Rujiravanit R, Theeramunkong S, Saito N (2017) Degradation of chitosan hydrogel dispersed in dilute carboxylic acids by solution plasma and evaluation of anticancer activity of degraded products. Jpn J Appl Phys 57:0102B5

Choudhury H, Maheshwari R, Pandey M, Tekade M, Gorain B, Tekade RK (2020) Advanced nanoscale carrier-based approaches to overcome biopharmaceutical issues associated with anticancer drug 'Etoposide'. Mater Sci Eng C 106:110275

de Santana SC, da Silva Filho RC, de Oliveira JA, de Macedo GR, Padilha FF, dos Santos ES (2015) Enhancing purification of chitosanase from Metarhiziumanisopliae by expanded bed adsorption chromatography using Doehlert design. Biocatal Agric Biotechnol 4:727–736

Elbehairi SE, Alfaifi MY, Shati AA, Alshehri MA, Elshaarawy RF, Hafez HS (2020) Role of Pd (II)–chitooligosaccharides–gboxinanalog in oxidative phosphorylation inhibition and energy depletion: targeting mitochondrial dynamics. Chem Biol Drug Design 96:1148–1161

El-Kenawi AE, El-Remessy AB (2013) Angiogenesis inhibitors in cancer therapy: mechanistic perspective on classification and treatment rationales. Br J Pharmacol 170:712–729

Eom TK, Senevirathne M, Kim SK (2012) Synthesis of phenolic acid conjugated chitooligosaccharides and evaluation of their antioxidant activity. Environ Toxicol Pharmacol 34:519–527

Farkona S, Diamandis EP, Blasutig IM (2016) Cancer immunotherapy: the beginning of the end of cancer? MNC Med 14:1–8

Fernandes JC, Borges M, Nascimento H, Bronze-da-Rocha E, Ramos OS, Pintado ME, Malcata FX, Santos-Silva A (2011) Cytotoxicity and genotoxicity of chitooligosaccharides upon lymphocytes. Int J Biol Macromol 49:433–438

Fernandes JC, Sereno J, Garrido P, Parada B, Cunha MF, Reis F, Pintado ME, Santos-Silva A (2012) Inhibition of bladder tumor growth by chitooligosaccharides in an experimental carcinogenesis model. Mar Drugs 10:2661–2675

Hamed I, Ozogul F, Regenstein JM (2016) Industrial applications of crustacean by-products (chitin, chitosan, and chitooligosaccharides): a review. Trends Food Sci Technol 48:40–50

Han FS, Cui BH, You XF, Xing YF, Sun XW (2015) Anti-proliferation and radiosensitization effects of chitooligosaccharides on human lung cancer line HepG2. Asian Pac J Trop Med 8:757–761

Han FS, Yang SJ, Lin MB, Chen YQ, Yang P, Xu JM (2016) Chitooligosaccharides promote radiosensitivity in colon cancer line SW480. World J Gastroenterol Hepatol Endosc 22:5193

Hong S, Ngo DN, Kim MM (2016) Inhibitory effect of aminoethyl-chitooligosaccharides on invasion of human fibrosarcoma cells. Environ Toxicol Pharmacol 45:309–314

Hu X, Chen S, Yin H, Wang Q, Duan Y, Jiang L, Zhao L (2020) Chitooligosaccharides-modified PLGA nanoparticles enhance the antitumor efficacy of AZD9291 (osimertinib) by promoting apoptosis. Int J Biol Macromol 162:262–272

Huang R, Mendis E, Kim SK (2005) Improvement of ACE inhibitory activity of chitooligosaccharides (COS) by carboxyl modification. Bioorg Med Chem 13:3649–3655

Huang R, Mendis E, Rajapakse N, Kim SK (2006) Strong electronic charge as an important factor for anticancer activity of chitooligosaccharides (COS). Life Sci 78:2399–2408

Huang X, Huang X, Jiang XH, Hu FQ, Du YZ, Zhu QF, Jin CS (2012) In vitro antitumour activity of stearic acid-g-chitosan oligosaccharide polymeric micelles loading podophyllotoxin. J Microencapsul 29:1–8

Jain KK (2005) Nanotechnology-based drug delivery for cancer. Technol Cancer Res Treatment 4:407–416

Jang A, Lee NY, Lee BD, Kim TH, Son JH, An BJ, Jo C (2009) Biological functions of a synthetic compound, octadeca-9, 12-dienyl-3, 4, 5-hydroxybenzoate, from gallic acid–linoleic acid ester. Food Chem 112:369–373

Jiang M, Zhuge X, Yang Y, Gu X, Ding F (2009) The promotion of peripheral nerve regeneration by chitooligosaccharides in the rat nerve crush injury model. Neurosci Lett 454:239–243

Jiang C, Wang H, Zhang X, Sun Z, Wang F, Cheng J, Xie H, Yu B, Zhou L (2014) Deoxycholic acid-modified chitooligosaccharide/mPEG-PDLLA mixed micelles loaded with paclitaxel for enhanced antitumor efficacy. Int J Pharm 475:60–68

Jing B, Cheng G, Li J, Wang ZA, Du Y (2019) Inhibition of liver tumor cell metastasis by partially acetylated chitosan oligosaccharide on a tumor-vessel microsystem. Mar Drugs 17:415

Kang HS, Kim GY, Jung I, Oh SD, Kim CH, Shim BS, Park KH, Oh SJ (2007) The effect of the compound of tomato extract to the prostatic cancer cell and the prostate of the rat model of benign prostatic hyperplasia. Korean J Pharmacogn 38:197–203

Karagozlu MZ, Kim JA, Karadeniz F, Kong CS, Kim SK (2010) Anti-proliferative effect of amino derivatized chitooligosaccharides on AGS human gastric cancer cells. Process Biochem 45:1523–1528

Kerch G (2015) The potential of chitosan and its derivatives in prevention and treatment of age-related diseases. Mar Drugs 13:2158–2182

Kerch G, Sabovics M, Kruma Z, Kampuse S, Straumite E (2011) Effect of chitosan and chitooligosaccharide on vitamin C and polyphenols contents in cherries and strawberries during refrigerated storage. Eur Food Res Technol 233:351–358

Kim EJ, Park MK, Byun HJ, Kang GJ, Yu L, Kim HJ, Shim JG, Lee H, Lee CH (2018) YdjC-chitooligosaccharide deacetylase homolog induces keratin reorganization in lung cancer cells: involvement of interaction between YDJC and CDC16. Oncotarget 9:22915

Kim EJ, Park MK, Kang GJ, Byun HJ, Kim HJ, Yu L, Kim B, Chae HS, Chin YW, Shim JG, Lee H (2019) YDJC induces epithelial-mesenchymal transition via escaping from interaction with CDC16 through ubiquitination of PP2A. J Oncol 2019:3542537

Kumara BN, Bhat SR, Prasad KS (2021) Why chitosan could be apt candidate for glaucoma drug delivery-an overview. Int J Biol Macromol 176:47–65

Li HF, Huang LF, Chen LH (2019) Chitooligosaccharides inhibit A549 lung cancer cell line proliferation by regulating cell autophagy. J Biol Regul Homeost Agents 33:1527–1532

Li R, Lyu Y, Luo S, Wang H, Zheng X, Li L, Ao N, Zha Z (2021) Fabrication of a multi-level drug release platform with liposomes, chitooligosaccharides, phospholipids and injectable chitosan hydrogel to enhance anti-tumor effectiveness. Carbohydr Polym 269:118322

Lin F, Jia XG, Lei WX, Li ZJ, Zhang TY (2009) Spectra analyses of chitosans degraded by hydrogen peroxide under optimal conditions. Spectrosc Spectr Anal 29:43–47

Liu X (2020) Anti-melanoma effect of intratumoral injection of PD-L1-siRNA-loaded phenylboronic acid-modified chitooligosaccharide nanoparticles: a preliminary study. Int J Pharm Res 47:638–644

Lu X, Guo H, Zhang Y (2012) Protective effects of sulphated chitooligosaccharides against hydrogen peroxide-induced damage in MIN6 cells. Int J Biol Macromol 50:50–58

Lu X, Guo H, Sun L, Zhang L, Zhang Y (2013) Protective effects of sulfatedchitooligosaccharides with different degrees of substitution in MIN6 cells. Int J Biol Macromol 52:92–98

Luo Z, Dong X, Ke Q, Duan Q, Shen L (2014) Downregulation of CD147 by chitooligosaccharide inhibits MMP-2 expression and suppresses the metastatic potential of human gastric cancer. Oncol Lett 8:361–366

Luo Y, Deng L, Deng QJ, Wen L (2016) Comparative study of the chitooligosaccharides effect on the proliferation inhibition and radiosensitization of three types of human gastric cancer cell line. Asian Pac J Trop Med 9:601–605

Manivasagan P, Oh J (2016) Marine polysaccharide-based nanomaterials as a novel source of nanobiotechnological applications. Int J Biol Macromol 82:315–327

Mitelman F, Johansson B, Mertens F (2007) The impact of translocations and gene fusions on cancer causation. Nat Rev Cancer 7:233–245

Mondal S, Swamy MJ (2020) Purification, biochemical/biophysical characterization and chitooligosaccharide binding to BGL24, a new PP2-type phloem exudate lectin from bottle gourd (*Lagenaria siceraria*). Int J Biol Macromol 164:3656–3666

Morris VB, Neethu S, Abraham TE, Pillai CK, Sharma CP (2009) Studies on the condensation of depolymerized chitosans with DNA for preparing chitosan-DNA nanoparticles for gene delivery applications. J Biomed Mater Res Part B Appl Biomater 89:282–292

Ngo DN, Qian ZJ, Je JY, Kim MM, Kim SK (2008) Aminoethyl chitooligosaccharides inhibit the activity of angiotensin converting enzyme. Process Biochem 43:119–123

Ngo DH, Qian ZJ, Ngo DN, Vo TS, Wijesekara I, Kim SK (2011) Gallylchitooligosaccharides inhibit intracellular free radical-mediated oxidation. Food Chem 128:974–981

Ngo DH, Ngo DN, Vo TS, Ryu B, Van Ta Q, Kim SK (2012) Protective effects of aminoethyl-chitooligosaccharides against oxidative stress and inflammation in murine microglial BV-2 cells. Carbohydr Polym 88:743–747

Ngo DH, Ngo DN, Kim SK, Vo TS (2019) Antiproliferative effect of aminoethyl-chitooligosaccharide on human lung A549 cancer cells. Biomolecules 9:195

Oh GW, Kim SC, Kim TH, Jung WK (2021) Characterization of an oxidized alginate-gelatin hydrogel incorporating a COS-salicylic acid conjugate for wound healing. Carbohydr Polym 252:117145

De Oliveira PN, Moussa A, Milhau N, Bini RD, Prouillac C, de Oliveira BF, Dias GS, Santos IA, Morfin I, Sudre G, Alcouffe P (2020) In situ synthesis of Fe3O4 nanoparticles coated by chito-oligosaccharides: physico-chemical characterizations and cytotoxicity evaluation for biomedical applications. Nanotechnol 31:175602

Ouchi T, Banba T, Matsumoto T, Suzuki S, Suzuki M (1990) Synthesis and antitumor activity of conjugates of 5-fluorouracil and chito-oligosaccharides involving a hexamethylene spacer group and carbamoyl bonds. Drug Des and Dev 6:281–287

Pan Z, Cheng DD, Wei XJ, Li SJ, Guo H, Yang QC (2021) Chitooligosaccharides inhibit tumor progression and induce autophagy through the activation of the p53/mTOR pathway in osteosarcoma. Carbohydr Polym 258:117596

Park BK, Kim MM (2010) Applications of chitin and its derivatives in biological medicine. Int J Mol Sci 11:5152–5164

Qin C, Du Y, Xiao L, Li Z, Gao X (2002) Enzymic preparation of water-soluble chitosan and their antitumor activity. Int J Biol Macromol 31:111–117

Qu Y, Chu BY, Peng JR, Liao JF, Qi TT, Shi K, Zhang XN, Wei YQ, Qian ZY (2015) A biodegradable thermo-responsive hybrid hydrogel: therapeutic applications in preventing the post-operative recurrence of breast cancer. NPG Asia Materials 7:e207

Quan H, Zhu F, Han X, Xu Z, Zhao Y, Miao Z (2009) Mechanism of anti-angiogenic activities of chitooligosaccharides may be through inhibiting heparanase activity. Med Hypotheses 73:205–206

Rajapakse N, Kim MM, Mendis E, Huang R, Kim SK (2006) Carboxylatedchitooligosaccharides (CCOS) inhibit MMP-9 expression in human fibrosarcoma cells via down-regulation of AP-1. Biochimica et Biophysica Acta (BBA)-General Subjects 1760:1780–1788

Rosenblum D, Joshi N, Tao W, Karp JM, Peer D (2018) Progress and challenges towards targeted delivery of cancer therapeutics. Nat Commun 9:1–2

Ryu B, Himaya SW, Napitupulu RJ, Eom TK, Kim SK (2012) Sulfatedchitooligosaccharide II (SCOS II) suppress collagen degradation in TNF-induced chondrosarcoma cells via NF-κB pathway. Carbohydr Res 350:55–61

Ryu B, Kim SY, Vo TS, Kim WS, Kim DG, Kim SK (2017) Characterization of the in vitro effects of gallic acid-grafted-chitooligosaccharides in the suppression of AGS human gastric cancer cell proliferation. RSC Adv 7:24561–24568

Sato K, Saimoto H, Morimoto M, Shigemasa Y (2003) Depolymerization of chitin and chitosan under hydrothermal conditions. Sen-I Gakkaishi 59:104–109

Schimpl M, Rush CL, Betou M, Eggleston IM, Recklies AD, Van Aalten DM (2012) Human YKL-39 is a pseudo-chitinase with retained chitooligosaccharide-binding properties. Biochem J 446:149–157

Shen KT, Chen MH, Chan HY, Jeng JH, Wang YJ (2009) Inhibitory effects of chitooligosaccharides on tumor growth and metastasis. Food Chem Toxicol 47:1864–1871

Sudhakar A (2009) History of cancer, ancient and modern treatment methods. J Cancer Sci Ther 1:1

Sun C, Shen C, Zhang Y, Hu C (2021) LncRNA ANRIL negatively regulated chitooligosaccharide-induced radiosensitivity in colon cancer cells by sponging miR-181a-5p. Adv Clin Exp Med 30:55–65

Wang Z, Zheng L, Yang S, Niu R, Chu E, Lin X (2007) N-acetylchitooligosaccharide is a potent angiogenic inhibitor both in vivo and in vitro. Biochem Biophys Res Commun 357:26–31

Wang SL, Lin HT, Liang TW, Chen YJ, Yen YH, Guo SP (2008) Reclamation of chitinous materials by bromelain for the preparation of antitumor and antifungal materials. Bioresour Technol 99:4386–4393

Wu H, Yao Z, Bai X, Du Y, Lin B (2008) Anti-angiogenic activities of chitooligosaccharides. Carbohydr Polym 73:105–110

Wu T, Zivanovic S, Hayes DG, Weiss J (2008) Efficient reduction of chitosan molecular weight by high-intensity ultrasound: underlying mechanism and effect of process parameters. J Agric Food Chem 56:5112–5119

Wu H, Yao Z, Bai X, Du Y, Ma X (2010) Chitooligosaccharides inhibit nitric oxide mediated migration of endothelial cells in vitro and tumor angiogenesis in vivo. Carbohydr Polym 82:927–932

Wu H, Aam BB, Wang W, Norberg AL, Sorlie M, Eijsink VG, Du Y (2012) Inhibition of angiogenesis by chitooligosaccharides with specific degrees of acetylation and polymerization. Carbohydr Polym 89:511–518
Wu S, Powers S, Zhu W, Hannun YA (2016) Substantial contribution of extrinsic risk factors to cancer development. Nature 529:43–47
Wu M, Li J, An Y, Li P, Xiong W, Li J, Yan D, Wang M, Zhong G (2019) Chitooligosaccharides prevents the development of colitis-associated colorectal cancer by modulating the intestinal microbiota and mycobiota. Front Microbiol 10:2101
Xie Y, Zhou N, Cao J, Ding B, Liu D, Liu J (2008) Antigrowth effects of chitosan and its derivatives on human hepatocellular carcinoma cell line SMMC7721. Chinese J Tissue Eng Res 12:4579–4582
Xin C, Yao X, Du B, Yang W, Wang L, Ma L, Weng W (2019) Stearic acid-grafted chitooligosaccharidenanomicelle system with biocleavable gadolinium chelates as a multifunctional agent for tumor imaging and drug delivery. Pharm Res 36:1–4
Xing R, Liu S, Yu H, Guo Z, Wang P, Li C, Li Z, Li P (2005) Salt-assisted acid hydrolysis of chitosan to oligomers under microwave irradiation. Carbohydr Res 340:2150–2153
Xiong C, Wu H, Wei P, Pan M, Tuo Y, Kusakabe I, Du Y (2009) Potent angiogenic inhibition effects of deacetylated chitohexaose separated from chitooligosaccharides and its mechanism of action in vitro. Carbohydr Res 344:1975–1983
Xu W, Jiang C, Kong X, Liang Y, Rong M, Liu W (2012) Chitooligosaccharides and N-acetyl-D-glucosamine stimulate peripheral blood mononuclear cell-mediated antitumor immune responses. Mol Med Rep 6:385–390
Yin X, Chi Y, Guo C, Feng S, Liu J, Sun K, Wu Z (2017) Chitooligosaccharides modified reduction-sensitive liposomes: enhanced cytoplasmic drug delivery and osteosarcomas-tumor inhibition in animal models. Pharm Res 34:2172–2184
Yin X, Feng S, Chi Y, Liu J, Sun K, Guo C, Wu Z (2018) Estrogen-functionalized liposomes grafted with glutathione-responsive sheddable chotooligosaccharides for the therapy of osteosarcoma. Drug Deliv 25:900–908
Yoksan R, Akashi M, Miyata M, Chirachanchai S (2004) Optimal γ-ray dose and irradiation conditions for producing low-molecular-weight chitosan that retains its chemical structure. Radiat Res 161:471–480
Yuan X, Zheng J, Jiao S, Cheng G, Feng C, Du Y, Liu H (2019) A review on the preparation of chitosan oligosaccharides and application to human health, animal husbandry and agricultural production. Carbohydr Polym 220:60–70
Zhai X, Li C, Ren D, Wang J, Ma C, Abd El-Aty AM (2021) The impact of chitooligosaccharides and their derivatives on the in vitro and in vivo antitumor activity: a comprehensive review. Carbohydr Polym 266:118132
Zhang J, Xia W, Liu P, Cheng Q, Tahi T, Gu W, Li B (2010) Chitosan modification and pharmaceutical/biomedical applications. Mar Drugs 8:1962–1987
Zhang Y, Sha W, Zhang X, Cheng M, Wu Q, Wang W, Yuan Z (2019) Zwitterionic chitooligosaccharide-modified ink-blue titanium dioxide nanoparticles with inherent immune activation for enhanced photothermal therapy. Biomater Sci 7:5027–5034
Zhao M, Gu L, Li Y, Chen S, You J, Fan L, Wang Y, Zhao L (2019) Chitooligosaccharides display anti-tumor effects against human cervical cancer cells via the apoptotic and autophagic pathways. Carbohydr Polym 224:115171
Zhu C, Zhao M, Fan L, Cao X, Xia Q, Zhou J, Yin H, Zhao L (2021) Chitopentaose inhibits hepatocellular carcinoma by inducing mitochondrial mediated apoptosis and suppressing protective autophagy. Bioresour Bioprocess 8:1–2
Zimmermann S, Dziadziuszko R, Peters S (2014) Indications and limitations of chemotherapy and targeted agents in non-small cell lung cancer brain metastases. Cancer Treat Rev 40:716–722

Disease Preventing Bioactivities of Chitooligosaccharides: Current Status and Future Trends

Sujata Sinha and Pushplata Tripathi

Abbreviations

DPPH	2, 2-Diphenyl-1-picrylhydrazyl
QCMCOS	Quaternised carboxymethyl chitooligosaccharide
DNA	Deoxyribonucleic acid
COS	Chitooligosaccharides
DD	Degree of deacetylation
DP	Degree of Polymerisation
DNA	Deoxyribonucleic acid
KDa	Kilo Dalton
LMWC	Low mol weight Chitosan
MIC	Minimum inhibitory concentration
SEM	Scanning electron microscope
TEM	Transmission electron microscope

1 Introduction

Chitooligosaccharides (COS) are derived from chitosan which is the most abundantly available polymer in nature after cellulose, generally they have molecular weight of 3.9 KDa or less and degree of polymerisation (DP) is lesser than 20

S. Sinha (✉)
Indian Red Cross Society Building, Department of Health Research, Red Cross Road, New Delhi 110001, India
e-mail: sujata.sinha@nic.in

P. Tripathi
School of Sciences, Indira Gandhi National Open University, Maidan Garhi, Saket, Delhi 110068, India

S.-K. Kim (ed.), *Chitooligosaccharides*,
https://doi.org/10.1007/978-3-030-92806-3_9

(Mourya et al. 2011). COS are water-soluble and non-cytotoxic to organisms, readily absorbed through the intestine, and can be easily eliminated through the kidneys. All these properties make them promising compounds for several medical and food applications (Jafari et al. 2020). Broadly, COS can be categorised into homo and heterooligomers. Homo oligomers contain only one type monomer while hetero-types may contain monomer as well as various other COS types like variable molecular weight as well as having various DAC (degree of acetylation). Solubility of hetero-types of COS decreases above 10 and depends on pH of the solution and degree of acetylation. Chitosan hydrolysis to COS can be achieved by chemical, physical or bio-catalytic method (Aam et al. 2010). Gamma-irradiation and ultrasonics wave has also been used for reducing the viscosity of chitosan solution and production of various oligomers of chitosan from 2–11, with maximum concentration (Popa-Nita et al. 2009; Choi et al. 2002) of trimer has been achieved. Chemical hydrolysis using strong acids like hydrochloric acids, acetic acids, phosphoric acids, fluoric acids, nitrous acids, etc., and milder acids like formic/lactic/trichloroacetic acid have been studied for their degradative action on chitosan (Ando and Kataoka 1980; Yamaguchi et al. 1982; Il'ina and Varlamov 2004). However, chemical methods are not considered feasible as they are not environmentally friendly and may lead to spurious and secondary product formation most of the time. Specific and non-specific biocatalysis is the most preferred method for COS production either by hydrolytic method or using trans glycosylation method. Biocatalytic methods are easy to control and minimise the formation of side reactions.

COS are known to possess many biological activities like anti-microbial, anti-tumour, antio-xidant, immunoenhancing, anti-arthritis, lowering of high blood pressure and protective effects against infection (Xia et al. 2011; Kim and Rajapakse 2005). Biological activities of COS depend on many factors like the degree of acetylation, degree of polymerisation, nature of the chemical modification and charge distribution (Muzzarelli 1996). Unlike high molecular weight chitosan, COS is easily absorbed through the intestine, quickly gets into the blood flow and has systemic biological effects on the organism. In the food industry, COS attracts a greater interest as an anti-microbial agent, anti-oxidants and enhancers of the nutritional quality of food (Shahidi et al. 1999). COS production from various sources and using various methods (physical, chemical, biological) and their bioactivity and disease preventing activities have been summarised in Fig. 1.

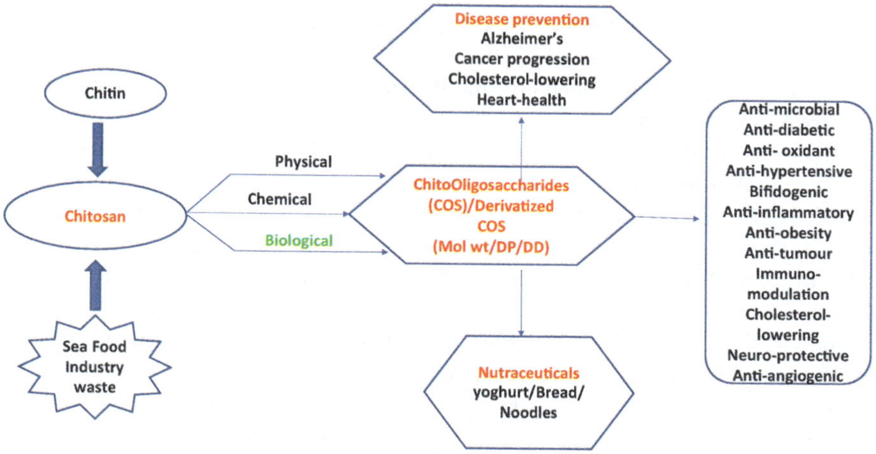

Fig. 1 Chitooligosaccharides production from various sources and their biological and disease prevention activity. Chitooligosaccharides with a range of molecular weight, degree of polymerisation and deacetylation decides various biological and disease preventing activity

2 Bioactivities of Chitosan Oligosaccharides

2.1 *Anti-microbial*

The anti-microbial activity of COS is influenced by factors such as DP (Park et al. 2004), DAC (Chung et al. 2004), type of microbes (Uchida et al. 1989) and some other physicochemical properties. The number of amino groups has proven to play a major role in antibacterial activity and several mechanisms have been given to describe this activity (Chen et al. 2002). The most widely accepted mechanism is that COS can alter permeability characteristics of microbial cell membrane and further prevent the entry of materials or cause leakage of cell constituents that finally leads to death of bacteria (Sudharshan et al. 1992). A study assessing the antibacterial activity of a mixture of COS and low molecular weight chitosan (230 kDa and degree of deacetylation: 91.5%) against clinically important bacterial pathogen, *Actinobacillus actinomycetemcomitans* was done, SEM and TEM images of those treated bacterial cells exhibited a change of their morphology from spherical shape to irregular condensed masses with bleb-like structures which indicated a separation of cytoplasmic membrane from cell wall and cytosolic components coagulation (Choi et al. 2001). It was suggested that the COS action site is probably the bacterial envelope and organism killing could be due to disruption of membrane. RNA transcription blocking by adsorption of penetrated chitosans to bacterial DNA (Kim et al. 2003) is another suggested mechanism for killing of microbes. Molecular weight of COS must be less than a critical value for furnishing this mechanism, which enables penetration of molecules into bacterial cells. Positive charge favours antibacterial activity of COS, so it was expected that modification of amino group of glucosamine at C-2 position by potent groups who are positively charged will enhance the antibacterial activity. N-conjugation (Jeon and Kim 2001) was used for introduction of asparagine to the C-2 position of glucosamine and resultant Asn-COS showed improved bactericidal activity. Lower MIC of Asn-COS on *E. coli* was recorded compared to that of COS. In addition to COS positive charge, cell wall charge distribution also seems to play a considerable role in antibacterial properties. Study of cell surface characteristics revealed close relationship of negative charge distribution and hydrophilicity (Chung et al. 2004). In case of higher negative charge distribution in gram-negative bacterial cell surface leads to greater hydrophobicity and leads to more adsorption of COS on the surface and is higher in negatively charged gram-negative bacteria to less negative or gram positive strains. This explains clearly that why most gram-negative bacteria are COS sensitive and distribution of charge on the cell surfaces is determining factor for this activity. Polycationic nature of COS allows reaction with fungal negatively charged groups and thereby showing antifungal activity on yeast and molds. COS inhibition mechanism for fungi is similar to that of bacterial species and has already been explained previously. Polyelectrolyte complexes formation between COS and groups on negatively charged cell surface interferes with the growth and normal physiological functions of fungi, indicating charge distribution of COS has a direct correlation with its activity against fungi (Hirano and Nagao 1989). COS are also effective in preventing several phage infections and depends on molecular weight, molecular structure, DD and also type of bacteriophages (Kochkina and Chirkov 2000). COS and LMWC are involved in the inhibition of bacteriophages replication by several different mechanisms. Interestingly, electron microscopic observations revealed that COS containing lower DP could change the structure of phage particles. Furthermore, these changes make the phage inactive directly through receptor-recognising structures on the phage particles.

2.2 *Anti-diabetic Activity*

Anti-oxidant activity of COS on pancreatic islet cells in streptozotocin induced diabetic rats have been detected (Yuan 2009). Optical microscopy, colorimetric assay and gel electrophoresis has been used for this purpose. It was concluded that COS was able to prohibit the apoptosis in pancreatic islet cells, all types of COS in all concentration improved the total anti-oxidant capacity, superoxide dismutase enzyme activity and decrease the malondialdehyde concentration drastically. Morphological investigation also proved that COS resulted in reduction of islet cells, loss of pancreatic cells and resulted in nuclear pyknosis in these cells. Reduction of in vivo postprandial blood glucose level has been reported using COS (Jeong et al. 2019) in a short-term cross over study. Consumption of COS (GO2KA1) with a supplemented meal effectively reduced postprandial glucose level in blood and is related to diabetes prevention. Reduction in fasting glucose and improved dyslipidemia was observed in diabetic mice with the help of COS (Zheng et al. 2018). Anti-diabetic effect of COS in type 2 diabetes mellitus mice was found to be due to maintenance of gut integrity/microbiota by inhibiting ρ-p38 and increasing AMPK level. COS also reversed the insulin resistance and was accompanied by suppression of inflammation mediators, down regulation of lipogenesis and adipocyte differentiation suppression in white adipose tissue. It also inhibited the occludin reduction and relieved gut dysbiosis by promoting certain microbial groups.

2.3 *Anti-oxidant or Radical Scavenging Activity*

Anti-oxidant or radical scavenging properties of COS depends primarily on their molecular weights and DD. Electron spin trapping experiment suggested that LMWC (1–3 kDa) have high potential to scavenge free radicals (Park et al. 2003) and at 0.5 mg ml^{-1} concentration greater than 80% of superoxide radical scavenging activity was observed. COS with DD (90%) have been shown to scavenge radicals like DPPH, -OH, carbon-centred and superoxide radicals (Je et al. 2004). Exact mechanism of anti-oxidant/radical scavenging activity of COS is not very known but attributed to amino and hydroxyl groups attached to C-2/3/6 of the pyranose ring. COS reacts with unstable free radicals and form a stable macromolecule. Some discrepancies in the -OH radical scavenging activities of COS and their derivatised version have been reported and reportedly metal ion uptake of COS has a great influence (Huang et al. 2006) it is partly attributed to transition Fe^{2+} chelating ability, molecular charge properties and proton donation through hydroxyl and amino groups. Metal ion uptake by COS can be through chelation via lone electron pairs in amino groups and ion exchange mechanisms of protonated amino groups (Guzman et al. 2003). Therefore, the above information further strengthens the effect of degree of deacetylation on radical scavenging which is directly correlated with protonation of amine groups. Hoverer, up-to-date information about the relationship between ion chelation and anti-oxidant properties of COS is lacking. Biocatalytic Production of Hetero-COS as an anti-oxidant has been reported and summarised recently (Jaiswal et al. 2019).

2.4 *Bifidogenic Activities*

COS with DP 2–8 have shown growth promoting effect on *Bifidobacterium bifidum* and *Lactobacillus sp.* (Lee et al. 2002). COS was prepared by enzymatic hydrolysis of fully deacetylated chitosan polymers and effect of COS on bacterial growth was studied. COS was found to be lethal against two prebiotic strains, however all others tested strains showed good growth till 5% COS. This growth stimulating effect was compared with that of FOS (fructo-oligosaccharides), an established prebiotics. In another study, increased growth rate of *B bifidum, B catenulatum* and *B infantis* was observed in presence of COS (Simunek et al. 2010). Proper explanation of bifidogenic properties of COS has not been given in any of the studies. Modulation of composition of gut microbiota by increasing bacteriodetes and by decreasing proteobacteria and

Actinobacteria and by lowering Firmicutes/Bacterioidetes ratio (Liu et al. 2020) has also been reported recently. COS reportedly promotes the generation of beneficial bacteria and suppresses the pathogens and is thus considered to be potential prebiotics in food products.

2.5 *Anti-tumour/Anti-cancer*

Medium molecular weight (1.5–10 KDa) COS have shown anti-tumour activity against Sarcoma cells (Jeon and Kim 2002). Impact of COS and their derivatives and related reports on In vitro and In vivo anti-tumour activity has been reviewed in detail (Zhai et al. 2021). It has been emphasised that COS derivatisation improves physicochemical properties and expands applications in gene and drug delivery. COS as drug carrier itself can inhibit tumour cell proliferation and metastasis which is an attribute to their anti-angiogenic activity as well as their ability to stimulate immune response.

2.6 *Cholesterol-Lowering Properties*

Several investigators in their earlier studies demonstrated that chitosan is capable of binding dietary fats and it can prevent their absorption through the gut and has ability to absorb fat from the body (Kanauchi et al. 1995). Fat up to five times of its weight have been absorbed reportedly and it lowered the level of low-density lipoprotein cholesterol while boosting the high-density lipoprotein level. In another study, as an effective hypo-cholesterolemic agent it can decrease blood cholesterol levels up to more than 50% (Jameela et al. 1994). COS applications also lead to control of blood cholesterol level especially in the liver and prevent development of fatty liver caused by the action of hepatotropic poisons. Unlike high molecular weight chitosan, COS application does not lead to enhancement compensatory cholesterol synthesis, decrease in essential fatty acids, fat soluble vitamins and microelements from the organisms (Muzzarelli 1997). Despite few researches carried out to search for the ability of chitosan and COS to bind with bile salts and lipids, up-to-date their exact mechanism on lowering blood cholesterol level has not been completely elucidated. However, several hypotheses have been suggested to explain the possible actions of COS in reducing blood cholesterol levels. One hypothesis suggests that ionic binding of COS with bile salts and bile acids may inhibit micelle formation during lipid digestion in the digestive track (Remunan-Lopez et al. 1998). Another hypothesis suggests that chitosan and its oligomers can directly trap lipids and fatty acids (Tanaka et al. 1997). However, contradictory results have been observed with regard to ionic interactions of COS, and thus the fat binding and cholesterol lowering effect of chitosan cannot be explained only using ion binding hypothesis (Sugano et al. 1988). Moreover, some other evidences from animal studies suggest that the effective lowering of cholesterol level can be explained in relation to the ability of COS to increase the excretion of neutral sterol and indigestion of dietary fats.

2.7 *Other Bioactivities*

Angiotensin I converting enzyme (ACE) inhibitory activity of COS has also been reported (Park et al. 2003) which has been dependent on DD. Chitosan trimer has been found most effective among all chitooligomers for this purpose (Hong et al. 1998). In vitro anticoagulant activity has been reported from sulphated COS (Park et al. 2004). COS derivatives have also been shown to have immune-stimulating properties (Matsuo and Miyazano 1993). COS and some of the related biological activities (Table 1) and their disease prevention activities (Table 2) have been summarised in the given tables.

3 COS Derivatives or Conjugates for Disease Prevention: Latest Research

Not only COS but their derivatives/conjugates have also shown various bioactivities in different biological systems (Table 3). Lycopene-COS complex of size ranging from 75 to 400 nm was prepared by mechanochemical (sonication, freeze drying and kneading) methods and were shown to have anti-oxidant activity (Bockuviene et al. 2021). Spectroscopy method was used for showing interaction and physico-chemical properties of complex was investigated, stability and water solubility of this complex was enhanced. This complexation was proposed for its application in the functional food development. Chemically synthesised phenolic acid conjugates of COS have been shown to have 1,1-diphenyl-2-picrylhydrazyl (DPPH) radical scavenging activity, Hydroxyl radical scavenging activity and nitric oxide radical scavenging activity (Eom et al. 2012a, b). Functional groups like -OH and -NH_2 groups in the chitosan monomers have been modified by bio-catalytic approaches to improve its application as it changes physicochemical properties. Phenolic and gallic acid conjugates have been prepared by amide coupling reaction (Liaqat and Rengin 2018). COS conjugated gallic acid has also shown anti-oxidant capacity (Vo et al. 2017). Quaternisation, alkylation, thiolation, hydroxyalkylation and carboxyalkylation have also been used for COS modification (Mourya and Inamdar 2009). In a study on functionalised COS, it was observed that COS has better anti-oxidant activity than their O- and N-carboxymethyl substituted counterparts, while reducing power was greatest in O-carboxymethyl substituted COS. In case of QCMCOS, anti-oxidant activity was directly related to degree of substitution. Substitution with quaternary ammonium and carboxymethyl group also enhanced their thermal stability and degree of crystallinity (Li et al. 2012).

Table 1 COS and reported bioactivities related to food and pharmaceutical industries

COS preparation	Polymerisation/ Mol wt	Bioactivity	Reference
Enzymatic hydrolysis of chitosan and ultrafiltration concentration	COS (1,2,3,4)	Anti-microbial against *B cereus*, *C albicans*, *P aeruginosa*	(El-Sayed et al. 2019)
COS produced enzymatically *Trichoderma harzianum*	(monomer to hexamer)	Anti-microbial, anti-cancer	(Rafael et al. 2019)
COS prepared by two methods (P1 and P2), P1: Enzymatic; P2: Chemical and Enzymatic	COS	Anti-microbial (*E coli and L monocytogenes*)	(Sanchez et al. 2017)
Three different enzyme treatment	COS and Chitosan	Anti-oxidant and anti-microbial	(Laokuldilok et al. 2017)
Two COS mixture and LMWC	COS and LMWC	Anti-oxidant	(Fernandes et al. 2010)
Preparation by enzymatic digestion	Low mol wt Chitosan OS (<1 Kda)	Anti-inflammatory	(Chung et al. 2012)
Acid hydrolysis with column chromatographic separation	COS with DP (2–12)	Anti-oxidant	(Li et al. 2012)
Enzyme-membrane coupling reactor system	COS with DP (5–12)	Anti-tumour	(Zhao et al. 2019)
Commercially procured	High mol wt chitosan (740 Kda, DD 91%) and COS (719 Da, DD 100%)	Hypolipidemic and anti-obesity	(Liu et al. 2021)
Commercially purchased and were from crustacean shell	High (740 Kda) and low mol wt chitosan (80 Kda)	Anti-hypercholesterol (Activation and inhibition of hepatic and intestinal cholesterol-modulators)	(Chiu et al. 2020)

4 COS as Functional Food or Nutraceuticals

Like dietary supplements, prebiotics, and bio-preservatives available in the market, COS are also soluble in water and is easily blended in the food products like dairy and beverages. Bread and dairy products can be considered ideal for food fortification and these can be used as functional foods (Table 4). In a study, shelf life of bread was extended reportedly due to chitosan and COS of higher molecular weight (30–120 Kda) coatings (No and Meyers 2007) which was explained by antimicrobial properties as well as by reducing starch degradation. Subject who used this COS coated bread every day for 12 weeks got decreased level of their mean LDL-cholesterol and increased level of good cholesterol. Chitosan has been approved as a

Table 2 Studies on Chitooligosaccharides related to prevention and treatment of diseases

Chitosan oligosaccharides	Disease	Mechanism /remarks	Reference
Deacetylated COS (biose/triose/tetrose)	Neuroprotective and anti-inflammatory	Comparable with FOS (fructo-oligosaccharides)	(Santos-Moriano et al. 2017)
COS from chitosanase enzyme	Anti-obesity	Reduction of serum lipid level	(Zhou et al. 2020)
COS	Lowering cholesterol level in healthy adults	Low-density lipoprotein level decreased	(Choi et al. 2011)
CHOS	Calcium fortifiers in calcium deficiency in Osteoporosis	Bone formation and suppressing bone resorption	(He and Wang et al. 2015)
CHOS (Pentamer/Hexamer)	Promoting Immuno-modulatory both In vitro and In vivo	Gene transcription and protein secretion of cytokines were promoted	(Wei et al. 2009)
COS (DP, 2–6)	Anti-oxidant activity	50% anti-oxidant activity as compared to Ascorbic acid	(Sinha et al. 2012)
COS	Anti-oxidant activity, anti-microbial, prebiotics	Sea waste utilisation	(Sinha et al. 2014)
CHOS	Skin tissue healing and regeneration	Wound dressing biomaterial	(Jafari et al. 2020)
COS-D3-7 and A5	Viability of breast cancer cell decreased upon treatment with COS, inhibited colony forming ability	Induced Apoptosis in breast cancer cells, inhibitory activity of COS on EGFR kinases promotes Apoptosis	(Mallakuntala et al. 2021)
NA-COS	NA-COS attenuated Amyloid β-generated cytotoxicity on PC-12 cells	Showed potential as functional food ingredient to prevent Alzheimer's disease	(Jiang et al. 2019)
Chitobiose and Chitotriose	Probiotic effect	Shifted microecology in Non-Alcoholic fatty liver diseases patients	(Ji et al. 2021)

food additive in Korea and Japan and USFDA approved chitosan as generally recognised as safe (GRAS) status, so it can be used as a food additive. Chitosan use in dairy products have been reported previously (Kerch et al. 2015). Chitosan coating microencapsulation of probiotic bacteria improve their survival rate as compared to free cells. COS has inhibitory effect on fermentation of sour cream and post-acidification during cold storage of set yoghurt and it also affects viscosity of sour cream according to their added concentration (Kerch et al. 2015). In fermented food, COS use and incorporation is limited due to their anti-microbial properties. However, in a study it was shown that COS at concentration higher than 0.1% only diminishes

Table 3 Summary of conjugated/derivatised COS and reported health-related bioactivities

COS derivatives/conjugates	Disease	Mechanism/remarks	Reference
Caffeic acid conjugated COS	Alzheimer's disease	β-Secretase inhibitor	(Eom et al. 2012a)
CMC-COS NP SC-COS NP	Immuno-modulatory	Regulate both humoral and cell-mediated immune response	(Lin et al. 2014)
AE-COS	Anti-oxidant and anti-inflammatory activity in murine microglial cells (BV-2)	Can be used as therapeutic agent for treatment of neuroinflammatory diseases	(Ngo et al. 2019)
AE-COS	Inhibitor against cell invasion associated with metastasis	Can be used for Chemoprevention	(Hong et al. 2016)
Carboxylated-COS (low mol wt)	MMP-9 expression control due to down regulation	Inhibition of tumour invasiveness	(Rajapakse et al. 2006)
Gallyl COS (G-COS)	Inhibitory effect on intracellular free radical generation	Anti-oxidant	(Ngo et al. 2011)
Gallyl COS (G-COS)	Free radical scavenging	Anti-oxidant	(Ngo et al. 2011)
Gallyl COS (G-COS)	Inhibited the release of Histamine	Anti-allergic	(Vo et al. 2012)
Sulfated-COS	Studied on H2O2-induced oxidative damage in Chang cells	Protective effect on liver and DNA against oxidative stress	(Lee et al. 2009)
Sulfated-COS	Binding on active site of ACE and competing with natural substrates	ACE inhibitory enzyme	(Qian et al. 2010)

bacterial acidification (Gurovic et al. 2015) thus indicating maximum concentration which could be added to the yoghurt. Significant difference between nutritional composition of control and yoghurt supplemented with COS up to 0.1% was not found, but milk acidification decreased at 0.5% and yoghurt formation failed at the 3.0% COS concentration without change in viable count. Chemical digestion was no different in the COS supplemented yoghurt as fatty acid profiles and CLA (conjugated linoleic acid) was found to be same in both control and COS supplemented yoghurt. Anti-inflammatory activity of COS was confirmed and was recommended as potential nutraceuticals and as functional foods (Santos-Moriano et al. 2019). Anti-inflammatory activities of three various mixtures of COS against TNF-α in murine macrophages (RAW 264.7) was checked after stimulation with mixture of lipopolysaccharides. COS with more than 90% DAC displayed (5–10 Kda) higher anti-inflammatory effect than those of 50%. COS from shrimps, crabs using method

Table 4 Chitooligosaccharides as functional foods or nutraceuticals in various food products

COS bioactivity	COS processing/form	Food products	Reference
COS enzymatically produced using *Bacillus cereus* chitosanase Prebiotic/anti-oxidant	Nanoparticle/microcapsule	Yoghurt	(Ismail et al. 2020)
COS (mol wt 1,5,30, 120 Kda) anti-microbial/anti-oxidant	COS dissolved in lactic acid added to dough at various concentration	Bread	(Lee et al. 2002)
Chitosan of various mol wt, anti-microbial	Coating of COS dissolved in acetic acid	Egg	(Caner. 2005)
Chitosan of various Mol wt, nutraceuticals/anti-microbial	Coating of COS dissolved in acetic acid with calcium, Vit E	Fruits and Vegetables	(Han et al. 2004)
Acetic acid hydrolysed water-soluble chitosan Anti-microbial/anti-oxidant	Decreased turbidity and controlled acidity	Fruit Juice (apple, grapes, lemon and orange juice)	(Chatterjee et al. 2004)
Chitosan (37 Kda) dissolved in Acetic acid Anti-microbial/anti-oxidant	Improved shelf life and quality	Noodle	(Lee and No 2002)

of deacetylation and chemical/enzymatic (papain and lysozyme) depolymerisation are produced and commercialised by several companies like Qingdao BZ-Oligo Co, Ltd, BioCHOS, AMSBIO.

Qingdao BZ-Oligo Co., Ltd produces COS from DP 2–10. Enzymatic hydrolysis with chemical derivatisation and column chromatography is used for obtaining COS (Muzzarelli et al. 2012). Controlled enzymatic degradation of chitosan has been used for preparation of BioCHOS (Dilokpimol et al. 2011). AMSBIO is chitosan oligomer from dimer to hexamer prepared from crab shell. Chromatographically pure COS (not less than 98%) were confirmed by HPLC (Jeon and Kim 2000).

5 Overview and Conclusion

COS functions and applications depends generally on size, DP and DA. It has already been established that molecular weight of generated COS depends upon substrate-enzyme synergism (Gutierrez-Roman et al. 2014). Three chitolytic enzymes were tested alone and in combination and additionally, three enzymes were tested in synergism with chitobiase and a non-catalytic binding protein. All combinations gave COS which were different in composition, however highest concentration was obtained when four enzymes were tested along with non-catalytic binding protein. Based on this study, it was suggested that studies on genetic modifications are required for over-production of enzymes along with non-catalytic binding protein which may have great impact on productivity of process along with improved quality of oligomers obtained. Finding novel source of enzymes through isolation and screening from extremophiles organism is one of the strategies which is being suggested for using agro-industrial or marine waste as starting material for COS which can work under non-physiological conditions such as high temperature, salinity and in low water activity conditions. Bioreactor designing using immobilised enzymes and packed bed reactor can ensure continuous production, high productivity and lower cost of product while looking for food grade COS for industrial level production. Lot of progress has been made in the research related to "COS acting as an anti-oxidant" but still industries have not used them in food, pharma, and cosmetics at larger scale. COS as promising anti-oxidant have been overshadowed by their non-availability on bigger scale. Obtaining COS in highly purified conditions in bulk is still appears to be difficult. Getting purified fraction of COS which has been produced bio-catalytically in terms of defined DP and exact known sequence of monomer (both acetylated and deacetylated) should have not been found feasible till date. Hetero-COS with single polymerisation (DP) still may be a mixture of various kinds of sequences and is difficult to determine like reported bioactivity is due to presence of which structure present in the mixture. However, recently chemical and enzymatic synthesis of these molecules along with separation techniques have been reviewed along with biological activities (Hao et al. 2021).

Studies to establish anti-oxidant activity and relation to degree of deacetylation/polymerisation and sequence in order to predict the connection between biological activity and their structure are needed. Most of In vitro and in vivo activity of COS have been done in COS mixture and they are not characterised properly, which is required for their establishment as an anti-oxidant. Biocatalytic COS production with improved yield and cost effectiveness are desirable for large-scale COS production. Downstream processing leading to purification/separation of COS with single DP and their In vitro and In vivo investigations are important for establishing as anti-oxidants. Anti-microbial, anti-oxidant activity, neuro-protective effect and anti-tumour effect of COS has been implicated in their uses as nutraceuticals and functional foods (Park et al. 2004; Huang et al. 2015; El-Sayed et al. 2017). Recently COS nanoparticles and their conjugates have been shown a promising drug delivery agent (Lu et al. 2015; Xu et al. 2016), but more investigations are needed for evaluation of anti-oxidant potential of nanoparticulate forms of COS.

Acknowledgements Corresponding author acknowledges Department of Science & Technology, Government of India for financial support vide reference no (SR/WOS-A/LS-1004/2015) and (SR/WOS-A/LS-129/2009) under Women Scientist Scheme.

Conflicts of Interest The author declares no conflict of interest.

References

Aam BB, Heggset EB, Norberg AL, Sørlie M, Vårum KM, Eijsink VGH (2010) Production of chitooligosaccharides and their potential applications in medicine. Mar Drugs 8:1482–1517. https://doi.org/10.3390/md8051482

Ando T, Kataoka S (1980) Acylations of chitin with acid anhydrides in trichloroacetic acid systems. Kobunshi Ronbunshu 37:1–7. https://doi.org/10.1295/koron.37.1

Bockuviene A, Zalneravicius R, Sereikaite J (2021) Preparation, characterization and stability investigation of lycopene-chitooligosaccharides complexes. Food Biosc 40:100854

Caner C (2005) The effect of edible eggshell coatings on egg quality and consumer perception. J Sci Food Agric 85:1897–1902

Chatterjee S, Chatterjee S, Chatterjee BP, Guha AK (2004) Clarification of fruit juice with chitosan. Process Biochem 39(12):2229–2232

Chen XG Liu CS Liu CG et al (2002) Manufacture of chitin microspheres. Chinese Patent No. ZL 96115714.3

Chiu CY, Yen TE, Liu SH, Chiang MT (2020) Comparative effects and mechanisms of chitosan and its derivatives on Hypercholesterolemia in high-fat diet-fed rats. Int J of Mol Sci 21(1):92

Choi BK, Kim KY, Yoo YJ, Oh SJ, Choi JH, Kim CY (2001) *In vitro* antimicrobial activity of a chitooligosacchride mixture against *Actinobacillus actinomycetemcomitans* and *Streptococcus mutans*. Int J Antimicrobial Agent 18(6):553–557

Choi CR, Kim EK, Kim YS, Je JY, An SH, Lee JD, Wang JH, Ki SS, Jeon BT, Moon SH, Park PJ (2011) Chitooligosaccharides decreases plasma lipid levels in healthy men. Int J Food Sci Nutri 63:103–106

Choi WS, Ahn KJ, Lee DW, Byun MW, Park HJ (2002) Preparation of chitosan oligomers by irradiation. Polym Degrad Stability 78(3):533–538. https://doi.org/10.1016/S0141-3910(02)00226-4

Chung Y, Su Y, Chen C, Jia G, Wang H, Wu JC, Lin J (2004) Relationship between antibacterial activity of chitosan and surface characteristic of cell wall. Acta Pharmaco Sinica 25(7):932–936

Chung MJ, Park JK, Park Y (2012) Anti-inflammatory effects of low-molecular weight chitosan oligosaccharides in IgE-antigen complex -stimulated RBL-2H3 cells and asthma model mice. Int Immunopharmacol 12(2):453–459

Dilokpimol A, Nakai H, Gotfredsen CH, Appledoorn M, Baumann MJ, Nakai N, Schols HA, Hachem MA, Svensson B (2011) Enzymatic synthesis of b-xylosyl-oligosaccharides by transxylosylation using two b-xylosidases of glycoside hydrolase family 3 from *Aspergillus nidulans* FGSCA4. Carohydr Res 346:421–429. https://doi.org/10.1016/j.carres.2010.12.010

El-Sayed ST, Omar NI, El Sayed ESM, Shousha WG (2017) Evaluation antioxidant and cytotoxic activities of novel chitooligosaccharides prepared from chitosan via enzymatic hydrolysis and ultrafiltration. J Appl Pharm Sci 7:50–55. https://doi.org/10.7324/JAPS.2017.71107

El-Sayed ST, Ali AM, Omar NI (2019) A comparative evaluation of antimicrobial activity of chitooligosaccharides with broad spectrum antibiotics on growth of some pathogenic microorganisms. Biocatal Agric Biotechnol 22:101382

Eom TK, Senevirathne M, Kim SK (2012a) Synthesis of phenolic acid conjugated chitooligosaccharides and evaluation of their antioxidant activity. Environ Toxicol Pharmacol 34:519–527. https://doi.org/10.1016/j.etap.2012.05.004

Eom TK, Ryu BM, Lee JK, Byun HG, Park SJ, Kim SK (2012b) β-Secretase inhibitory activity of phenolic acid conjugated chitooligosaccharides. J Enz Inhibi Med Chem 28:214–217

Fernandes JC, Eaton P, Nascimento H, Giao MS, Ramos OS, Belo L, Silva AS, Pintado ME, Malcata FX (2010) Antioxidant activity of chitooligosaccharides upon two biological systems: erythrocytes and bacteriophages. Carbohyr Polym 79(4):1101–1106

Gurovic MSV, Staffolo MD, Montero M, Debbaudt A, Albertengo L, Rodriguez MS (2015) Chitooligosaccharides as novel ingredients of fermented foods. Food Funct 6:3437–3443

Gutierrez-Roman MI, Dunn MF, Tinoco-Valencia R, Holguin-Melendez F, Huerta-Palacios G, Guillen-Navarro K (2014) Potentiation of the synergistic activities of chitinases Chi A, Chi B

and Chi C from *Serratia marcescens* CFFSUR-B2 by chitobiase (Chb) and chitin binding protein (CBP). World J Microbiol Biotechnol 30:33–42. https://doi.org/10.1007/s11274-013-1421-2
Guzman J, Saucedo I, Revilla J, Navarro R, Guibal E (2003) Copper sorption by chitosan in the presence of citrate ions: influence of metal speciation on sorption mechanism and uptake capacities. Int J Biol Macromol 33:57–65
Han C, Zhao Y, Leonard SW, andTraber MG, (2004) Edible coatings to improve storability and enhance nutritional value of fresh and frozen strawberries (Fragaria × ananassa) and raspberries (Rubus ideaus). Postharvest Biol Technol 33:67–78
Hao W, Li K and Li P(2021) Review: advances in preparation of chitooligosaccharides with heterogeneous sequences and their bioactivity. Carbohydr Polym 252:117206
He B, Wang J (2015) Chitooligosaccharides prevent osteopenia by promoting bone formation and suppressing bone resorption in overiectomised rats: possible involvement of COX-2. Nat Prod Res 29(4):359–362. https://doi.org/10.1080/14786419.2014.942301
Hirano S, Nagao N (1989) Effects of chitosan, pectic acid, lysozyme and chitinase on the growth of several phytopathogens. Agric Biological Chem 53:3065–3066
Hong SP, Kim MH, Oh SW, Han CH, Kim YH (1998) ACE inhibitory and antihypertensive effect of chitosan oligosaccharides in SHR. Korean J Food Sci Technol 30:1476–1479
Hong S, Ngo DN, Kim MM (2016) Inhibitory effect of Aminoethyl-chitooligosaccharides on invasion of human fibrosarcoma cells. Environ Toxicol Pharmacol 45:309–314
Huang R, Rajapakse N, Kim SK (2006) Structural factors affecting radical scavenging activity of chitooligosaccharides (COS) and its derivatives. Carbohydr Polym 63:122–129. https://doi.org/10.1016/j.carbpol.2005.08.022
Huang HC, Hong L, Chang P, Zhang J, Lu SY, Zheng BW, Jiang ZF (2015) Chitooligosaccharides attenuate Cu2 ± induced cellular oxidative damage and cell apoptosis involving Nrf2 activation. Neurotox Res 27:411–420. https://doi.org/10.1007/s12640-014-9512-x
https://doi.org/10.1023/B:ABIM.0000025956.98250.30
Ismail SA, El-Sayed HS, Fayed B (2020) Production of prebiotic chitooligosacchrides and its nano/microencapsulation for the production of functional yoghurt. Carbohydr Polym 234:115941
Jafari H, Bernaerts KV, Dodi G, Shavandi A (2020) Chitooligosaccharides for wound healing biomaterials engineering. Mat Sci Eng 117:111266
Jaiswal S, Tripathi P, Sinha S (2019) Biocatalytic production of hetero-chitosan oligosaccharides as an anti-oxidant. Sustain Agric Rev 35:125–146. Springer Nature
Jameela SR, Misra A, Jayakrishnan A (1994) Cross-linked chitosan microspheres as carriers for prolonged delivery of macromolecular drugs. J Biomater Sci Polym Ed 6:621–632. https://doi.org/10.1163/156856294x00563
Je JY, Park PJ, Kim SK (2004) Free radical scavenging properties of hetero-chitooligosaccharides using an ESR spectroscopy. Food Chem Toxicol 42:381–387. https://doi.org/10.1016/j.fct.2003.10.001
Jeon YJ, Kim SK (2000) Production of chitooligosaccharides using an ultrafiltration membrane reactor and their anti-bacterial activity. Carbohydr Polym 41:133–141
Jeon YJ, Kim SK (2001) Effect of antimicrobial activity by chitosan oligosaccharides N-conjugated with asparagine. J Microbio and Biotechnol 11(2):281–286
Jeon YJ, Kim SK (2002) Antitumor activity of chitosan oligosaccharides produced in an ultrafiltration membrane reactor system. J Microbiol Biotechnol 12:503–507
Jeong S, Cho JM, Kwon Y, Kim SC, Shin DY, Lee JH (2019) Chitosan oligosaccharides (GO2KA1) improves postpardinal glycemic response in subjects with impaired glucose tolerance and impaired fasting glucose and in healthy subjects: a crossover, randomized controlled trial. Nutr Diabetes 9:31
Ji X, Chen M, Zhao M, Song Y, Lin Y, Yin H, Zhao L (2021) Effect of chitooligosaccharides on the rebalance of gut microorganisms and their metabolites in patients with nonalcoholic fatty liver disease. J Func Food 77:104333

Jiang Z, Liu G, Yang Y, Shao K, Wang Y, Liu W, Han B (2019) N-Acetyl chitooligosaccharides attenuate amyloid β-induced damage in animal and cell models of Alzheimer's disease. Process Biochem 84:161–171

Kanauchi O, Deuchi K, Imasato Y, Shizukuishi M, Kobayashi E (1995) Mechanism for the inhibition of fat digestion by chitosan and for the synergistic effect of ascorbate. Biosci Biotechnol Biochem 59:786–790

Kerch G, Zicans J, Merijs Meri R, Stunda-Ramava A, Jakobsons E (2015) The use of thermal analysis in assessing the effect of bound water content and substrate rigidity on prevention of platelet adhesion. J Therm Anal Calorim 120:533–539

Kim SK, Rajapakse N (2005) Review: enzymatic production and biological activities of chitosan oligosaccharides (COS). Carbohydr Polym 62:357–368. https://doi.org/10.1016/j.foodchem.2006.01.038

Kim JY, Lee JK, Lee TS (2003) Synthesis of chitooligosaccharide derivative with quaternary ammonium group and its antimicrobial activity against streptococcus mutans. Int J Biol Macromol 32(1–2):23–27

Kochkina ZM, Chirkov SN (2000) Effect of chitosan derivatives on the development of phage infection in culture *Bacillus thuringiensis*. Microbiologiia 69:266–269

Laokuldilok T, Potivas T, Kanha N, Surawang S, Seesuriyachan P, Wangtueai S, Phimolsiripol Y, Regenstein JM (2017) Physicochemical, antioxidant and antimicrobial properties of chitooligosaccharides produced using three different enzyme treatments. Food Biosc 18:28–33

Lee MH, No HK (2002) Effect of chitosan on shelf-life and quality of wet noodle. J Chitin Chitosan 7(1):14–17

Lee HY, Kim SM, Kim JY, Youn SK, Choi JS, Park SM, Ahn DH (2002) Effect of addition of chitosan on improvement for shelf-life of bread. J Korean Soc Food Sci Nutr 31(3):445–450

Lee SJ, Kim EK, Kim J, Hwang W (2009) Antioxidative effects of sulfated chitooligosaccharides on oxidative injury. J Chitin Chitosan Sci 14(4):192–196

Li X, Liu BO, Wang X, Han Y, Su H, Zeng X (2012) Synthesis, characterization and antioxidant activity of quaternized carboxymethyl chitosan oligosaccharides. J Macromol Sci Part A 49:861–868. https://doi.org/10.1080/10601325.2012.714679

Liaqat F, Rengin E (2018) Review: chitooligosaccharides and their biological activities. Carbohydr Polym 184:243–259. https://doi.org/10.1016/j.carbpol.2017.12.067

Lin YS, Wu MF, Takamori Y, Okamoto Y, Minami S (2014) *In vivo* modulatory effects of chitooligosaccharides nanoparticles on mouse serum cytokines and splenocytes. Jl of Exp Nanosci 9:860–870

Liu SH, Chen RY, Chiang MT (2021) Effects and mechanism of chitosan and chitosan oligosaccharides on hepatic lipogenesis and lipid peroxidation, adipose lipolysis, and intestinal lipid adsorption in rats with high fat diet induced obesity. Int J Mol Sci 22:1139

Liu W, Li X, Zhao Z, Pi X, Meng Y, Fei D, Liu D, Wang X (2020) Effect of chitooligosaccharides on human gut microbiota and antiglycation. Carb Polym 242:116413

Lu C, Park MK, Lu C, Lee YH, Chai KY (2015) A mussel-inspired chitooligosaccharide based multidentate ligand for highly stabilized nanoparticles. J Mater Chem B 3:3730–3737. https://doi.org/10.1039/c5tb00114e

Mallakuntala M K, Penugurti V, Manavathi B, Podile AR (2021) Chitooligosaccharides induce apoptosis in human breast cancer cells. Carb Polymer Tech Appli 2:100077

Matsuo K, Miyazono I (1993) The influence of long-term administration of peptidoglycan on disease resistance and growth of juvenile rainbow trout. Nippon Suisan Gakkaishi 59:1377–1379

Mourya VK, Inamdar NN (2009) Trimethyl chitosan and its applications in drug delivery. J Mater Sci Mater Med 20:1057–1079. https://doi.org/10.1007/s10856-008-3659-z

Mourya VK, Inamdar NN, Choudhari YM (2011) Chitooligosaccharides: synthesis, characterization and applications. Polym Sci Ser A 53:583–612. https://doi.org/10.1134/S0965545X11070066

Muzzarelli RAA (1996) Chitosan based dietary food. Carbohydr Polym 29(4):309–316

Muzzarelli RAA (1997) Human enzymatic activities related to the therapeutical administration of chitin derivatives. Cell Mol Life Sci 53:131–140

Muzzarelli RAA, Boudrant J, Meyer D, Manno N, De Marchis M, Paoletti MG (2012) Current views on fungal chitin/chitosan human chitinases, food preservation, glucans, pectins and inulin: a tribute to Henri Braconnot, precursor of the carbohydrate polymer sciences, on the chitin bicentennial. Carbohydr Polym 87:995–1012. https://doi.org/10.1016/j.carbpol.2011.09.063

Ngo DH, Ngo DN, Kim SK, Vo TS (2019) Antiproliferative effect of aminoethyl-chitooligosacchride on human lung A549 cancer cells. Biomolecules 9(5):195

Ngo DH, Qian ZJ, Ngo DN, Vo TS, Vo IW, Kim S K (2011) Gallyl chitooligosaccharides inhibit intracellular free radical-mediated oxidation .Food Chem 128(4):974–981

No HK, Meyers SP, Prinyawiwatkul W, Xu Z (2007) Applications of chitosan for improvement of quality and shelf life of foods A review. J Food Sci 72:R87–R100

Santos-Moriano P, Fernandez-Arrojo L, Mengibar M, Belmonte-Reche E, Peñalver P, Acosta FN et al. (2017) Enzymatic production of fully deacetylated chitooligosaccharides and their neuroprotective and anti-inflammatory properties. Biocatal Biotranform 36:57–67

Santos-Moriano P, Kidibule P, Míguez N, Fernandez-Arrojo L, Ballesteros AO, Fernández-Lobato M, Plou FJ (2019) Tailored enzymatic synthesis of chitooligosaccharides with different deacetylation degrees and their anti-inflammatory activity. Catalysts 9:405. https://doi.org/10.3390/catal9050405

Park PJ, Je JY, Kim SK (2003) Free radical scavenging activity of chitooligosaccharides by electron spin resonance spectrometry. J Agric Food Chem 51:4624–4627. https://doi.org/10.1021/jf034039

Park PJ, Lee HK, Kim SK (2004) Preparation of hetero-chitooligosaccharides and their antimicrobial activity on *Vibrio parahaemolyticus*. J Microbiol Biotechnol 14:41–47

Popa-Nita S, Lucas JM, Ladaviere C, David L, Domard A (2009) Mechanisms involved during the ultrasonically induced depolymerization of Chitosan: characterization and control. Biomacromolecules 10(5):1203–1211. https://doi.org/10.1021/bm8014472

Qian ZJ, Eom TK, Ryu B (2010) Angiotensin I-converting enzyme inhibitory activity of sulfated chitooligosaccharides with different molecular weights. J Chitin Chitosan Sci 15(2):75–79

Rafael OHD, Fernándo ZGL, Abraham PT, Alberto VLP, Guadalupe GS, Pablo PG (2019) Production of chitosan-oligosaccharides by the chitin-hydrolytic system of Trichoderma harzianum and their antimicrobial and anticancer effects. Carbohydr Res 486

Rajapakse N, Kim MM, Mendis E, Kim SK (2006) Inhibition of free radical-mediated oxidation of cellular biomolecules by carboxylated chitooligosaccharides. Bioorg Med Chem 15(2):997–1003

Remunan-Lopez C, Portero A, Vila-Jato JL, Alonso MJ (1998) Design and evaluation of chitosan/ethylcellulose mucoadhesive bilayered devices for buccal drug delivery. J Control Release 55:143–152. https://doi.org/10.1016/s0168-3659(98)00044-3

Sanchez A, Mengibar M, Rodriguez GR, Moerchbacher B, Acosta N, Heras A (2017) The effect of preparation processes on the physicochemical characteristics and antibacterial activity of chitooligosaccharides. Carbohydr Polym 157:251–257

Shahidi F, Arachchi JKV, Jeon YJ (1999) Food application of chitin and chitosanas. Trend Food Sci Tech 10:37–51

Simunek J, Koppova I, Filip L, Tishchenko G, Belezcki, (2010) The antimicrobial action of low molecular mass chitosan, chitosan derivatives and chitooligosaccharides on Bifidobacteria. Folia Microbiol 55(4):379–382

Sinha S, Tripathi P, Chand S (2012) A new bifunctional chitosanase enzyme from *Streptomyces* sp. and its application in production of antioxidant chitooligosaccharides. Appl Biochem Biotechnol 167:1029–1039. https://doi.org/10.1007/s12010-012-9546-6

Sinha S, Chand S, Tripathi P (2014) Microbial degradation of chitin waste for production of chitosanase and food related bioactive compounds. Appl Biochem Microbiol 50(2):125–133

Sugano M, Watanabe S, Kishi A, Izume M, Ohtakara A (1988) Hypocholesterolemic action of chitosans with different viscosity in rats. Lipids 23:187–191

Tanaka Y, Tanioka S, Tanaka M, Tanigawa T, Kitamura Y, Minami S (1997) Effects of chitin and chitosan particles on BALB/c mice by oral and parenteral administration. Biomaterials 18:591–595. https://doi.org/10.1016/s0142-9612(96)00182-2

Uchida Y, Izume M and Ohtakara A (1989) Preparation of chitosan oligomers with purified chitosanase and its application. In: Skjak-Bræk G, Anthonsen T, Sandford P (Eds) Chitin and Chitosan Elsevier Essex 373–382

Vo TS, Ngo DH, Kim SK (2012) Gallic acid-grafted chitooligosaccharides suppress antigen-induced allergic reactions in RBL-2H3 mast cells. Eur J Pharma Sci 47(2):527–533

Vo TS, Ngo DH, Bach LG, Ngo DN, Kim SK (2017) The free radical scavenging and anti-inflammatory activities of gallate-chitooligosaccharides in human lung epithelial A549 cells. Process Biochem 54

Wei X, Wang Y, Zhu Q, Xiao J, Xia W (2009) Effect of chitosan pentamer and chitosan hexamer *in vivo* and *in vitro* on gene expression and secretion of cytokines. Food Agric Immunol 20:269–280

Xia W, Liu P, Zhang J, Chen J (2011) Biological activities of chitosan and chitooligosaccharides. Food Hydrocoll 25(2):170–179

Xu Y, Wang L, Li YK, Wang CQ (2016) Reduction and pH dual-responsive nanoparticles based chitooligosaccharide-based graft copolymer for doxorubicin delivery. Coll Surf Physicochem Eng Asp 497:8–15. https://doi.org/10.1016/j.colsurfa.2016.01.049

Yamaguchi R, Arai Y, Itoh T (1982) A microfibril formation from depolymerized chitosan by n-acetylation. Agric Biol Chem 46:2379–2381. https://doi.org/10.1080/00021369.1982.10865442

Yuan W (2009) Antioxidant activity of chito-oligosaccharides on pancreatic islet cells in streptozotocin-induced diabetes in rats. World J Gastroenterol 15(11):1339. https://doi.org/10.3748/wjg.15.1339

Zhai X, Li C, Ren D, Wang J, Ma C, El-Aty A MA (2021) The impact of chitooligosaccharides and their derivatives on the *in vitro* and *in vivo* antitumour activity: a comprehensive review. Carbohydr Polym 266:118132

Zhao M, Gu L, Li Y, Chen S, You J, Fan L, Wang Y, Zhao L (2019) Chitooligosaccharides display anti-tumor effects against human cervical cancer cells via the apoptotic and autophagic pathways. Carbohydr Polym 224:115171. https://doi.org/10.1016/j.carbpol.2019.115171

Zheng J, Yuan X, Cheng G, Jiao S, Feng C, Zhao X, Yin H, Du Y, Liu H (2018) Chitosan oligosaccharides improves the disturbance in glucose metabolism and reverse the dysbiosis of gut microbiota in diabetic mice. Carbohydr Polym 190:77–86

Zhou Y, Li S, Li D, Wang S, Zhao W, Lv Z, Li X, Li H, Han Y (2020) Enzymatic preparation of chitooligosaccharides and their anti-obesity application. Food Nutri Sci 84:1460–1466

Chitosan Oligosaccharide-Based Nanoparticle Delivery Systems for Medical Applications

Long Binh Vong, Nhu-Thuy Trinh, Van Toi Vo, and Dai-Nghiep Ngo

1 Introduction

Nanoparticle-based drug delivery systems have recently been developed to overcome the drawbacks of low molecular weight drugs by improving the stability, bioavailability, specific delivery, and controlled release of cargo molecules at the targeted tissues. Recently, the biomedical applications of nanoparticles have expanded into a broad range and the list of approved nanoparticles used in clinical practice has dramatically increased (Patra et al. 2018; Mitchell et al. 2021). Versatile nanoparticles including lipid-based nanoparticles, polymeric nanoparticles, and inorganic nanoparticles have been developed in medical applications for the past decades. Among them, polymeric nanoparticles have gained attraction since the wide range of characteristics and structures of various polymers from both nature and synthesis. In addition, the polymers can be easily modified to achieve the desired features as delivery nano-carriers. Possessing a hydrophilic surface and size of 10–100 nm, the polymeric micelles can escape the engulfment by the mononuclear phagocyte system to prolong blood circulation and systemically deliver a drug to the targeted tissues (Gaucher et al. 2005; Sung and Kim 2020). The polymers can also be grafted and blocked with hydrophobic molecules or polymers to create the versatile amphiphilic copolymers, which form stable nanoparticles composed of a hydrophobic core and a hydrophilic shell in an aqueous environment (Vong et al 2012). Poorly water-soluble drugs can be encapsulated in the internal hydrophobic cavity, which can improve their solubility and bioavailability (Vong et al. 2017; Nguyen et al. 2021). Polymeric

L. B. Vong (✉) · N.-T. Trinh · V. T. Vo
School of Biomedical Engineering, International University, Ho Chi Minh 700000, Vietnam
e-mail: vblong@hcmiu.edu.vn

L. B. Vong · N.-T. Trinh · V. T. Vo · D.-N. Ngo
Vietnam National University-Ho Chi Minh City (VNU-HCM), Ho Chi Minh 700000, Vietnam

D.-N. Ngo
Faculty of Biology and Biotechnology, University of Science, Ho Chi Minh 703000, Vietnam

S.-K. Kim (ed.), *Chitooligosaccharides*,
https://doi.org/10.1007/978-3-030-92806-3_10

nanoparticles are demonstrating high potential as nanomedicines capable of controlling the distribution and function of loaded bioactive agents in the body, effectively overcoming biological barriers, and various formulations are engaged in intensive preclinical and clinical testing (Cabral et al. 2018; Kataoka et al. 2001).

Chitosan is the linear polysaccharide obtained by the *N*-deacetylation of chitin, which is the second most abundant natural polymer. Chitosan is widely studied and applied in various fields including agriculture, tissue engineering, medicine, cosmetics, environment, etc. (Morin-Crini et al. 2019; Ghadi et al. 2014). Chitosan is a cationic polymer of glucosamine monomer with amino and hydroxyl functional groups, which have been used for chemical modification to generate numerous chitosan derivatives with different characteristics. However, due to low water solubility, poor mechanical property, and high viscosity, the biomedical applications of chitosan in drug delivery are significantly limited (Hu et al. 2013; Li et al. 2018). Chitosan oligomers (CSO) or chitooligosaccharides (COS) are the degraded products of chitosan and prepared by enzymatic or chemical hydrolysis of chitosan. Due to low molecular weight (about 0.5–5 kDa), CSO possesses an excellent water solubility and low viscosity, indicating that CSO and its derivatives have recently attracted considerable interest in biomedical applications and drug delivery (Fig. 1) (Lodhi et al. 2014). Under mild reaction conditions, the primary amino groups on the backbone of CSO provide reactive sites for further chemical modifications. CSO exhibits favorable pharmacological properties such as good pharmacokinetics and highly potential therapeutic activities in various diseases including inflammation, cancer, obesity, diabetes, etc. (Muanprasat and Chatsudthipong 2017; Huang et al. 2015; Kim et al. 2014a, b; Zhao et al. 2019). It has also been reported that pharmacokinetics and bioactivities of CSO are strongly associated with its structure and molecular weight (Zhou et al. 2021). In this chapter, recently CSO-based drug and gene delivery nano-carriers are reviewed and discussed.

2 CSO-Based Drug Delivery Nanoparticles for Cancer Therapy

Cancer is an intractable disease and a leading cause of mortality globally. The studies of drug delivery nano-carriers have been extensively employed in recent years as a promising approach for the treatment of cancer. It has been reported that nanoparticles with the diameter size of several tens to hundreds of nanometers tend to passively accumulate in the solid tumors, so-called enhanced permeability and retention (EPR) effect (Maeda et al. 2000; Rosenblum et al. 2018). However, due to the heterogeneity of cancer tissues, the EPR effect of the nanoparticle is highly impacted in different tumors and stages of tumors (Chauhan and Jain 2013). On the other hand, active targeting approaches involve using the affinity ligands conjugated on the surface of nanoparticles to improve the retention and cellular uptake. In addition, polymeric nano-carriers can trigger the release of encapsulated cargos by the response with

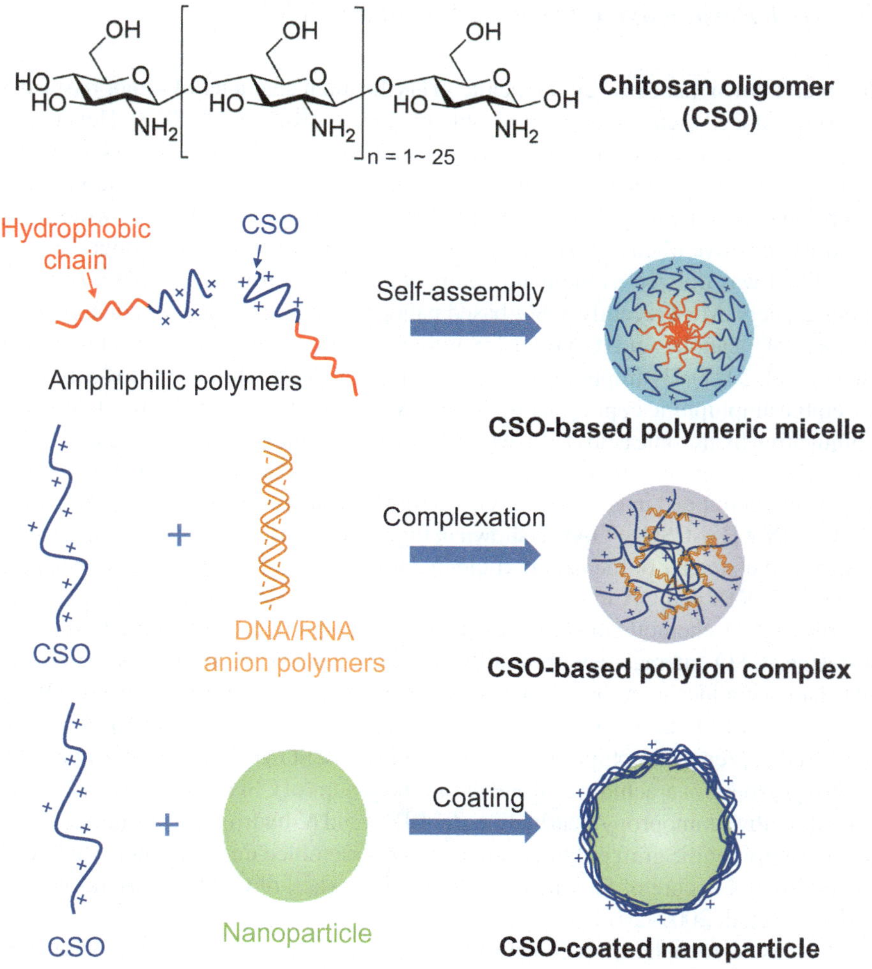

Fig. 1 Chitosan oligomer (CSO) and formations of CSO-based nanoparticles for drug and gene delivery

tumor microenvironmental stimuli to control drug distribution. Here, we discussed several stimuli-responsive CSO-based nanoparticles been designed and investigated in the application of cancer treatment.

2.1 pH-Responsive CSO-Based Nanoparticles

It is well known that extracellular acidic pH is the main character of tumor tissue due to overproduced lactic acid via anaerobic glycolysis (Kato et al. 2013). Hence, pH-responsive nanoparticles have been widely studied in cancer therapy, which significantly improves the release of drugs intratumorally. Since the amino group of CSO is protonated at acidic pH conditions, the surface charge of CSO can switch from neutral to positive when the pH changes from neutral or alkaline to acid values. Therefore, CSO was utilized as the main component for the formation of pH-responsive nano-carriers. Most recently, CSO-based nanoparticle was designed as dual prodrug delivery of oleanolic acid (OA) and doxorubicin (DOX) for targeting cancer treatment (Chen et al. 2021). Hydrophobic OA was grafted with hydrophilic CSO to form a self-assemble amphiphilic copolymer (CSO-*g*-OA). Targeting tumors can be obtained by conjugating hyaluronic acid (HA), which has a high affinity with overexpressed CD44 in various cancer cells (Chen et al. 2019; Luo et al. 2020). Prodrug HA-DOX was physically entrapped in HA-CSO-*g*-OA to form a dual prodrug delivery nano-carrier (HA-DOX@HA-CSO-*g*-OA), as shown in Fig. 2a. Due to the EPR effect and acid pH-responsive character of the nanoparticle, it could effectively control the intracellular release of DOX and OA in the tumor tissues (Fig. 2b), suppressing the non-specific distribution of chemotherapeutic agents and adverse effects in healthy tissues. As a design, CSO interacted with vanillin to form benzoic imine bonds, where the pH change could trigger the release of hydrophobic anticancer drug genistein (Wang et al. 2019). In addition, other fatty acids such as arachidic acid and linoleic acid were also used to prepare amphiphilic graft polymer with CSO via interaction between the carboxyl groups of arachidic acid and the amino groups of CSO in presence of 1-ethyl-3-(3-dimethylaminopropyl) carbodiimide (EDC) and *N*-hydroxysuccinimide (NHS), and the amphiphilic graft polymer can form self-assembled micelle nanoparticle with the hydrophobic core of fatty acid and hydrophilic shell of CSO (Termsarasab et al. 2013; Du et al. 2009, 2011a).

Besides, α-tocopherol succinate (α-TS), a derivative of vitamin E with anti-cancer potentials via induction of DNA synthesis arrest and apoptosis, was also used for grafting with CSO forming micelle for intracellular codelivery of PTX. The result showed that α-TS/PTX-CSO micelles exhibited higher anticancer activity as compared to PTX-loaded CSO micelles due to the synergistic effect of α-TS and PTX (Emami et al. 2015; Chen et al. 2017). Folic acid-modified CSO encapsulating DOX and gold nanorod was demonstrated as an effective combination of chemotherapy and photothermal therapy (Manivasagan et al. 2019). In this nano-system, folic acid was modified to enhance tumor targeting effect, CSO improved the stability and biocompatibility of nano-carrier, while nanorod was used for photoacoustic imaging, the laser-triggered release of DOX, and photothermal therapy. CSO also stabilized gold nanoparticles loading paclitaxel (PTX) and induced toxicity against cancer cells via the generation of reactive oxygen species (Manivasagan et al. 2016). The multifunctional nano-carriers were considered as a promising approach to improve anticancer efficacy and minimizing side effects.

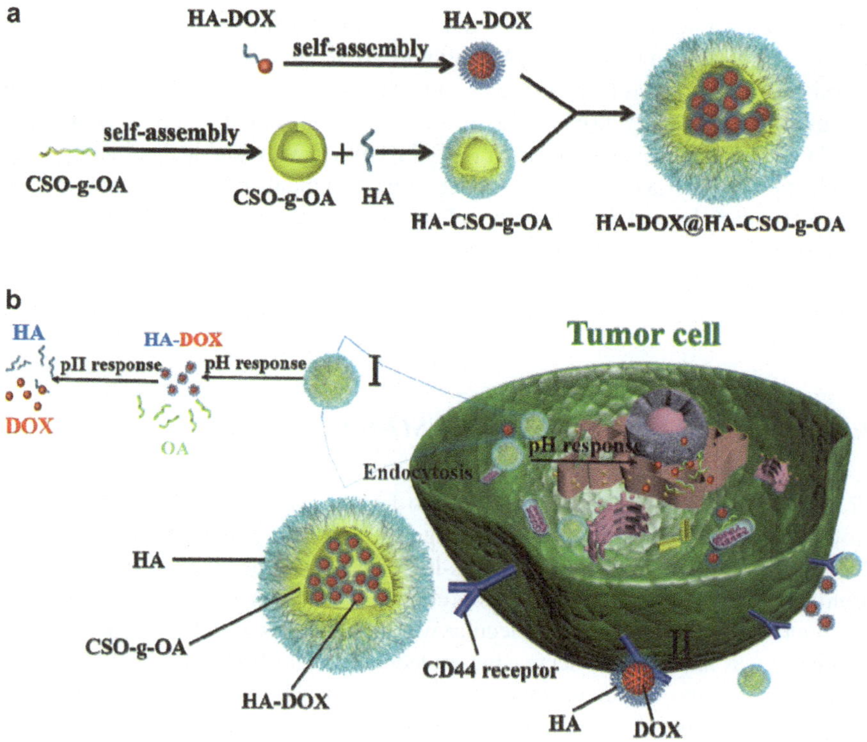

Fig. 2 Dual-prodrug delivery system CD44 targeted tumor cell therapy using CSO-based nanoparticles for controlled drug release and reducing the non-specific distribution. a Self-assembly of HA-DOX and CSO-OA to form HA-DOX-loaded HA-CSO-OA. **b** Nanoparticles target to CD44 receptor in cancer cells, followed by the pH-responsive release intracellularly (adapted from Chen et al. 2021)

2.2 *Redox-Responsive CSO-Based Nanoparticles*

In addition, the redox-responsive CSO-based nanosized micelle was also developed for the treatment of cancer. DOX was conjugated with stearic acid-graft CSO (CSO-SA) via disulfide bonds, which can be cleaved into thiols via intracellular high concentration of glutathione, particularly in cancer cells (Su et al. 2015; Lee et al. 2013). CSO-SA micelle exhibited a slow rate of drug release due to the slow degradation of the amide linkages between the hydrophobic core and hydrophilic shell in the neutral blood environment. When the micelle accumulates in tumor tissue via EPR effect and SA surface segment of micelle enhances the cellular internalization, followed by the degradation of micelle and DOX release via glutathione induced-disulfide bonds cleavage. Early studies reported that proteins bearing *O*-linked single or multiple *N*-acetylglucosamine (GlcNAc) moieties were particularly abundant in the cytoplasmic and nucleoplasmic (but not lumenal) sides of nuclear membranes (Holt et al. 1987; Holt and Hart 1986). As the monomer structure of chitosan is identical to the GlcNAc, we assumed that CSO-SA may hold the potential to gather into the cell nuclei. In fact, CSO-SA copolymer-based micelles have been applied as promising drug nano-carriers for the treatment of various cancers in vitro and in vivo (Hu et al. 2006, 2008a, 2008b 2009; You et al. 2007). Similarly, a folic acid conjugated CSO polylactic acid copolymer (FA-CSO-PLA) with double disulfide linkage, which rapidly destructed under a reductive environment, leading to encapsulated drug release (Yang et al. 2015). In the other work, the amphiphilic polycaprolactone graft CSO polymer possessing disulfide linkers has also been developed as a redox-responsive micelle for intracellular delivery of DOX in the cancer cells (Guerry et al. 2014).

2.3 Sustained Drug Release of CSO-Based Nanoparticles

Cross-linking methods have also been used to develop sustained or slow-release CSO-based nano-carriers, using crosslinking agents including glutaraldehyde (Hu et al. 2008b; Du et al. 2010b). The shell cross-linked nanoparticles were demonstrated to prolong the release of anticancer drugs as compared to non-cross-linking micelles. On the other hand, when different molecular weight CSO was used for coating, the rate of release was slower with decreasing molecular weight CSO coating on nanoparticles. This can be explained that low molecular weight CSO exhibited stronger folding ability and tight binding on the surface of nanoparticles (Yinga et al. 2011). Furthermore, CSO could also covalently conjugate with anticancer agents forming nanodrug. For example, CSO-conjugating methotrexate, a cytotoxicity agent for the treatment of a variety of malignant cancers, could form the micelle structure via a simple sonication process (Fattahi et al. 2015). The covalent bonds of a drug to polymer backbone of micelle nanoparticles has been reported to prevent the drug leakage during the delivery, and the concept was named self-assembled nanodrugs (Vong et al. 2020, 2019).

2.4 Enhancement of Tumor Targeting and Uptake

Besides the tumor accumulation of nanoparticles via passive targeting effect or EPR effect, to enhance the active cellular uptake of nanoparticles in cancer tissues, CSO could be used to conjugate with integrin-binding ligand peptide RGD (arginine-glycine-aspartic acid) (El-Sayed et al. 2016; Yuan et al. 2018). A dual coating of poly (ethylene glycol)-poly (D,L-lactic acid) (PEG-PDLLA) and CSO was used to form a colloidal stable nanoparticle via nanoprecipitation approach. When an appropriate ratio of CSO and PEG-PDLLA was coated on the surface of nanoparticles, it could achieve a significantly improved blood circulation and desirable interstitial penetration of nanoparticles in the tumor tissue by improved interaction of CSO with interstitial matrix in tumor tissue (Wang et al. 2016). RGD-modified CSO-coated palladium nanoparticle was also developed to improve the accumulation in the breast cancer cells to obtain an effective photothermal therapy (Bharathiraja et al. 2018). CSO coating significantly improved the colloidal stability and biocompatibility of palladium nanoparticles. For hepatocellular carcinoma cell-targeted delivery of ATP, galactosylated chitosan oligosaccharide nanoparticles were designed because of the galactose group as the specific ligand to the asialoglycoprotein receptor of hepatocellular carcinoma cells (Zhu et al. 2013). This single nanoplatform combining diagnostic and therapeutic functions has a great potential to revolutionize the diagnosis and treatment of diseases (Rizzo et al. 2013). Galactose-grafted CSO was synthesized via interaction between CSO and lactobionic acid using EDC-mediated condensation. Besides, CSO-coated nanorods could be easily modified by targeting monoclonal antibodies against epidermal growth factor receptor, which is overexpressed in human oral adenosquamous carcinoma cells (Charan et al. 2012). By introducing the ligand on the surface, CSO-based nanoparticles significantly improved the tumor targeting effect and deeper penetration in the tumor tissues, resulting in higher anticancer therapeutic efficacy.

3 CSO-Based Nanoparticles for Oral Drug Delivery

Oral drug administration is the most widely used route due to patient convenience and compliance, although numerous challenges are remained such as harsh pH gastric environment, low permeability, and retention in the gastrointestinal (GI) tract (Vong and Nagasaki 2017). Drug-encapsulated nanoformulation can improve the drug stability, bioavailability, and sustained release in GI tract (Date et al. 2016). It has been reported that CSO-coated nanoparticles possess a highly positive charge surface. Since the mucosa layer of the intestinal tract presents the negatively charged sialic acids of mucin, orally administered CSO-coated nanoparticles significantly improved the retention and permeability in the intestine via the interaction nanoparticle and mucosa layer of epithelial cells (Dyawanapelly et al. 2016). For oral drug administration of the gastric ulcer, CSO-based nanoparticles loading apocynin, a phytochemical antioxidant, was prepared using tripolyphosphate as cross-linkers. CSO-nanoparticles remarkably increased the mucin-binding efficacy, resulting in a significant improvement of gastric retention and apocynin therapeutic effect on gastric ulcer (Anter et al. 2019). Besides, surface coating with positively charged CSO also induced the opening of tight junctions in intestinal epithelial cells, increasing the internalization of drugs/proteins into the bloodstream (Chen et al. 2013). In fact, CSO-coated nanostructured lipid carrier (NLC) also improved the in vivo ocular mucoadhesion and retention (Fig. 3) (Pai and Vavia 2020). The retention in the ocular surface of CSO-coated NLC was 4 h, which was significantly higher than that of CSO-uncoated NLC. On the other hand, it would be better to stabilize the CSO-based nanoparticles against acid pH in the gastric environment for the delivery of drug to intestinal and colon areas. Eudragit® L100-55 (EL), an anionic copolymer based on

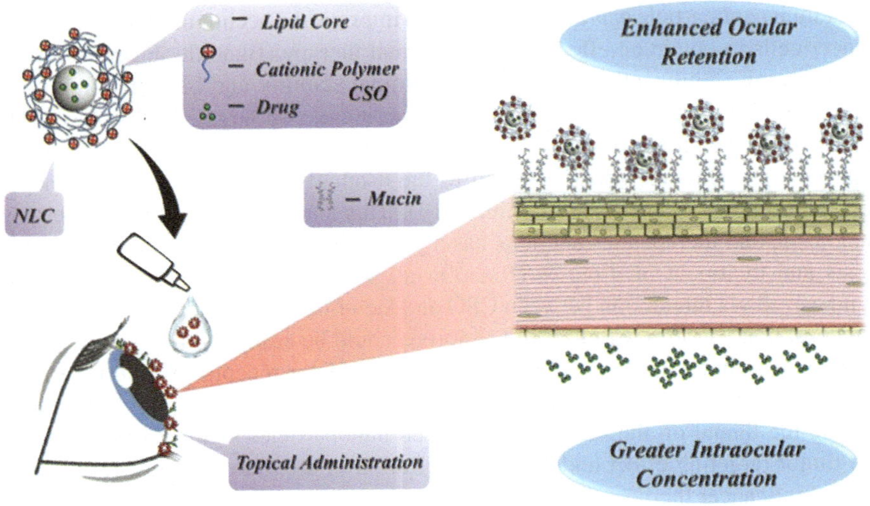

Fig. 3 Mucoadhesive property of CSO-coated nanostructured lipid carrier (NLC). CSO-coated NLC interacts with mucin to enhance the ocular retention (2020 adapted from Pai and Vavia)

methacrylic acid and ethylacrylate, was used to stabilize the alginate-CSO microparticles in acid pH by the formation of CSO-EL polyelectrolyte complex (Calija et al. 2013). Higher content of CSO was used to form complex, slower release of nonsteroidal anti-inflammatory drug naproxen was obtained. When the pH raised higher (intestinal and colon conditions), the particles started to swell and the rate of drug release was also higher. These reports indicated that using CSO-based nanoparticles, the release of a drug could be controlled depending on the various pH values in the GI tract along with mucosa adhesive property.

4 COS-Based Complexes for Gene Delivery

Gene therapy using DNA and RNA has enormous therapeutic potential for the treatment of various intractable diseases (Dunbar et al. 2018). However, due to nature characters, DNA and RNA exhibit extremely low stability in the serum environment, a rapid elimination by the kidney, and a low plasma membrane permeability, leading to low accumulation at the targeted sites and low therapeutic efficacy. The development of gene delivery systems has become a promising approach to deliver a sufficient amount of DNA and RNA to the targeted cells. Various natural and synthetic polymers have been studied for complexation with DNA and RNA to protect from degradation and control the release of DNA and RNA intracellularly (Pack et al. 2005). Possessing a positively charged backbone of the amino group, which can be complexed with negatively charged DNA and RNA molecules, chitosan and CSO have been studied as non-viral gene vectors for gene delivery (Chae et al. 2005).

4.1 DNA Delivery

Although chitosan has been widely used as vectors for the delivery of genes with low toxicity, the high molecular weight of chitosan exhibits a low solubility at physiological pH and high viscosity, limiting the complexation processes with DNA and RNA molecules. In addition, the chitosan complexes often tend to form aggregates and high molecular weight chitosan can form strong complexation inhibiting the release of DNA molecules from the complex (Issa et al. 2006). To overcome the drawbacks of chitosan, the highly water-soluble CSOs with different molecular weights have become potential cationic oligomers for studying the complexation with DNA molecules. It has been reported that CSO-based complex dissociated more easily than those of high molecular weight chitosan, leading to higher gene expression both in vitro and in vivo after administration (Köping-Höggård et al. 2003; Miao et al. 2019). In addition, the stability of the CSO-DNA-complexes at lower pH conditions (lower pH 6) was greater as compared at higher pH values (above pH 7) due to the protonation or deprotonation of the amino group in different pH, resulting in high transfection efficiency (Strand et al. 2005).

It was reported that various molecular weight CSOs from 1.2 kDa to 4.7 kDa (6 to 24 units) form different physical shapes of complexes, which are strongly associated with stability and transfection efficiency of the complexes as gene delivery systems (Köping-Höggård et al. 2004). Using the COS-7 cell line, CSO-DNA nanocomplex showed slightly higher transfection efficiency as compared to the polyethylenimine (PEI)-DNA nanoparticles, in which PEI is well known as a highly toxic cationic polymer for the non-viral vector of gene delivery (Moghimi et al. 2005; Jia et al. 2013). It should be noticed that high molecular weight chitosan-DNA nanoparticles exhibited a very low transgene expression in COS-7 cells. As the result in vivo testing, CSO-DNA nanoparticles induced luciferase gene expression in rat corneas 5.4 times higher than in rat treated with PEI-DNA nanoparticles (Klausner et al. 2010). In addition, different DNA plasmids used for complexation with CSO also strongly affected the transfection efficiency, in which CpG-free plasmid DNA led to the highest transgene expression in rat corneas in comparison with other tested plasmid DNA (Klausner et al. 2012). In another work, CSO-SA complexed with DNA plasmid did not remarkably enhance the transfection efficiency on HEK293 and MCF-7 cell lines (Du et al. 2011b). However, the addition of protamine, a small peptide with high content of arginine, significantly improved the gene expression efficiency of DNA-CSO-SA complex. The cationic CSO-SA formed a complex with plasmid DNA via electrostatic interaction and the stability of the complex was highly associated with the graft ratio of SA, the molecular weight of CSO, ionic strength, and pH conditions (Du et al. 2010a). It was also reported that branched glycosylated-CSO substituted with a trisaccharide containing *N*-acetylglucosamine (AAM) possessed higher transfection efficiency as compared to unsubstituted and substituted linear CSO. It can be explained by the balance between the stability of the complexes and facile release of DNA from the complexes, where the random substitution of the uncharged AAM ligand to the CSO reduced the affinity for DNA, leading to weaker DNA-CSO complexes, while the substitution of AAM branching in CSO backbone enhanced the colloidal stability via the sterical hindrance effect (Strand et al. 2008).

4.2 siRNA Delivery

Short interfering RNA (siRNA) has great potential for gene silencing therapy; however, low stability in the physiological environment is a great challenge (Davis et al. 2010). Fatty acyl peptide-modified CSO was designed to form a complex with siRNA. The stability of the complex was significantly improved by hydrophobic and electrostatic interactions and effectively protected siRNA from serum enzyme degradation (El-Sayed et al. 2018). Furthermore, the molecular weight and degree of acetylation of CSO along with charged ratio were reported to highly affect the stability and dissociation of the CSO-siRNA complexes (Delas et al. 2019; Alameh et al. 2018). Recently, FA-PEG-CSO lactate nanoparticle was developed to deliver siRNA to suppress the expression of hypoxia-inducible factor-1 alpha (HIF-1α) in ovarian cancer treatment. Treatment with siRNA-loaded FA-PEG-CSO lactate nanoparticles remarkably reduced the expression of HIF-1α mRNA and protein, leading to the proliferation inhibition of OVK#18 ovarian cancer cells (Li et al. 2014). In addition, the aforementioned siRNA complex also effectively inhibited the invasion and metastasis of pancreatic cancer cells in vivo (Taniuchi et al. 2019). Despite of numerous promising results in vitro, the comprehensive research on CSO-based gene delivery systems investigating in vivo animal models and clinical trials is limited.

5 Conclusions

CSO exhibits great potential in nanoparticle-based drug and gene delivery. Highly water-soluble of CSO is an important feature for designing various nano-carriers, which might not suitable with insoluble high molecular weight chitosan. This chapter presented a review on research of CSO-based nanoparticles for drug and gene delivery to apply in the treatment of cancer and other diseases. Since CSO can be easily modified to form various types of nanoparticles, it is able to effectively encapsulate and protect bioactive agents against physiological degradation with the controlled release manner and targeted delivery. As the result, CSO-based nanoparticles significantly improved the therapeutic efficacy in comparison with the free drugs or gene treatments. Although numerous studies on the formation and in vitro evaluation of drug-loaded CSO-based nanoparticles have been studied, the in vivo results have still not been well investigated, limiting the application of CSO-based nanoparticles.

Taken together, CSO-based nanoparticles exhibit a as great promising biomaterials for drug and gene delivery systems, and further investigations should be carefully considered and evaluated to bring these nano-systems to clinical applications.

Acknowledgements This work was also supported by Vietnam National University Ho Chi Minh City (VNU-HCM) under grant numbers NCM2020-28-01.

Conflicts of Interest The authors declare that they have no competing financial interests.

References

Alameh M, Lavertu M, Tran-Khanh N et al (2018) siRNA delivery with chitosan: Influence of chitosan molecular weight, degree of deacetylation, and amine to phosphate ratio on in vitro silencing efficiency, hemocompatibility, biodistribution, and in vivo efficacy. Biomacromol 19(1):112–131

Anter HM, Abu Hashim II, Awadin W et al (2019) Novel chitosan oligosaccharide-based nanoparticles for gastric mucosal administration of the phytochemical "apocynin." Int J Nanomedicine 14:4911–4929

Bharathiraja S, Bui NQ, Manivasagan P et al (2018) Multimodal tumor-homing chitosan oligosaccharide-coated biocompatible palladium nanoparticles for photo-based imaging and therapy. Sci Rep 8(1):500

Cabral H, Miyata K, Osada K et al (2018) Block copolymer micelles in nanomedicine applications. Chem Rev 118(14):6844–6892

Calija B, Cekić N, Savić S et al (2013) pH-sensitive microparticles for oral drug delivery based on alginate/oligochitosan/Eudragit(®) L100–55 "sandwich" polyelectrolyte complex. Coll Surf B Biointerface 110:395–402

Chae SY, Son S, Lee M et al (2005) Deoxycholic acid-conjugated chitosan oligosaccharide nanoparticles for efficient gene carrier. J Control Release 109(1–3):330–344

Charan S, Sanjiv K, Singh N et al (2012) Development of chitosan oligosaccharide-modified gold nanorods for in vivo targeted delivery and noninvasive imaging by NIR irradiation. Bioconjug Chem 23(11):2173–2182

Chauhan VP, Jain RK (2013) Strategies for advancing cancer nanomedicine. Nat Mater 12(11):958–962

Chen MC, Mi FL, Liao ZX et al (2013) Recent advances in chitosan-based nanoparticles for oral delivery of macromolecules. Adv Drug Deliv Rev 65(6):865–879

Chen Y, Feng S, Liu W et al (2017) Vitamin E succinate-grafted-chitosan oligosaccharide/RGD-conjugated TPGS mixed micelles loaded with paclitaxel for U87MG tumor therapy. Mol Pharm 14(4):1190–1203

Chen Y, Chen Q, Zhu Q et al (2019) Small molecular theranostic assemblies functionalized by doxorubicin-hyaluronic acid-methotrexate prodrug for multiple tumor targeting and imaging-guided combined chemo-photothermal therapy. Mol Pharm 16(6):2470–2480

Chen X, Bremner DH, Ye Y et al (2021) A dual-prodrug nanoparticle based on chitosan oligosaccharide for enhanced tumor-targeted drug delivery. Coll Surf A Physicochem Eng Asp 619:126512

Date AA, Hanes J, Ensign LM (2016) Nanoparticles for oral delivery: Design, evaluation and state-of-the-art. J Control Release 240:504–526

Davis ME, Zuckerman JE, Choi CH et al (2010) Evidence of RNAi in humans from systemically administered siRNA via targeted nanoparticles. Nature 464(7291):1067–1070

Delas T, Mock-Joubert M, Faivre J et al (2019) Effects of chain length of chitosan oligosaccharides on solution properties and complexation with siRNA. Polymers (basel) 11(8):1236

Du YZ, Wang L, Yuan H et al (2009) Preparation and characteristics of linoleic acid-grafted chitosan oligosaccharide micelles as a carrier for doxorubicin. Coll Surf B Biointerface 69(2):257–263

Du YZ, Ying XY, Wang L et al (2010) Sustained release of ATP encapsulated in chitosan oligosaccharide nanoparticles. Int J Pharm 392(1–2):164–169

Du YZ, Lu P, Zhou JP et al (2010) Stearic acid grafted chitosan oligosaccharide micelle as a promising vector for gene delivery system: factors affecting the complexation. Int J Pharm 391(1–2):260–266

Du YZ, Lu P, Yuan H et al (2011) Quaternary complexes composed of plasmid DNA/protamine/fish sperm DNA/stearic acid grafted chitosan oligosaccharide micelles for gene delivery. Int J Biol Macromol 48(1):153–159

Du YZ, Wang L, Yuan H et al (2011) Linoleic acid-grafted chitosan oligosaccharide micelles for intracellular drug delivery and reverse drug resistance of tumor cells. Int J Biol Macromol 48(1):215–222

Dunbar CE, High KA, Joung JK et al (2018) Gene therapy comes of age. Science 359(6372):eaan4672

Dyawanapelly S, Koli U, Dharamdasani V et al (2016) Improved mucoadhesion and cell uptake of chitosan and chitosan oligosaccharide surface-modified polymer nanoparticles for mucosal delivery of proteins. Drug Deliv Transl Res 6(4):365–379

El-Sayed NS, Shirazi AN, El-Meligy MG et al (2016) Design, synthesis, and evaluation of chitosan conjugated GGRGDSK peptides as a cancer cell-targeting molecular transporter. Int J Biol Macromol 87:611–622

El-Sayed NS, Sharma M, Aliabadi HM et al (2018) Synthesis, characterization, and in vitro cytotoxicity of fatty acyl-CGKRK-chitosan oligosaccharides conjugates for siRNA delivery. Int J Biol Macromol 112:694–702

Emami J, Rezazadeh M, Rostami M et al (2015) Co-delivery of paclitaxel and α-tocopherol succinate by novel chitosan-based polymeric micelles for improving micellar stability and efficacious combination therapy. Drug Dev Ind Pharm 41(7):1137–1147

Fattahi A, Asgarshamsi M, Hasanzadeh F et al (2015) Methotrexate-grafted-oligochitosan micelles as drug carriers: synthesis and biological evaluations. J Mater Sci Mater Med 26(2):119

Gaucher G, Dufresne MH, Sant VP et al (2005) Block copolymer micelles: preparation, characterization and application in drug delivery. J Control Release 109(1–3):169–188

Ghadi A, Mahjoub S, Tabandeh F et al (2014) Synthesis and optimization of chitosan nanoparticles: potential applications in nanomedicine and biomedical engineering. Caspian J Intern Med 5(3):156–161

Guerry A, Cottaz S, Fleury E et al (2014) Redox-stimuli responsive micelles from DOX-encapsulating polycaprolactone-g-chitosan oligosaccharide. Carbohydr Polym 112:746–752

Holt GD, Snow CM, Senior A et al (1987) Nuclear pore complex glycoproteins contain cytoplasmically disposed O-linked N-acetylglucosamine. J Cell Biol 104(5):1157–1164

Holt GD, Hart GW (1986) The subcellular distribution of terminal N-acetylglucosamine moieties. Localization of a novel protein-saccharide linkage, O-linked GlcNAc. J Biol Chem 261(17):8049–8057

Hu FQ, Ren GF, Yuan H et al (2006) Shell cross-linked stearic acid grafted chitosan oligosaccharide self-aggregated micelles for controlled release of paclitaxel. Coll Surf B Biointerface 50(2):97–103

Hu FQ, Meng P, Dai YQ et al (2008) PEGylated chitosan-based polymer micelle as an intracellular delivery carrier for anti-tumor targeting therapy. Eur J Pharm Biopharm 70(3):749–757

Hu FQ, Wu XL, Du YZ et al (2008) Cellular uptake and cytotoxicity of shell crosslinked stearic acid-grafted chitosan oligosaccharide micelles encapsulating doxorubicin. Eur J Pharm Biopharm 69(1):117–125

Hu FQ, Liu LN, Du YZ et al (2009) Synthesis and antitumor activity of doxorubicin conjugated stearic acid-g-chitosan oligosaccharide polymeric micelles. Biomaterials 30(36):6955–6963

Hu L, Sun Y, Wu Y (2013) Advances in chitosan-based drug delivery vehicles. Nanoscale 5(8):3103–3111

Huang L, Chen J, Cao P et al (2015) Anti-obese effect of glucosamine and chitosan oligosaccharide in high-fat diet-induced obese rats. Mar Drugs 13(5):2732–2756

Issa MM, Köping-Höggård M, Tømmeraas K et al (2006) Targeted gene delivery with trisaccharide-substituted chitosan oligomers in vitro and after lung administration in vivo. J Control Release 115(1):103–112

Jia LJ, Li ZY, Zhang DR et al (2013) Redox-responsive catiomer based on PEG-ss-chitosan oligosaccharide-ss-polyethylenimine copolymer for effective gene delivery. Polym Chem 4:156–165

Kataoka K, Harada A, Nagasaki Y (2001) Block copolymer micelles for drug delivery: design, characterization and biological significance. Adv Drug Deliv Rev 47(1):113–131

Kato Y, Ozawa S, Miyamoto C et al (2013) Acidic extracellular microenvironment and cancer. Cancer Cell Int 13(1):89

Kim JH, Kim YS, Hwang JW et al (2014) Sulfated chitosan oligosaccharides suppress LPS-induced NO production via JNK and NF-κB inactivation. Molecules 19(11):18232–18247

Kim HJ, Ahn HY, Kwak JH et al (2014) The effects of chitosan oligosaccharide (GO2KA1) supplementation on glucose control in subjects with prediabetes. Food Funct 5(10):2662–2669

Klausner EA, Zhang Z, Chapman RL et al (2010) Ultrapure chitosan oligomers as carriers for corneal gene transfer. Biomaterials 31(7):1814–1820

Klausner EA, Zhang Z, Wong SP et al (2012) Corneal gene delivery: chitosan oligomer as a carrier of CpG rich, CpG free or S/MAR plasmid DNA. J Gene Med 14(2):100–108

Köping-Höggård M, Mel'nikova YS, Vårum KM et al (2003) Relationship between the physical shape and the efficiency of oligomeric chitosan as a gene delivery system in vitro and in vivo. J Gene Med 5(2):130–141

Köping-Höggård M, Vårum KM, Issa M et al (2004) Improved chitosan-mediated gene delivery based on easily dissociated chitosan polyplexes of highly defined chitosan oligomers. Gene Ther 11(19):1441–1452

Lee MH, Yang Z, Lim CW et al (2013) Disulfide-cleavage-triggered chemosensors and their biological applications. Chem Rev 113(7):5071–5109

Li TS, Yawata T, Honke K (2014) Efficient siRNA delivery and tumor accumulation mediated by ionically cross-linked folic acid-poly(ethylene glycol)-chitosan oligosaccharide lactate nanoparticles: for the potential targeted ovarian cancer gene therapy. Eur J Pharm Sci 52:48–61

Li J, Cai C, Li J et al (2018) Chitosan-based nanomaterials for drug delivery. Molecules 23(10):2661

Lodhi G, Kim YS, Hwang JW et al (2014) Chitooligosaccharide and its derivatives: preparation and biological applications. Biomed Res Int 2014:654913

Luo S, Feng J, Xiao L et al (2020) Targeting self-assembly peptide for inhibiting breast tumor progression and metastasis. Biomaterials 249:120055

Maeda H, Wu J, Sawa T et al (2000) Tumor vascular permeability and the EPR effect in macromolecular therapeutics: a review. J Control Release 65(1–2):271–284

Manivasagan P, Bharathiraja S, Bui NQ et al (2016) Paclitaxel-loaded chitosan oligosaccharide-stabilized gold nanoparticles as novel agents for drug delivery and photoacoustic imaging of cancer cells. Int J Pharm 511(1):367–379

Manivasagan P, Jun SW, Nguyen VT et al (2019) A multifunctional near-infrared laser-triggered drug delivery system using folic acid conjugated chitosan oligosaccharide encapsulated gold nanorods for targeted chemo-photothermal therapy. J Mater Chem B 6:3811–3825

Miao J, Yang XQ, Gao Z et al (2019) Redox-responsive chitosan oligosaccharide-SS-Octadecylamine polymeric carrier for efficient anti-Hepatitis B Virus gene therapy. Carbohydr Polym 212:215–221

Mitchell MJ, Billingsley MM, Haley RM et al (2021) Engineering precision nanoparticles for drug delivery. Nat Rev Drug Discov 20(2):101–124

Moghimi SM, Symonds P, Murray JC et al (2005) A two-stage poly(ethylenimine)-mediated cytotoxicity: implications for gene transfer/therapy. Mol Ther 11(6):990–995

Morin-Crini N, Lichtfouse E, Torri G et al (2019) Applications of chitosan in food, pharmaceuticals, medicine, cosmetics, agriculture, textiles, pulp and paper, biotechnology, and environmental chemistry. Environ Chem Lett 17:1667–1692

Muanprasat C, Chatsudthipong V (2017) Chitosan oligosaccharide: biological activities and potential therapeutic applications. Pharmacol Ther 170:80–97

Nguyen TT, Trinh NT, Tran HN et al (2021) Improving silymarin oral bioavailability using silica-installed redox nanoparticle to suppress inflammatory bowel disease. J Control Release 331:515–524

Pack DW, Hoffman AS, Pun S et al (2005) Design and development of polymers for gene delivery. Nat Rev Drug Discov 4(7):581–593

Pai RV, Vavia PR (2020) Chitosan oligosaccharide enhances binding of nanostructured lipid carriers to ocular mucins: effect on ocular disposition. Int J Pharm 577:119095

Patra JK, Das G, Fraceto LF et al (2018) Nano based drug delivery systems: recent developments and future prospects. J Nanobiotechnol 16(1):71

Rizzo LY, Theek B, Storm G et al (2013) Recent progress in nanomedicine: therapeutic, diagnostic and theranostic applications. Curr Opin Biotechnol 24(6):1159–1166

Rosenblum D, Joshi N, Tao W et al (2018) Progress and challenges towards targeted delivery of cancer therapeutics. Nat Commun 9(1):1410

Strand SP, Danielsen S, Christensen BE et al (2005) Influence of chitosan structure on the formation and stability of DNA-chitosan polyelectrolyte complexes. Biomacromol 6(6):3357–3366

Strand SP, Issa MM, Christensen BE et al (2008) Tailoring of chitosans for gene delivery: novel self-branched glycosylated chitosan oligomers with improved functional properties. Biomacromol 9(11):3268–3276

Su Y, Hu Y, Du Y et al (2015) Redox-responsive polymer-drug conjugates based on doxorubicin and chitosan oligosaccharide-g-stearic acid for cancer therapy. Mol Pharm 12(4):1193–1202

Sung YK, Kim SW (2020) Recent advances in polymeric drug delivery systems. Biomater Res 24:12. https://doi.org/10.1186/s40824-020-00190-7

Taniuchi K, Yawata T, Tsuboi M et al (2019) Efficient delivery of small interfering RNAs targeting particular mRNAs into pancreatic cancer cells inhibits invasiveness and metastasis of pancreatic tumors. Oncotarget 10(30):2869–2886

Termsarasab U, Cho HJ, Kim DH et al (2013) Chitosan oligosaccharide-arachidic acid-based nanoparticles for anti-cancer drug delivery. Int J Pharm 441(1–2):373–380

Vong LB, Tomita T, Yoshitomi T et al (2012) An orally administered redox nanoparticle that accumulates in the colonic mucosa and reduces colitis in mice. Gastroenterology 143(4):1027-1036.e3

Vong LB, Kimura S, Nagasaki Y (2017) Newly designed silica-containing redox nanoparticles for oral delivery of novel TOP2 catalytic inhibitor for treating colon cancer. Adv Healthc Mater 6(20):1700428

Vong LB, Ibayashi Y, Lee Y et al (2019) Poly(ornithine)-based self-assembling drug for recovery of hyperammonemia and damage in acute liver injury. J Control Release 310:74–81

Vong LB, Trinh NT, Nagasaki Y (2020) Design of amino acid-based self-assembled nano-drugs for therapeutic applications. J Control Release 326:140–149

Vong LB, Nagasaki Y (2017) Development of redox nanomedicine for gastrointestinal complications via oral administration route. In: Ito Y (ed) Advances in bioinspired and biomedical materials, vol 2. ACS Symposium Series 1253, pp 47–67

Wang G, Chen Y, Wang P et al (2016) Preferential tumor accumulation and desirable interstitial penetration of poly(lactic-co-glycolic acid) nanoparticles with dual coating of chitosan oligosaccharide and polyethylene glycol-poly(D, L-lactic acid). Acta Biomater 29:248–260

Wang Y, Khan A, Liu Y et al (2019) Chitosan oligosaccharide-based dual pH responsive nano-micelles for targeted delivery of hydrophobic drugs. Carbohydr Polym 223:115061

Yang Q, He C, Xu Y et al (2015) Chitosan oligosaccharide copolymer micelles with double disulphide linkage in the backbone associated by H-bonding duplexes for targeted intracellular drug delivery. Polym Chem 6:1454–1464

Yinga XY, Cui D, Yub L et al (2011) Solid lipid nanoparticles modified with chitosan oligosaccharides for the controlled release of doxorubicin. Carbohydr Polym 84:1357–1364

You J, Hu FQ, Du YZ et al (2007) Polymeric micelles with glycolipid-like structure and multiple hydrophobic domains for mediating molecular target delivery of paclitaxel. Biomacromol 8(8):2450–2456

Yuan Z, Yuan Y, Han L et al (2018) Bufalin-loaded vitamin E succinate-grafted-chitosan oligosaccharide/RGD conjugated TPGS mixed micelles demonstrated improved antitumor activity against drug-resistant colon cancer. Int J Nanomedicine 13:7533–7548

Zhao M, Gu L, Li Y et al (2019) Chitooligosaccharides display anti-tumor effects against human cervical cancer cells via the apoptotic and autophagic pathways. Carbohydr Polym 224:115171

Zhou J, Wen B, Xie H (2021) Advances in the preparation and assessment of the biological activities of chitosan oligosaccharides with different structural characteristics. Food Funct 12(3):926–951

Zhu XL, Du YZ, Yu RS et al (2013) Galactosylated chitosan oligosaccharide nanoparticles for hepatocellular carcinoma cell-targeted delivery of adenosine triphosphate. Int J Mol Sci 14(8):15755–15766

Biocomposites-Based on Chitooligosaccharides for Biomedical Applications

Sesha Subramanian Murugan, Se-Kwon Kim, Pandurang Appana Dalavi, Jayachandran Venkatesan, and Gi Hun Seong

1 Introduction

Chitosan is a cationic polymer produced through partly deacetylated chitin. Chitin is commonly observed in crab shells. Chitin is the second most common polymer in nature. It is composed of β (1 → 4) linked N-acetyl-D glucosamine molecules (Aam et al. 2010). Chitin and chitosan have significant potential to be used as a biomaterial since it has the properties of biodegradable, non-toxic, and non-allergenic. It has shorter ester linkages and free amino groups through D-glucosamine units. Chitooligosaccharide (COS) is a derivative of chitosan and is used in a variety of biomedical applications. COS is made from chitosan by acid hydrolysis or enzymatic hydrolysis. Chitosanases are often utilized enzyme to produce COS using the enzymatic method (Fig. 1). COS has three types of reactive sites at the C-2, C-3, and C-6 positions: an amino/acetamido group, primary hydroxyl functional group, and secondary hydroxyl group (Phil et al. 2018). COS has a positively charged surface that binds tightly to negatively charged components and triggers a wide range of biological functions (Kurita 1998). Several research studies have found that chitosan and its derivatives of COS possess antibacterial, antifungal, anticancer, anti-HIV, and antioxidant effects (Kim and Rajapakse 2005; Kumar and Kumar 2020). Numerous

S. S. Murugan · P. A. Dalavi · J. Venkatesan (✉)
Biomaterials Research Laboratory, Yenepoya Research Centre, Yenepoya (Deemed to be University), Deralakatte, Mangalore 575018, Karnataka, India
e-mail: jvenkatesan@yenepoya.edu.in

S.-K. Kim
Department of Marine Science and Convergence Engineering, College of Science and Technology, Hanyang University, Ansan 426-791, South Korea

J. Venkatesan · G. H. Seong (✉)
Department of Bionano Engineering, Center for Bionano Intelligence Education and Research, Hanyang University, Ansan 426-791, South Korea
e-mail: ghseong@hanyang.ac.kr

S.-K. Kim (ed.), *Chitooligosaccharides*,
https://doi.org/10.1007/978-3-030-92806-3_11

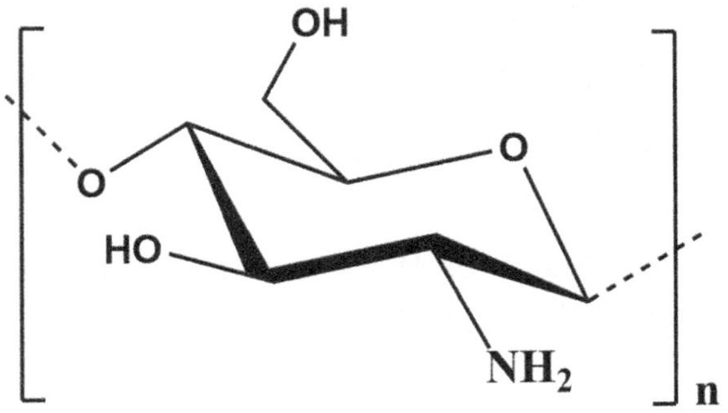

Fig. 1 Basic chemical structure of N–deacetylated COS. The figure is redrawn from the reference research article (Liaqat and Eltem 2018)

investigations have found that COS has potent anticancer properties (Huang et al. 2006). It was shown that chitosan with at least 70% deacetylation activated the macrophages and natural killer (NK) cells and improved the cytotoxic effects of immune cells that produce interleukins and interferons (Kumar and Kumar 2020). Chitin, chitosan, and COS were synthesized and fabricated nanoparticles, gels, films, microparticles, and patches. These polymeric derivatives have been used in tissue engineering, wound healing, drug delivery, gene delivery, antitumor, and antimicrobial biomedical applications (Jung and Park 2014; Jafari et al. 2020; Hao et al. 2020; Mo et al. 2020; Tabassum et al. 2021; Xiang et al. 2021; Pan et al. 2021; Chapelle et al. 2021; Zhai et al. 2021; Singh et al. 2021).

2 COS-Based Biocomposites in Antimicrobial Applications

Bacterial and viral infections cause a threat to human health. Hence, researchers are focusing on developing innovative materials to prevent microbial and viral infections. Therefore, polymer or polymer-based biocomposites with antibacterial potential are extensively used (Kasinathan et al. 2020; Paladini et al. 2015; Chen et al. 2014). Chitosan possesses excellent antibacterial potential but is limited in its application due to poor solubility in neutral solutions. This drawback makes it challenging to utilize in the industrial and biomedical fields. On the other hand, COS having antibacterial potential and is soluble in water (El-Sayed et al. 2017). Lillo et al. (2008) developed COS by partial acid hydrolysis of chitosan, and aminoglycosylated derivative COS was developed by the reductive alkylation method. Further, they have proven the antibacterial potential of the developed materials against *staphylococcus aureus* (Lillo et al. 2008). Wang et al. (2007) have developed COS from chitosanase. The degree of deacetylation and degree of polymerization of COS on the antibacterial activity was investigated. The higher degree of deacetylation provides better antimicrobial activity than the lower degree of acetylation (Wang et al. 2007). Tsai et al. (2000) have developed a biocomposite of COS containing shrimp chitosan with cellulose. Further, they have proven the antibacterial potential of this bio composition against different pathogens. Aside from that, they've suggested that the discovered bio composition can be employed to preserve milk (Tsai et al. 2000). Fernandes and his co-workers have proven the antibacterial efficiency of chitosan and COS in the food components containing pathogens (Fernandes et al. 2008). These findings results reveal that COS has potential applications to be used in the therapeutic industries such as the food and medical sectors. In another study, COS was combined with the bacterial cellulose, and its antibacterial potential was checked against *Escherichia coli* and *S. aureus* (Yin et al. 2020). Hui et al. have proved the antibacterial potential of the quaternary ammonium COS containing zinc oxide and palygorskite nanocomposite (Hui et al. 2020). Dalavi et al. (2020) have developed COS-coated silver nanoparticles and subsequently characterized them using different analytical techniques. COS-coated silver nanoparticles show efficient antimicrobial activity. Further, COS-coated silver nanoparticles were combined with calcium hydroxide for dentistry application (Dalavi et al. 2020; Rajasekaran et al. 2021).

3 COS-Based Biocomposites in Antitumor Applications

Chitosan and COS derivatives have anticancer effects dependent on molecular weight (Salah et al. 2013). Mallakuntla et al. (2021) reported that low molecular weight COS has a more decisive anticancer action and a greater surface charge, which considerably inhibits tumor cell proliferation. COS has anti-carcinogenic effects on the human colon adenocarcinoma cancer cell HT-29, according to an in-vitro study. Furthermore, COS treatment suppressed the long-term proliferation capability of breast cancer cells, as COS significantly decreased colony formation among breast cancer cells in a dose-dependent manner (Mallakuntla et al. 2021). Also, through activating the p53/mTOR pathway, COS has been found to decrease tumor development and induce autophagy in osteosarcoma (Pan et al. 2021). COS suppresses the production of matrix metalloproteinase-9 in human breast cancer cells (MDA-MB-231) (Shen et al. 2009).

4 Tissue Engineering

Tissue engineering is a branch of bioengineering that uses biology, materials science, and engineering concepts. The main aim of tissue engineering is to restore, repair, enhance, or replace damaged tissue. Biomaterials are essential components in tissue engineering concepts because they provide a new architectural framework analogous to the natural extracellular matrix, allowing cells to proliferate and differentiate, resulting in tissue regeneration. Biomaterials are involved in cell adhesion, proliferation and provide the extracellular matrix for the normal tissue. Biomaterials should be biocompatible, biodegradable, stable, and have mechanical properties. Biomaterials are classified as synthetic and natural materials. Metals, ceramics, polymers, and composite materials are the commonly used biomaterials for tissue engineering applications. These biomaterials were used to fabricate the heart, neuron, and skin tissues (Hubbell 1995).

4.1 COS-Based Biocomposites in Cardiac Tissue Engineering

Arteriosclerosis, stroke, myocardial infarction, renal failure, and coronary artery disease are caused by high blood pressure. Coronary heart disease was caused by circulating cholesterol, which builds on arterial walls, blocks the arteries, and reduces blood flow to the heart (Lawler et al. 2017). COS has been discovered to block angiotensin-converting enzymes and lower blood pressure (Hong et al. 1998). It also has excellent biological characteristics, effective antibacterial and antioxidant activity (Jiang et al. 2019). The usage of the COS was demonstrated to be effective in limiting gastrointestinal lipid absorption (Liu et al. 2018). COS has anti-atherosclerosis active components that may be formulated into a tablet that lowers atherosclerotic and hyperlipidemic risk while controlling blood lipid levels (Yang et al. 2019).

4.2 COS-Based Biocomposites for Coronary Heart Diseases

Nerve crush and nerve injury are common physical injuries to peripheral nerves (Roglio et al. 2008). Some of the most recent advances in regenerative medicine that have fascinating therapeutic implications in artificial neural grafts include enhanced biomaterials, customized scaffold topologies, and the integration of diverse tissue cells and growth hormones (Gu et al. 2011). Wang et al. (2016) reported that COS activates peripheral nerve regeneration in rats (Fig. 2). COS stimulates the proliferation of Schwann cells during nerve regeneration, and it was shown that COS enhances miR-27a gene expression while decreasing FOXO1, promoting the cell cycle to accelerate neuron regeneration (Wang et al. 2016). In addition, Jiang et al. (2014) discovered that COS treatment increases Schwan cell proliferation and survival in a dose-dependent manner (0.25–1.0 mg/mL). It was assessed by a real-time cell analyzer-based assay, cell proliferation assay, and cell division analysis. The protein activity of N-cadherin and -catenin was likely boosted by cultivating the Schwann cells with COS (Jiang et al. 2014).

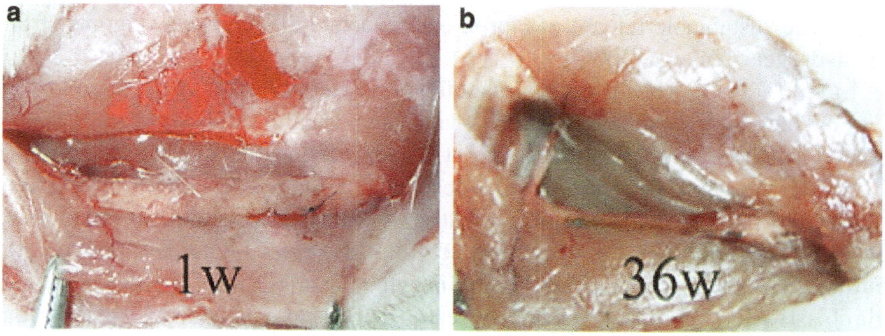

Fig. 2 Chitosan degradation metabolites promote nerve regeneration and Schwann cell proliferation. Gross view **a** 1 week and **b** 36 weeks after chitosan scaffold been transplanted into 10-mm rat sciatic nerve injury. The figure is adapted with permission from the reference (Wang et al. 2016)

4.3 COS-Based Biocomposites in Skin Tissue Engineering and Wound Healing Application

Skin is frequently injured tissue by physical and chemical factors. When skin is harmed, the tissues initiate a complex sequence of actions to restore its integrity and function. In several acute and chronic conditions of partial, complete, or severe injuries, the restore function has been taken regularly by a sequence of complex reactions (Jayarama Reddy et al. 2013). Skin transplantation is one of the conventional procedures to restore skin tissue. However, shortage and scarring of donor sites, inflammation, and infection at the injured area also make transplantation more challenging. The tissue repair process depended on different healing stages of hemostasis, proliferation, inflammation, and remodeling. They are regulated by the immunomodulatory actions of healing materials and inflammatory cells, cytokines, and growth factors (Dixit et al. 2017).

Polymer-based biomaterials have received a lot of attention for skin tissue regeneration and wound healing applications. Polymeric composites serve as an ideal material for wound healing applications that prevent infections and heal tissues (Negut et al. 2020; Rădulescu et al. 2016; Moura et al. 2017). Chitosan and COS polymers have received much interest for skin tissue wound healing applications due to their therapeutic characteristics, including antibacterial activity, antifungal activity, antioxidant activity, immunostimulatory activity, and anti-inflammatory activity. Also, they encourage cell adhesion and proliferation. However, the wound healing capability of COS is influenced by its molecular weight and degree of acetylation (DA) (Jafari et al. 2020; Singh et al. 2017).

Yu et al. (2006) have developed hydrogels containing chitosan oligomer, poly (*N*-vinyl pyrrolidone) (PVP), and poly (vinyl alcohol) (PVA) for wound dressing applications. It has been observed that, with the addition of the PVA and chitosan oligomer, the elongation ratio and swelling ratio were improved significantly. Further, ciprofloxacin lactate was incorporated into the hydrogels. In-vitro drug release study shows the sustainable release of ciprofloxacin lactate and chitosan oligomer from the hydrogels (Yu et al. 2006).

Li et al. (2013) have fabricated nanofibres containing COS, AgNPs, and PVA. Further, they have proven in-vitro and in-vivo wound healing ability and antibacterial potential of the developed nanofibres (Li et al. 2013). A composite scaffold containing COS, alginate, and fish collagen was designed for skin regeneration application. Substantial porosity, in-vitro biodegradation swelling ratio, elongation ratio were observed. Besides this, the developed scaffold possesses higher tensile strength. Further, biological assays show scaffold is cytocompatible with the human dermal fibroblasts-neonatal cells. It has been observed that because of the COS molecule, the swelling ratio, elongation ratio, and cytocompatibility were significantly improved (Chandika et al. 2015).

5 COS-Based Biocomposites in Drug Delivery Applications

The Drug Delivery System (DDS) has been widely used in the biomedical industry to deliver therapeutical drugs in the desired place to treat various diseases. DDS systems can improve drug efficiency, make it safer to use, and save healthcare costs. Furthermore, by employing biocompatible DDS, the drug's toxicity can be reduced. Recently, polymers have been widely used in DDS because of their biocompatible, biodegradable, and non-toxic properties. In addition, polymers can be molded into desirable geometries, and their physicochemical properties can be tuned (Bamrungsap et al. 2012; Pillai and Panchagnula 2001; Chacko et al. 2012; Allen and Cullis 2004).

COS has excellent potential to be used in DDS due to its non-toxic, biocompatible, biodegradable, and water-soluble properties (Qin and Zhao 2019; Park et al. 2018). Besides this, COS is comparatively inexpensive, and it has readily available from natural sources. In addition to this, fabrication of the COS-based DDS can be achieved easily due to fictional groups' presence (Hamman 2010). The development of COS-based biocomposites for drug delivery applications has gained significant importance over the years. DDS containing single-walled carbon nanotubes conjugated with COS was developed, and Gliotoxin (GTX) drug was incorporated. Targeting molecules including folic acid, p53, and lysozyme were added to the DDS system of COS. The effectiveness of the developed DDS was tested on the breast cancer (MCF-7) cells and cervical cancer (HeLa) cells. Finding results reveals that created DDS containing p53 showing excellent results, including sustainable drug release and higher cytotoxicity with HeLa cells (Bhatnagar et al. 2014). Mahato et al. (2020) have developed PVA and COS-based hydrogel systems for drug delivery applications. Several analytical techniques confirmed the formation of the PVA-COS biocomposite. In addition, the Lomefloxacin drug was incorporated into the hydrogels, and sustainable release of the drug was observed (Fig. 3) (Mahato et al. 2020). Qu et al. (2015) developed a biodegradable heat-sensitive hydrogel system of gold

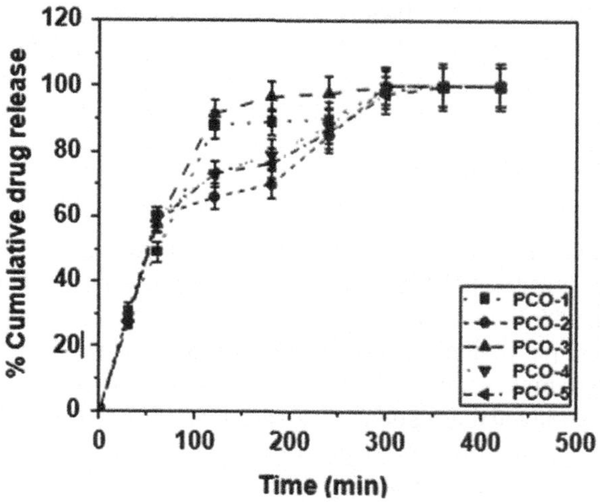

Fig. 3 Drug release profile from PCO composite hydrogels. The figure is adopted from the reference (Mahato et al. 2020)

nanorods and doxorubicin incorporated in a methoxylpoly (ethylene glycol)-poly(-caprolactone)-acryloyl chloride (PECA)/glycidyl methacrylate chitooligosaccharide (COS-GMA)/N-isopropyl acrylamide. A near-infrared (NIR) laser was employed to initiate the delivery of a packed Doxorubicin drug (Qu et al. 2015).

6 COS-Based Biocomposites in Gene Delivery

Gene delivery is the process of delivering genetic material such as DNA, plasmids, RNA, and siRNA to specific receptors of cells, possibly encapsulated within and enclosed to nanoparticles, activate or suppress protein biosynthesis to cure or treat a variety of diseases (Jayakumar et al. 2010). Many biomaterials have been employed in gene transfer. However, due to its efficacy, chitosan has been used. Chitosan is employed as a carrier of gene transfer to mucosal organs such as the lung and intestinal epithelium because it creates stable connections with plasmid DNA (MacLaughlin et al. 1998; Köping-Höggård et al. 2001; Chae et al. 2005). Furthermore, utilizing high molecular weight chitosan has some drawbacks, including limited solubility at physiological pH, substantial viscosity, and the possibility of chitosan complex clumping. To overcome these drawbacks, a COS was used. Şenel (2019) have developed COS containing nanosized lipid-based nanoparticles as a gene carrier system. The biocompatibility of the gene carrier system was tested on the MDA-MB-231, NIH 3T3, and A549 using the MTT assay, and the results showed no toxicity (Şenel 2019). In-vitro and in-vivo tests proved that deacetylated COS developed persistent compounds on plasmid DNA, proved effective vehicles for gene delivery (Köping-Höggård et al. 2003, 2004). Furthermore, Liu et al. (2021) used phenylboronic acid-modified COS as the carrier to transport the survivin-targeted siRNA for the melanoma therapy study. The results show that these particles inhibited B16F10 cell growth in-vitro

(Liu et al. 2021). Interestingly, ultrapure COS have been identified as potent nonviral corneal and retinal gene delivery (Agirre et al. 2015).

7 Conclusion

COS and its derivatives in composites have been used in antimicrobial and antitumor applications. COS-based composites materials are biocompatible, biodegradable, and non-toxic. Significant research works have been carried out on COS due to the unique biological properties of COS. As a result, COS is abundant, inexpensive, and highly water-soluble in neutral conditions. COS has been utilized in tissue engineering and drug delivery applications. Cardiac, nerve, and skin tissue engineering. Thus, COS-based biocomposites can be potentially useful for various biological and biomedical applications.

Acknowledgements This research was supported by the Basic Science Research Program through the National Research Foundation (NRF) of Korea (2018R1A6A1A03024231 and 2021R1A2C1003566).

References

Aam BB, Heggset EB, Norberg AL, Sørlie M, Vårum KM, Eijsink VG (2010) Production of chitooligosaccharides and their potential applications in medicine. Mar Drugs 8:1482–1517

Agirre M, Zarate J, Puras G, Ojeda E, Pedraz JL (2015) Improving transfection efficiency of ultrapure oligochitosan/DNA polyplexes by medium acidification. Drug Deliv 22:100–110

Allen TM, Cullis PR (2004) Drug delivery systems: entering the mainstream. Science 303:1818–1822

Bamrungsap S, Zhao Z, Chen T, Wang L, Li C, Fu T, Tan W (2012) Nanotechnology in therapeutics: a focus on nanoparticles as a drug delivery system. Nanomedicine 7:1253–1271

Bhatnagar I, Venkatesan J, Kim S-K (2014) Polymer functionalized single walled carbon nanotubes mediated drug delivery of gliotoxin in cancer cells. J Biomed Nanotechnol 10:120–130

Chacko RT, Ventura J, Zhuang J, Thayumanavan S (2012) Polymer nanogels: a versatile nanoscopic drug delivery platform. Adv Drug Deliv Rev 64:836–851

Chae SY, Son S, Lee M, Jang M-K, Nah J-W (2005) Deoxycholic acid-conjugated chitosan oligosaccharide nanoparticles for efficient gene carrier. J Control Release 109:330–344

Chandika P, Ko S-C, Oh G-W, Heo S-Y, Nguyen V-T, Jeon Y-J, Lee B, Jang CH, Kim G, Park WS (2015) Fish collagen/alginate/chitooligosaccharides integrated scaffold for skin tissue regeneration application. Int J Biol Macromol 81:504–513

Chapelle C, David G, Caillol S, Negrell C, Desroches Le Foll M (2021) Advances in chitooligosaccharides chemical modifications. Biopolymers e23461

Chen J, Wang F, Liu Q, Du J (2014) Antibacterial polymeric nanostructures for biomedical applications. Chem Commun 50:14482–14493

Dalavi PA, Prabhu A, Shastry RP, Venkatesan J (2020) Microspheres containing biosynthesized silver nanoparticles with alginate-nano hydroxyapatite for biomedical applications. J Biomater Sci Polym Ed 31:2025–2043

Dixit S, Baganizi DR, Sahu R, Dosunmu E, Chaudhari A, Vig K, Pillai SR, Singh SR, Dennis VA (2017) Immunological challenges associated with artificial skin grafts: available solutions and stem cells in future design of synthetic skin. J Biol Eng 11:1–23

El-Sayed ST, Ali AM, El-Sayed E, Shousha WG, Omar NI (2017) Characterization and potential antimicrobial effect of novel chitooligosaccharides against pathogenic microorganisms. J Appl Pharm Sci 7:6–12

Fernandes JC, Tavaria FK, Soares JC, Ramos OS, Monteiro MJ, Pintado ME, Malcata FX (2008) Antimicrobial effects of chitosans and chitooligosaccharides, upon Staphylococcus aureus and Escherichia coli, in food model systems. Food Microbiol 25:922–928

Gu X, Ding F, Yang Y, Liu J (2011) Construction of tissue engineered nerve grafts and their application in peripheral nerve regeneration. Prog Neurobiol 93:204–230

Hamman JH (2010) Chitosan based polyelectrolyte complexes as potential carrier materials in drug delivery systems. Mar Drugs 8:1305–1322

Hao W, Li K, Li P (2020) Advances in preparation of chitooligosaccharides with heterogeneous sequences and their bioactivity. Carbohydr Polym 117206

Hong S-P, Kim M-H, Oh S-W, Han C-K, Kim Y-H (1998) ACE inhibitory and antihypertensive effect of chitosan oligosaccharides in SHR. Korean J Food Sci Technol 30:1476–1479

Huang R, Mendis E, Rajapakse N, Kim S-K (2006) Strong electronic charge as an important factor for anticancer activity of chitooligosaccharides (COS). Life Sci 78:2399–2408

Hubbell JA (1995) Biomaterials in tissue engineering. Bio/technology 13:565–576

Hui A, Yan R, Wang W, Wang Q, Zhou Y, Wang A (2020) Incorporation of quaternary ammonium chitooligosaccharides on ZnO/palygorskite nanocomposites for enhancing antibacterial activities. Carbohydr Polym 247:116685

Jafari H, Bernaerts KV, Dodi G, Shavandi A (2020) Chitooligosaccharides for wound healing biomaterials engineering. Mat Sci Eng C 111266

Jayakumar R, Chennazhi K, Muzzarelli R, Tamura H, Nair S, Selvamurugan N (2010) Chitosan conjugated DNA nanoparticles in gene therapy. Carbohydr Polym 79:1–8

Jayarama Reddy V, Radhakrishnan S, Ravichandran R, Mukherjee S, Balamurugan R, Sundarrajan S, Ramakrishna S (2013) Nanofibrous structured biomimetic strategies for skin tissue regeneration. Wound Repair Regen 21:1–16

Jiang M, Cheng Q, Su W, Wang C, Yang Y, Cao Z, Ding F (2014) The beneficial effect of chitooligosaccharides on cell behavior and function of primary Schwann cells is accompanied by up-regulation of adhesion proteins and neurotrophins. Neurochem Res 39:2047–2057

Jiang T, Xing X, Zhang L, Liu Z, Zhao J, Liu X (2019) Chitosan oligosaccharides show protective effects in coronary heart disease by improving antioxidant capacity via the increase in intestinal probiotics. Oxidative medicine and cellular longevity

Jung W-J, Park R-D (2014) Bioproduction of chitooligosaccharides: present and perspectives. Mar Drugs 12:5328–5356

Kasinathan K, Murugesan B, Pandian N, Mahalingam S, Selvaraj B, Marimuthu K (2020) Synthesis of biogenic chitosan-functionalized 2D layered MoS2 hybrid nanocomposite and its performance in pharmaceutical applications: in-vitro antibacterial and anticancer activity. Int J Biol Macromol 149:1019–1033

Kim S-K, Rajapakse N (2005) Enzymatic production and biological activities of chitosan oligosaccharides (COS): a review. Carbohydr Polym 62:357–368

Köping-Höggård M, Tubulekas I, Guan H, Edwards K, Nilsson M, Vårum KM, Artursson P (2001) Chitosan as a nonviral gene delivery system. Structure–property relationships and characteristics compared with polyethylenimine in vitro and after lung administration in vivo. Gene Ther 8:1108–1121

Köping-Höggård M, Mel'nikova YS, Vårum KM, Lindman B, Artursson P (2003) Relationship between the physical shape and the efficiency of oligomeric chitosan as a gene delivery system in vitro and in vivo. J Gene Med Cross-Discipl J Res Sci Gene Transf Clin Appl 5:130–141

Köping-Höggård M, Vårum K, Issa M, Danielsen S, Christensen B, Stokke B, Artursson P (2004) Improved chitosan-mediated gene delivery based on easily dissociated chitosan polyplexes of highly defined chitosan oligomers. Gene Ther 11:1441–1452
Kumar A, Kumar A (2020) The virtuous potential of chitosan oligosaccharide for promising biomedical applications. J Mater Res 35:1123–1134
Kurita K (1998) Chemistry and application of chitin and chitosan. Polym Degrad Stab 59:117–120
Lawler PR, Akinkuolie AO, Ridker PM, Sniderman AD, Buring JE, Glynn RJ, Chasman DI, Mora S (2017) Discordance between circulating atherogenic cholesterol mass and lipoprotein particle concentration in relation to future coronary events in women. Clin Chem 63:870–879
Li C, Fu R, Yu C, Li Z, Guan H, Hu D, Zhao D, Lu L (2013) Silver nanoparticle/chitosan oligosaccharide/poly (vinyl alcohol) nanofibers as wound dressings: a preclinical study. Int J Nanomed 8:4131
Liaqat F, Eltem R (2018) Chitooligosaccharides and their biological activities: a comprehensive review. Carbohydr Polym 184:243–259
Lillo L, Alarcon J, Cabello G, Céspedes C, Caro C (2008) Antibacterial activity of chitooligosaccharides. Zeitschrift Für Naturforschung C 63:644–648
Liu S-H, Chiu C-Y, Shi C-M, Chiang M-T (2018) Functional comparison of high and low molecular weight chitosan on lipid metabolism and signals in high-fat diet-fed rats. Mar Drugs 16:251
Liu X, Chen L, Zhang Y, Xin X, Qi L, Jin M, Guan Y, Gao Z, Huang W (2021) Enhancing anti-melanoma outcomes in mice using novel chitooligosaccharide nanoparticles loaded with therapeutic survivin-targeted siRNA. Eur J Pharm Sci 158:105641
MacLaughlin FC, Mumper RJ, Wang J, Tagliaferri JM, Gill I, Hinchcliffe M, Rolland AP (1998) Chitosan and depolymerized chitosan oligomers as condensing carriers for in vivo plasmid delivery. J Control Release 56:259–272
Mahato KK, Sabbarwal S, Misra N, Kumar M (2020) Fabrication of polyvinyl alcohol/chitosan oligosaccharide hydrogel: physicochemical characterizations and in vitro drug release study. Int J Polym Anal Charact 25:353–361
Mallakuntla MK, Penugurti V, Manavathi B, Podile AR (2021) Chitooligosaccharides induce apoptosis in human breast cancer cells. Carbohydr Polym Technol Appl 2:100077
Mo IV, Feng Y, Dalheim MØ, Solberg A, Aachmann FL, Schatz C, Christensen BE (2020) Activation of enzymatically produced chitooligosaccharides by dioxyamines and dihydrazides. Carbohydr Polym 232:115748
Moura D, Souza M, Liverani L, Rella G, Luz G, Mano J, Boccaccini A (2017) Development of a bioactive glass-polymer composite for wound healing applications. Mater Sci Eng C 76:224–232
Negut I, Dorcioman G, Grumezescu V (2020) Scaffolds for wound healing applications. Polymers 12:2010
Paladini F, Pollini M, Sannino A, Ambrosio L (2015) Metal-based antibacterial substrates for biomedical applications. Biomacromolecules 16:1873–1885
Pan Z, Cheng D-d, Wei X-j, Li S-j, Guo H, Yang Q-c (2021) Chitooligosaccharides inhibit tumor progression and induce autophagy through the activation of the p53/mTOR pathway in osteosarcoma. Carbohydr Polym 258:117596
Park H-H, Ko S-C, Oh G-W, Jang Y-M, Kim Y-M, Park WS, Choi I-W, Jung W-K (2018) Characterization and biological activity of PVA hydrogel containing chitooligosaccharides conjugated with gallic acid. Carbohydr Polym 198:197–205
Phil L, Naveed M, Mohammad IS, Bo L, Bin D (2018) Chitooligosaccharide: an evaluation of physicochemical and biological properties with the proposition for determination of thermal degradation products. Biomed Pharmacother 102:438–451
Pillai O, Panchagnula R (2001) Polymers in drug delivery. Curr Opin Chem Biol 5:447–451
Qin Z, Zhao L (2019) The History of Chito/chitin oligosaccharides and its monomer. In: Oligosaccharides of chitin and chitosan. Springer, pp 3–14
Qu Y, Chu BY, Peng JR, Liao JF, Qi TT, Shi K, Zhang XN, Wei YQ, Qian ZY (2015) A biodegradable thermo-responsive hybrid hydrogel: therapeutic applications in preventing the post-operative recurrence of breast cancer. NPG Asia Mater 7:e207–e207

Rădulescu M, Holban AM, Mogoantă L, Bălşeanu T-A, Mogoşanu GD, Savu D, Popescu RC, Fufă O, Grumezescu AM, Bezirtzoglou E (2016) Fabrication, characterization, and evaluation of bionanocomposites based on natural polymers and antibiotics for wound healing applications. Molecules 21:761

Rajasekaran S, Rao SS, Dalavi PA, Prabhu A, Anil S, Venkatesan J, Bhat SS (2021) Rapid microwave-assisted biosynthesis of chitooligosaccharide coated silver nanoparticles: assessments of antimicrobial activity for paediatric pulp therapy. Adv Nat Sci Nanosci Nanotechnol 11:045018

Roglio I, Giatti S, Pesaresi M, Bianchi R, Cavaletti G, Lauria G, Garcia-Segura L-M, Melcangi RC (2008) Neuroactive steroids and peripheral neuropathy. Brain Res Rev 57:460–469

Salah R, Michaud P, Mati F, Harrat Z, Lounici H, Abdi N, Drouiche N, Mameri N (2013) Anticancer activity of chemically prepared shrimp low molecular weight chitin evaluation with the human monocyte leukaemia cell line, THP-1. Int J Biol Macromol 52:333–339

Şenel B (2019) In vitro preliminary studies of chitooligosaccharide coated nanostructured lipidic nanoparticles for efficient gene delivery. J Res Pharm 23:671–681

Shen K-T, Chen M-H, Chan H-Y, Jeng J-H, Wang Y-J (2009) Inhibitory effects of chitooligosaccharides on tumor growth and metastasis. Food Chem Toxicol 47:1864–1871

Singh R, Shitiz K, Singh A (2017) Chitin and chitosan: biopolymers for wound management. Int Wound J 14:1276–1289

Singh A, Mittal A, Benjakul S (2021) Chitosan, chitooligosaccharides and their polyphenol conjugates: preparation, bioactivities, functionalities and applications in food systems. Food Rev Int1–23

Tabassum N, Ahmed S, Ali MA (2021) Chitooligosaccharides and their structural-functional effect on hydrogels: a review. Carbohydr Polym 117882

Tsai G-J, Wu Z-Y, Su W-H (2000) Antibacterial activity of a chitooligosaccharide mixture prepared by cellulase digestion of shrimp chitosan and its application to milk preservation. J Food Prot 63:747–752

Wang Y, Zhao Y, Sun C, Hu W, Zhao J, Li G, Zhang L, Liu M, Liu Y, Ding F (2016) Chitosan degradation products promote nerve regeneration by stimulating schwann cell proliferation via miR-27a/FOXO1 axis. Mol Neurobiol 53:28–39

Wang Y, Zhou P, Yu J, Pan X, Wang P, Lan W, Tao S (2007) Antimicrobial effect of chitooligosaccharides produced by chitosanase from Pseudomonas CUY8. Asia Pacif J Clin Nutr 16

Xiang J, Wang J, Xie H, Liu Y, Bai Y, Che Q, Cao H, Huang G, Guo J, Su Z (2021) Protective effect and mechanism of chitooligosaccharides on acetaminophen-induced liver injury. Food Funct

Yang D, Hu C, Deng X, Bai Y, Cao H, Guo J, Su Z (2019) Therapeutic effect of chitooligosaccharide tablets on lipids in high-fat diets induced hyperlipidemic rats. Molecules 24:514

Yin N, Du R, Zhao F, Han Y, Zhou Z (2020) Characterization of antibacterial bacterial cellulose composite membranes modified with chitosan or chitooligosaccharide. Carbohydr polym 229:115520

Yu H, Xu X, Chen X, Hao J, Jing X (2006) Medicated wound dressings based on poly (vinyl alcohol)/poly (N-vinyl pyrrolidone)/chitosan hydrogels. J Appl Polym Sci 101:2453–2463

Zhai X, Li C, Ren D, Wang J, Ma C, Abd El-Aty A (2021) The impact of chitooligosaccharides and their derivatives on the in vitro and in vivo antitumor activity: a comprehensive review. Carbohydr Polym 118132

Chitooligosaccharides as Wound Healing Agent

Manish Kumar, V. Vivekanand, and Nidhi Pareek

1 Introduction

Chitooligosaccharides (COS) are the degradation products of polymer chitin or chitosan (CHS). Chitin is a linear polymer of N-acetyl-D-glucosamine (GlcNAc) residues linked through β(1 → 4) glycoside bonds. In contrast, CHS is deacetylated form of chitin that is a heteropolymer of GlcNAc and D-glucosamine (GlcN) residues. So, COS can be defined as the homo- or hetero-oligomers of GlcNAc and GlcN units in varying proportions. COS can be classified into two groups, i.e., homo-COS and hetero-COS. Homo-COS contain GlcNAc or GlcN units, whereas hetero-COS comprise both GlcN and GlcNAc units (Barroca-Aubry et al. 2010). Moreover, COS have a low molecular weight (MW) (<3.9 kDa), degree of polymerization (DP) (<20), lower viscosity, and superior solubility in aqueous solutions (Mourya et al. 2011). COS have several biological activities: antimicrobial, antitumor, antioxidant, immunostimulatory, and elicitors of plant immunity (Liaqat and Eltem 2018). These biological properties of COS depend upon DP and DA/DD (Kim and Rajapakse 2005). COS have three types of reactive functional groups: an amino/acetamido group at C-2, a primary hydroxyl group at C-3, and a secondary hydroxyl group at C-6 positions (Zou et al. 2016). The wide range of biological activities of COS has applications in several sectors such as agriculture, medicine, food, cosmetics, and wastewater treatment (Hamed et al. 2016). However, the most significant and applications of COS are in human health care. COS can be used as a vector for

M. Kumar · N. Pareek (✉)
Department of Microbiology, School of Life Sciences, Central University of Rajasthan, Ajmer 305817, Rajasthan, India
e-mail: nidhipareek@curaj.ac.in

V. Vivekanand
Centre for Energy and Environment, Malaviya National Institute of Technology, Jaipur 302017, Rajasthan, India

S.-K. Kim (ed.), *Chitooligosaccharides*,
https://doi.org/10.1007/978-3-030-92806-3_12

gene and drug delivery, in tumor suppression, in wound healing, in asthma treatment, in improving bone strength, and preventing malaria (Aam et al. 2010; Liaqat and Eltem 2018). The suitable physiological properties such as water solubility and low MW, degree of acetylation/deacetylation (DA/DDA), and pattern of acetylation/deacetylation (PA/PDA) and the presence of amino groups are responsible for the various biological activities of COS (Xia et al. 2011; Jafari et al. 2020). However, the mechanism for the biological activities of COS is still not much precise due to the lack of purity, quality, and well-characterized COS. In this chapter, the various biological properties of COS responsible for wound healing have been discussed. Furthermore, detailed insight into the different COS-based materials for wound healing has been presented to understand COS's current scenario and prospects in wound healing.

2 COS Production

The vast range of applications of COS in the biomedical field has resulted in its production demand. Depolymerization of chitin or CHS by physical or chemical and enzymatic means can yield COS. Hydrothermal treatment, microwave irradiation, ultrasonication, and rays are examples of physical processes of COS production (Wu et al. 2017; Bockuviene et al. 2021; Rajasekaran et al. 2021). However, the physical processes are not used in commercial practice due to their ineffectiveness. Furthermore, the production of COS from chitin and CHS can also be carried out by acid hydrolysis and oxidative degradation method (Kumar et al. 2020). Acids such as hydrochloric acid, sulfuric acid, acetic acid, lactic acid, trichloroacetic acid, and formic acid are commonly used in the chemical production of COS (Li et al. 2019b). However, Chemical procedures have various drawbacks, such as the development of toxic secondary compounds, low specificity, and uncontrolled reactions (Mourya et al. 2011). The issue related to the chemical COS production through conventional chemicals can be checked by the application of less toxic, ionic liquids. The utilization of ionic liquids in the production of well-defined COS with improved biological activities is still in its primitive stage (Li et al. 2019a).

Moreover, COS can also be generated through enzymatic hydrolysis. The enzymatic mean of COS generation has advancement over chemical production in terms of eco-toxicity, specificity and ease of control. However, an enzymatic process has limitations like low yield and high production cost (Jung and Park 2014). The enzymatic production of COS has been of much interest due to the agricultural, medical, pharmaceutical, and nutritional values of the produced COS. COS synthesis by enzymes improved DP and DA control while having no negative environmental impact. An assortment of carbohydrate-degrading enzymes, primarily chitinolytic and chitosanolytic enzymes, such as chitinase, chitosanase, lysozyme, cellulases, pectinase, protease, lipase, and pepsin, showed affinity to the degradation of chitin (Sashiwa et al. 2003; Poshina et al. 2018). However, chitozymes have shown a significant role in the production of COS with improved biological activities. Chitinases and chitosanases are the most important biocatalysts for the biotransformation of chitin and CHS to COS (Kumar et al. 2018). Nowadays, the synergistic approach consisting of chemical and enzymatic means has shown high potential for COS production (Sivaramakrishna et al. 2020). The chemoenzymatic approach consists of mild chemical pretreatment of chitin or CHS followed by enzymatic hydrolysis has resulted in a noteworthy level of COS (Kumar et al. 2021). Thus, the involvement of the chemoenzymatic approach by utilizing ionic liquids and chitinolytic enzymes seems to be effective in the production of high titer of COS with improved biological activities.

3 Wound Healing Properties of COS

The wound healing process is completed through a "cascade of healing" that consists of four overlapping phases, viz. hemostasis phase, defensive/inflammatory phase, proliferative phase, and maturation/remodeling phase (Enoch and Leaper 2008) (Fig. 1). During the hemostatic phase, mainly the activation of the blood clotting system and blockage of the drainage occurs, leading to the stoppage of bleeding. Further, the defensive/inflammatory phase deals with the destruction of bacteria and the removal of debris required to grow new tissue. In the third phase of wound

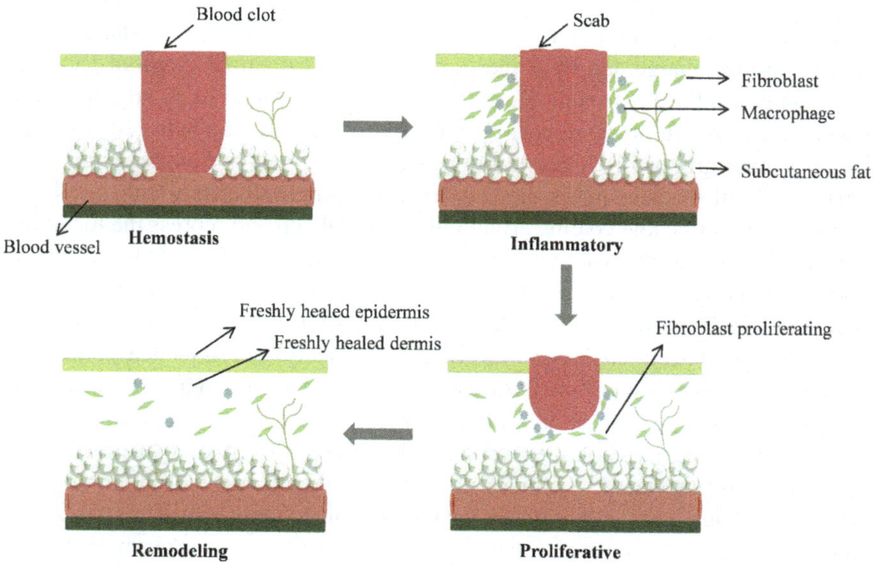

Fig. 1 Illustration of different phases of wound healing. (Hemostasis: coagulation by the initiation of blood clotting system; Inflammatory: destruction of microorganisms and eradication of debris followed by the preparation of base for new tissue growth; Proliferative: wound filling and contraction of the wound margins followed by epithelialization; Remodeling: maturation of the new tissues along with acquisition strength and flexibility)

healing, i.e., proliferative phase, the covering and filling of the wound took place. This phase complete in three stages: (i) wound filling, (ii) contraction of the wound margins, and (iii) wound covering (Wallace et al. 2017). After the proliferative phase, the maturation phase occurs, where the new tissue slowly gains strength and flexibility. The employment of COS can augment the wound healing process due to its significant wound healing biological properties, viz. antimicrobial, antioxidant, anti-inflammatory, and immunostimulatory (Jafari et al. 2020).

3.1 Antimicrobial

The wounds are always vulnerable to microbial infection. The microbial infection leads to the formation of unpleasant exudates and toxins in the infected wounds. The infection of wounds results in the simultaneous death of regenerating cells and delay of the healing process. Bacterial species such as *Staphylococcus aureus*, *Streptococcus pyrogenes*, *Enterococci*, and *Pseudomonas aeruginosa* are the most common agents causing infection to wounds. The application of COS can check microbial infection due to its antimicrobial activity. The antimicrobial property of COS can accelerate the wound healing process through tissue granulation and collagenase activity. Thus, the wound healing potential of COS can be attributed to its antimicrobial activity. The antimicrobial activity of COS differs in the different forms of bacteria, i.e., Gram-positive and Gram-negative. Gram-negative bacteria have a negative cell surface charge and thus, the positively charged amine group of COS can significantly inhibit the growth. However, COS are not found to be so effective in inhibiting the growth of Gram-positive bacteria. The interaction between positively charged amine groups of COS and the negatively charged carboxylic acid groups of bacterial cell surfaces paves the way for generating polyelectrolyte complexes, forming an impermeable coating around the bacterial cell and repress the metabolic activity. The antimicrobial activity of COS depends on its DP, MW, DA, and pattern of acetylation (PA), and the type of organism. According to a recent study, acetylated sequences in COS structure are required for antibacterial activity, and COS with a higher number of acetylated sequences (and less deacetylated sequences) and lesser free amino groups has superior antimicrobial activity (Sánchez et al. 2017). Another study found that COS with a high number of deacetylated units have antibacterial properties and can cause colonic microbiota imbalance by reducing bacterial population, whereas acetylated COS have no negative effects on faecal microbiota and can increase *Lactobacillus/Enterococcus* population (Mateos-Aparicio et al. 2016). Findings of the study suggested that the acetylated COS has no detrimental effect on fecal microbiota, however, deacetylated COS has antibacterial action. Generally, a decrease in the MW of COS has shown the increment of its antibacterial and antifungal properties. However, the impact of MW/DP and DA of COS has been found to be irregular in different studies. The use of different microorganisms and techniques of analysis, as well as the quality, purity, and features of COS examined, could be the cause of these discrepancies. The antimicrobial activity of COS can also be enhanced by incorporating with other nanomaterials. Recently, a study investigated the impact on the antimicrobial activity of quaternary ammonium COS (QACOS) incorporated onto the ZnO/palygorskite (ZnO/PAL) nanocomposite (Hui et al. 2020). The study demonstrated that QACOS/ZnO/PAL had superior MIC values for inhibiting bacteria (0.5 mg/mL for *E. coli* and 1 mg/L for *S. aureus*) than ZnO/PAL and QACOS. Better contact and interface interaction between QACOS/ZnO/PAL and bacteria results in increased antibacterial activity, making it easier to destroy bacteria's structural integrity.

3.2 *Antioxidant*

The antioxidant activity of COS can be attributed to the scavenging activity of reactive oxygen species (ROS) (Yang et al. 2006). ROS has a significant role in the wound healing process when present at a limited level. However, the undue level of ROS can have adverse effects on wound healing due to stimulation of processes like inflammation and fibrotic scarring (Steen et al. 2020). COS possess strong antioxidant activity. Hence, the application of COS with ROS scavenging activity in wound dressing materials could quicken the wound healing process. The molecular mechanism of COS' scavenging activity is not known clearly. But, the antioxidant activity of COS is supposed to be due to presence of hydroxyl and free amino groups (Li et al. 2012). Due to the short-chain length, COS have weaker intermolecular and intramolecular hydrogen bonding. It results in more hydroxyl groups and activated free amine groups, which can react with free radicals and thus promotes antioxidant activity (Li et al. 2012). There is also another concept present in literature for the scavenging mechanism of COS. It suggests that the scavenging mechanism of COS is due to the reaction of hydroxyl and superoxide anion radicals with active hydrogen atoms present in COS, which lead to the generation of stable macromolecule radicals (Jafari et al. 2020).

Further, DP and MW have also shown a significant effect on the antioxidant activity of COS (Zhao et al. 2013, 2016). The antioxidant activity of COS is inversely proportional to the DP. A study reported an increment in the hydroxyl scavenging activity of COS from 34.97 to 48.19% when the DP was decreased from 10 to 3 (Li et al. 2012). However, the same samples exhibited the strongest superoxide radical scavenging ability (Ic_{50}: 0.68 mg ml^{-1}), with COS having DP 10. Furthermore, DA also has a noticeable effect on the antioxidant activity of COS. The ratio of acetylated/deacetylated units in the COS could play a significant role in the antioxidant activity. A study conducted by Mengíbar et al. (2013) revealed that COS exhibited the best antioxidant activity when >70% of the acetylated/deacetylated ratio lie in between 0 and 1. The study observed the highest 2,2-diphenyl-1-picrylhydrazyl (DPPH) radical scavenging activity (90%) when the COS had a 91% acetylated/deacetylated ratio below 1 was used. The DPPH radical scavenging activity has also shown its role in wound closure. In a study, the wound dressing with 54% DPPH radical scavenging activity was used and an improvement in wound closure (>90%) after 10 days was observed. The improved wound closure indicated the significance of intracellular ROS production suppression in tissue regeneration (Le Thi et al. 2020). Qu and Han (2016) investigated the capacity of COS to scavenge free radicals in vitro and in vivo in a high-fat diet mouse model, and COS was revealed to have powerful antioxidant properties that can protect mice from oxidative stress. The study showed that catalase, superoxide dismutase, and glutathione peroxidase activity in the stomach, liver, and serum of mice were significantly increased when COS was combined with a high-fat diet. This suggests that COS may have antioxidant properties and could be able to restore the activity of enzymes that have been affected by a high-fat diet.

3.3 Anti-inflammatory

An inflammatory response is a primitive action of host protection against different stimuli such as physical damage, microbial invasion, ultraviolet irradiation, and immune responses. The human body experiences several diseases such as chronic asthma, multiple sclerosis, inflammatory bowel disease, rheumatoid arthritis, and cancer, due to severe or extended inflammatory action (Ngo et al. 2015). However, acute inflammation can be resolved by involving different cell types and inflammatory mediators such as cytokines (Medzhitov 2008). COS has been well reported for its anti-inflammatory activity. COS have shown to regulate the inflammatory activity in macrophages. Moreover, COS has also been reported to be a potential therapy for inflammatory bowel disease and allergic asthma (Chung et al. 2012; Yousef et al. 2012). COS' anti-inflammatory actions are dependent on their physicochemical features. The inflammatory effect and the possible mechanism of CHS oligomers with DP (2–6) on the lipopolysaccharides (LPS) stimulated RAW 264.7 macrophage cells was studied by Ma et al. (2011). The study suggested an inhibitory effect of the studied COS on the LPS-induced interleukin-6 (IL-6) and tumor necrosis factor α (TNF-α) expression in macrophages. The study proposed the inhibitory mechanism that was based on the down-regulation of the phosphorylated levels of mitogen-activated protein kinase (MAPK) and P13K/Akt signaling pathways and subsequent nuclear factor kappa B (NF-$_k$B) and AP-1 activation (Ma et al. 2011). The effect of MW on the anti-inflammatory activity of COS is not well established as compared to antioxidant and antibacterial activities. There are several contradictory research findings related to the effect of MW on the anti-inflammatory activity of COS. Some studies suggested that high MW (>3 kDa) CHS, COS, and its derivatives were not efficient in the inhibition of nitric oxide (NO) production and the cytokines expression of IL-1, IL-6, or TNF-α in the RAW 264.7 macrophages (Porporatto et al. 2003; Wu et al. 2015). However, COS with MW (<20 kDa) were reported to extraordinarily inhibited the production of NO in LPS induced macrophages (Yang et al. 2010). Thus, the expression of proinflammatory cytokines (IL-1, IL-6, or TNF-α) is mainly effective in the inflammatory phage of the wound healing process. Hence, acceleration in wound healing and reduction in scar formation is highly dependent on the production modulation of the inflammatory mediators (Deldar et al. 2018). Hajiali et al. (2016) developed a wound dressing with the combination of sodium alginate and lavender essential oil having ability to down-regulate cytokine production (IL-1, IL-6, or TNF-α) had shown a significant effect in promoting wound healing of burned skin induced by midrange ultraviolet radiation. Moreover, the similar type of contradiction can be observed for the anti-inflammatory activity of COS due to DA, PA, and the COS production methods. COS' anti-inflammatory properties entice scientists to investigate its potential against inflammatory diseases. COS should be studied further in the food and pharmaceutical industries to see if they may be used as an anti-inflammatory agent.

3.4 Immunostimulatory

Immunostimulation is the process of stimulating immune system cells with various immunostimulants. Immunostimulants are chemicals that boost the immune system's ability to combat infection and disease, such as drugs, vitamins, and immune system components. The immune system is in charge of detecting and eliminating foreign harmful substances in the body. Macrophages play an essential role in host defense because their activation drives adaptive immunity by producing accessory and costimulatory molecules and interactions with T cells (Cheng et al. 2008; Sun et al. 2019). Indeed, macrophages are a key player in immunomodulation, playing a crucial role in wound healing, angiogenesis, cell proliferation, and the exit from the inflammatory phase (Han et al. 2005; Larouche et al. 2018). COS has been shown to exhibit immunostimulatory properties, including activation of macrophage motility, in vivo or in vitro cytokine secretion, and NO release in several investigations (Seo et al. 2000; Allavena et al. 2005).

Moreover, in cultured human monocytes, CHOS may trigger the production of tumor necrosis factor-alpha (TNF-α), a pro-inflammatory known to have a role in cardiovascular problems (Otterlei et al. 1994; Hanna and Frangogiannis 2020). The immunostimulatory activity of COS is dependent on its physicochemical properties such as MW, DP, and DA, similar to other biological properties. COS with a lower MW can up-regulate mRNA expression by activating RAW 264.7 macrophages (Wu et al. 2015; Sumayya and Muraleedhara Kurup 2021). In a study, the COS with MW of 3 kDa was found to have a better expression of TNF-α, IFN-γ, and IL-6 from RAW 264.7 cells compared to the COS with MW 50 kDa (Chen et al. 2008). The immunostimulatory activities of COS also depend on the various preparation methods. COS obtained from three different procedures, including chemical hydrolysis, microwave irradiation, and enzymatic hydrolysis, was compared in vitro and in vivo immunomodulatory capabilities (Xing et al. 2017). The immune system was significantly stimulated by COS obtained from the three methods by acting through cellular and humoral immunity. In comparison to the other two COS, COS produced by microwave irradiation had the most potent immunomodulatory effect, increasing the spleen index and stimulating deacetylated-type hypersensitivity. All of these studies show that COS can aid improve immunostimulatory activity, resulting in faster wound healing.

4 Application of COS-Based Biomaterials in Wound Healing

The role of COS in wound healing has been linked with their above-mentioned biological properties, i.e., antimicrobial, antioxidant, immunostimulatory, and anti-inflammatory. However, immunostimulating activities have a direct impact on the wound healing process. It leads to increased macrophage production and cytokine secretion, both of which are important for wound healing. In a rat wound model, COS showed superior fibroblast activation and break strength than the parent polymer chitin (Minagawa et al. 2007). Furthermore, it has been suggested that COS' wound-healing activity is due to their ability to induce fibroblast formation via controlling fibroblast growth factors. Moreover, the effectiveness of COS in polyurethane membrane modification was also investigated (Luo et al. 2017). COS-modified membranes were found to be very beneficial in retaining moisture content, speeding wound healing, and boosting antibacterial activity, all of which are critical features for wound dressings. Thus, COS can be applied in the different wound healing materials to speed up the healing process. The demand for the employment of COS-based biomaterials with superior biological activities has been rising fast. Due to its biological activities, COS has emerged as a potential therapeutic agent and has shown a positive effect on wound healing (Jafari et al. 2020, 2021). COS has shown its beneficial potential in protecting infection of injured tissues, improving the permeability of air and moisture, supporting cell adhesion and promoting cell proliferation (Venugopal et al. 2008; Sarabahi 2012). However, COS cannot be used directly in wound healing due to its low MW. But, the COS with incorporation with other materials has shown improved biological or physicochemical properties (Table 1). Thus, the wound healing properties of COS can be explored by using it in combination with other biomaterials, viz. fibers, sponges, hydrogels, films, and nanoparticles.

Hydrogels are appealing choices for wound dressings because they can keep wounds moist while also acting as a microbial barrier (Peppas and Hoffman 2020). It is a three-dimensional (3D) network of hydrophilic polymers such as poly(vinyl alcohol) (PVA) or poly(ethylene glycol) (PEG) that can swell in water and store a large quantity of water while preserving structural integrity thanks to chemical or physical crosslinking between individual polymer chains (Murray et al. 2019). Hydrogel wound dressings are beneficial because they can maintain a moist environment in the wound site, aid in removing wound exudates, preventing infection, and promoting tissue regeneration (Kamoun et al. 2015). The advances in hydrogel polymerization processes have also enabled the use of injectable hydrogels that may

Table 1 COS-based material preparation methods and their biological properties

COS complex	Preparation method	Biological/Physicochemical properties	References
Gallic acid-grafted COS	Precipitation	Anti-inflammatory (inhibition of TNF-α; elevation of cytokine IL-8) Antioxidant [DPPH, 70% at a conc. of 200 μg ml^{-1}]	Vo et al. (2017)
Hesperidin-COS	Spay-drying	Increased water solubility Antioxidant (in μM TE g^{-1}) [DPPH & ABTS radical scavenging activities: 446.0 ± 5.7 & 43.0 ± 7.6]	Cao et al. (2018)
β-carotene-COS	Mechanochemical	Enhanced water solubility	Bockuviene and Sereikaite (2019)
Quaternary ammonium COS-ZnO/Palygorskite	Electrostatic self-assembly	Antibacterial (MIC mg ml^{-1}) [*E. coli:* 0.5; *S. aureus*: 1.0]	Hui et al. (2020)
COS-epigallocatechin gallate	Free radical grafting	Antioxidant (in mM TE g^{-1}) [DPPH & ABTS radical scavenging activities: 41.69 ± 5.60 & 123.03 ± 15.47; Ferric reducing antioxidant power, 390.29 ± 17.98; Oxygen radical absorbance capacity: 12.17 ± 1.40] Antimicrobial (MIC mg ml^{-1}) [*E. coli:* 0.01 ± 0.01; *Listeria monocytogenes*: 1.00 ± 0.12]	Singh et al. (2020)
Apple polyphenols-COS microcapsule	Spray-drying	Anti-inflammatory (inhibition of NO & TNF-α; elevation of cytokine IL-10)	Zheng et al. (2020)
Segmented poly(ester-urethane) film-COS	Surface grafting	Surface biocompatibility (blood compatibility)	Liu et al. (2021)

be given directly to a patient's wound, allowing for complete and tailored coverage (Overstreet et al. 2012). Hydrogels can be of various types based on source, ionic charge, preparation, crosslinking, and physical properties (Aswathy et al. 2020). Hydrogels with antimicrobial, antioxidant, and anti-inflammatory activities could enhance the effectiveness of wound dressing (Chai et al. 2017). As COS possesses these biological properties, a COS-based hydrogel may be advantageous in wound healing.

Hydrogel prepared by combining glycidyl methacrylated COS, di-acrylated pluronic, and a recombinant human epidermal growth factor (rhEGF) has been an effective wound care product (Choi and Yoo 2010). The hydrogel was applied to dorsal burn lesions in healing-impaired animals and photo-irradiated in the study. The researchers concluded that including COS and rhEGF into hydrogels could improve epidermal differentiation by increasing local rhEGF concentrations at wound locations while sustaining keratinocytic differentiation (Choi and Yoo 2010). The incorporation of COS with salicylic acid (SA), alginate and gelatin can also be applied in wound healing due to increased antioxidant activity. In a study, H_2O_2-induced grafting polymerization was used to make COS and SA conjugates. Schiff's base reaction was used to synthesize oxidized alginate (OA) and gelatin hydrogels (Oh et al. 2021). The developed hydrogel exhibited improved antioxidant activity. The study investigated wound healing capacity (contraction and closing) of the developed conjugates using a circular-excision wound model on the back of 8-week-old ICR mice. The results indicated that the hydrogels with COS-SA conjugates have negligible cytotoxicity with significant wound healing activity. Moreover, COS could also be conjugated with gallic acid (GA) through hydrogen peroxide-mediated method (Park et al. 2018). Further, the fabrication was done with PVA to form PVA hydrogel with GA-COS. The developed hydrogel inhibited intracellular formation of ROS and exerted improved antimicrobial activity against *Propinibacterium acenes*, which is responsible for the onset of inflammation leading to acne. The study observed a strong effect of GA on the scavenging activity with a minor effect against *P. acenes*. However, the GA-COS manifested a stronger antibacterial activity against *P. acenes*, suggesting the potential of the developed hydrogel conjugates in treating acne.

Nowadays, the investigation for the development of injectable hydrogels is getting attention due to its improved and instant effects. COS-mediated hydrogel has been better than the carboxymethyl CHS (CM-CHS)-based hydrogel (Lv et al. 2019). The hydrogel developed by a combination of COS and alginate resulted in a two-stage gelling process. The primary gelling process was identical to that of a CM-CHS-based hydrogel. Still, about 20 min later, the secondary gelling process developed, resulting in substantially more robust hydrogels with a storage modulus of nearly 1 MPa, which was 10^4 times higher (Lv et al. 2019). The study reported that the hydrogels containing 0.5% COS significantly increased the proliferation of human umbilical cord mesenchymal stem cells (HUMSCs) from 80 to 160% and improved wound healing in an SD mouse skin defect model. The rise in the thickness and integrity of epidermal tissue augmented development of collagen fibers and enhanced expression of vascular endothelial growth factor indicated the potentiality of COS-based injectable hydrogels in wound healing.

Moreover, COS has been observed as an excellent candidate for nanoparticle (NPs) preparation because of its free amine groups and cationic nature and its solubility, biocompatibility, and biodegradability (Chen and Zhao 2019). COS NPs can be prepared using crosslinking, precipitation, ionic gelation, and emulsion-droplet coalescence (Liu et al. 2007). The traditional synthesis of NPs from metals such as silver (Ag) results in chemical reducing agents, leading to increased toxicity. COS can be applied to stabilize Ag-NPs to reduce the risk of toxicity. In a study, the incorporation of COS-Ag NPs into a porcine acellular dermal matrix (pADM) was performed and checked for its effect on the cell viability of 1.929 fibroblasts. The study observed no adverse effect on the cell viability of COS-Ag NPs. However, increasing the concentration of Ag-NPs from 0.4% to 1.2% reduced the cell viability of 1.929 fibroblasts and HUVECs, from 90.8 ± 3% to 86.3 ± 3.3% and 91.6 ± 4.1% to 82.1 ± 5.1%, respectively (Chen et al. 2019). COS-based nanoparticles could boost antibacterial activity, which can check colonized wounds from infection and thus promote wound healing (Kang et al. 2016). In a study, silver chloride nanoparticles were stabilized with chitosan oligomers (COS-AgCl NPs) and their effect on burn wound healing was investigated in a rat model. The result suggested that COS-AgCl NPs-based ointment could heal entirely after 14 days of treatment and was found similar to normal skin. The combination of curcumin (CUR) with COS showed better colloidal stability due to the improved charge distribution in COS compared to CHS (Nguyen et al. 2019). The developed CUR-COS nanoplex was investigated for cytotoxicity towards human keratinocytes cells. The developed nanoplex was simple yet effective, with a CUR utilization rate of over 100%. The CUR-OCHI nanoplex outperformed the native CUR in wound healing efficacy, with wound closure of >90% after 7 days versus 9 days for the native CUR, resulting in reduced scars, owing to its ability to generate high CUR concentration in wound sites (Nguyen et al. 2019).

Furthermore, COS-based nanofibers have emerged as an alternate for the traditional wool or polyamide-based dressings due to improved moisture vapor permeability, biological compatibility, and biodegradability (Chen and Zhao 2019). The electrospinning method is the most studied method for the preparation of COS nanofibers. Because of its effects on skin cell fibroblast proliferation and migration, COS-based nanofibers have been identified as a promising choice for developing wound dressing materials (Li et al. 2013). The study developed a nanofiber from the mixture of PVA and COS by electrospinning with Ag-NPs. The developed nanofiber accelerated the early stage of wound healing compared to gauze. A silver nanoparticles/CHS oligosaccharides/poly(vinyl alcohol) PVA/COS-AgNPs nanofiber was made by electrospinning in another study, and its influence on wound healing was investigated (Li et al. 2016). The study suggested enhanced wound healing and up-regulated expression levels of cytokines related with the transforming growth factor β1 (TGFβ1)/Smad signaling pathways such as TGFβ1, TGFβR1, TGFβRII, collagen I, collagen III, pSmad2, and pSmad3. Thus, it can be deduced that the TGF1/Smad signaling pathway is involved in the wound healing process of the PVA/COS-AgNPs nanofiber. A study of the composite electrospun nanomembrane of fish scale collagen peptides (FSCP) and COS were combined to prepare nanofibers and investigated fibroblast cell viability (Wang et al. 2011). The prepared FSCP/COS nanofibers exhibited significant increment in antibacterial activity against *Staphylococcus aureus*, biocompatibility, and in vitro proliferation of fibroblasts.

Moreover, a nanocomposite based on halloysite and COS (H-COS) can treat chronic wounds (Sandri et al. 2017). The study utilized a composite with 98.6% w/w H and 1.4% COS composition to study the biocompatibility and skin reepithelization capacity in a murine model. The result suggested that the developed H-COS composite could allow better skin reepithelization and reorganization than H or COS alone. COS-based sponges have also been proven to be beneficial as a wound healing accelerator in the early stages of wound healing. In a study conducted by You et al. (2004), after 8 days of initial wounding, the wound treated with the

COS-loaded PVA sponge was almost re-epithelialized, and granulation tissues in the wound healing were significantly replaced by fibrosis. By increasing the COS content in the PVA sponge from 10 to 50%, the average wound area reduced from 90.8 to 27.7%. The study concluded that the influence of COS on wound closure could be attributed to its homeostatic effect and antibacterial activity during the early stages of wound healing. Furthermore, COS can improve the wound healing assets of fish collagen/alginate scaffolds by improving the human dermal fibroblasts-neonatal (NHDF-neo) (Chandika et al. 2015). COS (1–3 kDa) dramatically enhanced cell survival and proliferation, demonstrating, a reduction in MW can aid wound healing.

5 Conclusion

The chapter details the biological properties of COS involved in wound healing and an insight into the different COS-based wound healing materials. COS can accelerate the wound healing process by enhancing biological properties such as antimicrobial, antioxidant, anti-inflammatory, and immunostimulatory. The antimicrobial properties of COS promote wound healing by checking infection from pathogens. COS-induced down-regulation of pro-inflammatory molecules such as IL-1, IL-6, and TNF-α can resolve the inflammation phase of wound healing. Furthermore, COS can speed up the proliferation and remodeling phase of wound healing by refining the proliferation and migration of fibroblasts sententiously with angiogenesis and new blood formation.

Moreover, the impact of the physicochemical properties such as MW, DA, DDA, and PA of COS on wound healing biological properties of COS has been sensed in recent studies. However, the studies are in their primitive stage and there is a need for much more investigation to understand the complete molecular and cellular mechanism of COS behind wound healing. Moreover, the unique properties of COS, such as improved solubility and biocompatibility, have advantages for various wound healing material preparation. Due to the heterogeneous nature of wounds, different wound healing materials are required to heal different wounds. Thus, much more exploitation of COS for the development of wound healing materials such as hydrogel, nanofibers, membrane, sponge, and composite is required. Therefore, the better utilization of the remarkable wound healing bioactivities of COS can be achieved after understanding the complete cellular and molecular mechanism behind it along with the preparation of well-defined COS.

References

Aam BB, Heggset EB, Norberg AL, Sørlie M, Vårum KM, Eijsink VG (2010) Production of chitooligosaccharides and their potential applications in medicine. Mar Drugs 8:1482–1517

Allavena P, Signorelli M, Chieppa M, Erba E, Bianchi G, Marchesi F, Olimpio CO, Bonardi C, Garbi A, Lissoni A (2005) Anti-inflammatory properties of the novel antitumor agent yondelis (trabectedin): inhibition of macrophage differentiation and cytokine production. Cancer Res 65:2964–2971

Aswathy S, Narendrakumar U, Manjubala I (2020) Commercial hydrogels for biomedical applications. Heliyon 6:e03719

Barroca-Aubry N, Pernet-Poil-Chevrier A, Domard A, Trombotto S (2010) Towards a modular synthesis of well-defined chitooligosaccharides: synthesis of the four chitodisaccharides. Carbohydr Res 345:1685–1697

Bockuviene A, Sereikaite J (2019) Preparation and characterisation of novel water-soluble β-carotene-chitooligosaccharides complexes. Carbohydr Polym 225:115226

Bockuviene A, Zalneravicius R, Sereikaite J (2021) Preparation, characterization and stability investigation of lycopene-chitooligosaccharides complexes. Food Biosci 40:100854

Cao R, Zhao Y, Zhou Z, Zhao X (2018) Enhancement of the water solubility and antioxidant activity of hesperidin by chitooligosaccharide. J Sci Food Agric 98:2422–2427

Chai Q, Jiao Y, Yu X (2017) Hydrogels for biomedical applications: their characteristics and the mechanisms behind them. Gels 3:6

Chandika P, Ko S-C, Oh G-W, Heo S-Y, Nguyen V-T, Jeon Y-J, Lee B, Jang CH, Kim G, Park WS (2015) Fish collagen/alginate/chitooligosaccharides integrated scaffold for skin tissue regeneration application. Int J Biol Macromol 81:504–513

Chen Q, Zhao L (2019) The application of chitooligosaccharides on biomaterials. Oligosaccharides of chitin and chitosan. Springer, pp 275–288

Chen C-L, Wang Y-M, Liu C-F, Wang J-Y (2008) The effect of water-soluble chitosan on macrophage activation and the attenuation of mite allergen-induced airway inflammation. Biomat 29:2173–2182

Chen Y, Dan N, Dan W, Liu X, Cong L (2019) A novel antibacterial acellular porcine dermal matrix cross-linked with oxidized chitosan oligosaccharide and modified by in situ synthesis of silver nanoparticles for wound healing applications. Mater Sci Eng: C 94:1020–1036

Cheng A, Wan F, Wang J, Jin Z, Xu X (2008) Macrophage immunomodulatory activity of polysaccharides isolated from *Glycyrrhiza uralensis* fish. Int Immunopharmacol 8:43–50

Choi JS, Yoo HS (2010) Pluronic/chitosan hydrogels containing epidermal growth factor with wound-adhesive and photo-crosslinkable properties. J Biomed Mater Res Part A 95:564–573

Chung MJ, Park JK, Park YI (2012) Anti-inflammatory effects of low-molecular weight chitosan oligosaccharides in IgE–antigen complex-stimulated RBL-2H3 cells and asthma model mice. Int Immunopharmacol 12:453–459

Deldar Y, Pilehvar-Soltanahmadi Y, Dadashpour M, Montazer Saheb S, Rahmati-Yamchi M, Zarghami N (2018) An in vitro examination of the antioxidant, cytoprotective and anti-inflammatory properties of chrysin-loaded nanofibrous mats for potential wound healing applications. Artif Cells Nanomed Biotechnol 46:706–716

Enoch S, Leaper DJ (2008) Basic science of wound healing. Surg Infect (larchmt) 26:31–37

Hajiali H, Summa M, Russo D, Armirotti A, Brunetti V, Bertorelli R, Athanassiou A, Mele E (2016) Alginate–lavender nanofibers with antibacterial and anti-inflammatory activity to effectively promote burn healing. J Mater Chem: B 4:1686–1695

Hamed I, Özogul F, Regenstein JM (2016) Industrial applications of crustacean by-products (chitin, chitosan, and chitooligosaccharides): a review. Trends Food Sci Technol 48:40–50

Han Y, Zhao L, Yu Z, Feng J, Yu Q (2005) Role of mannose receptor in oligochitosan-mediated stimulation of macrophage function. Int Immunopharmacol 5:1533–1542

Hanna A, Frangogiannis NG (2020) Inflammatory cytokines and chemokines as therapeutic targets in heart failure. Cardiovasc Drugs Ther 1–15

Hui A, Yan R, Wang W, Wang Q, Zhou Y, Wang A (2020) Incorporation of quaternary ammonium chitooligosaccharides on ZnO/palygorskite nanocomposites for enhancing antibacterial activities. Carbohydr Polym 247:116685

Jafari H, Bernaerts K V, Dodi G, Shavandi A (2020) Chitooligosaccharides for wound healing biomaterials engineering. Matr Sci Eng: C 111266

Jafari H, Delporte C, Bernaerts K V, De Leener G, Luhmer M, Nie L, Shavandi A (2021) Development of marine oligosaccharides for potential wound healing biomaterials engineering. Chem Eng J Adv 7:100113

Jung W-J, Park R-D (2014) Bioproduction of chitooligosaccharides: present and perspectives. Mar Drugs 12:5328–5356

Kamoun EA, Chen X, Eldin MSM, Kenawy E-RS (2015) Crosslinked poly (vinyl alcohol) hydrogels for wound dressing applications: a review of remarkably blended polymers. Arab J Chem 8:1–14

Kang YO, Jung J-Y, Cho D, Kwon OH, Cheon JY, Park WH (2016) Antimicrobial silver chloride nanoparticles stabilized with chitosan oligomer for the healing of burns. Materials 9:215

Kim S-K, Rajapakse N (2005) Enzymatic production and biological activities of chitosan oligosaccharides (COS): a review. Carbohydr Polym 62:357–368

Kumar M, Brar A, Vivekanand V, Pareek N (2018) Bioconversion of chitin to bioactive chitooligosaccharides: amelioration and coastal pollution reduction by microbial resources. Mar Biotechnol 20:269–281

Kumar M, Rajput M, Soni T, Vivekanand V, Pareek N (2020) Chemoenzymatic production and engineering of chitooligosaccharides and N-acetyl glucosamine for refining biological activities. Front Chem 8:469

Kumar M, Madhuprakash J, Balan V, Singh AK, Vivekanand V, Pareek N (2021) Chemoenzymatic production of chitooligosaccharides employing ionic liquids and *Thermomyces lanuginosus* chitinase. Bioresour Technol 125399

Larouche J, Sheoran S, Maruyama K, Martino MM (2018) Immune regulation of skin wound healing: mechanisms and novel therapeutic targets. Adv Wound Care 7:209–231

Le Thi P, Lee Y, Tran DL, Thi TTH, Kang JI, Park KM, Park KD (2020) *In situ* forming and reactive oxygen species-scavenging gelatin hydrogels for enhancing wound healing efficacy. Acta Biomater 103:142–152

Li K, Xing R, Liu S, Li R, Qin Y, Meng X, Li P (2012) Separation of chito-oligomers with several degrees of polymerization and study of their antioxidant activity. Carbohydr Polym 88:896–903

Li C, Fu R, Yu C, Li Z, Guan H, Hu D, Zhao D, Lu L (2013) Silver nanoparticle/chitosan oligosaccharide/poly (vinyl alcohol) nanofibers as wound dressings: a preclinical study. Int J Nanomed 8:4131

Li C-w, Wang Q, Li J, Hu M, Shi S-j, Li Z-w, Wu G-l, Cui H-h, Li Y-y, Zhang Q (2016) Silver nanoparticles/chitosan oligosaccharide/poly (vinyl alcohol) nanofiber promotes wound healing by activating TGFβ1/Smad signaling pathway. Int J Nanomed 11:373

Li M, Han J, Xue Y, Dai Y, Liu J, Gan L, Xie R, Long M (2019a) Hydrogen peroxide pretreatment efficiently assisting enzymatic hydrolysis of chitosan at high concentration for chitooligosaccharides. Polym Degrad Stab 164:177–186

Li J, Huang W-C, Gao L, Sun J, Liu Z, Mao X (2019b) Efficient enzymatic hydrolysis of ionic liquid pretreated chitin and its dissolution mechanism. Carbohydr Polym 211:329–335

Liaqat F, Eltem R (2018) Chitooligosaccharides and their biological activities: a comprehensive review. Carbohydr Polym 184:243–259

Liu C, Tan Y, Liu C, Chen X, Yu L (2007) Preparations, characterizations and applications of chitosan-based nanoparticles. J Ocean Uni China 6:237–243

Liu Y, Liu Z, Gao Y, Gao W, Hou Z, Zhu Y (2021) Facile method for surface-grafted chitooligosaccharide on medical segmented poly (ester-urethane) film to improve surface biocompatibility. Membranes 11:37

Luo C, Liu W, Luo B, Tian J, Wen W, Liu M, Zhou C (2017) Antibacterial activity and cytocompatibility of chitooligosaccharide-modified polyurethane membrane via polydopamine adhesive layer. Carbohydr Polym 156:235–243

Lv X, Liu Y, Song S, Tong C, Shi X, Zhao Y, Zhang J, Hou M (2019) Influence of chitosan oligosaccharide on the gelling and wound healing properties of injectable hydrogels based on carboxymethyl chitosan/alginate polyelectrolyte complexes. Carbohydr Polym 205:312–321

Ma P, Liu H-T, Wei P, Xu Q-S, Bai X-F, Du Y-G, Yu C (2011) Chitosan oligosaccharides inhibit LPS-induced over-expression of IL-6 and TNF-α in RAW264. 7 macrophage cells through blockade of mitogen-activated protein kinase (MAPK) and PI3K/Akt signaling pathways. Carbohydr Polym 84:1391–1398

Mateos-Aparicio I, Mengíbar M, Heras A (2016) Effect of chito-oligosaccharides over human faecal microbiota during fermentation in batch cultures. Carbohydr Polym 137:617–624

Medzhitov R (2008) Origin and physiological roles of inflammation. Nature 454:428–435

Mengíbar M, Mateos-Aparicio I, Miralles B, Heras Á (2013) Influence of the physico-chemical characteristics of chito-oligosaccharides (COS) on antioxidant activity. Carbohydr Polym 97:776–782

Minagawa T, Okamura Y, Shigemasa Y, Minami S, Okamoto Y (2007) Effects of molecular weight and deacetylation degree of chitin/chitosan on wound healing. Carbohydr Polym 67:640–644

Mourya V, Inamdar N, Choudhari YM (2011) Chitooligosaccharides: synthesis, characterization and applications. Polym Sci Ser A 53:583–612

Murray RZ, West ZE, Cowin AJ, Farrugia BL (2019) Development and use of biomaterials as wound healing therapies. Burns Trauma 7

Ngo D-H, Vo T-S, Ngo D-N, Kang K-H, Je J-Y, Pham HN-D, Byun H-G, Kim S-K (2015) Biological effects of chitosan and its derivatives. Food Hydrocoll 51:200–216

Nguyen M-H, Lee SE, Tran T-T, Bui C-B, Tran T-T, Nguyen T-T, Hadinoto K (2019) A simple strategy to enhance the *in vivo* wound-healing activity of curcumin in the form of self-assembled nanoparticle complex of curcumin and oligochitosan. Mat Sci Eng: C 98:54–64

Oh G-W, Kim S-C, Kim T-H, Jung W-K (2021) Characterization of an oxidized alginate-gelatin hydrogel incorporating a COS-salicylic acid conjugate for wound healing. Carbohydr Polym 252:117145

Otterlei M, Vårum KM, Ryan L, Espevik T (1994) Characterization of binding and TNF-α-inducing ability of chitosans on monocytes: the involvement of CD14. Vaccine 12:825–832

Overstreet DJ, Dutta D, Stabenfeldt SE, Vernon BL (2012) Injectable hydrogels. J Polym Sci Part B: Polym Phy 50:881–903

Park H-H, Ko S-C, Oh G-W, Jang Y-M, Kim Y-M, Park WS, Choi I-W, Jung W-K (2018) Characterization and biological activity of PVA hydrogel containing chitooligosaccharides conjugated with gallic acid. Carbohydr Polym 198:197–205

Peppas NA, Hoffman AS (2020) Hydrogels. Elsevier Biomat Sci 153–166

Porporatto C, Bianco ID, Riera CM, Correa SG (2003) Chitosan induces different L-arginine metabolic pathways in resting and inflammatory macrophages. Biochem Biophys Res Commun 304:266–272

Poshina DN, Raik SV, Poshin AN, Skorik YA (2018) Accessibility of chitin and chitosan in enzymatic hydrolysis: a review. Polym Degrad Stab 156:269–278

Qu D, Han J (2016) Investigation of the antioxidant activity of chitooligosaccharides on mice with high-fat diet. Rev Bras Zootec 45:661–666

Rajasekaran S, Rao SS, Dalavi PA, Prabhu A, Anil S, Venkatesan J, Bhat SS (2021) Rapid microwave-assisted biosynthesis of chitooligosaccharide coated silver nanoparticles: assessments of antimicrobial activity for paediatric pulp therapy. Adv Nat Sci: Nanosci Nanotechnol 11:045018

Sánchez Á, Mengíbar M, Rivera-Rodríguez G, Moerchbacher B, Acosta N, Heras A (2017) The effect of preparation processes on the physicochemical characteristics and antibacterial activity of chitooligosaccharides. Carbohydr Polym 157:251–257

Sandri G, Aguzzi C, Rossi S, Bonferoni MC, Bruni G, Boselli C, Cornaglia AI, Riva F, Viseras C, Caramella C (2017) Halloysite and chitosan oligosaccharide nanocomposite for wound healing. Acta Biomat 57:216–224

Sarabahi S (2012) Recent advances in topical wound care. Indian J Plast Surg 45:379–387

Sashiwa H, Fujishima S, Yamano N, Kawasaki N, Nakayama A, Muraki E, Sukwattanasinitt M, Pichyangkura R, Aiba S-i (2003) Enzymatic production of N-acetyl-D-glucosamine from chitin. Degradation study of N-acetylchitooligosaccharide and the effect of mixing of crude enzymes. Carbohydr Polym 51:391–395

Seo W-G, Pae H-O, Kim N-Y, Oh G-S, Park I-S, Kim Y-H, Kim Y-M, Lee Y-h, Jun C-D, Chung H-T (2000) Synergistic cooperation between water-soluble chitosan oligomers and interferon-γ for induction of nitric oxide synthesis and tumoricidal activity in murine peritoneal macrophages. Cancer Lett 159:189–195

Singh A, Benjakul S, Huda N, Xu C, Wu P (2020) Preparation and characterization of squid pen chitooligosaccharide–epigallocatechin gallate conjugates and their antioxidant and antimicrobial activities. RSC Adv 10:33196–33204

Sivaramakrishna D, Bhuvanachandra B, Mallakuntla MK, Das SN, Ramakrishna B, Podile AR (2020) Pretreatment with KOH and KOH-urea enhanced hydrolysis of α-chitin by an endo-chitinase from *Enterobacter cloacae* subsp. cloacae. Carbohydr Polym 235:115952

Steen EH, Wang X, Balaji S, Butte MJ, Bollyky PL, Keswani SG (2020) The role of the anti-inflammatory cytokine interleukin-10 in tissue fibrosis. Adv Wound Care 9:184–198

Sumayya A, Muraleedhara Kurup G (2021) In vitro anti-inflammatory potential of marine macro-molecules cross-linked bio-composite scaffold on LPS stimulated RAW 264.7 macrophage cells for cartilage tissue engineering applications. J Biomat Sci Polym 32:1040–1056

Sun L, Liu W, Zhang L-j (2019) The role of toll-like receptors in skin host defense, psoriasis, and atopic dermatitis. J Immunol Res 2019

Venugopal J, Low S, Choon AT, Ramakrishna S (2008) Interaction of cells and nanofiber scaffolds in tissue engineering. J Biomed Mat Res Part B: Appl Biomat 84:34–48

Vo T-S, Ngo D-H, Bach LG, Ngo D-N, Kim S-K (2017) The free radical scavenging and anti-inflammatory activities of gallate-chitooligosaccharides in human lung epithelial A549 cells. Process Biochem 54:188–194

Wallace HA, Basehore BM, Zito PM (2017) Wound healing phases

Wang Y, Zhang C-l, Zhang Q, Li P (2011) Composite electrospun nanomembranes of fish scale collagen peptides/chito-oligosaccharides: antibacterial properties and potential for wound dressing. Int J Nanomed 6:667

Wu N, Wen Z-S, Xiang X-W, Huang Y-N, Gao Y, Qu Y-L (2015) Immunostimulative activity of low molecular weight chitosans in RAW264. 7 macrophages. Mar Drugs 13:6210–6225

Wu T, Shao Q, Zhao W (2017) Microwave hydrothermal degradation of chitosan. J Shandong Univ Sci Technol (Nat Sci) 02

Xia W, Liu P, Zhang J, Chen J (2011) Biological activities of chitosan and chitooligosaccharides. Food Hydrocoll 25:170–179

Xing R, Liu Y, Li K, Yu H, Liu S, Yang Y, Chen X, Li P (2017) Monomer composition of chitooligosaccharides obtained by different degradation methods and their effects on immunomodulatory activities. Carbohydr Polym 157:1288–1297

Yang Y, Shu R, Shao J, Xu G, Gu X (2006) Radical scavenging activity of chitooligosaccharide with different molecular weights. Eur Food Res Technol 222:36–40

Yang E-J, Kim J-G, Kim J-Y, Kim SC, Lee NH, Hyun C-G (2010) Anti-inflammatory effect of chitosan oligosaccharides in RAW 264.7 cells. Cent Eur J Biol 5:95–102

You Y, Park WH, Ko BM, Min B-M (2004) Effects of PVA sponge containing chitooligosaccharide in the early stage of wound healing. J Mat Sci: Mat Med 15:297–301

Yousef M, Pichyangkura R, Soodvilai S, Chatsudthipong V, Muanprasat C (2012) Chitosan oligosaccharide as potential therapy of inflammatory bowel disease: therapeutic efficacy and possible mechanisms of action. Pharmacol Res 66:66–79

Zhao D, Wang J, Tan L, Sun C, Dong J (2013) Synthesis of N-furoyl chitosan and chito-oligosaccharides and evaluation of their antioxidant activity in vitro. Int J Biol Macromol 59:391–395

Zhao X, Yu Z, Wang T, Guo X, Luan J, Sun Y, Li X (2016) The use of chitooligosaccharide in beer brewing for protection against beer-spoilage bacteria and its influence on beer performance. Biotechnol Lett 38:629–635

Zheng H-Z, Cui C-L, Jeong W-S, Chung S-K (2020) Anti-inflammatory effect of unripe apple polyphenols-chitooligosaccharides microcapsule against LPS-induced RAW 264.7 cells. Appl Biol Chem 63:1–9

Zou P, Yang X, Wang J, Li Y, Yu H, Zhang Y, Liu G (2016) Advances in characterisation and biological activities of chitosan and chitosan oligosaccharides. Food Chem 190:1174–1181

Chitooligosaccharides: Preparation and Applications in Food and Nutraceuticals

Soottawat Benjakul, Avtar Singh, and Ajay Mittal

1 Introduction

Chitooligosaccharide (COS) is the degraded product of chitosan (CS) extracted from shrimp and crab shells, squid pen, etc. (Benhabiles et al. 2012; Rakkhumkaew and Pengsuk 2018; Singh et al. 2019a, 2020b). CS is known to possess a variety of bioactivities such as antibacterial, antioxidant, anti-aging, anti-cancer, drug delivery, etc. (Elieh-Ali-Komi and Hamblin 2016). However, aqueous acid solubility limits the use of CS in various fields. To produce water-soluble CS derivatives, hydrolysis via physical, chemical, and enzymatic methods has been used (Singh et al. 2021d; Liang et al. 2018). Physical methods such as irradiation and ultrasonication, etc., are not suitable to hydrolyze or degrade CS to low molecular weight (MW) (Lodhi et al. 2014). On the other hand, uncontrolled and harsh chemical hydrolysis is unable to produce the desired COS (Lodhi et al. 2014). Enzymatic hydrolysis has gained interest in the production of COS with a controlled degree of depolymerization (DDP) (Lee et al. 2008; Lodhi et al. 2014). Specific enzyme (chitosanase) and non-specific enzymes (carbohydrases, proteases, cellulase, lipase, etc.) have been used to hydrolyze CS for the production of COS (Lodhi et al. 2014). However, utilization of specific enzymes could lead to increase in production cost of COS.

COS has been used widely as antioxidant and antibacterial agents (Rhoades et al. 2006; Singh et al. 2019a) and vectors in gene therapy (Köping-Höggård et al. 2003, 2004). Furthermore, it has been used to reduce metastasis and tumor growth in cancer (Shen et al. 2009; Nam et al. 2007), and to increase bone strength for preventing osteoporosis (Klokkevold et al. 1996; Ratanavaraporn et al. 2009; Jung et al. 2006). It also showed immune modulatory effects (Long et al. 2018), anti-fungal activities

S. Benjakul (✉) · A. Singh · A. Mittal
International Center of Excellence in Seafood Science and Innovation, Faculty of Agro-Industry, Prince of Songkla University, Hat Yai, Songkhla 90110, Thailand
e-mail: soottawat.b@psu.ac.th

S.-K. Kim (ed.), *Chitooligosaccharides*,
https://doi.org/10.1007/978-3-030-92806-3_13

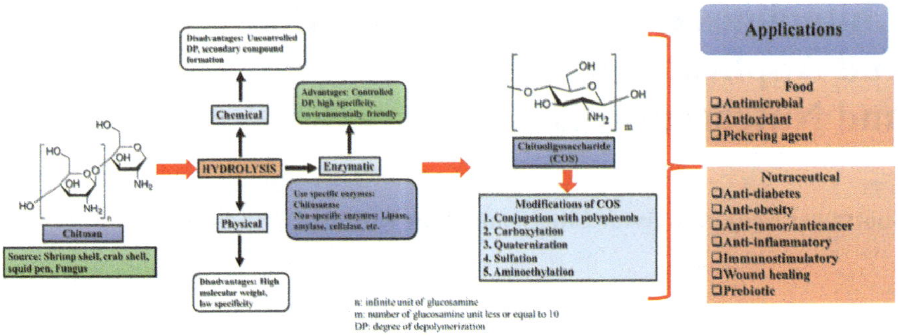

Fig. 1 Overview of chitooligosaccharide production and applications

(Mei et al. 2015; Moon et al. 2017), and lowering effect on serum glucose levels in diabetics (Liu et al. 2007).

Additionally, COS has been gaining the attention of food scientists due to its excellent antioxidant and antibacterial activities. Although CS has been employed in food, especially as films or emulsifying agents, etc., few studies have been available for COS in foods. Recently, COS from squid pen has been used to extend the shelf life and improved the quality of surimi gel and fillets from tuna and Asian sea bass (Singh et al. 2019a, b, 2021a, b, c). The objective of this book chapter is to provide the updated information regarding the bioactivities and application of COS in food as well as its nutraceutical aspect. Furthermore, several preparative methods of COS have been also revisited (Fig. 1).

2 Preparation and Characterization of Chitooligosaccharide (COS)

2.1 *Chemical Methods*

CS is made up of glucosamine unit linked via unstable glycosidic bonds, which are cleavable by hydrolyzing agents such as acids, enzymes, etc., to produce CS oligomers or COS with variable degree of polymerization (DPs) or degree of depolymerization (DDP) (Singh et al. 2019a; Liang et al. 2018; Lodhi et al. 2014). CS with DP less than 20 and an average MW less than 3.9 kDa are called COS, CS oligomers or chitooligomers (Il'ina and Varlamov 2004; Lodhi et al. 2014). Additionally, COS or CS oligomer having MW higher than 3–4 kDa (range of 30–75 kDa) has been reported (Singh et al. 2021d; Boonviset and Pirak 2020; Singh et al. 2019a). Non-enzymatic hydrolysis of CS includes chemicals and physical methods. Hydrolysis of CS can be performed using mineral acid such as HCl, or a mixture of acids and electrolytes including nitrous acid, phosphoric acid, and hydrofluoric acid, etc. (Mourya et al. 2011). Moreover, hydrogen peroxide or persulfate have also been used for CS hydrolysis (Mourya et al. 2011). Lactic, trichloroacetic, formic, and acetic acids were also used for hydrolytic reactions (Mourya et al. 2011; Il'ina and Varlamov 2004). During the cleavage of glycosidic bond (β-1,4-linkage) of CS, a new alcohol hydroxyl is formed at the C4 atom chain fragment (Struszczyk et al. 1999; Il'ina and Varlamov 2004). In addition, a free glycoside hydroxyl (an aldehyde group) can be formed at C1 of fragment, which results in the production of a mixture of *N*-acetylglucosamine and glucosamine units (Il'ina and Varlamov 2004; Einbu et al. 2007). The rate of hydrolysis depends on the concentration of the catalyst, substrate, temperature, duration of the reaction, and degree of deacetylation (DDA) of CS (Il'ina and Varlamov 2004). Generally, the mechanism consists of the formation of conjugate via hydration of the glycosidic oxygen atom, which is followed by the breakdown of exocyclic O-5 to C-1 bond and generates a cyclic carbonium–oxonium ion. Reaction with water then gives the protonated reducing sugar (Einbu et al. 2007). Einbu et al. (2007) stated that hydrolysis of chitin (*N*-acetyl-D-glucosamine; $GlcNAc_4$) and CS (D-glucosamine; $GlcN_4$) tetramers using HCl (12.07 M) is a non-random depolymerization process. The glycosidic bond next to one of the end residues is hydrolyzed at 2.5 and 2.0 times faster as compared to the other glycosidic linkages in the chitin and CS chain, respectively. Moreover, the glycosidic bond residing next to the non-reducing end had the fastest rate of hydrolysis. Furthermore, rate constant of glycosidic bond hydrolysis in $GlcNAc_4$ was 50 time higher than that in $GlcN_4$, which is associated with the catalytic role of the *N*-acetyl group and presence of the positively charged amino-group on the *N*-deacetylated sugar residue (Einbu et al. 2007). Various studies have been reported on the acid hydrolysis of CS (Table 1). Due to safety concern and uncontrolled hydrolysis, chemical methods are limited. In addition, the secondary products formed during hydrolysis are difficult to remove from the mixture (Lodhi et al. 2014). However, chemical methods such as oxidative hydrolysis using hydrogen peroxide alone or in combination with ascorbic acid, ferric chloride, etc., have been used widely to produce water-soluble CS oligomers or COS (Li et al. 2019; Zoldners et al. 2005). Although these methods reduced the MW and viscosity of CS, a few studies have been conducted on the effect of oxidative hydrolysis on the bioactivities of COS or water-soluble CS.

2.2 *Physical Methods*

Physical methods such as microwave (Xing et al. 2005; Omari et al. 2012), lambda radiation (Ulański and Rosiak 1992; Hai et al. 2003; Zhong et al. 1993; Yoksan et al. 2004), and high-intensity ultrasonication (Wu et al. 2008; Popa-Nita et al. 2009) have been employed to hydrolyze CS. However, those methods are unable to reduce the MW of CS efficiently (Table 1). The main principle behind the physical methods is the absorption of energy by CS molecules, which might cause the scission of chemical bonds (Mourya et al. 2011). However, due to high energy consumption and lower

Table 1 Applications and bioactivities of chitooligosaccharides

COSs	MW of COS (kDa) and DP	Applications and Bioactivities	References
Squid pen COS	74.5 kDa	Inhibit discoloration and extend the shelf life of yellowfin tuna slices at 4 °C	Singh et al. (2021b)
		Inhibit discoloration and extend the shelf life of yellowfin tuna slices at 4 °C in combination with oxygen-based MAP	Singh et al. (2021c)
		1. Antimicrobial activity against *Pseudomonas aeruginosa* PSU.SCB.16S.11, *Listeria monocytogenes* F2365, *Vibrio parahaemolyticus* PSU.SCB.16S.14, *Staphylococcus aureus* DMST 4745, *Salmonella enterica* serovar Enteritidis S5–371 2. Antioxidant activities 3. Inhibit lipid oxidation and extend the shelf life of refrigerated sardine surimi gel	Singh et al. (2019)
		Shelf life extension of Asian sea bass slices stored at 4 °C for 12 days in combination with high-voltage cold atmospheric plasma	Singh and Benjakul (2020)
		Shelf life extension of Asian sea bass slices inoculated with *Pseudomonas aeruginosa* stored at 4 °C for 12 days in combination with high-voltage cold atmospheric plasma	Singh et al. (2021a)
COS	99.77 kDa	Shelf life extension of reduced fat-low sugar Chinese pork sausage for 14 days at room temperature	Boonviset and Pirak (2020)
COS	3.8 kDa	Enhanced stability of β-carotene-COS complexes	Bockuviene and Sereikaite (2020)
Shrimp shell COS	13 kDa	Shelf life extension of bread for 4 days at 30 °C	Rakkhumkaew and Pengsuk (2018)
COS	>100 to <1 kDa	Antimicrobial activity against *Bacillus cereus*, *Pseudomonas aeruginosa*, and *Candida albicans*	El-Sayed et al. (2019)

(continued)

Table 1 (continued)

COSs	MW of COS (kDa) and DP	Applications and Bioactivities	References
COS	5.06 kDa	1. Antioxidant activities 2. Antimicrobial activity against *Escherichia coli*, *Salmonella typhimurium*, and *Salmonella enteritidis*	Laokuldilok et al. (2017)
COS	DP: 2–7	1. Antioxidant activities 2. Antimicrobial activity against *Salmonella enterica*, *Pseudomonas aeruginosa*, *Enterobacter* sp., *Staphylococcus aureus*, *Micrococcus luteus*, *Listeria monocytogenes*, and *Bacillus cereus* 3. Antifungal activity against *Fusarium solani* and *Rhizoctonia solani*	Affes et al. (2020)
COS	DP: 2–6	Anti-tumor activity against human osteosarcoma MNNG, MG63, and U2OS cell lines	Pan et al. (2021)
COS nanoparticle	2.005 kDa	1. Act as prebiotic for *Lactobacillus plantarum* 4496, *Lactobacillus rhamnosus* 306, *Lactobacilluscasei* FEGY 9973, *Lactobacillus acidophilus* CH-2, *Bifidobacteriumlactis* BB12, *Streptococcus thermophilus* CH-1, *Lactobacillus delbrueckii* subsp. *bulgaricus* Lb-12 DRI-VAC 2. Fortified in yogurt	Ismail et al. (2020)
Food grade COS from crab	≤3000 kDa	Shelf life extension of grass carp (*Ctenopharyngodon idellus)* for 11 days assisted with ultrasonication	Yu et al. (2021a, b)
COS-epigallocatechin gallate conjugate	74.5 kDa	1. Antimicrobial activity against *Listeria monocytogenes* and *Escherichia coli* 2. Antioxidant activities	Singh et al. (2020a, b)
COS-polyphenol conjugates	3–5 kDa	Antioxidant activities	Eom et al. (2012)
COS-gallic acid conjugate	3–5 kDa	Antioxidant activities (mouse macrophage RAW264.7 cells)	Ngo et al. (2011)
		Free radical scavenging and anti-inflammatory activities (human lung epithelial A549 cells)	Vo et al. (2017)
COS-lycopene complex	3.8 kDa	Antioxidant activities	Bockuviene et al. (2021)

(continued)

Table 1 (continued)

COSs	MW of COS (kDa) and DP	Applications and Bioactivities	References
COS	DP: 2–3	Alleviate the hepatic steatosis in high-fat diet-fed mice and inhibition of cluster of differentiation 36 (CD36)	Zhao et al. (2019b)
COS	DP: 2–12; 2490 Da	Anti-tumor activity (human cervical cancer cells)	Zhao et al. (2019a)
COS	DP: 2–6	Inhibitory effects against hepatocellular carcinoma	Zou et al. (2019)
COS	DP: 3–7	Reduce the viability of human breast cancer cell lines, BT-474 and SUM-159	Mallakuntla et al. (2021)
Curcumin-loaded liposomes coated with COS	DP: 2–7; ≤1 kDa	Inhibitory effect of MCF-7 cells	Li et al. (2021)
COS	DP: 2–5; 489.73 kDa	Anti-obesity activity (mice)	Zhou et al. (2020)
COS	≤1 kDa	1. Prebiotic 2. Antiglycation activity 3. Reduce serum Nε-(carboxymethyl) lysine level, related to anti-obesity activity	Liu et al. (2020)
COS tablets	≤1 kDa	1. Upregulate gene expressions of cholesterol excretion in mice 2. Prevent and treat metabolic hyperlipidaemia syndrome	Yang et al. (2019a, b)
COS-citronellol conjugate	1 kDa	Anti-inflammatory activity (mice)	Mao et al. (2021)
COS	8 kDa	Anti-inflammatory activity (mice and mouse macrophage RAW264.7 cells)	Sanchez et al. (2018)
COS	DP: 2–4	Neuroprotective (human SH-S5Y5 neurons) and anti-inflammatory (in RAW macrophages)	Santos-Moriano et al. (2018)
Polyvinyl alcohol/methacrylate κ-carrageenan/COS hydrogel	1–3 kDa	Wound healing properties	Chandika et al. (2021a)
Fish collagen/poly (ε-caprolactone)/COS			Chandika et al. (2021b)

COS: chitooligosaccharide; MW: molecular weight; DP: degree of polymerization; kDa: kilodalton

DDA, physical methods have not been commonly implemented for depolymerization of CS.

2.3 Enzymatic Method

Enzymatic hydrolysis of CS can be carried out by using specific enzymes (chitinases or chitosanases) and non-specific enzymes such as lipase, carbohydrase, proteases, etc. (Table 1) (Lodhi et al. 2014; Singh et al. 2019a). Jeon et al. (2001) prepared COS using a continuous ultrafiltration membrane reactor system and an immobilized chitosanase from *Bacillus* sp. adsorbed on chitin as a carrier. Similarly, chitosanase immobilized onto agar or agarose gel was used to prepare pentamer and hexamer of COS (Kuroiwa et al. 2002). Kidibule et al. (2021) extracted chitinase Chit42 from fungus (*Trichoderma harzianum*) and immobilized on magnetic nanoparticles and CS beads to produce COS from chitin and CS from various sources. Three different types of COS, fully deacetylated (fdCOS), partially acetylated (paCOS), and fully acetylated (faCOS) of high bioactivities were prepared from chitin or CS (varying DDA) as substrates using a combination of chitinase or chitosanase without and with immobilization (Miguez et al. 2021). Various studies have been reported on the usage of specific enzymes (chitosanase) for depolymerization of CS extracted from different sources (Ichikawa et al. 2002; Sakai et al. 1991; Fukamizo et al. 1995; De Medeiros Dantas et al. 2020; Kidibule et al. 2020). However, the high cost and availability of specific enzymes, such as chitosanase and chitinase, limit their use in an industrial scale. Therefore, various non-specific enzymes have been used to produce COS or CS oligomers of varying MW instead of chitosanase (Roncal et al. 2007). Roncal et al. (2007) reported that cellulase, pepsin, and lipase A were able to produce COS from CS with characteristic equivalent to that accomplished by chitosanase. Out of three enzymes, pepsin (1:100, w/w) produced a higher yield (52%) of COS (DP: 16) after 20 h at 40 °C and pH 4. Similarly, Lee et al. (2008) and Gohi et al. (2017) successfully produced COS via CS hydrolysis with the help of lipase and pepsin, respectively. Similarly, carbohydrases such as α- and β-amylase were able to hydrolyze CS (Rokhati et al. 2013). In general, the pH of the CS solution was varied from 4 to 5, except pepsin (2–3.5), which can work well under acidic conditions (Gohi et al. 2017; Lee et al. 2008; Rokhati et al. 2013). Normally, hydrolysis of polysaccharides by enzymes could be achieved via hydrolyzing an interior (endoaction) or terminal (exoaction) glycosidic bond, which resulted in the formation of monomer units (Rokhati et al. 2013; Roncal et al. 2007). CS consists of four types of glycosidic linkages (-GlcN-GlcN-, -GlcN-GlcNAc-, -GlcNAc-GlcN-, and -GlcNAc-GlcNAc-) (Roncal et al. 2007; Singh et al. 2019a), therefore, it can be cleaved at four different linkages by the enzymes related to their specificities at different pHs and DDAs (Lee et al. 2008). Singh et al. (2019b) also obtained COS with different DDP of squid pen CS when hydrolyzed using three different enzymes (lipase, amylase, and pepsin) at pH 5 for the same hydrolysis time. Therefore, different enzymes can produce COS of different DDP and MW.

3 Bioactivities and Applications of COS in Foods

3.1 Antioxidative Activity

Lipid oxidation has been known to produce off-flavor and off-odor of fatty foods, especially fatty fish, which is associated with oxidative destruction of polyunsaturated fatty acid (PUFA), thus causing short shelf life (Mao and Wu 2007). Synthetic antioxidants including butylated hydroxytoluene and butylated hydroxyanisole have been employed to inhibit lipid oxidation in the various foods. However, concerning health-related issue with the synthetic antioxidants, food industries are focusing on natural antioxidants (Hamed et al. 2016). COS, a natural antioxidant, which can provide H-ion to free radicals with the help of free amino group at the glucosamine unit, indicating the potential of COS to scavenge free radicals (Laokuldilok et al. 2017; Senphan and Benjakul 2014). High MW COS has lower antioxidant activity as compared to the lower MW counterpart (Kim and Thomas 2007), which is associated with the higher exposure of free amino group of the latter to scavenge radicals and acts as hydrogen donor. COS with different MWs in the range of 5–50 kDa showed the increases in DPPH radical scavenging activity, metal chelating activity, and reducing power as compared to the CS (Laokuldilok et al. 2017). Singh et al. (2019a) hydrolyzed squid pen CS using lipase and the obtained COS showed excellent radical (DPPH and ABTS) scavenging activities, reducing power, and metal chelating activity. The COS has been modified using various methods to enhance the antioxidant activity. Singh et al. (2020a) prepared COS and epigallocatechin gallate (EGCG) using free radical method, in which higher antioxidant activity was obtained for COS-EGCG conjugate as compared to COS. In contrast, Bockuviene et al. (2021) observed lower DPPH radical scavenging activity of water-soluble lycopene-COS complexes prepared using kneading method.

3.2 Antimicrobial Activity

COS has been reported to inhibit the bacterial growth in food, plant, animal, as well as to cease viral growth. Generally, antimicrobial potential of COS is more likely associated with leakage of intracellular constituents caused by altering the membrane integrity when attacked by charged COS (Kumirska et al. 2011; No et al. 2007). Moreover, penetration of COS into the cells resulted in destruction of mRNA and protein translation machinery of bacteria and interaction between the microbial DNA. Chelation of essential nutrients, metals, and spore elements could be another reason for microbial inactivation (No et al. 2007). The antimicrobial activity of COS is mainly varied with its MW, DP, and chemical modification. In general, COS with lower MW showed the higher antimicrobial activity than the medium- and high-MW COS. However, on contrary, Jeon et al. (2001) noticed higher antimicrobial for COS having high MW than the lower MW COS, which might be due to the differences in sources of COS and preparative procedures. COS (~8 kDa) prepared with the help of crude chitinase isolated from *Astrosphaeriella* sp. BUSK 55–1 had higher inhibitory activity against *Staphylococcus aureus, Escherichia coli, Pseudomonas aeruginosa,* and *Bacillus subtilis* as compared to those produced from chitinase prepared from *Oxydothis* sp. BUSK43-2 (Pilantanapak et al. 2017). Fernandes et al. (2010) documented that COS (<3 and <5 kDa) had a greater inhibitory effect against *Klebsiella pneumoniae, Escherichia. coli, and Pseudomonas aeruginosa* than CS (MW: 628, 591, and 107 kDa). Additionally, COS has an excellent ability to create novel antifungal drugs. The result was confirmed by Mei et al. (2015), in which 0.5 and 1% COS significantly suppressed the growth of dermatophytic fungus *Trichophyton rubrum.* Although COS is efficient to inhibit the bacterial growth, various methods have been employed to enhance those inhibitory activities. For example, *N*-quaternization of COS enhanced antimicrobial activity, and higher efficiency of antimicrobial activity was noticed when alkyl chain length of the quaternary amine was increased (Rúnarsson et al. 2007, 2010). Generally, quaternary ammonium salt increases the electropositivity and water solubility of COS (Hui et al. 2020). Therefore, highly electropositive COS could bind with negatively charged bacterial surface to a higher extent. Additionally, conjugation of various phenolic compounds (PC) with CS successfully enhanced the antibacterial properties of CS-PC conjugate (Mittal et al. 2020a, b; Attia et al. 2021; Yi et al. 2021). However, rare information has been available for COS-PC conjugate, especially in terms of antimicrobial properties (Oh et al. 2021; Singh et al. 2020a). For example, squid pen COS conjugated with EGCG showed drastic increase in antibacterial activity against both pathogenic and foodborne bacteria as determined by the minimum inhibitory and bactericidal concentration methods (Singh et al. 2020a).

4 Applications of COS in Foods

CS has been mostly explored in terms of gelling or film-forming properties and as a coating agent in various kinds of foods. However, applications of COS in foods are limited (Table 1), which might be related to tedious preparative methods, lower yield, and high cost of specific/non-specific enzymes for CS hydrolysis. COS and CS from white leg shrimp shell were used as substitute of flour in bread formulation (1 g/100 g total weight basis) and stored at 30 °C for 4 days. Bread added with COS did not show the growth of *Bacillus cereus* and *Rhizopus* sp. throughout the storage, whereas bread containing CS showed microbial growth after day 3. Additionally, COS was able to delay the fruity odor to a higher extent (4 days) than the CS (3 days) (Rakkhumkaew and Pengsuk 2018). Similarly, squid pen COS enhanced the gel properties and shelf life of gel prepared from sardine surimi stored for 10 days at 4 °C (Singh et al. 2019a, b). Singh (2021a) reported the inhibition of *Pseudomonas aeruginosa* inoculated on Asian sea bass slices when treated with squid pen COS in combination with high-voltage atmospheric plasma (HV-CAP). Additionally, COS was able to reduce lipid oxidation of seabass slices caused by

HV-CAP. Similarly, COS at appropriate levels was able to lower the lipid or protein oxidation and inhibited the growth of various pathogenic and spoilage bacteria including psychrophilic bacteria, *Clostridium perfringens*, lactic acid bacteria, Enterobacteriaceae, *Pseudomonas*, and hydrogen sulfide (H_2S)-producing bacteria of HV-CAP-treated slices during the first 15 days of storage at 4 °C (Singh and Benjakul 2020). Moreover, COS lowered metmyoglobin formation and color discoloration in yellowfin tuna slices when treated with COS at different concentrations (200–800 ppm), irrespective of modified atmospheric packaging (Singh et al. 2021b, c, d). COS could also prolonged the shelf life of minced pork via inhibiting Gram-positive and Gram-negative microbes (Chantarasataporn et al. 2014). In addition to antimicrobial or antioxidative properties, Yu et al. (2021b) explored COS as a thickener in Pickering emulsion. Glycated whey protein isolate (gWPI)–COS nanoparticles were assembled through electrostatic interactions. Those prepared nanoparticles successfully increased the viscoelasticity and stability of the curcumin-loaded Pickering emulsions than in the emulsions prepared using WPI or gWPI or WPI–COS alone. This was more likely associated with the interactions of COS with the proteins molecules and helped in the interactions between neighboring proteins, thus favoring the formation of aggregates (Yu et al. 2021b).

5 Chitooligosaccharides as a Nutraceutical

High biocompatibility of COS leads to its exploitation in treatment of various kinds of diseases related to cardiovascular, hypertension, inflammation, obesity, cancer, etc. (Table 1) (Zhai et al. 2021). Moreover, it has been known to enhance immune or antibody response (Yang et al. 2019a, b).

5.1 Anti-obesity Activity

The sedentary lifestyle with consumption of junk food resulted in the predominance of obesity, regardless of age and gender, which in turn causes various cardiovascular diseases, digestive problems, type 2 diabetes mellitus (T2DM), etc. (Singh et al. 2021e; Lavie et al. 2009). CS has been considered as effective agent in controlling plasma lipids and body weight in both animals and humans (Muanprasat and Chatsudthipong 2017). Considering the limitations of CS, COS has shown higher absorption ability in intestine. Various studies have been conducted to determine the COS potential to lower body weight gain, serum levels of triglyceride/cholesterol or lipid accumulation in the liver and adipose tissues (Choi et al. 2012; Huang et al. 2015; Muanprasat and Chatsudthipong 2017). Moreover, when obese rats were fed with COS, the high-density lipoproteins or cardioprotective lipid-containing particle, which removes excessive amount of cholesterol from tissues to the liver, were found at higher levels than control rats (treated without COS) (Huang et al. 2015; Kang et al.

2012). Although various theories have been raised for anti-obesity activity of COS, no exact mechanism has been well elucidated. COS may have the ability to inhibit enzyme (mainly lipase) or to bind with bile acids, which lower the fat absorption and enhance fecal fat excretion (Kang et al. 2012). Moreover, at the cellular levels, several cell lines studies confirmed the ability of COS to cease adipocyte cell differentiation, triglyceride accumulation, and the expression of adipogenic markers in 3T3-L1 cells (Cho et al. 2008). The anti-adipogenic effect of COS also downregulate peroxisome proliferator-activated receptor gamma (PPAR-γ) or de-methylation of leptin, which plays important roles in adipocyte differentiation (Bahar et al. 2013). Also, COS has been known to suppress the urge to eat via reducing ghrelin (hunger hormone) in the stomach (Kao et al. 2012). Additionally, COS has also been shown to reduce the triglyceride level in obese diabetic mice via inhibiting digestive enzymes, especially lipase. Kang et al. (2012) observed inhibition of pancreatic lipase activity (in vitro) in the presence of COS. However, a few reports have been available on the inhibition of those digestive enzymes.

5.2 *Anti-diabetic Activity*

When beta-islet cells (pancreas) are unable to produce enough insulin, or the inability of body to utilize insulin efficiently occurs, the diabetes, especially hyperglycemia can be found (Karadeniz and Kim 2014). Even though T2DM patients contained healthy β-cells for many years after disease onset, the high level of glucose can cause cellular damage (Karadeniz and Kim 2014). Therefore, to improve the health status of T2DM patients, β-cells and insulin secretion needed to be control. COS showed potential to protect β-cells against high glucose via accelerating the proliferation of pancreatic β-cells, which resulted in higher secretion of insulin to lower the glucose (Karadeniz et al. 2010). Liu et al. (2007) reported that COS treatment could improve the general condition and diabetic symptoms of diabetic rats, decrease the blood glucose levels, and normalize the impaired insulin sensitivity. Moreover, COS activities such as enzyme inhibition, antioxidant, immunomodulation, etc., could help non-obese diabetic mice from developing T2DM, which advocated the protection of pancreatic β-cells and normalize the crucial insulin secretion (Srinivasan et al. 2005). The T2DM can be controlled via inhibiting digestive enzymes such as α-amylase and α-glucosidase, which could lead to the lower amount of calories intakes in the form of glucose (Awosika and Aluko 2019). In general, these digestive enzymes are able to break down dietary carbohydrates into smaller units called monosaccharides (glucose). Subsequently, those mono units are absorbed into blood via intestinal tract and causing diabetes.

5.3 Other Bioactivities/Nutraceutical Properties

COS has been explored widely in terms of prebiotic, anti-inflammatory, immunostimulatory, antibody enhancing, anti-tumor and wound healing activities (Table 1). Prebiotics are known to provide various health benefits including treatment of inflammatory bowel disease, enhancement of mineral absorption, prevention of colorectal cancer, weight management, improvement of insulin sensitivity, and lowering serum lipid concentration (Charalampopoulos and Rastall 2012). Ismail et al. (2020) also determined prebiotic activity of COS against five probiotic strains, namely, *Lactobacillus rhamnosus*, *Lactobacillus plantarum*, *Lactobacillus casei*, *Lactobacillus acidophilus*, and *Bifidobacterium lactis*. COS showed the highest prebiotic activity (enhanced the growth of the probiotics) than inulin, CS, and glucose. Various researchers have determined the prebiotic efficiency of COS against various *Lactobacillus* sp. (Harti et al. 2015; Fernandes et al. 2012; Liang et al. 2013). Recently, Ismail et al. (2020) determined the stability of COS nanoparticles and microparticles in yoghurt for 3 weeks at 4 °C. COS was not stable in yoghurt, which was more likely due to metabolization of the oligosaccharide during storage by the stater culture (*Lactobacillus bulgaricus*) present in samples, thus diminishing COS. In general, inflammation plays a significant role in the pathology of numerous diseases such as cancer, autoimmune diseases, and cardiovascular disease (Liang et al. 2018). COS showed an anti-inflammatory activity by inhibiting various inflammation markers, including nitric oxide (NO), alpha tumor necrosis factor-α and interleukins-1β (Zhu et al. 2015). COS has also been known to enhance the various immune responses such as reactive oxygen species (ROS), generation of pro-inflammatory cytokines, NO production, macrophages phagocytosis, etc. (Deng et al. 2020). The dietary supplementation of COS was proved to show strong immune enhancing activity in animals (Mei et al. 2013; Nguyen et al. 2017). The anti-tumor activity of COS has been determined in vitro using HeLa, Hep3B, and SW480 cell. The effects of different chain length, charge density, MW, modifications, etc., of COS on its anti-tumor activity have been studied under both in vivo and in vitro conditions (Huang et al. 2006; Li et al. 2011; Salah et al. 2013; Hong et al. 2016). The results suggest that COS of lower MW or DP showed excellent anti-tumor activity. Moreover, anti-tumor activity can be enhanced via modification, such a conjugation of COS with phenolics, carboxylation, etc. For wound healing activity, the encapsulated COS was incorporated into composite hydrogel, in which a polyvinyl alcohol/methacrylate kappa-carrageenan composite hydrogel showed closure of full-thickness excision skin wound. This was more likely associated with the wound healing, antibacterial, and antioxidant of COS and high fluid absorption capacity of kappa-carrageenan (Chandika et al. 2021a). The wound healing activity of COS was due to its immunostimulant potential, which could have positive impacts on inflammatory cells, proinflammatory cytokines, and growth factor responsible for wound healing (Jafari et al. 2020). Moreover, COS also improved the water absorption, flexibility, and mechanical properties of dressing (Salamone et al. 2016). Therefore, COS possessed vast range of bioactivities and nutraceutical properties.

6 Conclusion

Chitooligosaccharide (COS) has high water solubility and biocompatibility along with excellent antioxidant and antimicrobial activities. Moreover, nutraceutical values of COS made it become a potential replacer for the synthetic additives and drugs. COS could be used to inhibit oxidation of lipids/proteins and microbial growth in foods. In addition to those preservative properties, it has ability to act as an anti-diabetic, anti-obesity, digestive enzyme inhibitory agent. A little report has been available on the use of COS as a food additive and its effects on the human health. Moreover, high cost associated with COS productions could be a limiting factor for its employment at large scale in both food and health-related industries. Considering the excellent potential of COS to act as nutraceutical or food additives, the potential preparative methods of COS should be further optimized for better benefit.

References

Affes S, Maalej H, Aranaz I et al (2020) Enzymatic production of low-Mw chitosan-derivatives: characterization and biological activities evaluation. Int J Biol Macromol 144:279–288

Attia MS, El-Sayyad GS, Abd Elkodous M et al (2021) Chitosan and EDTA conjugated graphene oxide antinematodes in Eggplant: toward improving plant immune response. Int J Biol Macromol 179:333–344

Awosika TO, Aluko RE (2019) Inhibition of the *in vitro* activities of α-amylase, α-glucosidase and pancreatic lipase by yellow field pea (*Pisum sativum* L.) protein hydrolysates. Int J Food Sci Technol 54:2021–2034

Bahar B, O'doherty JV, O'doherty AM et al (2013) Chito-oligosaccharide inhibits the de-methylation of a 'CpG'island within the leptin (LEP) promoter during adipogenesis of 3T3-L1 cells. PloS One 8:e60011

Benhabiles MS, Salah R, Lounici H et al (2012) Antibacterial activity of chitin, chitosan and its oligomers prepared from shrimp shell waste. Food Hydrocoll 29:48–56

Bockuviene A, Sereikaite J (2020) New β-Carotene-Chitooligosaccharides complexes for food fortification: Stability study. Foods 9:765

Bockuviene A, Zalneravicius R, Sereikaite J (2021) Preparation, characterization and stability investigation of lycopene-chitooligosaccharides complexes. Food Biosci 40:100854

Boonviset P, Pirak T (2020) Physicochemical and sensory characteristics of reduced fat-low sugar Chinese pork sausage as produced by chitooligosaccharide using commercial pectinase hydrolysis. Int J Food Prop 23:22–33

Chandika P, Kim M-S, Khan F et al (2021a) Wound healing properties of triple cross-linked poly (vinyl alcohol)/methacrylate kappa-carrageenan/chitooligosaccharide hydrogel. Carbohydr Polym 269:118272

Chandika P, Oh G-W, Heo S-Y et al (2021b) Electrospun porous bilayer nano-fibrous fish collagen/PCL bio-composite scaffolds with covalently cross-linked chitooligosaccharides for full-thickness wound-healing applications. Mater Sci Eng C 121:111871

Chantarasataporn P, Tepkasikul P, Kingcha Y et al (2014) Water-based oligochitosan and nanowhisker chitosan as potential food preservatives for shelf life extension of minced pork. Food Chem 159:463–470

Charalampopoulos D, Rastall RA (2012) Prebiotics in foods. Curr Opin Biotechnol 23:187–191

Cho E-J, Rahman A, Kim S-W et al (2008) Chitosan oligosaccharides inhibit adipogenesis in 3T3-L1 adipocytes. J Microbiol Biotechn 18:80–87

Choi EH, Yang HP, Chun HS (2012) Chitooligosaccharide ameliorates diet-induced obesity in mice and affects adipose gene expression involved in adipogenesis and inflammation. Nutr Res 32:218–228

De Medeiros Dantas JM, Da Silva NS, De Araújo Padilha CE et al (2020) Enhancing chitosan hydrolysis aiming chitooligosaccharides production by using immobilized chitosanolytic enzymes. Biocatal Agric Biotechnol 28:101759

Deng J-J, Li Z-Q, Mo Z-Q et al (2020) Immunomodulatory effects of N-acetyl chitooligosaccharides on RAW264.7 macrophages. Mar Drugs 18:421

Einbu A, Grasdalen H, Vårum KM (2007) Kinetics of hydrolysis of chitin/chitosan oligomers in concentrated hydrochloric acid. Carbohydr Res 342:1055–1062

Elieh-Ali-Komi D, Hamblin MR (2016) Chitin and chitosan: production and application of versatile biomedical nanomaterials. Int J Adv Res 4:411

El-Sayed ST, Ali AM, Omar NI (2019) A comparative evaluation of antimicrobial activity of chitooligosaccharides with broad spectrum antibiotics on growth of some pathogenic microorganisms. Biocatal Agric Biotechnol 22:101382

Eom T-K, Senevirathne M, Kim S-K (2012) Synthesis of phenolic acid conjugated chitooligosaccharides and evaluation of their antioxidant activity. Environ Toxicol Pharmacol 34:519–527

Fernandes JC, Eaton P, Franco I et al (2012) Evaluation of chitoligosaccharides effect upon probiotic bacteria. Int J Biol Macromol 50:148–152

Fernandes JC, Tavaria FK, Fonseca SC et al (2010) In vitro screening for anti-microbial activity of chitosans and chitooligosaccharides, aiming at potential uses in functional textiles. J Microbiol Biotechnol 20:311–318

Fukamizo T, Honda Y, Goto S et al (1995) Reaction mechanism of chitosanase from *Streptomyces* sp. N174. Biochem J 311:377–383

Gohi BFCA, Zeng H-Y, Pan AD et al (2017) pH dependence of chitosan enzymolysis. Polym 9:174

Hai L, Diep TB, Nagasawa N et al (2003) Radiation depolymerization of chitosan to prepare oligomers. Nucl Instrum Methods Phys Res B 208:466–470

Hamed I, Özogul F, Regenstein JM (2016) Industrial applications of crustacean by-products (chitin, chitosan, and chitooligosaccharides): a review. Trends Food Sci Technol 48:40–50

Harti AS, Haryati DS, Setyaningsih W et al (2015) The potential chito-oligosaccharide (COS) as natural prebiotic and preservatives on synbiotic tofu in Indonesia. Int J Pharma Med Biol Sci 4:204

Hong S, Ngo D-N, Kim M-M (2016) Inhibitory effect of aminoethyl-chitooligosaccharides on invasion of human fibrosarcoma cells. Environ Toxicol Pharmacol 45:309–314

Huang R, Mendis E, Rajapakse N et al (2006) Strong electronic charge as an important factor for anticancer activity of chitooligosaccharides (COS). Life Sci 78:2399–2408

Huang L, Chen J, Cao P et al (2015) Anti-obese effect of glucosamine and chitosan oligosaccharide in high-fat diet-induced obese rats. Mar Drugs 13:2732–2756

Hui A, Yan R, Wang W et al (2020) Incorporation of quaternary ammonium chitooligosaccharides on ZnO/palygorskite nanocomposites for enhancing antibacterial activities. Carbohydr Polym 247:116685

Ichikawa S, Takano K, Kuroiwa T et al (2002) Immobilization and stabilization of chitosanase by multipoint attachment to agar gel support. J Biosci Bioeng 93:201–206

Il'ina AV, Varlamov VP (2004) Hydrolysis of chitosan in lactic acid. Appl Biochem Microbiol 40:300–303

Ismail SA, El-Sayed HS, Fayed B (2020) Production of prebiotic chitooligosaccharide and its nano/microencapsulation for the production of functional yoghurt. Carbohydr Polym 234:115941

Jafari H, Bernaerts KV, Dodi G et al (2020) Chitooligosaccharides for wound healing biomaterials engineering. Mater Sci Eng C 111266

Jeon Y-J, Park P-J, Kim S-K (2001) Antimicrobial effect of chitooligosaccharides produced by bioreactor. Carbohydr Polym 44:71–76

Jung W-K, Moon S-H, Kim S-K (2006) Effect of chitooligosaccharides on calcium bioavailability and bone strength in ovariectomized rats. Life Sci 78:970–976

Kang N-H, Lee WK, Yi B-R et al (2012) Modulation of lipid metabolism by mixtures of protamine and chitooligosaccharide through pancreatic lipase inhibitory activity in a rat model. Lab Anim Res 28:31

Kao C-H, Hsiang C-Y Ho T-Y (2012) Assessment of chitosan-affected metabolic response by peroxisome proliferator-activated receptor bioluminescent imaging-guided transcriptomic analysis. PloS One 7:e34969

Karadeniz F, Kim S-K (2014) Antidiabetic activities of chitosan and its derivatives: a mini review. Adv Food Nutr Res 73:33–44

Karadeniz F, Artan M, Kong C-S et al (2010) Chitooligosaccharides protect pancreatic β-cells from hydrogen peroxide-induced deterioration. Carbohydr Polym 82:143–147

Kidibule PE, Santos-Moriano P, Plou FJ et al (2020) Endo-chitinase Chit33 specificity on different chitinolytic materials allows the production of unexplored chitooligosaccharides with antioxidant activity. Biotechnol Rep 27:e00500

Kidibule PE, Costa J, Atrei A et al (2021) Production and characterization of chitooligosaccharides by the fungal chitinase Chit42 immobilized on magnetic nanoparticles and chitosan beads: selectivity, specificity and improved operational utility. RSC Adv 11:5529–5536

Kim KW, Thomas RL (2007) Antioxidative activity of chitosans with varying molecular weights. Food Chem 101:308–313

Klokkevold PR, Vandemark L, Kenney EB et al (1996) Osteogenesis enhanced by chitosan (poly-N-acetyl glucosaminoglycan) in vitro. J Periodontol 67:1170–1175

Köping-Höggård M, Mel'nikova YS, Vårum KM et al (2003) Relationship between the physical shape and the efficiency of oligomeric chitosan as a gene delivery system in vitro and *in vivo*. J Gene Med 5:130–141

Köping-Höggård M, Vårum K, Issa M et al (2004) Improved chitosan-mediated gene delivery based on easily dissociated chitosan polyplexes of highly defined chitosan oligomers. Gene Ther 11:1441–1452

Kumirska J, Weinhold MX, Thöming J et al (2011) Biomedical activity of chitin/chitosan based materials influence of physicochemical properties apart from molecular weight and degree of N-acetylation. Polym 3:1875–1901

Kuroiwa T, Ichikawa S, Hiruta O et al (2002) Factors affecting the composition of oligosaccharides produced in chitosan hydrolysis using immobilized chitosanases. Biotechnol Prog 18:969–974

Laokuldilok T, Potivas T, Kanha N et al (2017) Physicochemical, antioxidant, and antimicrobial properties of chitooligosaccharides produced using three different enzyme treatments. Food Biosci 18:28–33

Lavie CJ, Milani RV, Ventura HO (2009) Obesity and cardiovascular disease: risk factor, paradox, and impact of weight loss. J Am Coll Cardiol 53:1925–1932

Lee D-X, Xia W-S, Zhang J-L (2008) Enzymatic preparation of chitooligosaccharides by commercial lipase. Food Chem 111:291–295

Li X, Wang J, Chen X et al (2011) Effect of chitooligosaccharides on cyclin D1, bcl-xl and bcl-2 mRNA expression in A549 cells using quantitative PCR. Sci Bull 56:1629

Li M, Han J, Xue Y et al (2019) Hydrogen peroxide pretreatment efficiently assisting enzymatic hydrolysis of chitosan at high concentration for chitooligosaccharides. Polym Degrad Stab 164:177–186

Li R, Lyu Y, Luo S et al (2021) Fabrication of a multi-level drug release platform with liposomes, chitooligosaccharides, phospholipids and injectable chitosan hydrogel to enhance anti-tumor effectiveness. Carbohydr Polym 269:118322

Liang T-W, Liu C-P, Wu C et al (2013) Applied development of crude enzyme from *Bacillus cereus* in prebiotics and microbial community changes in soil. Carbohydr Polym 92:2141–2148

Liang S, Sun Y, Dai X (2018) A review of the preparation, analysis and biological functions of chitooligosaccharide. Int J Mol Sci 19:2197

Liu B, Liu W-S, Han B-Q et al (2007) Antidiabetic effects of chitooligosaccharides on pancreatic islet cells in streptozotocin-induced diabetic rats. World J Gastroenterol 13:725

Liu W, Li XQ, Zhao ZL et al (2020) Effect of chitooligosaccharides on human gut microbiota and antiglycation. Carbohydr Polym 242:116413
Lodhi G, Kim Y-S, Hwang J-W et al (2014) Chitooligosaccharide and its derivatives: preparation and biological applications. BioMed Res Int 2014:1–14
Long T, Yu J, Wang J et al (2018) Orally administered chitooligosaccharides modulate colon microbiota in normal and colitis mice. Int J Pharmacol 14:291–300
Mallakuntla MK, Penugurti V, Manavathi B et al (2021) Chitooligosaccharides induce apoptosis in human breast cancer cells. Carbohydr Polym Technol Appl 2:100077
Mao L, Wu T (2007) Gelling properties and lipid oxidation of kamaboko gels from grass carp (*Ctenopharyngodon idellus*) influenced by chitosan. J Food Eng 82:128–134
Mao SF, Wang B, Yue L et al (2021) Effects of citronellol grafted chitosan oligosaccharide derivatives on regulating anti-inflammatory activity. Carbohydr Polym 262:117972
Mei YX, Chen HX, Zhang J et al (2013) Protective effect of chitooligosaccharides against cyclophosphamide-induced immunosuppression in mice. Int J Biol Macromol 62:330–335
Mei Y-X, Dai X-Y, Yang W et al (2015) Antifungal activity of chitooligosaccharides against the dermatophyte Trichophyton rubrum. Int J Biol Macromol 77:330–335
Miguez N, Kidibule P, Santos-Moriano P et al (2021) Enzymatic synthesis and characterization of different families of chitooligosaccharides and their bioactive properties. Appl Sci 11:3212
Mittal A, Singh A, Aluko RE et al (2020a) Pacific white shrimp (*Litopenaeus vannamei*) shell chitosan and the conjugate with epigallocatechin gallate: Antioxidative and antimicrobial activities. J Food Biochem 45:e13569
Mittal A, Singh A, Benjakul S et al (2020b) Composite films based on chitosan and epigallocatechin gallate grafted chitosan: characterization, antioxidant and antimicrobial activities. Food Hydrocoll 111:106384
Moon C, Seo D-J, Song Y-S et al (2017) Antifungal activity and patterns of N-acetyl-chitooligosaccharide degradation via chitinase produced from Serratia marcescens PRNK-1. Microb Pathog 113:218–224
Mourya V, Inamdar N, Choudhari YM (2011) Chitooligosaccharides: synthesis, characterization and applications. Polym Sci Ser A 53:583–612
Muanprasat C, Chatsudthipong V (2017) Chitosan oligosaccharide: biological activities and potential therapeutic applications. Pharmacol Ther 170:80–97
Nam K, Kim M, Shon Y (2007) Inhibition of proinflammatory cytokine-induced invasiveness of HT-29 cells by chitosan oligosaccharide. J Microbiol Biotechnol 17:2042
Ngo DH, Qian ZJ, Vo TS et al (2011) Antioxidant activity of gallate-chitooligosaccharides in mouse macrophage RAW264.7 cells. Carbohydr Polym 84:1282–1288
Nguyen ND, Van Dang P, Le AQ et al (2017) Effect of oligochitosan and oligo-β-glucan supplementation on growth, innate immunity, and disease resistance of striped catfish (*Pangasianodon hypophthalmus*). Biotechnol Appl Biochem 64:564–571
No H, Meyers S, Prinyawiwatkul W et al (2007) Applications of chitosan for improvement of quality and shelf life of foods: a review. J Food Sci 72:R87–R100
Oh G-W, Kim S-C, Kim T-H et al (2021) Characterization of an oxidized alginate-gelatin hydrogel incorporating a COS-salicylic acid conjugate for wound healing. Carbohydr Polym 252:117145
Omari KW, Besaw JE, Kerton FM (2012) Hydrolysis of chitosan to yield levulinic acid and 5-hydroxymethylfurfural in water under microwave irradiation. Green Chem 14:1480–1487
Pan Z, Cheng DD, Wei X J et al (2021) Chitooligosaccharides inhibit tumor progression and induce autophagy through the activation of the p53/mTOR pathway in osteosarcoma. Carbohydr Polym 258:117596
Pilantanapak A, Waiprib Y, Eadtem P et al (2017) Production of chitooligosaccharides with antibacterial potential via crude chitinase enzymes from marine fungi. Chiang Mai J Sci 44:1224–1230
Popa-Nita S, Lucas J-M, LadavièRe C et al (2009) Mechanisms involved during the ultrasonically induced depolymerization of chitosan: characterization and control. Biomacromol 10:1203–1211

Rakkhumkaew N, Pengsuk C (2018) Chitosan and chitooligosaccharides from shrimp shell waste: characterization, antimicrobial and shelf life extension in bread. Food Sci Biotechnol 27:1201–1208

Ratanavaraporn J, Kanokpanont S, Tabata Y et al (2009) Growth and osteogenic differentiation of adipose-derived and bone marrow-derived stem cells on chitosan and chitooligosaccharide films. Carbohydr Polym 78:873–878

Rhoades J, Gibson G, Formentin K et al (2006) Inhibition of the adhesion of enteropathogenic Escherichia coli strains to HT-29 cells in culture by chito-oligosaccharides. Carbohydr Polym 64:57–59

Rokhati N, Widjajanti P, Pramudono B et al (2013) Performance comparison of α-and β-amylases on chitosan hydrolysis. Int Sch Res Notices. https://doi.org/10.1155/2013/186159

Roncal T, Oviedo A, De Armentia IL et al (2007) High yield production of monomer-free chitosan oligosaccharides by pepsin catalyzed hydrolysis of a high deacetylation degree chitosan. Carbohydr Res 342:2750–2756

Rúnarsson ÖV, Holappa J, Nevalainen T et al (2007) Antibacterial activity of methylated chitosan and chitooligomer derivatives: synthesis and structure activity relationships. Eur Polym J 43:2660–2671

Rúnarsson ÖV, Holappa J, Malainer C et al (2010) Antibacterial activity of N-quaternary chitosan derivatives: Synthesis, characterization and structure activity relationship (SAR) investigations. Eur Polym J 46:1251–1267

Sakai K, Katsumi R, Isobe A et al (1991) Purification and hydrolytic action of a chitosanase from *Nocardia orientalis*. Biochimica ET Biophysica Acta-Protein Struct Mol Enzymol 1079:65–72

Salah R, Michaud P, Mati F et al (2013) Anticancer activity of chemically prepared shrimp low molecular weight chitin evaluation with the human monocyte leukaemia cell line, THP-1. Int J Biol Macromol 52:333–339

Salamone JC, Salamone AB, Swindle-Reilly K et al (2016) Grand challenge in biomaterials-wound healing. Regen Biomater 3:127–128

Sanchez A, Mengibar M, Fernandez M et al (2018) Influence of preparation methods of chitooligosaccharides on their physicochemical properties and their anti-inflammatory effects in mice and in RAW264.7 macrophages. Mar Drugs 16

Santos-Moriano P, Fernandez-Arrojo L, Mengibar M et al (2018) Enzymatic production of fully deacetylated chitooligosaccharides and their neuroprotective and anti-inflammatory properties. Biocatal Biotransfor 36:57–67

Senphan T, Benjakul S (2014) Antioxidative activities of hydrolysates from seabass skin prepared using protease from hepatopancreas of Pacific white shrimp. J Funct Foods 6:147–156

Shen K-T, Chen M-H, Chan H-Y et al (2009) Inhibitory effects of chitooligosaccharides on tumor growth and metastasis. Food Chem Toxicol 47:1864–1871

Singh A, Benjakul S (2020) The combined effect of squid pen chitooligosaccharides and high voltage cold atmospheric plasma on the shelf life extension of Asian sea bass slices stored at 4 °C. Innov Food Sci Emerg Technol 64:102339

Singh A, Benjakul S, Prodpran T (2019a) Chitooligosaccharides from squid pen prepared using different enzymes: characteristics and the effect on quality of surimi gel during refrigerated storage. Food Prod Process Nutr 1:1–10

Singh A, Benjakul S, Prodpran T (2019b) Effect of chitooligosaccharide from squid pen on gel properties of sardine surimi gel and its stability during refrigerated storage. Int J Food Sci Technol 54:2831–2838

Singh A, Benjakul S, Huda N et al (2020a) Preparation and characterization of squid pen chitooligosaccharide–epigallocatechin gallate conjugates and their antioxidant and antimicrobial activities. RSC Adv 10:33196–33204

Singh A, Mittal A, Benjakul S (2020b) Full utilization of squid meat and its processing by-products: revisit. Food Rev Int. https://doi.org/10.1080/87559129.2020.1734611

Singh A, Benjakul S, Olatunde OO et al (2021a) The combined effect of squid pen chitooligosaccharide and high voltage cold atmospheric plasma on the quality of Asian sea bass slices inoculated with *Pseudomonas aeruginosa*. Turk J Fish Aquat Sci 21:41–50

Singh A, Benjakul S, Peng Z et al (2021b) Effect of squid pen chitooligosaccharide and epigallocatechin gallate on discoloration and shelf life of yellowfin tuna slices during refrigerated storage. Food Chem 351:129296

Singh A, Benjakul S, Zhang B et al (2021c) Effect of squid pen chitooligosaccharide on discoloration and shelf life of yellowfin tuna slices packed under different modified atmospheric packaging during refrigerated storage. Food Cont 125:108013

Singh A, Benjakul S, Odunayo O, Ahmet O, Yesilsu F (2021d) The combined effect of squid pen chitooligosaccharide and high voltage cold atmospheric plasma on the quality of Asian sea bass slices inoculated with *Pseudomonas aeruginosa*. Turk J Fish Aquat Sci 21(01) 41–50. https://doi.org/10.4194/1303-2712-v21_1_05

Singh A, Mittal A, Benjakul S (2021e) Chitosan, chitooligosaccharides and their polyphenol conjugates: preparation bioactivities functionalities and applications in food systems. Food Rev Int 1–23. https://doi.org/10.1080/87559129.2021.1950176

Singh A, Mittal A, Benjakul S (2021f) Chitosan nanoparticles: preparation food applications and health benefits. Sci Asia 47(1). https://doi.org/10.2306/scienceasia1513-1874.2021.020

Singh A, Mittal A, Benjakul S Undesirable discoloration in edible fish muscle: Impact of indigenous pigments chemical reactions processing and its prevention. Compr Rev Food Sci Food Saf. https://doi.org/10.1111/1541-4337.12866

Srinivasan K, Viswanad B, Asrat L et al (2005) Combination of high-fat diet-fed and low-dose streptozotocin-treated rat: a model for type 2 diabetes and pharmacological screening. Pharmacol Res 52:313–320

Struszczyk M, Peter M Loth F (1999) Progress on chemistry and application of chitin and its derivatives. In: Struszczyk H (ed). Lodz, Poland, p 168

Ulański P, Rosiak J (1992) Preliminary studies on radiation-induced changes in chitosan. Int J Radiat Appl Instrum Part C Radiat Phys Chem 39:53–57

Vo TS, Ngo DH, Bach LG et al (2017) The free radical scavenging and anti-inflammatory activities of gallate-chitooligosaccharides in human lung epithelial A549 cells. Process Biochem 54:188–194

Wu T, Zivanovic S, Hayes DG et al (2008) Efficient reduction of chitosan molecular weight by high-intensity ultrasound: underlying mechanism and effect of process parameters. J Agric Food Chem 56:5112–5119

Xing R, Liu S, Yu H et al (2005) Salt-assisted acid hydrolysis of chitosan to oligomers under microwave irradiation. Carbohydr Res 340:2150–2153

Yang D, Hu CJ, Deng XY et al (2019a) Therapeutic effect of chitooligosaccharide tablets on lipids in high-fat diets induced hyperlipidemic rats. Molecules 24:514

Yang Y, Xing R, Liu S et al (2019b) Immunostimulatory effects of chitooligosaccharides on RAW 264.7 mouse macrophages via regulation of the MAPK and PI3K/Akt signaling pathways. Mar Drugs 17:36

Yi J, Huang H, Wen Z et al (2021) Fabrication of chitosan-gallic acid conjugate for improvement of physicochemical stability of β-carotene nanoemulsion: Impact of Mw of chitosan. Food Chem 362:130218

Yoksan R, Akashi M, Miyata M et al (2004) Optimal γ-ray dose and irradiation conditions for producing low-molecular-weight chitosan that retains its chemical structure. Radiat Res 161:471–480

Yu DW, Zhao WY, Yang F et al (2021a) A strategy of ultrasound-assisted processing to improve the performance of bio-based coating preservation for refrigerated carp fillets (*Ctenopharyngodon idellus*). Food Chem 345:128862

Yu J, Wang Q, Zhang H et al (2021b) Increased stability of curcumin-loaded pickering emulsions based on glycated proteins and chitooligosaccharides for functional food application. LWT-Food Sci Technol 148:111742

Zhai X, Li C, Ren D et al (2021) The impact of chitooligosaccharides and their derivatives on the *in vitro* and *in vivo* antitumor activity: a comprehensive review. Carbohydr Polym 266:118132

Zhao MY, Gu LM, Li Y et al (2019a) Chitooligosaccharides display anti-tumor effects against human cervical cancer cells via the apoptotic and autophagic pathways. Carbohydr Polym 224:115171

Zhao MY, Shen X, Li XD et al (2019b) Chitooligosaccharide supplementation prevents the development of high fat diet-induced non-alcoholic fatty liver disease (NAFLD) in mice via the inhibition of cluster of differentiation 36 (CD36). J Funct Foods 57:7–18

Zhong X, Yu L, Zhao W et al (1993) Estimation of the radiation induced damage in PTFE by depression of the melting and crystallization temperatures. Polym Degrad Stab 41:223–227

Zhou Y, Li SY, Li DD et al (2020) Enzymatic preparation of chitooligosaccharides and their anti-obesity application. Biosci Biotech Bioch 84:1460–1466

Zhu J, Zhang Y, Wu G et al (2015) Inhibitory effects of oligochitosan on TNF-α, IL-1β and nitric oxide production in lipopolysaccharide-induced RAW264.7 cells. Mol Med Rep 11:729–733

Zoldners J, Kiseleva T, Kaiminsh I (2005) Influence of ascorbic acid on the stability of chitosan solutions. Carbohydr Polym 60:215–218

Zou P, Yuan S, Yang X et al (2019) Structural characterization and antitumor effects of chitosan oligosaccharides against orthotopic liver tumor via NF-κB signaling pathway. J Funct Foods 57:157–165

Anti-inflammatory Activity of Well-Defined Chitooligosaccharides (CHOS) Derived from Enzymatic Hydrolysis of Chitosan

Thae Thae Min and Montarop Yamabhai

1 Chitin, Chitosan and Chitooligosaccharides (Chitosan Oligomers or COS or CHOS)

Chitin is the second most abundant natural polymer on earth and the global annual turnover of chitin is about 10^{11} tons. Chitin and its derivatives are biocompatible, biodegradable, and non-toxic (Schmitz et al. 2019). It is found in the exoskeletons of insects, crustaceans such as crabs and shrimps, and the cell walls of fungi. Chitin is made up of randomly distributed glucosamine (GlcN) and much more abundant N-acetylglucosamine (GlcNAc) units connected by β(1–4) linkage. It is insoluble in water and has few industrial applications (Khoushab and Yamabhai 2010; Elieh-Ali-Komi and Hamblin 2016). The degree of acetylation of chitin is >90%. The degree of polymerization is about 5,000–10,000 and its molecular weight can be 1,000–2,500 kDa (Kaczmarek et al. 2019). Chitosan is the deacetylated form of chitin and about 70–95% of its units are deacetylated. The molecular weight of chitosan can be 10–1000 kDa (Moura et al. 2011). Chitosan is soluble in acid solutions such as acetic, formic, lactic, citric acids, and solvents such as dimethylsulfoxide by structural modification; therefore, chitosan is a more suitable substrate for enzymatic bioconversion into COS than chitin, which must be dissolved in harsh acidic condition (Roy et al. 2017). Chitooligosaccharides (chitosan oligomers, COS or CHOS) are short chitosan polymers. Its degree of polymerization varies from 2 to 20 units and the average molecular weight is <3.9 kDa. In general, COS is fully soluble in water, partially soluble in methanol and dimethylsulfoxide, and insoluble in acetone, butanol, ethanol, ethyl acetate, propanol, and pyridine (Phil et al. 2018; Liaqat and Eltem 2018; Lodhi et al. 2014). The bioactive properties of both

T. T. Min · M. Yamabhai (✉)
Molecular Biotechnology Laboratory, School of Biotechnology, Institute of Agricultural Technology, Suranaree University of Technology, 111 University Avenue, Nakhon Ratchasima 30000, Thailand
e-mail: montarop@g.sut.ac.th

S.-K. Kim (ed.), *Chitooligosaccharides*,
https://doi.org/10.1007/978-3-030-92806-3_14

chitosan and COS depend on DA (degree of acetylation), PA (pattern of acetylation), and DP (degree of polymerization). Based on the correlation between the pattern of acetylation and biologic properties, COS has been proposed as information-carrying molecules (Basa et al. 2020).

2 Bioconversion of Chitosan into COS

To produce COS, the glycosidic bonds of chitosan must be cleaved by acid hydrolysis, enzymatic hydrolysis, oxidative degradation, or ultrasonic degradation methods. Depending on the preparation method, the DP and PA of COS products vary (Thomas et al. 2015). Subsequently, the bioactivities of COS differ depending on the DP, DA, and PA of its constituent oligomers. Because of this, the homogeneity and reproducibility of each batch of manufactured COS is critical for the quality control; hence reliability and efficacy of the products. When the chitosan substrate of 89 kDa and DA 17% (Novamatrix, Norway) was hydrolyzed by chitosanase from *Streptomyces griseus* (Sigma-Aldrich, USA) into COS (42% fully deacetylated and 54% monoacetylated oligomers of 8 kDa, DA 13%, PA 0.8), that COS could attenuate the LPS-induced inflammation in RAW264.7 macrophages as well as wild type C57BL/6 mice. When the same chitosan substrate was depolymerized first with nitrous acid and then hydrolyzed by chitosanase, COS (50% fully deacetylated and 27% monoacetylated oligomers) of 5 kDa, DA 11%, PA 0.9 was obtained. Not only this COS could not attenuate the inflammatory response, but it also promoted inflammation in both RAW264.7 macrophages and C57BL/6 mice (Sanchez et al. 2018).

2.1 *Chitosanase and Its Role in COS Production*

Chitosan is made up of four types of glycosidic bonds which are: A-A (between two GlcNAc units), A-D (between GlcNAc and GlcN units), D-A (between GlcN and GlcNAc units), and D-D (between two GlcN units) (Kim and Rajapakse 2005; Einbu 2007; Thomas et al. 2015). Enzymes that can hydrolyze the glycosidic bonds of chitosan are known as chitosanases which have been grouped into six glycoside hydrolase families based on their sequence similarities; namely, GH-5, GH-7, GH-8, GH-46, GH-75, and GH-80 (Johnsen et al. 2010; Viens et al. 2015). Chitosanases are classified into two major categories: endochitosanases and exochitosanases, depending on their mode of hydrolysis (Miguez et al. 2021).

Moreover, chitosanases can be divided into four main classes: class I, II, III, and IV, based on their affinity for glycosidic bonds of chitosan. In general, all chitosanases are capable of cleaving D-D linkage. In addition to that, Class I members, e.g., chitosanase of *Streptomyces* sp. N174, can hydrolyze A-D linkage. Class III members like chitosanase from *Bacillus circulans* MH-K1 can hydrolyze both D-D and D-A linkages. Class IV members, such as chitosanase of *Pseudomonas* sp. A-01 can cleave

D-D, A-D, and D-A bonds. On the other hand, Class II enzymes, e.g., chitosanase of *Bacillus* sp. No. 7-M can act only on D-D bond (Weikert et al. 2017; Singh et al. 2019). Unlike above chitosanases, that of *Streptomyces coelicolor* A3(2) can cleave all four types of glycosidic bonds of chitosan where the specific rate of degradation of substrate ranges from 0.006/min to 816/min, depending on the DA of chitosan (Heggset et al. 2009, 2010).

Recombinant chitosanases have been commonly used to hydrolyze chitosan to COS. For example, recombinant chitosanase originated from *Bacillus* sp. MET 1299 from Korean soil was expressed in TOP10 *Escherichia coli* which hydrolyzed chitosan into COS ranging from DP1 monomer to DP6 hexamer (Kim et al. 2004). In some studies, native chitosanases were directly used for COS production. Native chitosanase produced by *Bacillus subtilis* Ege-B-1.19 screened from vineyard soil was purified and used to hydrolyze chitosan into glucosamine monomers, dimers, and trimers which have antifungal activities (Oztopuz et al. 2019). Likewise, the chitosanase of *Bacillus cereus* P16 soil isolate was expressed and used for COS production (Yu-Young et al. 2003). Chitosanases are also commercially available in both liquid and lyophilized forms. Some are of *Streptomyces* sp. origin, and some are from *Bacillus* sp. In general, the specific enzyme activity of commercially available chitosanases ranges between 0.694–50 units/mg protein as shown in the Table 1.

Although commercially not available yet, optimizations to maximize the production of certain chitosanases have been documented. For example, the yield of chitosanase (*Bs*Csn46A) originated from *B. subtilis* strain 168 and expressed in TOP10 *E. coli* in shake-flask culture was increased from 0.4 to 18.5 mg/L by substituting native signal peptide with OmpA signal peptide of *E. coli*. The signal peptide

Table 1 Examples of commercially available chitosanases

Chitosanase	Source	Company	References
Chitosanase in aqueous solution (≥15 U/mg protein)	*Streptomyces* sp.	Sigma-Aldrich Pte Ltd	Sigma (2019)
Chitosanase in aqueous solution (≥25 U/mg protein)	*Streptomyces* sp. N174	Merck	VWR (2021)
Chitosanase Lyophilized powder (>50 U/mg protein)	*Streptomyces griseus*	Merck	Sigma-Aldrich (2021)
Chitosanase Lyophilized powder (>50 U/mg protein)	*S. griseus*	Creative Enzymes	Creative-Enzymes (2021)
Chitosanase Lyophilized powder (35 U/mg of protein)	*Bacillus* sp.	Amicogen Co., Jinju, Korea	Park et al. (2003), Amicogen (2018), Kim and Kim (2006)
Chitosanase (0.694 U/mg protein)	*Bacillus pumilus* BN-262	Wako Chemical Industries, Ltd., Japan	Jeon and Kim (2000), Jeon et al. (2001)

substitution didn't affect the specific activity of the enzyme which was about 650 U/mg of protein (Pechsrichuang et al. 2016). This *Bacillus* chitosanase (*Bs*Csn46A) can digest chitosan with F_A (fraction of acetylation) 0.3, DP >1000, and MW 200–400 kDa into COS with the initial specific activity of 8.4×10^3/min and has become one of the fastest chitosanases described so far (Pechsrichuang et al. 2018). To produce food-grade chitosanase, a *Lactobacillus* expression platform for chitosanase has been established. In that system, alanine racemase (alr) gene was used as a selection marker to produce the enzyme from an alr-deficient *L. plantarum*, using pSIP-based expression system. The yield and secretion efficiency of chitosanase (*Bs*Csn46A) in *Lactobacillus* expression system were 79 kU/L and 56%, respectively, at 20 h after induction with 12.5 ng/mL of inducing peptide in MRS medium supplied with 20 g/L glucose (Sak-Ubol et al. 2016). The bioconversion of chitosan to COS is influenced not only by the properties of chitosan substrate and chitosanase but also by the temperature, pH, and duration of the reaction. As for the chitosanase *Bs*Csn46A, the optimal pH is 5.0–6.0 and the optimal temperature is between 40–50 °C under standard assay conditions. The presence of chitosan has been shown to enhance the thermostability of *Bs*Csn46A (Pechsrichuang et al. 2013).

2.2 COS Production by Enzymatic Hydrolysis

The general outline of COS production by enzymatic hydrolysis of chitosan is depicted in Fig. 1 below. For enzyme-based COS production, shake-flasks, batch bioreactors or continuous dual reactor systems have been utilized. Researchers like Fukamizo et al. (1994) used chitosanase from Wako Chemical Industries, Ltd., Japan, and sodium acetate buffer to dissolve chitosan. Jung et al. (2006) used chitosanase from Amicogen Co., Jinju, Korea, and lactic acid to dissolve chitosan. Pechsrichuang et al. (2018) used chitosanase *Bs*Csn46A, and sodium acetate buffer to dissolve chitosan. Sodium azide or bovine serum albumin (BSA) can be added into the reaction (Fukamizo et al. 1994; Pechsrichuang et al. 2018). Certain research groups such as Santos-Moriano et al. (2019) used Neutrase 0.8 L and chitinase from *Trichoderma harzianum* expressed in *Pichia pastoris* to hydrolyze chitosan. The Li et al. (2018) group used chitosanase and exo-β-D-glucosaminidase enzymes in bi-enzyme single-step hydrolysis of chitosan. The temperature and duration of hydrolysis depend on the specification of chitosanase used and the desired COS product mixture.

The hydrolysis has to be stopped by boiling and after that, the undigested impurities can be removed by pH change or centrifugation (Fukamizo et al. 1994; Santos-Moriano et al. 2019). The fragmentation of COS products can be performed by chromatography or ultrafiltration using charcoal column, CM-Sephadex C-25 column or ultrafiltration membrane reactor system (Millipore Co., USA). Depending on the molecular weight cut off (MWCO) of the filteration membranes, the COS products can generally be grouped into high molecular weight COS (HMWCOS), medium molecular weight COS (MMWCOS), and low molecular weight COS (LMWCOS) (Fukamizo et al. 1994; Santos-Moriano et al. 2019; Jung et al. 2006; Jeon et al. 2001).

Fig. 1 General outline of enzymatic hydrolysis of chitosan

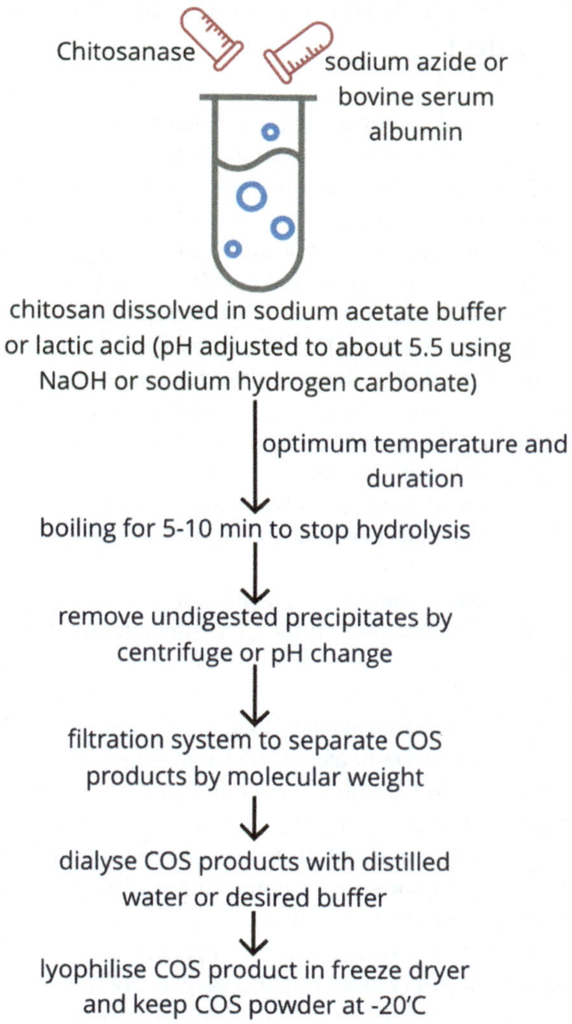

After separation, COS products could be dialyzed against distilled water for 3 days, followed by lyophilization in a freeze-dryer for 5 days (Park et al. 2004). In order to recycle the enzyme, chitosanase was immobilized on chitin carriers and an immobilized enzyme column reactor was introduced (Jeon et al. 2001). Later, a dual reactor system that can hydrolyze chitosan in equivalent to 11 batches of the batch reactor using the same amount of chitosanase has been developed for continuous production of COS. Immobilized chitosanase is packed in an enzyme column reactor and is connected to an ultrafiltration membrane reactor (Kim and Je 2010). The final COS products from that controlled-hydrolysis were lyophilized and used in cosmetic formula for preventing skin aging. The COS formula has been patented in South Korea and USA (Kim et al. 2011).

3 Biologic Activities of COS and Its Anti-inflammatory Effect

COS is water-soluble and its potential to be used for human well-being has been proven in several aspects. It has been shown to possess a wide range of biologic functions such as antimicrobial, antioxidant, anti-inflammatory effect, anti-tumor/anticancer, immunostimulatory, wound healing and tissue regeneration, antidiabetic, and hypocholesterolemic activities. In addition, COS has been investigated as drug carrier, food additive, and elicitor of plant defense (Jeon and Kim 2002; Marmouzi et al. 2019; Oztopuz et al. 2019; Phil et al. 2018; Guan et al. 2019; Miguez et al. 2021; Muanprasat and Chatsudthipong 2017). It has also been well-accepted as a potential drug for the prevention and treatment of Alzheimer's disease because of its anti-neuroinflammatory, antioxidant and acetylcholinesterase inhibitory effects, as well as its ability to adsorb copper ions (Ouyang et al. 2017). Among these, the property that has the highest impact on human health and extensively studied is the anti-inflammatory effect, which will be the focus of this chapter. Recent research works on the anti-inflammatory activity of COS are described in Table 2 and below.

To demonstrate the anti-inflammatory effect of COS, the proinflammatory cytokines productions from LPS-stimulated mammalian cell lines were observed in the absence or presence of COS-pretreatment. In the study of Jitprasertwong et al. (2021), COS (0.4–1.2 kDa, DP2-DP6) could significantly attenuate the LPS-induced production of interleukin-1β (IL-1β), IL-6 and tumor necrosis factor-α (TNF-α) by vitamin-D3 treated THP-1 monocytes of human origin. COS concentrations from 5 to 200 μg/mL were found to have anti-inflammatory activity in a dose-dependent manner. Notably, a high concentration of COS, 200 μg/mL, caused a slight increase in IL-1β, IL-6, and TNF-α levels in LPS-untreated control cells, indicating the importance of optimization of COS dosage for application in various healthcare products. Similarly, the effect of COS (0.2–1.2 kDa) on TNF-α production in LPS-exposed murine macrophages (RAW 264.7) has been studied. Out of the three COS preparations tested, a statistically significant anti-inflammatory effect was seen in cells treated with 250 ng/well of fully deacetylated and fully acetylated COS (Santos-Moriano et al. 2019). Likewise, COS (5–10 kDa) could inhibit the release of inflammation-related mediators such as inducible nitric oxide synthase (iNOS), nitric oxide (NO), TNF-α, IL-6, cyclooxygenase-2 (COX-2) and prostaglandin-E_2 (PGE_2) in LPS-stimulated RAW264.7 macrophage cells at a dose-dependent manner (Lee et al. 2009). In addition to the above macrophage cell lines, human lung epithelial A549 cells have been used to observe the anti-inflammatory effect of COS (3–5 kDa) or gallic acid conjugated COS at 50–200 μg/mL dose range. Both preparations of COS could inhibit the production of TNF-α, IL-8, COX-2, and PGE_2 in LPS-stimulated A549 cells in a dose-dependent manner. But the inhibitory effect was more prominent in gallic acid-COS-treated cells than those of COS-treated ones (Vo et al. 2017).

Table 2 Recent studies on anti-inflammatory activity of COS

COS	Treatment regime	Test system	References
0.4–1.2 kDa, DP2–DP6	5–200 μg/mL	Human THP-1 macrophage	Jitprasertwong et al. (2021)
0.2–1.2 kDa, DP1–DP5	100–500 ng/well	Murine macrophage (RAW 264.7)	Santos-Moriano et al. (2019)
<1–10 kDa	1–5 mg/mL	Murine macrophage (RAW 264.7)	Lee et al. (2009)
3–5 kDa COS conjugated with gallic acid	50–200 μg/mL	Human lung epithelial A549 cells	Vo et al. (2017)
1.2 and 5.3 kDa, 80–85% deacetylated	Orally (50–1,000 mg/kg body weight)	Balb/c male mice	Fernandes et al. (2010)
GlcN (80.0%)-COS (9.8%)	Orally (40–160 mg/kg/day)	C57BL/6 male mice	Li et al. (2018)
COS	2.0 g/kg	Juvenile turbot fish	Gu et al. (2021)
1.5 kDa, 90% deacetylated	Orally (125–500 mg/kg/day)	C57BL/6J female mice	Tong et al. (2016)
5–10 kDa, >90% deacetylated	20–100 mg/kg/day	Human colonic epithelial T84 cells, ICR male mice	Yousef et al. (2012)
5–14 kDa, >90% deacetylated	10–100 μg/mL, 20–500 mg/kg/day	human colonic epithelial T84 cells, human colon adenocarcinoma HT-29 and Caco-2 cells, C57/BL6 male mice	Muanprasat et al. (2015)
COS-A (10–20 kDa), COS-C (1–3 kDa)	COS-A (0.1–0.4%), COS-C (0.05–0.2%)	Murine macrophage (RAW 264.7), healthy female Korean volunteers	Yang et al. (2010)

In-vivo anti-inflammatory effect of COS has been tested in animal models such as mice and fish (turbots). Fernandes et al. (2010) treated Balb/c male mice with COS (1.2 kDa) or (5.3 kDa) preparations orally and the effect was observed by carrageenan-induced paw edema method. The anti-inflammatory effect of COS (1.2 kDa) given at 500 mg/kg body weight was comparable to that of indomethacin and oral LD50 value for both COS in mice was above 1,000 mg/kg body weight. Li et al. (2018) treated C57BL/6 male mice osteoarthritis models with GlcN (80.00%)-COS (9.80%) orally at 40–160 mg/kg body weight for 15 days. There was significant down-regulation in serum levels of IL-1β, IL-6, TNF-α, and C-reactive protein, and significant increase in the level of anti-inflammatory cytokine, IL-2. Another study group, Tong et al. (2016), used female C57BL/6 J mice. The oral administration of

COS (125–500 mg/kg) could prevent dextran sulfate sodium-induced body weight loss and bloody diarrhea event rate and reduce the histological damage scores by suppressing the Toll-like receptor-4 (TLR4)/ nuclear factor kappa B (NF-κB)/COX-2 signaling cascade. In addition to mammals, the anti-inflammatory effect of COS or chitosan supplement diet on juvenile turbot fish has been investigated for 8 weeks by Gu and the group in 2021. Although the chitosan and COS used were not clearly specified, both chitosan and COS could improve the general growth performance and feed efficiency ratio, significantly prevent the histological alterations due to local inflammation caused by soyabean meal, as well as suppress the transcriptional levels of IL-1β, IL-8, TNF-α in addition to major pathway effectors such as NF-κB and p38 mitogen-activated protein kinase (Gu et al. 2021).

In the study of Yousef et al. (2012), oral administration of COS (5–10 kDa) at 10–20 mg/kg/day could protect the ICR male mice from mortality and intestinal inflammation due to acute colitis induced by 5% dextran sulfate sodium or chronic colitis by rectal instillation of 4% acetic acid. In COS-treated mice, suppression of NF-κB activation, and IL-6 and TNF-α levels in colonic tissues were observed. Moreover, in COS-treated human colonic epithelial T84 cells, NF-κB activation, and IL-6 and TNF-α expression were down-regulated. As a continuation of the above study, Muanprasat et al. (2015) tested the effect of COS (5–14 kDa) on human colonic epithelial T84 cells, human colon adenocarcinoma HT-29 cells, and Caco-2 cells, as well as C57/BL6 male mice model and closed mouse ileal loop models. COS treatment in cell lines was associated with AMP-activated protein kinase (AMPK) activation caused by calcium-sensing receptor-mediated release of calcium from endoplasmic reticulum. COS could also augment the tight junction assembly in T84 cells. The oral administration of COS (500 mg/kg) could prevent the development of aberrant crypt foci and colitis-associated colorectal cancer in mouse model by AMPK activation-induced β-catenin suppression and by caspase-3 cleavage-mediated apoptosis of cancer cells. In closed mouse ileal loop models, COS treatment, as low as 100 μg/mL, could reduce the cholera toxin-induced intestinal fluid secretion by 30%, proving that COS can effectively prevent cholera toxin-induced intestinal fluid secretion (Muanprasat et al. 2015).

For cosmeceutical application of COS, so far only the human skin primary irritation test for COS has been performed on human subjects by Yang et al. (2010). Prior to testing on human skin, investigators checked the effects of two COS formulas: COS-A (10–20 kDa) and COS-C (1–3 kDa) on LPS-stimulated RAW264.7 macrophages. Both COS-A and COS-C could significantly suppress the COX-2 and PGE_2 production, inducible nitric oxide synthase, and nitric oxide production, and IL-1β production, without cytotoxic effect on the cells. Next, 32 healthy female Korean volunteers (20–49 years) were tested with patches, containing 1% COS-A and COS-C in water, firmly secured to the skin of the back for 48 h. The subjects were not allowed to shower nor perform any work or exercise that might wet or loosen the patches. The site was evaluated at 30 min and 24 h after the patches were removed and the scores were analyzed according to the Cosmetic, Toiletry, and Fragrance Association guidelines. There were no adverse reactions of COS (Yang et al. 2010).

4 Conclusion and Future Perspectives

COS is a versatile bioactive product. Because of its water-soluble property, COS is superior to its parental chitin and chitosan. Although COS is generally defined as a short polymer composed of 2 to 20 units and has an average molecular weight of <3.9 kDa, polymers larger than 3.9 kDa are also described as COS in publications. COS production procedures are different from one research group to the other and so are the properties of the produced COS. Despite the lack of consensus definition of COS, the divert bioactivities of COS have been proved and COS is well-accepted as a promising biologic agent with anti-inflammatory effect. The global chitosan market was valued at USD 476.6 million in 2016 and is speculated to reach USD 1,088.0 million by 2022 and together with chitosan, the demand of both food-grade chitosanase and industrial chitosanase is uprising for medical and agricultural applications, and for the beauty product industries as well. Moreover, chitosanase-based dietary supplement enzyme formulas for humans are available in the market. Pure COS powder packages to be taken as an immune booster for humans and several COS-based dietary supplement formulas for animal farms are commercially available as well. On top of that, COS has rapidly gained interest to be used in the treatment of Alzheimer's disease, inflammatory bowel diseases including Crohn's disease and ulcerative colitis and many more inflammation-based pathologies. Good manufacturing practice of different well-defined COS is essential for safety and efficacy applications in the prevention and control of various diseases.

References

Amicogen (2018) Chimax-O. Amicogen, Korea

Basa S, Nampally M, Honorato T, Das SN, Podile AR, El Gueddari NE, Moerschbacher BM (2020) The pattern of acetylation defines the priming activity of chitosan tetramers. J Am Chem Soc 142:1975–1986. https://doi.org/10.1021/jacs.9b11466

Creative-Enzymes (2021) Native Streptomyces griseus chitosanase. Creative Enzymes, USA

Einbu A (2007) Characterisation of chitin and a study of its acid-catalysed hydrolysis. Norwegian University of Science and Technology

Elieh-Ali-Komi D, Hamblin MR (2016) Chitin and chitosan: production and application of versatile biomedical nanomaterials. Int J Adv Res (Indore) 4:411–427

Fernandes JC, Spindola H, de Sousa V, Santos-Silva A, Pintado ME, Malcata FX, Carvalho JE (2010) Anti-inflammatory activity of chitooligosaccharides in vivo. Mar Drugs 8:1763–1768. https://doi.org/10.3390/md8061763

Fukamizo T, Ohkawa T, Ikeda Y, Goto S (1994) Specificity of chitosanase from *Bacillus pumilus*. Biochem Biophys Acta 1205:183–188. https://doi.org/10.1016/0167-4838(94)90232-1

Gu M, Pan S, Li Q, Qi Z, Deng W, Bai N (2021) Chitosan and chitooligosaccharides attenuate soyabean meal-induced intestinal inflammation of turbot (*Scophthalmus maximus*): possible involvement of NF-small ka, CyrillicB, activator protein-1 and mitogen-activated protein kinases pathways. Br J Nutr:1–12. https://doi.org/10.1017/S0007114521000489

Guan G, Azad MAK, Lin Y, Kim SW, Tian Y, Liu G, Wang H (2019) Biological effects and applications of chitosan and chito-oligosaccharides. Front Physiol 10:516. https://doi.org/10.3389/fphys.2019.00516

Heggset EB, Hoell IA, Kristoffersen M, Eijsink VGH, Vårum KM (2009) Degradation of chitosans with chitinase G from *Streptomyces coelicolor* A3(2): production of chitooligosaccharides and insight into subsite specificities. Biomacromolecules 10:892–899. https://doi.org/10.1021/bm801418p

Heggset EB, Dybvik AI, Hoell IA, Norberg AL, Sørlie M, Eijsink VGH, Vårum KM (2010) Degradation of chitosans with a family 46 chitosanase from *Streptomyces coelicolor* A3(2). Biomacromolecules 11:2487–2497. https://doi.org/10.1021/bm1006745

Jeon YJ, Kim SK (2000) Production of chitooligosaccharides using an ultrafiltration membrane reactor and their antibacterial activity. Carbohydr Polym 41:133–141. https://doi.org/10.1016/S0144-8617(99)00084-3

Jeon YJ, Kim SK (2002) Antitumor activity of chitosan oligosaccharides produced in ultrafiltration membrane reactor system. J Microbiol Biotechnol 12:503–507

Jeon YJ, Park PJ, Kim SK (2001) Antimicrobial effect of chitooligosaccharides produced by bioreactor. Carbohydr Polym 44:71–76. https://doi.org/10.1016/S0144-8617(00)00200-9

Jitprasertwong P, Khamphio M, Petsrichuang P, Eijsink VGH, Poolsri W, Muanprasat C, Rangnoi K, Yamabhai M (2021) Anti-inflammatory activity of soluble chito-oligosaccharides (CHOS) on VitD3-induced human THP-1 monocytes. PLoS ONE 16:e0246381. https://doi.org/10.1371/journal.pone.0246381

Johnsen MG, Hansen OC, Stougaard P (2010) Isolation, characterization and heterologous expression of a novel chitosanase from *Janthinobacterium* sp. strain 4239. Microbial Cell Fact 9:5. https://doi.org/10.1186/1475-2859-9-5

Jung WK, Moon SH, Kim SK (2006) Effect of chitooligosaccharides on calcium bioavailability and bone strength in ovariectomized rats. Life Sci 78:970–976. https://doi.org/10.1016/j.lfs.2005.06.006

Kaczmarek MB, Struszczyk-Swita K, Li X, Szczesna-Antczak M, Daroch M (2019) Enzymatic modifications of chitin, chitosan, and chitooligosaccharides. Front Bioeng Biotechnol 7:243. https://doi.org/10.3389/fbioe.2019.00243

Khoushab F, Yamabhai M (2010) Chitin research revisited. Mar Drugs 8:1988–2012. https://doi.org/10.3390/md8071988

Kim SK, Je JY (2010) Continuous production of chitooligosaccharides by enzymatic hydrolysis. In: Kim S-K (ed) Chitin, chitosan, oligosaccharides and their derivatives-biological activities and applications. CRC Press, USA, pp 47–50

Kim MM, Kim SK (2006) Chitooligosaccharides inhibit activation and expression of matrix metalloproteinase-2 in human dermal fibroblasts. FEBS Lett 580:2661–2666. https://doi.org/10.1016/j.febslet.2006.04.015

Kim SK, Rajapakse N (2005) Enzymatic production and biological activities of chitosan oligosaccharides (COS): a review. Carbohyd Polym 62:357–368. https://doi.org/10.1016/j.carbpol.2005.08.012

Kim PI, Kang TH, Chung KJ, Kim IS, Chung KC (2004) Purification of a constitutive chitosanase produced by *Bacillus* sp. MET 1299 with cloning and expression of the gene. FEMS Microbiol Lett 240:31–39. https://doi.org/10.1016/j.femsle.2004.09.006

Kim S-K, Kong C-S, Bak S-S, Kim J-A, Byul-Nim A (2011) Cosmetic composition for preventing skin aging containing chitooligosaccharides. USA Patent

Lee SH, Senevirathne M, Ahn CB, Kim SK, Je JY (2009) Factors affecting anti-inflammatory effect of chitooligosaccharides in lipopolysaccharides-induced RAW264.7 macrophage cells. Bioorg Med Chem Lett 19:6655–6658. https://doi.org/10.1016/j.bmcl.2009.10.007

Li Y, Chen L, Liu Y, Zhang Y, Liang Y, Mei Y (2018) Anti-inflammatory effects in a mouse osteoarthritis model of a mixture of glucosamine and chitooligosaccharides produced by bi-enzyme single-step hydrolysis. Sci Rep 8:5624. https://doi.org/10.1038/s41598-018-24050-6

Liaqat F, Eltem R (2018) Chitooligosaccharides and their biological activities: a comprehensive review. Carbohydr Polym 184:243–259. https://doi.org/10.1016/j.carbpol.2017.12.067

Lodhi G, Kim YS, Hwang JW, Kim SK, Jeon YJ, Je JY, Ahn CB, Moon SH, Jeon BT, Park PJ (2014) Chitooligosaccharide and its derivatives: preparation and biological applications. Biomed Res Int 2014:654913. https://doi.org/10.1155/2014/654913

Marmouzi I, Ezzat SM, Salama MM, Merghany RM, Attar AM, El-Desoky AM, Mohamed SO (2019) Recent updates in pharmacological properties of chitooligosaccharides. Biomed Res Int 2019:4568039. https://doi.org/10.1155/2019/4568039

Miguez N, Kidibule P, Santos-Moriano P, Ballesteros AO, Fernandez-Lobato M, Plou FJ (2021) Enzymatic synthesis and characterization of different families of chitooligosaccharides and their bioactive properties. Appl Sci 11. https://doi.org/10.3390/app11073212

Moura CMd, Moura JMd, Soares NM, Pinto LAdA (2011) Evaluation of molar weight and deacetylation degree of chitosan during chitin deacetylation reaction: used to produce biofilm. Chem Eng Process 50:351–355. https://doi.org/10.1016/j.cep.2011.03.003

Muanprasat C, Chatsudthipong V (2017) Chitosan oligosaccharide: biological activities and potential therapeutic applications. Pharmacol Ther 170:80–97. https://doi.org/10.1016/j.pharmthera.2016.10.013

Muanprasat C, Wongkrasant P, Satitsri S, Moonwiriyakit A, Pongkorpsakol P, Mattaveewong T, Pichyangkura R, Chatsudthipong V (2015) Activation of AMPK by chitosan oligosaccharide in intestinal epithelial cells: mechanism of action and potential applications in intestinal disorders. Biochem Pharmacol 96:225–236. https://doi.org/10.1016/j.bcp.2015.05.016

Ouyang QQ, Zhao S, Li SD, Song C (2017) Application of chitosan, chitooligosaccharide, and their derivatives in the treatment of Alzheimer's disease. Mar Drugs 15. https://doi.org/10.3390/md15110322

Oztopuz O, Sarigul N, Liaqat F, Park R-D, Eltem R (2019) Chitinolytic *Bacillus subtilis* Ege-B-1.19 as a biocontrol agent against mycotoxigenic and phytopathogenic fungi. Turk J Biochem 44:323–331. https://doi.org/10.1515/tjb-2018-0144

Park PJ, Je J, Kim S-K (2003) Free radical scavenging activity of chitooligosaccharides by electron spin resonance spectrometry. J Agric Food Chem 51:4624–4627. https://doi.org/10.1021/jf034039+

Park P-J, Je J-Y, Jung W-K, Ahn C-B, Kim S-K (2004) Anticoagulant activity of heterochitosans and their oligosaccharide sulfates. Eur Food Res Technol 219:529–533. https://doi.org/10.1007/s00217-004-0977-3

Pechsrichuang P, Yoohat K, Yamabhai M (2013) Production of recombinant *Bacillus subtilis* chitosanase, suitable for biosynthesis of chitosan-oligosaccharides. Bioresour Technol 127:407–414. https://doi.org/10.1016/j.biortech.2012.09.130

Pechsrichuang P, Songsiriritthigul C, Haltrich D, Roytrakul S, Namvijtr P, Bonaparte N, Yamabhai M (2016) OmpA signal peptide leads to heterogenous secretion of *B. subtilis* chitosanase enzyme from *E. coli* expression system. Springerplus 5:1200. https://doi.org/10.1186/s40064-016-2893-y

Pechsrichuang P, Lorentzen SB, Aam BB, Tuveng TR, Hamre AG, Eijsink VGH, Yamabhai M (2018) Bioconversion of chitosan into chitooligosaccharides (CHOS) using family 46 chitosanase from *Bacillus subtilis* (BsCsn46A). Carbohydr Polym 186:420–428. https://doi.org/10.1016/j.carbpol.2018.01.059

Phil L, Naveed M, Mohammad IS, Bo L, Bin D (2018) Chitooligosaccharide: an evaluation of physicochemical and biological properties with the proposition for determination of thermal degradation products. Biomed Pharmacother 102:438–451. https://doi.org/10.1016/j.biopha.2018.03.108

Roy JC, Salaün F, Giraud S, Ferri A, Chen G, Guan J (2017) Solubility of chitin: solvents, solution behaviors and their related mechanisms. In: Solubility of polysaccharides. https://doi.org/10.5772/intechopen.71385

Sak-Ubol S, Namvijitr P, Pechsrichuang P, Haltrich D, Nguyen TH, Mathiesen G, Eijsink VG, Yamabhai M (2016) Secretory production of a beta-mannanase and a chitosanase using a *Lactobacillus plantarum* expression system. Microb Cell Fact 15:81. https://doi.org/10.1186/s12934-016-0481-z

Sanchez A, Mengibar M, Fernandez M, Alemany S, Heras A, Acosta N (2018) Influence of preparation methods of chitooligosaccharides on their physicochemical properties and their anti-inflammatory effects in mice and in RAW264.7 macrophages. Mar Drugs 16. https://doi.org/10.3390/md16110430

Santos-Moriano P, Kidibule P, Míguez N, Fernández-Arrojo L, Ballesteros AO, Fernández-Lobato M, Plou FJ (2019) Tailored enzymatic synthesis of chitooligosaccharides with different deacetylation degrees and their anti-inflammatory activity. Catalysts 9(5):405. https://doi.org/10.3390/catal9050405

Schmitz C, Auza LG, Koberidze D, Rasche S, Fischer R, Bortesi L (2019) Conversion of chitin to defined chitosan oligomers: current status and future prospects. Mar Drugs 17. https://doi.org/10.3390/md17080452

Sigma-Aldrich (2021) Chitosanase from Streptomyces griseus. Sigma-Aldrich, USA

Sigma (2019) Chitosanase

Singh R, Weikert T, Basa S, Moerschbacher BM (2019) Structural and biochemical insight into mode of action and subsite specificity of a chitosan degrading enzyme from *Bacillus spec.* MN. Sci Rep 9:1132. https://doi.org/10.1038/s41598-018-36213-6

Thomas NV, Venkatesan J, Manivasagan P, Kim S-K (2015) Production and biological activities of chitooligosaccharides (COS)—an overview. J Chitin Chitosan Sci 3:1–10. https://doi.org/10.1166/jcc.2015.1084

Tong Q, Yang Y, Xiong Z, Li Z, Yuan W, Wang T (2016) Chitooligosaccharides downregulate TLR4/NF-κB/COX-2 signaling cascade in dextran sulfate sodium-treated mice: a potential mechanism for the anti-colitis effect. Int J Pharmacol 12:720–728. https://doi.org/10.3923/ijp.2016.720.728

Viens P, Lacombe-Harvey ME, Brzezinski R (2015) Chitosanases from family 46 of glycoside hydrolases: from proteins to phenotypes. Mar Drugs 13:6566–6587. https://doi.org/10.3390/md13116566

Vo TS, Ngo DH, Bach LG, Ngo DN, Kim SK (2017) The free radical scavenging and anti-inflammatory activities of gallate-chitooligosaccharides in human lung epithelial A549 cells. Process Biochem 54:188–194. https://doi.org/10.1016/j.procbio.2017.01.001

VWR (2021) Chitosanase

Weikert T, Niehues A, Cord-Landwehr S, Hellmann MJ, Moerschbacher BM (2017) Reassessment of chitosanase substrate specificities and classification. Nat Commun 8:1698. https://doi.org/10.1038/s41467-017-01667-1

Yang E-J, Kim J-G, Kim J-Y, Kim S, Lee N, Hyun C-G (2010) Anti-inflammatory effect of chitosan oligosaccharides in RAW 264.7 cells. Open Life Sci 5:95–102. https://doi.org/10.2478/s11535-009-0066-5

Yousef M, Pichyangkura R, Soodvilai S, Chatsudthipong V, Muanprasat C (2012) Chitosan oligosaccharide as potential therapy of inflammatory bowel disease: therapeutic efficacy and possible mechanisms of action. Pharmacol Res 66:66–79. https://doi.org/10.1016/j.phrs.2012.03.013

Yu-Young J, Jo K-J, Jin Y-L, Jung W-J, Kuk J-H, Kim K-Y, Kim T-H, Park R-D (2003) Characterization of endochitosanases-producing *Bacillus cereus* P16. J Microbiol Biotechnol 13:960–968

Insect Enzymes in Chitin Turnover and Deacetylation

Mingbo Qu, Lin Liu, and Qing Yang

1 Introduction

Chitin acts as an important protective and supporting polysaccharide in insects. It plays essential roles in insect organs such as the epidermis, trachea, and midgut. However, it also restricts the growth and metamorphosis of insects because of its rigid structure. Since the amount of chitin in insects changes periodically corresponding to the molting cycle, they possess multiple enzymes to accomplish a turnover and modifications of chitin, such as chitinases (Chts), β-*N*-acetylhexosaminidases (Hexes), lytic polysaccharide monooxygenases (LPMOs), and chitin deacetylases (CDAs) during development. In this chapter, our focus is exclusively on the structures, mechanisms of action, and catalytic properties of these enzymes in insects.

2 Chitinases

Chitinases (Chts, EC 3.2.1.14) catalyze the breakdown of β-1, 4 glycosidic bonds in chitin and chitooligosaccharides (Zhu et al. 2016). Insect chitinases belong to the glycoside hydrolase family 18 (GH18) as mentioned in the Carbohydrate Active enZymes database (CAZy; http://www.cazy.org/), except for a GH48 chitinase from

M. Qu · L. Liu · Q. Yang (✉)
School of Bioengineering, Dalian University of Technology, Dalian 116024, China
e-mail: qingyang@dlut.edu.cn

Q. Yang
State Key Laboratory for Biology of Plant Diseases and Insect Pests, Institute of Plant Protection, Chinese Academy of Agricultural Sciences, Beijing 100193, China

Shenzhen Branch, Guangdong Laboratory of Lingnan Modern Agriculture, Genome Analysis Laboratory, Ministry of Agriculture and Rural Affairs, Agricultural Genomics Institute at Shenzhen, Chinese Academy of Agricultural Sciences, Shenzhen 518120, China

S.-K. Kim (ed.), *Chitooligosaccharides*,
https://doi.org/10.1007/978-3-030-92806-3_15

the leaf beetle *Gastrophysa atrocyanea* (Fujita et al. 2006). All the insect GH18 chitinases share a conserved GH18 catalytic domain (CAD). The overall structure of CAD comprises a classic $(\beta/\alpha)_8$ TIM-barrel with eight β-strands (β1–β8) tethered by eight α-helices (α1–α8). Four highly conserved regions or signature motifs in this structure have been identified, including motif 1 (KXXXXXGGW) in strand β3, motif 2 (FDGXDLDWEYP) in strand β4, motif 3 (MXYDXXG) in strand β6, and motif 4 (GXXXWXXDXDD) in strand β8 (Arakane and Muthukrishnan 2010). The critical amino acids that directly participate in the hydrolytic reaction are located in motif 2, while the other motifs provide sugar-binding sites and play a role in processivity based on the presence of aromatic residues and their location along the substrate-binding cleft.

2.1 Catalytic Mechanism of Insect Chitinases

Chitinases generate β-anomeric products during catalysis of β-1, 4 glycosidic bonds and hence use a molecular mechanism that retains the anomeric configuration at the cleavage site (Armand et al. 1994). Crystal structures of several family 18 chitinases with bound substrates or products indicated that the *N*-acetyl group of the sugar plays a role in catalysis, providing evidence for a mechanism known as substrate-assisted catalysis. This mechanism involves the formation of a positively charged oxazolinium-ion intermediate, which is stabilized by the carbonyl oxygen of the *N*-acetyl group of −1 sugar unit adjacent to the scissile bond (Brameld et al. 1998). The whole catalytic process includes three steps (van Aalten et al. 2001): (1) Binding of substrate in the cleft. Multiple sugar-binding sites in the substrate-binding cleft of GH18 chitinases can bind several GlcNAc units. Motif 2 (DXDXE) containing the catalytic triad (Asp144, Asp 146, and Glu148 in chitinase I from *Ostrinia furnacalis*, OfChtI) is involved in the binding of the −1 sugar unit. The side chain of Asp146 enables an interaction of hydrogen bonds between the hydrogen molecules of the acetamido group of the −1 sugar so as to fix it in a distorted conformation. (2) Formation of oxazolinium-ion intermediate and breakdown of glycosidic bond. Glu148 acts as a proton donor and interacts with the oxygen of the glycosidic bond. The carbonyl oxygen in the acetamido moiety of the distorted −1 sugar acts as a nucleophile; it attacks the anomeric center to form a covalent bicyclic oxazolinium-ion intermediate, resulting in a breakdown of the glycosidic bond. Asp146 stabilizes the oxazolinium intermediate during catalysis. (3) Dissociation of the product and ring opening. After the product at the reducing end dissociates from the active site, a water molecule as a general base attacks the intermediate. The ring of the oxazolinium-ion opens and cyclization occurs with inversion of the stereochemical structure; the overall reaction proceeds with a net retention of the stereochemical structure of the oxazolinium-ion (refer to Fig. 3 in Chen et al. 2020).

2.2 Modes of Action of Insect Chitinases

Although the GH18 chitinases employ a similar catalytic mechanism, different architectures of their sugar-binding sites indicate their different modes of action during chitin hydrolysis. Based on their cleavage type, chitinases are classified into exo- or endo-acting chitinases (Chen et al. 2020). Exo-acting chitinases hydrolyze the chitin chain from either the reducing or non-reducing end, primarily resulting in the production of chitobiose. The substrate-binding clefts are usually closed and deep. Conserved aromatic residues are present along the cleft to form hydrophobic stacking interactions with sugar moieties, which ensure that the enzymes are able to bind to the chitin chain and slide onward after one catalytic cycle (Chen et al. 2020). In general, these chitinases are processive enzymes, which mean that they are able to slide along the chitin fibrils and undergo hydrolysis without detachment after each catalytic event (Igarashi et al. 2014). Endo-acting chitinases catalyze internal hydrolysis of chitin chains at random positions. Usually, these chitinases do not exhibit processivity and possess structural characteristics with a shallow substrate-binding cleft (Chen et al. 2020).

The processivity of chitinases, as their important property, is difficult to evaluate, although various approaches have been developed. One classical approach is to determine the ratio of disaccharide units to the sum of monosaccharide and trisaccharide units produced (Fox et al. 2012), because processive chitinases primarily produce disaccharides during processive catalysis, while only producing odd-numbered saccharides when the first cleavage occurs. Another approach is to monitor the ratio of soluble to insoluble reducing ends simultaneously (Kurasin et al. 2015), as processive chitinases are assumed to primarily generate soluble products during processive catalysis, while non-processive catalysis primarily generates insoluble reducing ends. Recently, high-speed atomic force microscopy (HS-AFM) (Qu et al. 2020) and single-molecule fluorescence (Nakamura et al. 2018) have been applied to directly visualize the processive movement of chitinases on chitin fibrils. During each catalytic cycle, chitinases move forward by two sugar units on chitin fibrils. So the processivity of chitinases could be determined based on the moving distance of chitinase on chitin fibrils divided by the length of the product, chitobiose (1 nm). In addition, water-soluble chitosan (partially deacetylated chitin) can also be used as a substrate to estimate processivity (Horn et al. 2006). Because the acetamido group of the sugar unit at -1 subsite is required for productive binding, when processive chitinases confront a non-productive residue at the -1 subsite, they move forward by two sugar units each time to find an available acetamido group, which ultimately leads to catalysis. Thus, the processive chitinases generate products dominated by even-numbered oligomers, while non-processive chitinases dissociate from the chitin chain and rebind to the substrate randomly, leading to equal amounts of odd- and even-numbered products.

2.3 Modular Organization and Groups of Insect Chitinases

The differences in insect GH18 chitinases render them to be classified into 11 subgroups [ChtI to ChtX (10 subgroups) and a Chi-h subgroup] based on the phylogenetic relationship of the GH18 domain (Arakane and Muthukrishnan 2010; Zhu et al. 2016). They also possess various domain organizations, including 1–5 CADs (some of which may be catalytically inactive), 0–7 chitin-binding domains (CBDs), and linker regions that connect these domains (Fig. 1).

GH20, glycoside hydrolase family 20; GH18, glycoside hydrolase family 18; CBM14, carbohydrate-binding module family 14; AA15, auxiliary activity family 15; FnIII, fibronectin type-III; CRD, cysteine-rich domain.

The CADs present in most of the groups are thought to have chitinolytic activity, except for CADs in the Group V Cht-like proteins and several CADs in Group II chitinases. The structure and properties of the CADs are illustrated above. The CBDs of insect chitinases (except for Chi-h) belong to the carbohydrate-binding module family 14 (CBM14). Although the structure of CBM14 from insects has not been determined, the structure of CBM14 from chitinase HsCHIT1 has been resolved – it exhibits a distorted β-sandwich 3D fold (Fadel et al. 2016). One highly conserved tryptophan residue in a surface-exposed conformation is involved in chitin binding (Madland et al. 2019). The CBD in the Chi-h family is a fibronectin type-III (FnIII) domain, which is a typical immunoglobulin-like β-sandwich of two antiparallel β-sheets comprising three and four strands, respectively. Aromatic residues in the FnIII domain of SmChiA, a bacterial homolog of insect Chi-h, are exposed on the surface and are proven to be essential for substrate binding and hydrolysis efficiency (Uchiyama et al. 2001).

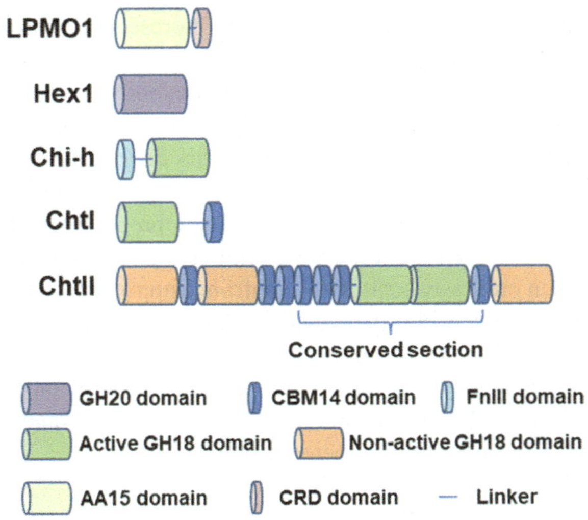

Fig. 1 Domain organization of chitinolytic enzymes in insects

Overall, the diversity of insect chitinases demonstrates that they may have functional specialization during chitin degradation, which is closely related to the domain organization and architecture of their catalytic domain.

2.3.1 Chi-hs Are Exo-Acting Processive Chitinases

Chi-hs belong to one subgroup of insect chitinases that are only found in lepidopteran insects. They possess one FnIII CBD at the N-terminus and one GH18 CAD at the C-terminus. They share more than 70% identity of amino acid sequences with their bacterial homologs, suggesting that Chi-hs may have been horizontally transferred from bacteria (Daimon et al. 2003). Chi-hs have been proven to be mostly related to cuticular chitin degradation, because the group of genes coding for Chi-hs is mainly upregulated during insect molting. A report suggested that knockdown of *SeChi-h* from the beet armyworm *Spodoptera exigua* caused molting deficiency and death (Zhang et al. 2012).

The properties of Chi-h from *O. furnacalis* (OfChi-h) have been extensively studied (Liu et al. 2017). Crystal structures of both OfChi-h and its complex with chitoheptaose ($(GlcN)_7$) revealed that there were 13 solvent-exposed aromatic residues at the non-reducing end, but only two at the reducing end, indicating an asymmetric substrate-binding cleft. Furthermore, two extra α-helices are present in the wall of the OfChi-h substrate-binding cleft, which narrows the substrate-binding cleft and may increase the binding affinity of OfChi-h for chitin chains (Liu et al. 2017). HS-AFM analysis at single-molecule level demonstrated that OfChi-h is a processive chitinase that moves on crystalline chitin at a speed of 56 nm/s (Qu et al. 2020). Enzymological studies demonstrated that compared to the Group I chitinase OfChtI, OfChi-h is more active toward insoluble substrates but shows lower activity toward the soluble substrate ethylene glycol chitin (Liu et al. 2017). The BmChi-h from *Bombyx mori* displays an exo-type substrate preference. Thus, Chi-h has been proven to be an exo-acting processive chitinase that degrades chitin chains from the reducing end, similar to its bacterial homolog (Daimon et al. 2005).

2.3.2 ChtIs Are Endo-Acting Non-Processive Chitinases

Group I chitinases, ChtIs (also known as Cht5), are the most widely studied insect chitinases. In most insects, only one ChtI is present, which possesses an N-terminal signal peptide, a GH18 CAD, a Thr/Ser-rich linker domain, and a C-terminal CBM14 CBD. However, in some species such as *Anopheles gambiae*, *Locusta migratoria*, and *Nilaparvata lugens*, ChtI is duplicated, but usually only one gene copy encodes the protein containing CBD (Muthukrishnan et al. 2019). The physiological role of ChtIs has been analyzed by RNAi in a variety of insects such *as Tribolium castaneum* (Zhu et al. 2008b), *Chilo suppressalis* (Su et al. 2016), *N. lugens* (Xi et al. 2015) and *L. migratoria* (Li et al. 2015). In general, these studies have indicated that ChtIs containing CBD participate in the turnover of old cuticle, especially at the pupal to adult stages.

The crystal structure of the CAD of OfChtI from *O. furnacalis* represents the first insect GH18 CAD structure with catalytic activity (Chen et al. 2014). The structure revealed that 10 aromatic residues were distributed symmetrically around the catalytic center along the substrate-binding cleft, indicating a long substrate-binding cleft. However, unlike the CAD of the exo-acting chitinase OfChi-h, which has a tunnel-like cleft, that of OfChtI possesses a groove-like substrate-binding cleft with an open end, suggesting that OfChtI is an endo-acting chitinase (Chen et al. 2014). Furthermore, a unique hydrophobic plane composed of four surface-exposed aromatic residues is situated adjacent to the entrance of the substrate-binding cleft, mutations of which greatly impair the binding ability and hydrolytic activity toward chitin, thus indicating that the surface hydrophobic plane may render OfChtI to possess the ability to anchor and bind chitin (Chen et al. 2014).

Biochemical analysis indicated that OfChtI could effectively convert colloidal chitin into chitobiose but had relatively low activity toward α-chitin (Wu et al. 2013). Some other Group I chitinases including TcCht5, DmCht5, and MsCht5 have been proven to be able to cleave both polymeric and oligomeric substrates (Zhu et al. 2008a). The CBD in MsCht5 enhances activity toward an insoluble polymer but not toward a soluble chitin oligosaccharide (Arakane et al. 2003). HS-AFM analysis at the single-molecule level demonstrated that OfChtI is a non-processive chitinase on crystalline chitin (Qu et al. 2020). Thus, these data indicate that Group I chitinases are endo-acting non-processive chitinases.

2.3.3 ChtIIs Are Endo-Acting Non-Processive Chitinases with Multiple Domains

Group II chitinases, ChtIIs, possess an N-terminal signal peptide, and 4–5 GH18 CADs interspersed with 4–7 CBM14 CBDs. In general, CADs located at the N-terminus have substitutions in the catalytic amino acids, which may result in inactivation of the CADs (Tetreau et al. 2015). Modular structure analysis indicated that the CAD located at the C-terminus is not conserved. It is either missing or inactive in some species. All the ChtIIs share a conserved structure organization with two tandem active CADs in the middle, one CBD at the C-terminus, and multiple CBDs at the N-terminus (Fig. 1).

The physiological role of ChtIIs has been investigated in various insect species. The genes of ChtIIs were shown to be upregulated during insect molting, and the time of the upregulation occurred earlier than that of Chi-h and ChtI in *O. furnacalis* (Qu et al. 2021) and *N. lugens* (Xi et al. 2015). Downregulating the expression of *TcChtII* in *T. castaneum* resulted in molting deficiency at every developmental stage as well as inhibition of egg hatching (Zhu et al. 2008b). Feeding or injecting the dsRNA for ChtII in the lepidoptera species resulted in a failure of ecdysis (He et al. 2013; Su et al. 2016). These results prove that ChtIIs are essential for insect molting.

The crystal structures of the two catalytically active CADs (OfChtII-C1 and OfChtII-C2) of OfChtII in both an unliganded form or complexed with chitooligosaccharide substrates have been determined (Chen et al. 2018). The structures demonstrate that the distribution of aromatic residues along the substrate-binding cleft of OfChtII is asymmetric, but the subsite occupancy is symmetrical because the most significant interactions occurring between the enzyme and substrate are localized in the area ranging from sugar subsites −2 to +2 (Chen et al. 2018). Compared to those of the endo-acting processive Chi-h, the two β-sheets present in the wall of the substrate-binding cleft of ChtII are short and form an open, flat tunnel. All these structural features demonstrate ChtIIs to be endo-acting chitinases.

Biochemical analyses showed that a truncated form of OfChtII containing one C-terminal CAD and three N-terminal CBDs (OfChtII-B4C1) exhibited a higher activity toward insoluble polymeric substrates than those of OfChtII-C1, OfChi-h, and OfChtI (Chen et al. 2018), indicating that the enzymatic activity of OfChtII toward the polymeric substrate is closely related to the number of carbohydrate-binding modules. OfChtII-B4C1 could loosen compacted chitin particles and produce holes in the cuticle surface, as evidenced by scanning electron microscopy (Qu et al. 2021). These results suggest that full-length ChtII may be an endo-acting non-processive chitinase that has a high potential for hydrolyzing polymeric chitin. The multiple inactive catalytic domains and CBMs may serve as anchors to ensure binding to the polymeric substrate.

2.3.4 ChtIIIs Are Endo-Acting Chitinases Active Toward Single-Chained Chitin

Group III chitinases, ChtIIIs, possess an N-terminal transmembrane domain, two active GH18 CADs in tandem, and a C-terminal CBM14 CBD. This group of chitinases is highly conserved in all the insects analyzed as well as in several other subphyla of arthropods, suggesting an ancient origin for this group of chitinases (Noh et al. 2018; Liu et al. 2018b). They are predicted to be membrane-anchored chitinases and verified by their cell membrane localization expressed in a baculovirus system (Noh et al. 2018).

The physiological roles of ChtIIIs have been investigated in different insect species. In *T. castaneum*, downregulation of *TcChtIII* by RNAi resulted in defective abdominal contraction and forewing expansion during the pupal stage, shortened elytra, and improper folding of hindwings during the adult stage (Zhu et al. 2008b). The newly synthesized cuticular layers were abnormal and contained amorphous fibers, as revealed by transmission electron microscopy (Noh et al. 2018). Similarly, in *Sogatella furcifera* and *Chilo suppresalis*, an RNAi of *ChtIII* affected molting and wing development (Su et al. 2016; Chen et al. 2017). Furthermore, in *O. furnacalis*, OfChtIII was observed to be co-localized with chitin synthase in the epidermal cell layer (Liu et al. 2018b).

Crystal structures of two CADs of OfChtIII, one in an unliganded form and the other complexed with chitooligosaccharide substrates, have been obtained (Liu et al.

2018b). Structural analysis indicated that both these CADs have shorter and shallower substrate-binding clefts than that of OfChtI. They also lack a surface hydrophobic plane, similar to that observed in OfChtI. Consequently, it can be inferred that OfChtIIIs have shallower substrate-binding clefts. Biochemical studies have shown that OfChtIIIs are active toward single-chained chitin substrates but are inactive toward insoluble chitin substrates (Liu et al. 2018b). These data suggest that ChtIIIs are endo-acting chitinases, which may be associated with the processing of nascent chitin chains into higher-order structures during chitin synthesis rather than with chitin degradation.

2.3.5 ChtIVs

Group IV chitinases, ChtIVs, possess an N-terminal signal peptide and a CAD. Several ChtIVs are found in many insects, and some of these contain an extra CBD. ChtIVs are generally expressed in the gut tissue at both larval and adult stages (Zhu et al. 2008b; Su et al. 2016; Khajuria et al. 2010). RNAis of *ChtIV*s have yielded mixed results concerning their physiological roles. In *O. nubilalis*, RNAi of *OnChtIV* resulted in a reduction in body weight and an increase in chitin content in the peritrophic matrix (PM) (Khajuria et al. 2010). In *C. suppresalisr*, injection of dsRNA targeting two different *CiChtIV*s resulted in the death of more than two-thirds of larvae (Su et al. 2016). In *D. melanogaster*, RNAi of *ChtIV* led to thinning of the procuticle and larval lethality (Pesch et al. 2016). These data indicate a connection between ChtIVs and chitin turnover in the PM. However, some studies suggest that ChtIVs may play an important role in protecting insects from fungal infections of the midgut.

The structure of OfChtIV from *O. furnacalis* has been resolved recently. It is structurally more similar to plant antifungal chitinases than insect chitinolytic chitinases such as OfChtI and OfChi-h. It lacks the elements for crystalline chitin binding and contains a few solvent-exposed aromatic residues in the substrate-binding cleft, suggesting that it is an endo-active chitinase that facilitates the degradation of chitin with less crystallinity (Liu et al. 2020). Biochemical analysis indicated that OfChtIV exhibits high mycelial hydrolytic activity (Liu et al. 2020). A ChtIV from *D. melanogaster* (DmChtIV) exhibited high activity toward a polymeric substrate but no or low activity toward an oligosaccharide substrate (Zhu et al. 2008a). The physiological role of ChtIVs and their enzymological properties need to be further investigated.

2.3.6 Other Chitinases

To date, there have been insufficient efforts to investigate the biochemical properties and physiological functions of chitinases from other families of chitinases. These chitinases seem to be less important than the chitinases summarized above. However, further studies are needed to elucidate their functions during insect development and their potential applications.

3 GH20 β-N-Acetylhexosaminidases

The glycoside hydrolase family 20 (GH20) β-*N*-acetylhexosaminidases (Hex, EC 3.2.1.52) are essential enzymes that catalyze the removal of β-linked *N*-acetylhexosamine from the non-reducing ends (http://www.cazy.org). They are involved in chitin degradation by hydrolyzing chitooligosaccharides to generate the monomer *N*-acetylglucosamine (GlcNAc). The crystal structure of GH20 obtained from bacteria was first reported in 1996 (Tews et al. 1996), while that of an insect GH20 Hex was first reported by Liu et al. in 2011 (Liu et al. 2011b). Most GH20 Hexes possess two closely connected domains, domain I and domain II. Domain I consists of highly conserved six-stranded antiparallel β-sheets. Domain II is a GH20 CAD, in which all the GH20 Hexes share a highly conserved structure of the $(\beta/\alpha)_8$-barrel (TIM-barrel). This CAD contains a highly conserved pair of catalytic residues, Asp-Glu, wherein the Glu residue represents a general acid/base involved in the catalytic mechanism of these enzymes.

3.1 Catalytic Mechanism of GH20 β -N-Acetylhexosaminidases

GH 20 Hexes also employ a substrate-assisted catalytic mechanism similar to GH18 chitinases (Slamova et al. 2010). The active site of GH20 Hexes contains a highly conserved pair of catalytic residues, Asp-Glu. The Glu residue acts as a proton donor and forms a hydrogen bond with glycosidic oxygen. The 2-acetamido group, GlcNAc, at the −1 position acts as a nucleophile instead of Asp in the active center; it attacks the glycosidic bond to form an oxazoline or oxazolinium-ion reaction intermediate. The Asp residue, the nucleophilic residue, is responsible for orienting the C2-acetamido group and stabilizing the oxazoline intermediate (refer to Fig. 3 in Prag et al. 2000). Glu then activates a water molecule, which hydrolyzes the intermediate.

3.2 The Groups of GH20 β-N-Acetylhexosaminidases

Although all the GH20 Hexes share the same catalytic mechanism, the loops on top of the catalytic amino acids that form the substrate-binding pocket determine substrate specificity (Liu et al. 2018a). Based on the different amino acid sequences, the insect GH20 Hexes could be classified into four subgroups that exhibit different enzymological and physiological properties.

3.2.1 Hex1s Are the Major Chitinolytic Hexes

Group I β-*N*-acetylhexosaminidase Hex1s have been shown to be the major chitinolytic enzymes. Hex1s from different insect species such as *B. mori* (Okada et al. 2007), *Manduca sexta* (Kramer et al. 1993), and *Choristoneura fumiferana* (Zheng et al. 2008) are mainly expressed in the epidermis during insect molting. Downregulation of these genes by RNAi resulted in molting defects in *T. castaneum* (Hogenkamp et al. 2008) and *O. furnacalis* (Liu et al. 2011c). The old cuticle of the larvae injected with ds *Hex1* could not be shed off. Studies on enzymatic properties of Hex1s from *D. melanogaster*, *B. mori*, and *O. furnacalis* demonstrated that Hex1s could efficiently hydrolyze β-1,4 linked chitin oligosaccharide with very high specificity but could not act on β-1,3 or β-1,2 linked substrates (Liu et al. 2009; Okada et al. 2007). These data suggest that Hex1s are specialized enzymes that are essential for cuticle degradation.

To date, the only crystal structure of insect Hex studied is that of OfHex1 from *O. furnacalis*. The crystal structure of OfHex1 in an unliganded form and complexed with N, N, N-trimethyl-D-glucosaminyl (TMG)-chitotriomycin, a mimic of the substrate $(GlcNAc)_4$, was determined (Liu et al. 2011c). Compared to the structures of GH20 Hexes from humans, OfHex1 has a deep boot-shaped substrate-binding pocket and two loops (Loop314–335 and Loop478–496) situated at the entrance to the active site. The residues Val327, Glu328, and Trp490 in these loops form a narrow entrance at the +1 subsite to the active pocket. Trp490 and Val327 sandwich the +1 GlcNAc, while Glu328 forms a hydrogen bond with the C3-OH of the + 1 GlcNAc. Mutation of Val327 to glycine changes the selectivity of the glycosidic bond in the substrate, whereas mutation of Trp490 and Glu328 to alanine leads to a severe decrease in the binding affinity for $(GlcNAc)_2$, indicating that these structural features are essential for the strict substrate specificity of OfHex1 (Liu et al. 2012b). The deep substrate-binding pocket and the loops at the entrance to the active site are also conserved in other chitinolytic GH20s, such as SmChb, SpHex, PsHex, ScHex, and AoHex from *Aspergillus oryzae*, suggesting an evolutionary conservation of structural elements in determining the specificity of chitin oligosaccharides (Liu et al. 2011a).

3.2.2 Other Groups of Hexes in Insects

The group II β-*N*-acetylhexosaminidases Hex2s show a high amino acid sequence similarity to human β-*N*-acetylhexosaminidase compared to the other groups of insect Hexes. The RNAi of *OfHex2* from *O. furnacalis* causes abnormalities in the larval abdomen, pupa, and adult appendages (Liu et al. 2012a). Studies on the enzymatic properties of Hex2s from *O. furnacalis*, *B. mori*, and *Spodoptera frugiperda* have shown that they are enzymes with a broad substrate spectrum, which hydrolyze β-*N*-acetylhexosamine from chitin oligosaccharides, N-glycan, and glycolipids (Liu et al. 2012a; Okada et al. 2007; Tomiya et al. 2006). Hex2s may be involved in the degradation of glycoconjugates similar to human β-*N*-acetylhexosaminidases.

The group III β-*N*-acetylhexosaminidases Hex3s have not been well studied so far. The RNAi of *TcHex3* from *T. castaneum* was able to cause abnormalities during insect molting (Hogenkamp et al. 2008). Hex3 was proved to be present in molting fluid where it interacted with Hex1. Enzymatic activity analysis indicated that OfHex3 from *O. furnacalis* was able to degrade chitooligosaccharides, but at a lower rate than that of OfHex1 (Qu et al. 2013). Hex3s, together with Hex1s and Hex4s, were also found in the plasma membrane of spermatozoa in dipteran insects, suggesting that they may be involved in sperm-egg recognition (Cattaneo et al. 2002).

The group IV β-*N*-acetylhexosaminidases Hex4s are also called fused lobes (fdl), because the mutant of *Hex4* gene caused a fused lobes phenotype of the mushroom body in *D. melanogaster* (Leonard et al. 2006). Biochemical studies of these enzymes from *O. furnacalis*, *B. mori*, and *D. melanogaster* suggest that they also possess a strict substrate specificity. They could exclusively hydrolyze the terminal β-1,2-GlcNAc residue from the α-1,3 branch instead of the α-1,6 branch of the substrate, N-glycan GnGn (Leonard et al. 2006; Tomiya et al. 2006; Nomura et al. 2010). The specific activity of Hex4s may be closely related to the two long loops located at the +1 substrate-binding sites, but so far, the crystalline structure of Hex4 has not been demonstrated. These results suggest that Hex4s are involved in the modification of N-glycans in insects.

3.3 The Cocktail for Insect Chitin Hydrolysis System

Proteomic analysis of molting fluids, which are secreted for the degradation of insect old cuticle during insect molting, uncovered a "211" cocktail of glycoside hydrolases for cuticular chitin degradation, including two insect-originated ChtI and ChtII, one bacteria-originated chitinase Chi-h and one β-*N*-acetylhexosaminidase (Qu et al. 2014). As illustrated above, a series of studies on these chitinases in *O. furnacalis* suggested that both OfChtI and OfChtII are endo-acting non-processive chitinases, while OfChi-h is an exo-acting processive chitinase. OfChtII is upregulated earlier than OfChtI and OfChi-h during the larva–pupa molting of *O. furnacalis* (Qu et al. 2021). It efficiently loosens compacted chitin, which may initiate chitin degradation, thereby preparing chitin structures for further degradation by OfChtI and OfChi-h when they are produced after OfChtII. Synergy between these three chitinases has been demonstrated. As revealed by HS-AFM, the endo-acting OfChtI and OfChtII promote substrate binding as well as processivity of exo-acting OfChi-h during hydrolysis of crystalline chitin (Qu et al. 2020). OfHex1 hydrolyzes $(GlcNAc)_2$ to GlcNAc to eliminate the production inhibition of chitinases.

Similar to the extensively studied chitin hydrolysis system of *Serratia marcescens,* that of *O. furnacalis* also involves three chitinases and one β-*N*-acetylhexosaminidase; however, some enzymes work in a different manner. The β-*N*-acetylhexosaminidases in both the species (OfHex1 and SmChb) possess deep substrate-binding pockets and conserved loops located at the entrance to the active site (Liu et al. 2011a). OfChi-h shares a very high sequence identity and exhibits similar substrate-binding sites as SmChiA (Liu et al. 2017). They both are exo-acting processive chitinases that prefer to act on the reducing end. OfChtI is an endo-acting non-processive chitinase that possesses an open and symmetric cleft, while SmChiB is an exo-acting processive chitinase that prefers to act on the non-reducing end (Igarashi et al. 2014). Both OfChtII and SmChiC are endo-acting chitinases that contain more than two domains. However, OfChtII is unique and possesses five CADs and seven CBDs. Compared to that of SmChiC, the CAD of OfChtII contains an extra $\alpha + \beta$ insertion domain that forms one "wall" of the substrate-binding cleft, resulting a longer and deeper substrate-binding cleft (Chen et al. 2018). The differences between the insect and bacterial chitin hydrolysis systems may be closely related to the conditions under which they achieve an effective turnover of chitin. The insect chitin degradation system may provide new candidates for chitin application in the field of biorefinery.

4 Insect Lytic Polysaccharide Monooxygenases

Lytic polysaccharide monooxygenases (LPMOs, EC 1.14.99.53/54/55/56) are copper-dependent enzymes that oxidize glycosidic bonds in polysaccharides (Forsberg et al. 2019). Their discovery circumvents the challenges faced by glycoside hydrolase during polysaccharide degradation (Vaaje-Kolstad et al. 2010). LPMOs generate new chain ends through their oxidative activity and decrease crystallinity of the polysaccharides. They create new starting points for GHs, so that the overall activity of the enzyme mixtures is boosted (Vaaje-Kolstad et al. 2005a, 2010). Today, LPMOs are classified in the CAZy database as the auxiliary activities (AA) families AA9-AA11 and AA13-AA17. They show substrate specificity toward different polysaccharides such as cellulose, chitin, starch (Lo Leggio et al. 2015), xylan (Couturier et al. 2018), and pectin (Sabbadin et al. 2021). Insect LPMOs were discovered in 2018. They are grouped as the AA15 family and have been shown to be active toward chitin and cellulose (Sabbadin et al. 2018). These LPMOs provide a new potential for polysaccharide applications; however, to date, their properties are not fully understood.

4.1 Structure and Catalytic Mechanism of Insect LPMO

The first structure of LPMO to be described was chitin-active SmLPMO10A (SmCBP21) from *S. marcescens* (Vaaje-Kolstad et al. 2005b). To date, more than 40 LPMO structures have been deposited in the CAZy database. All the LPMOs share a conserved structure of β-sandwich fold encompassing 8–9 β-strands, yielding an overall pyramidal shape. The structure is stabilized by disulfide bridges and hydrophobic interactions between residues. The base of the pyramid consists of loops and helices that constitute the substrate-binding surface. Mature LPMOs possess a conserved N-terminal His. The N-terminal amino group together with the side chain of the N-terminal His and the side chain of a more distal His form the "histidine brace" active site of LPMOs, which is located at the center of the substrate-binding surface (Vaaje-Kolstad et al. 2010). The active site is completed by a Tyr or Phe in the axial position and coordinates with one copper ion, which is classified as a type II copper site (Gudmundsson et al. 2014).

To date, the only structure of insect LPMO (TdAA15A) was obtained from *Thermobia domestica* (Sabbadin et al. 2018). It shares a conserved typical central β-sandwich fold structure and is stabilized by five disulfide bonds. The TdAA15A active site is also characterized by a ubiquitous LPMO histidine brace. His1 and His91 of the mature enzyme directly coordinate with the essential copper cofactor with a T-shaped geometry, as observed for all copper-bound LPMOs characterized to date. The TdAA15A active site is also characterized by the ubiquitous LPMO histidine brace (Fig. 4b). In addition to the conserved structural features shared by all the LPMOs, the insect AA15 LPMOs (TdAA15A as representative) structure reveals an unusual β-tongue-like protrusion on one side of the histidine brace and forms a part of the surface surrounding the active site. On the opposite side of the histidine brace, there are co-planar aromatic rings of Tyr166 and Tyr24 in TdAA15A, which mark the boundaries of the flat surface surrounding the active site and may be involved in substrate binding (Sabbadin et al. 2018).

During catalysis, the Cu(II) located at the catalytic site of LPMO is first reduced to Cu(I) by an electron donor. LPMO-bound Cu(I) then binds oxygen (or hydrogen peroxide) to form Cu-oxo (or Cu-oxyl) species (Forsberg et al. 2019). They oxidize the C-H bond of the C1 or C4 carbon of polysaccharides, followed by hydroxylation of the resulting substrate radical. This results in the formation of either a lactone when H1 is oxidized or a ketone when H4 is oxidized (Vaaje-Kolstad et al. 2017).

4.2 Groups of Insect LPMOs and Their Activities

Based on phylogenetic analysis, Qu et al. grouped insect AA15 LPMOs into four subgroups (Qu et al., unpublished data). Group I LPMOs (LPMO1s) process an N-terminal AA15 domain and a C-terminal extension. The C-terminal extension is highly conserved only in LPMO1s and is featured by 12 cysteines; thus, it is named as a cysteine-rich domain (CRD) (Fig. 1). Group II LPMOs possess a potential C-terminal transmembrane helix in addition to an AA15 domain, while group III and group IV LPMOs only contain an AA15 domain. Several group III genes have been identified in some species, such as *T. castaneum* and *T. domestica* (Sabbadin et al. 2018). Group III LPMOs do not include any representatives from the lepidopteran species, while group IV LPMOs are lepidopteran-specific.

So far, the biochemical activities of only several LPMOs have been analyzed. TdAA15A from *T. domestica* exhibits a predominant C1-oxidation pattern and generates C1-aldonic acids on both chitin and cellulose in the presence of an external electron donor. It is able to synergize with various hydrolases such as GH6 (cellobiohydrolase), GH7 (endoglucanase), GH9 (endoglucanase), GH1 (β-glucosidase), and GH18 (chitinase) families (Sabbadin et al. 2018). Compared to TdAA15A, TdAA15B only exhibits oxidative activity toward chitin, producing C1-oxidation products from both highly crystalline α-chitin and partially amorphous β-chitin (Sabbadin et al. 2018). Similarly, two recombinant LPMOs (CgAA15a and CgAA15b) from the lower termite, *Coptotermes gestroi*, exhibit oxidative activity toward chitin but not toward cellulose, xylan, xyloglucan, or starch (Franco Cairo et al. 2020).

The discovery of LPMO in insects suggests that insect chitin is mainly degraded by glycoside hydrolases. Insect LPMOs share less than 20% identity with bacterial enzymes. It provides a new candidate for the application of LPMO in the biorefinery field. However, the understanding of insect LPMOs is still very limited. More efforts are needed to investigate the enzymatic properties and synergistic mechanisms with other glycoside hydrolases during chitin degradation.

5 Insect Chitin Deacetylases

Chitin deacetylase (CDA; EC 3.5.1.41), an extracellular chitin-modifying metalloenzyme, catalyzes the removal of acetamido group in the *N*-acetylglucosamine units of chitin, thereby generating the glucosamine unit chitosan and acetic acid (AcOH) (Tsigos et al. 2000). CDAs belong to the carbohydrate esterase family 4 (CE4) in the CAZy database (http://www.cazy.org) together with peptidoglycan deacetylases (EC 3.5.1.104), MurNAc deacetylases (EC 3.5.1.-), poly-β1, 6-GlcNAc deacetylases (EC 3.5.1.-), and acetylxylan esterases (EC 3.1.1.72) (Li et al. 2021). Although CDAs have been widely identified in protists, diatoms, bacteria, fungi, nematodes, insects, and other arthropods, insects possess a greater number of CDAs than any other organism (Xi et al. 2014).

5.1 Crystal Structure of Insect Chitin Deacetylases

The CE4 CDAs contain a conserved CAD known as the NodB homologous domain because of its similarity to the NodB oligosaccharide deacetylase (John et al. 1993). The NodB homology domain is a $(\beta/\alpha)_7$ barrel, composed of seven parallel β-strands arranged in a barrel shape and surrounded by six α-helices. Notably, the $(\beta/\alpha)_7$ barrel contains one loop insertion between β5 and α5. The NodB domains are known to contain five conserved catalytic motifs (motifs 1–5) (Andres et al. 2014; Blair and van Aalten 2004; Blair et al. 2005). Motif 1 (TFDD) possesses two aspartic acid residues that interact with zinc or cobalt; the second residue binds to the acetate released from the substrate. Motif 2 (H[S/T]xxH) possesses two histidines that bind a metal ion and a serine or threonine and forms a hydrogen bond with the second histidine to stabilize the loop. Motif 3 (RxPY/F) constitutes one side of the active-site groove that is involved in acetate binding and zinc binding. Motif 4 (DxxDW/Y) constitutes the other side of the active-site groove, with tryptophan being the most critical residue. Motif 5 (LxH) constitutes a hydrophobic pocket that binds the acetate methyl group, and the histidine forms a hydrogen bond with the produced acetate.

The structure of the insect CDAs was first demonstrated in *B. mori* as a Group I CDA (BmCDA1-CAD; the catalytic domain of BmCDA1, PDB 5ZNS, and 5ZNT) (Liu et al. 2019b). Later, the structure of a midgut-specific CDA (BmCDA8) from *B. mori* was also determined (PDB 5Z34) (Liu et al. 2019b). Alignment of insect CDAs with other CE4 enzymes revealed the presence of two main structural elements among these CDAs that are not present in any of the other CE4 structures determined to date. Moreover, motif 4 (SMVDS/A) of insect CDAs is different from the conserved motif 4 (DxxDW/Y) in microbial CDAs, although the other four motifs are conserved. In addition, the C-terminal region consists of an extra α8 helix, a pair of antiparallel β-strands (β9 and β10), and several loops (Liu et al. 2019b). These structural characteristics contribute to the distinct architecture of insect CDAs by having a much longer and wider substrate-binding cleft than that of microbial CDAs. The longer substrate-binding cleft formed by unique loops appears to be an intentional feature to fit chitin fibers in vivo, suggesting that they may possess different substrate specificities and physiological functions. Structural base comparisons between BmCDA1 and BmCDA8 also revealed that the substrate-binding cleft of BmCDA1 is more open than that of BmCDA8, because of the lack of a series of surface-exposed aromatic amino acids (Liu et al. 2019b).

5.2 Catalytic Mechanism of Insect Chitin Deacetylases

The conserved NodB homologous domain of BmCDA1-CAD and CDA8 confers on insect CDAs a conserved metal-assisted acid/base catalytic mechanism using a His–His–Asp zinc-binding triad, which is very similar to the conserved catalytic base (Asp) and acid (His) mechanism of CE4 enzymes (Liu et al. 2019b). Regarding

CDA from *Vibrio cholerae* (VcCDA), His295 acts as a catalytic acid and Asp39 acts as a catalytic base in the catalytic process (Andres et al. 2014). First, a metal-bound water molecule is activated by Asp39 to form a tetrahedral oxyanion intermediate. The negative charge at the carbonyl oxygen atom is stabilized both by the main chain nitrogen atom of Tyr169 as well as metal coordination. Second, the nitrogen group of the intermediate is protonated by His295 to assist in the cleavage of a C–N bond, generating a free amine in the de-*N*-acetylated chitin and releasing acetate. This catalytic mechanism of microorganisms can provide a base for understanding the mode of action of insect CDAs, because some key amino acids involved in catalysis are conserved, although their crystal structures are slightly different (refer to Fig. 3 in (Li et al. 2021). Liu et al. studied the deacetylation mode of BmCDA8 from *B. mori*. The results showed that BmCDA8 was not able to deacetylate GlcNAc or $(GlcNAc)_2$ but could deacetylate $(GlcNAc)_3$. Deacetylation occurred at the first GlcNAc, which is situated at the non-reducing end. Thus, BmCDA8 activity requires substrates to occupy subsites 0, +1, and +2, where 0 is the catalytic site (Liu et al. 2019b).

5.3 *Subfamilies of Insect Chitin Deacetylases*

There are several CDAs in insects, which can be classified into five groups based on sequence similarity and domain diversity (Li et al. 2021; Dixit et al. 2008). Except group V CDAs, all other CDAs have a multidomain architecture composed of a CAD and several other domains, including CBM14 and a low-density lipoprotein receptor class A domain (LDLa). These CDAs are thought to possess different physiological and biochemical properties.

5.3.1 Group I Chitin Deacetylases

In insects, all group I CDAs possess an N-terminal CBD, an LDLa in the middle, and a C-terminal CAD. In each species, there are usually two CDAs in group I CDAs, CDA1 and CDA2 (the amino acid sequence identity between CDA1s and CDA2s is approximately 60%). In *D. melanogaster*, CDA Serpentine (Serp, CDA1) or Vermiform (Verm, CDA2) mutant embryos display elongated and tortuous tracheal tubes (Luschnig et al. 2006; Wang et al. 2006). In *T. castaneum*, *TcCDA1* and *TcCDA2* are essential for insect molting and survival (Arakane et al. 2009). In *L. migratoria*, *LmCDA2* is essential for molting (Yu et al. 2016). These data suggest that group I CDAs are involved in chitin turnover in tissues originating from the ectoderm.

Enzymatic activity of insect CDAs has been pursued for a long time, but these CDAs seem to be less active than CDAs of microorganisms. The enzymatic activity of insect CDA1s was first reported in 2008 using artificial substrates. Later, Toprak et al. reported the activity of McCDA1 from *Mamestra configurata* toward colloidal chitin using an in-gel assay to detect chitosan formation (Toprak et al. 2008, 2016). In addition, Liu et al. suggested that the deacetylase activity of BmCDA1 from *B.*

mori was detectable only when the cuticular chitin-binding protein CPAP3-A1 was added (Liu et al. 2019b).

5.3.2 Other Groups of Chitin Deacetylases

Group II CDAs are composed of CDA3s, which exhibit the same structural organization as group I CDAs. However, the overall amino acid sequence has only about 38% identity with group I CDAs. Groups III and IV CDAs include CDA4s and CDA5s, respectively. These CDAs lack the LDLa domain but contain an N-terminal CBD and a C-terminal CAD. They differ in the length of the linker sequence between the domains. Group V CDAs include several CDAs that possess only a CAD.

The functions of CDAs from groups II to V have not yet been well established. In *T. castaneum*, silencing of these genes by RNAi failed to produce any visible effects (Arakane et al. 2009). In *Diaphorina citri*, DcCDA3 was suggested to play an important role in the immune response (Yu et al. 2019). RNAi of Group IV CDA leads to mortality in *N. lugens* (Xi et al. 2014). Hemimetabolous insects belonging to the orders Hemiptera, Anopleura, and Orthopteta seem to lack Groups II and V CDAs (Xi et al. 2014; Yu et al. 2016; Dixit et al. 2008). In other insect species, Group V CDAs are predominantly expressed in the gut tissue and are also known as midgut CDAs. No obvious phenotypes were observed when dsRNAs for gut-specific CDAs (*TcCDA6–9*) from *T. castaneum* were co-injected at the larval stage, suggesting that gut CDAs may not be essential for survival (Arakane et al. 2009). In *Helicoverpa armigera*, the gene expression of group V CDA (*HaCDA5a*) is downregulated by baculovirus infection in larvae, suggesting that it may contribute to defense against pathogens (Han et al. 2015). Recombinant HaCDA5a could increase the permeability of PM in a concentration-dependent manner, suggesting it might be involved in the modification of endogenous chitin (Jakubowska et al. 2010).

Zhong et al. (2014) reported that group V BmCDA7 from *B. mori* catalyzed the conversion of the artificial substrate p-nitroacetanilide to p-nitroaniline (Zhong et al. 2014). In 2019, the deacetylase activity of insect CDAs toward native chitin was demonstrated. Two midgut CDAs (BmCDA7 and BmCDA8) from *B. mori* were active toward their physiological substrate, PM chitin. It is worth noting that another *B. mori* midgut CDA BmCDA6, which was expressed almost exclusively at the mid-molt stage, showed no activity toward PM chitin and slight activity toward colloidal chitin (Liu et al. 2019a). BmCDA6, BmCDA7, and BmCDA8 were shown to have a binding activity toward PM chitin and colloidal chitin (Liu et al. 2019a). Moreover, CDA in the midgut of *Trichoplusia ni* was tightly bound to the PM (Guo et al. 2005). Insect CDAs exhibit a broad substrate specificity in that they act on chitin polymers such as PM chitin, ethylene glycol chitin, and colloidal chitin, as well as on short-chain chitin oligosaccharides, and display a higher activity against long-chain chitin oligosaccharides.

6 Conclusions

Insects provide a pool of multiple enzymes for degradation and modification of chitin. These enzymes can achieve efficient degradation and modification of chitin during development. Some enzymes possess unique properties, which may provide new insights for designing novel chitinases that can be applied in the field of biorefinery. Synergism between these enzymes needs to be further investigated to fully understand the chitin degradation system in insects, which may provide useful information for industrial applications.

References

Andres E, Albesa-Jove D, Biarnes X, Moerschbacher BM, Guerin ME, Planas A (2014) Structural basis of chitin oligosaccharide deacetylation. Angew Chem Int Ed Engl 53(27):6882–6887. https://doi.org/10.1002/anie.201400220

Arakane Y, Muthukrishnan S (2010) Insect chitinase and chitinase-like proteins. Cell Mol Life Sci 67(2):201–216. https://doi.org/10.1007/s00018-009-0161-9

Arakane Y, Zhu Q, Matsumiya M, Muthukrishnan S, Kramer KJ (2003) Properties of catalytic, linker and chitin-binding domains of insect chitinase. Insect Biochem Mol Biol 33(6):631–648. https://doi.org/10.1016/s0965-1748(03)00049-3

Arakane Y, Dixit R, Begum K, Park Y, Specht CA, Merzendorfer H, Kramer KJ, Muthukrishnan S, Beeman RW (2009) Analysis of functions of the chitin deacetylase gene family in Tribolium castaneum. Insect Biochem Mol Biol 39(5–6):355–365. https://doi.org/10.1016/j.ibmb.2009.02.002

Blair DE, van Aalten DM (2004) Structures of Bacillus subtilis PdaA, a family 4 carbohydrate esterase, and a complex with N-acetyl-glucosamine. FEBS Lett 570(1–3):13–19. https://doi.org/10.1016/j.febslet.2004.06.013

Blair DE, Schuttelkopf AW, MacRae JI, van Aalten DM (2005) Structure and metal-dependent mechanism of peptidoglycan deacetylase, a streptococcal virulence factor. Proc Natl Acad Sci U S A 102(43):15429–15434. https://doi.org/10.1073/pnas.0504339102

Brameld KA, Shrader WD, Imperiali B, Goddard WA 3rd (1998) Substrate assistance in the mechanism of family 18 chitinases: theoretical studies of potential intermediates and inhibitors. J Mol Biol 280(5):913–923. https://doi.org/10.1006/jmbi.1998.1890

Cattaneo F, Ogiso M, Hoshi M, Perotti ME, Pasini ME (2002) Purification and characterization of the plasma membrane glycosidases of *Drosophila melanogaster* spermatozoa. Insect Biochem Mol Biol 32(8):929–941

Chen L, Liu T, Zhou Y, Chen Q, Shen X, Yang Q (2014) Structural characteristics of an insect group I chitinase, an enzyme indispensable to moulting. Acta Crystallogr D Biol Crystallogr 70(Pt 4):932–942. https://doi.org/10.1107/S1399004713033841

Chen C, Yang H, Tang B, Yang WJ, Jin DC (2017) Identification and functional analysis of chitinase 7 gene in white-backed planthopper, Sogatella furcifera. Comp Biochem Physiol B Biochem Mol Biol 208–209:19–28. https://doi.org/10.1016/j.cbpb.2017.03.002

Chen W, Qu M, Zhou Y, Yang Q (2018) Structural analysis of group II chitinase (ChtII) catalysis completes the puzzle of chitin hydrolysis in insects. J Biol Chem 293(8):2652–2660. https://doi.org/10.1074/jbc.RA117.000119

Chen W, Jiang X, Yang Q (2020) Glycoside hydrolase family 18 chitinases: The known and the unknown. Biotechnol Adv 43:107553. https://doi.org/10.1016/j.biotechadv.2020.107553

Couturier M, Ladeveze S, Sulzenbacher G, Ciano L, Fanuel M, Moreau C, Villares A, Cathala B, Chaspoul F, Frandsen KE, Labourel A, Herpoel-Gimbert I, Grisel S, Haon M, Lenfant N, Rogniaux H, Ropartz D, Davies GJ, Rosso MN, Walton PH, Henrissat B, Berrin JG (2018) Lytic xylan oxidases from wood-decay fungi unlock biomass degradation. Nat Chem Biol 14(3):306–310. https://doi.org/10.1038/nchembio.2558

Daimon T, Hamada K, Mita K, Okano K, Suzuki MG, Kobayashi M, Shimada T (2003) A Bombyx mori gene, BmChi-h, encodes a protein homologous to bacterial and baculovirus chitinases. Insect Biochem Mol Biol 33(8):749–759

Daimon T, Katsuma S, Iwanaga M, Kang WK, Shimada T (2005) The BmChi-h gene, a bacterial-type chitinase gene of Bombyx mori, encodes a functional exochitinase that plays a role in the chitin degradation during the molting process. Insect Biochem Mol Biol 35(10):1112–1123

Dixit R, Arakane Y, Specht CA, Richard C, Kramer KJ, Beeman RW, Muthukrishnan S (2008) Domain organization and phylogenetic analysis of proteins from the chitin deacetylase gene family of Tribolium castaneum and three other species of insects. Insect Biochem Mol Biol 38(4):440–451. https://doi.org/10.1016/j.ibmb.2007.12.002

Fadel F, Zhao Y, Cousido-Siah A, Ruiz FX, Mitschler A, Podjarny A (2016) X-Ray crystal structure of the full length human chitotriosidase (CHIT1) reveals features of its chitin binding domain. PLoS One 11(4):e0154190. https://doi.org/10.1371/journal.pone.0154190

Forsberg Z, Sorlie M, Petrovic D, Courtade G, Aachmann FL, Vaaje-Kolstad G, Bissaro B, Rohr AK, Eijsink VG (2019) Polysaccharide degradation by lytic polysaccharide monooxygenases. Curr Opin Struct Biol 59:54–64. https://doi.org/10.1016/j.sbi.2019.02.015

Fox JM, Levine SE, Clark DS, Blanch HW (2012) Initial- and processive-cut products reveal cellobiohydrolase rate limitations and the role of companion enzymes. Biochemistry 51(1):442–452. https://doi.org/10.1021/bi2011543

Franco Cairo JPL, Cannella D, Oliveira LC, Goncalves TA, Rubio MV, Terrasan CRF, Tramontina R, Mofatto LS, Carazzolle MF, Garcia W, Felby C, Damasio A, Walton PH, Squina F (2020) On the roles of AA15 lytic polysaccharide monooxygenases derived from the termite Coptotermes gestroi. J Inorg Biochem 216:111316. https://doi.org/10.1016/j.jinorgbio.2020.111316

Fujita K, Shimomura K, Yamamoto K, Yamashita T, Suzuki K (2006) A chitinase structurally related to the glycoside hydrolase family 48 is indispensable for the hormonally induced diapause termination in a beetle. Biochem Biophys Res Commun 345(1):502–507. https://doi.org/10.1016/j.bbrc.2006.04.126

Gudmundsson M, Kim S, Wu M, Ishida T, Momeni MH, Vaaje-Kolstad G, Lundberg D, Royant A, Stahlberg J, Eijsink VG, Beckham GT, Sandgren M (2014) Structural and electronic snapshots during the transition from a Cu(II) to Cu(I) metal center of a lytic polysaccharide monooxygenase by X-ray photoreduction. J Biol Chem 289(27):18782–18792. https://doi.org/10.1074/jbc.M114.563494

Guo W, Li G, Pang Y, Wang P (2005) A novel chitin-binding protein identified from the peritrophic membrane of the cabbage looper, Trichoplusia ni. Insect Biochem Mol Biol 35(11):1224–1234. https://doi.org/10.1016/j.ibmb.2005.06.003

Han G, Li X, Zhang T, Zhu X, Li J (2015) Cloning and tissue-specific expression of a chitin deacetylase gene from helicoverpa armigera (Lepidoptera: Noctuidae) and its response to bacillus thuringiensis. J Insect Sci 15. https://doi.org/10.1093/jisesa/iev076

He B, Chu Y, Yin M, Mullen K, An C, Shen J (2013) Fluorescent nanoparticle delivered dsRNA toward genetic control of insect pests. Adv Mater 25(33):4580–4584. https://doi.org/10.1002/adma.201301201

Hogenkamp DG, Arakane Y, Kramer KJ, Muthukrishnan S, Beeman RW (2008) Characterization and expression of the β-*N*-acetylhexosaminidase gene family of *Tribolium castaneum*. Insect Biochem Mol Biol 38(4):478–489. https://doi.org/10.1016/j.ibmb.2007.08.002

Horn SJ, Sorbotten A, Synstad B, Sikorski P, Sorlie M, Varum KM, Eijsink VG (2006) Endo/exo mechanism and processivity of family 18 chitinases produced by Serratia marcescens. FEBS J 273(3):491–503. https://doi.org/10.1111/j.1742-4658.2005.05079.x

Igarashi K, Uchihashi T, Uchiyama T, Sugimoto H, Wada M, Suzuki K, Sakuda S, Ando T, Watanabe T, Samejima M (2014) Two-way traffic of glycoside hydrolase family 18 processive chitinases on crystalline chitin. Nat Commun 5:3975. https://doi.org/10.1038/ncomms4975

Jakubowska AK, Caccia S, Gordon KH, Ferre J, Herrero S (2010) Downregulation of a chitin deacetylase-like protein in response to baculovirus infection and its application for improving baculovirus infectivity. J Virol 84(5):2547–2555. https://doi.org/10.1128/JVI.01860-09

John M, Röhrig H, Schmidt J, Wieneke U, Schell J (1993) Rhizobium NodB protein involved in nodulation signal synthesis is a chitooligosaccharide deacetylase. Proc Natl Acad Sci 90(2):625. https://doi.org/10.1073/pnas.90.2.625

Khajuria C, Buschman LL, Chen MS, Muthukrishnan S, Zhu KY (2010) A gut-specific chitinase gene essential for regulation of chitin content of peritrophic matrix and growth of Ostrinia nubilalis larvae. Insect Biochem Mol Biol 40(8):621–629. https://doi.org/10.1016/j.ibmb.2010.06.003

Kramer KJ, Corpuz L, Choi HK, Muthukrishnan S (1993) Sequence of a cDNA and expression of the gene encoding epidermal and gut chitinases of Manduca sexta. Insect Biochem Mol Biol 23(6):691–701

Kurasin M, Kuusk S, Kuusk P, Sorlie M, Valjamae P (2015) Slow off-rates and strong product binding are required for processivity and efficient degradation of recalcitrant chitin by family 18 chitinases. J Biol Chem 290(48):29074–29085. https://doi.org/10.1074/jbc.M115.684977

Leonard R, Rendic D, Rabouille C, Wilson IBH, Preat T, Altmann F (2006) The *Drosophila* fused lobes gene encodes an *N*-acetylglucosaminidase involved in N-glycan processing. J Biol Chem 281(8):4867–4875. https://doi.org/10.1074/jbc.M511023200

Li D, Zhang J, Wang Y, Liu X, Ma E, Sun Y, Li S, Zhu KY, Zhang J (2015) Two chitinase 5 genes from Locusta migratoria: molecular characteristics and functional differentiation. Insect Biochem Mol Biol 58:46–54. https://doi.org/10.1016/j.ibmb.2015.01.004

Li Y, Liu L, Yang J, Yang Q (2021) An overall look at insect chitin deacetylases: promising molecular targets for developing green pesticides. J Pestic Sci 46(1):43–52. https://doi.org/10.1584/jpestics.D20-085

Liu T, Liu F, Yang Q, Yang J (2009) Expression, purification and characterization of the chitinolytic β-*N*-acetyl-D-hexosaminidase from the insect *Ostrinia furnacalis*. Protein Expr Purif 68(1):99–103. https://doi.org/10.1016/j.pep.2009.06.004

Liu T, Zhang H, Liu F, Wu Q, Shen X, Yang Q (2011) Structural determinants of an insect beta-N-Acetyl-D-hexosaminidase specialized as a chitinolytic enzyme. J Biol Chem 286(6):4049–4058. https://doi.org/10.1074/jbc.M110.184796

Liu T, Zhang H, Liu F, Chen L, Shen X, Yang Q (2011) Active-pocket size differentiating insectile from bacterial chitinolytic β-*N*-acetyl-D-hexosaminidases. Biochem J 438:467–474

Liu T, Zhang H, Liu F, Wu Q, Shen X, Yang Q (2011) Structural determinants of an insect β-*N*-acetyl-D-hexosaminidase specialized as a chitinolytic enzyme. J Biol Chem 286(6):4049–4058

Liu T, Wu Q, Liu L, Yang Q (2012) Elimination of substrate inhibition of a β-*N*-acetyl-D-hexosaminidase by single site mutation. Process Biochem 48(1):103–108

Liu F, Liu T, Qu M, Yang Q (2012) Molecular and biochemical characterization of a novel β-*N*-acetyl-D-hexosaminidase with broad substrate-spectrum from the Aisan corn borer, *Ostrinia furnacalis*. Int J Biol Sci 8(8):1085–1096

Liu T, Chen L, Zhou Y, Jiang X, Duan Y, Yang Q (2017) Structure, catalysis, and inhibition of OfChi-h, the lepidoptera-exclusive insect chitinase. J Biol Chem 292(6):2080–2088. https://doi.org/10.1074/jbc.M116.755330

Liu T, Duan Y, Yang Q (2018) Revisiting glycoside hydrolase family 20 beta-N-acetyl-d-hexosaminidases: crystal structures, physiological substrates and specific inhibitors. Biotechnol Adv 36(4):1127–1138. https://doi.org/10.1016/j.biotechadv.2018.03.013

Liu T, Zhu W, Wang J, Zhou Y, Duan Y, Qu M, Yang Q (2018) The deduced role of a chitinase containing two nonsynergistic catalytic domains. Acta Crystallographica Section D Struct Biol 74(Pt 1):30–40. https://doi.org/10.1107/S2059798317018289

Liu L, Qu M, Liu T, Chen Q, Guo X, Yang J, Yang Q (2019) Biochemical characterization of three midgut chitin deacetylases of the Lepidopteran insect Bombyx mori. J Insect Physiol 113:42–48. https://doi.org/10.1016/j.jinsphys.2019.01.005

Liu L, Zhou Y, Qu M, Qiu Y, Guo X, Zhang Y, Liu T, Yang J, Yang Q (2019) Structural and biochemical insights into the catalytic mechanisms of two insect chitin deacetylases of the carbohydrate esterase 4 family. J Biol Chem 294(15):5774–5783. https://doi.org/10.1074/jbc.RA119.007597

Liu T, Guo X, Bu Y, Zhou Y, Duan Y, Yang Q (2020) Structural and biochemical insights into an insect gut-specific chitinase with antifungal activity. Insect Biochem Mol Biol 119:103326. https://doi.org/10.1016/j.ibmb.2020.103326

Lo Leggio L, Simmons TJ, Poulsen JC, Frandsen KE, Hemsworth GR, Stringer MA, von Freiesleben P, Tovborg M, Johansen KS, De Maria L, Harris PV, Soong CL, Dupree P, Tryfona T, Lenfant N, Henrissat B, Davies GJ, Walton PH (2015) Structure and boosting activity of a starch-degrading lytic polysaccharide monooxygenase. Nat Commun 6:5961. https://doi.org/10.1038/ncomms6961

Luschnig S, Batz T, Armbruster K, Krasnow MA (2006) serpentine and vermiform encode matrix proteins with chitin binding and deacetylation domains that limit tracheal tube length in Drosophila. Curr Biol 16(2):186–194. https://doi.org/10.1016/j.cub.2005.11.072

Madland E, Crasson O, Vandevenne M, Sorlie M, Aachmann FL (2019) NMR and fluorescence spectroscopies reveal the preorganized binding site in family 14 carbohydrate-binding module from human chitotriosidase. ACS Omega 4(26):21975–21984. https://doi.org/10.1021/acsomega.9b03043

Muthukrishnan S, Merzendorfer H, Arakane Y, Yang Q (2019) Chitin organizing and modifying enzymes and proteins involved in remodeling of the insect cuticle. In: Yang Q, Fukamizo T (eds) Targeting shitin-containing organisms. Springer Singapore, Singapore, pp 83–114. https://doi.org/10.1007/978-981-13-7318-3_5

Nakamura A, Okazaki KI, Furuta T, Sakurai M, Iino R (2018) Processive chitinase is Brownian monorail operated by fast catalysis after peeling rail from crystalline chitin. Nat Commun 9(1):3814. https://doi.org/10.1038/s41467-018-06362-3

Noh MY, Muthukrishnan S, Kramer KJ, Arakane Y (2018) A chitinase with two catalytic domains is required for organization of the cuticular extracellular matrix of a beetle. PLoS Genet 14(3):e1007307. https://doi.org/10.1371/journal.pgen.1007307

Nomura T, Ikeda M, Ishiyama S, Mita K, Tamura T, Okada T, Fujiyama K, Usami A (2010) Cloning and characterization of a beta-N-acetylglucosaminidase (BmFDL) from silkworm Bombyx mori. J Biosci Bioeng 110(4):386–391. https://doi.org/10.1016/j.jbiosc.2010.04.008

Okada T, Ishiyama S, Sezutsu H, Usami A, Tamura T, Mita K, Fujiyama K, Seki T (2007) Molecular cloning and expression of two novel β-*N*-acetylglucosaminidases from silkworm *Bombyx mori*. Biosci Biotechnol Biochem 71(7):1626–1635

Pesch YY, Riedel D, Patil KR, Loch G, Behr M (2016) Chitinases and Imaginal disc growth factors organize the extracellular matrix formation at barrier tissues in insects. Sci Rep 6:18340. https://doi.org/10.1038/srep18340

Prag G, Papanikolau Y, Tavlas G, Vorgias CE, Petratos K, Oppenheim AB (2000) Structures of chitobiase mutants complexed with the substrate di-N-acetyl-d-glucosamine: the catalytic role of the conserved acidic pair, aspartate 539 and glutamate 54011Edited by R. Huber. J Mol Biol 300(3):611–617. https://doi.org/10.1006/jmbi.2000.3906

Qu M, Ma L, Chen P, Yang Q (2014) Proteomic analysis of insect molting fluid with a focus on enzymes involved in chitin degradation. J Proteome Res 13(6):2931–2940. https://doi.org/10.1021/pr5000957

Qu M, Watanabe-Nakayama T, Sun S, Umeda K, Guo X, Liu Y, Ando T, Yang Q (2020) High-speed atomic force microscopy reveals factors affecting the processivity of chitinases during interfacial enzymatic hydrolysis of crystalline chitin. ACS Catal 10(22):13606–13615. https://doi.org/10.1021/acscatal.0c02751

Qu MB, Sun SP, Liu YS, Deng XR, Yang J, Yang Q (2021) Insect group II chitinase OfChtII promotes chitin degradation during larva-pupa molting. Insect Sci 28(3):692–704. https://doi.org/10.1111/1744-7917.12791

Qu M, Liu T, Chen P, Yang Q (2013) A sperm-plasma beta-N-acetyl-D-hexosaminidase interacting with a Chitinolytic beta-N-Acetyl-D-hexosaminidase in insect molting fluid. PLoS One 8(8):e71738. https://doi.org/10.1371/journal.pone.0071738

S. Armand HT, A. Heyraud, C. Gey, T. Watanabe, B. Henrissat, (1994) Stereochemical course of the hydrolysis reaction catalyzed by chitinases Al and D from Bacillus circulans WL-12. FEBS Lett 343(2):177–180

Sabbadin F, Hemsworth GR, Ciano L, Henrissat B, Dupree P, Tryfona T, Marques RDS, Sweeney ST, Besser K, Elias L, Pesante G, Li Y, Dowle AA, Bates R, Gomez LD, Simister R, Davies GJ, Walton PH, Bruce NC, McQueen-Mason SJ (2018) An ancient family of lytic polysaccharide monooxygenases with roles in arthropod development and biomass digestion. Nat Commun 9(1):756. https://doi.org/10.1038/s41467-018-03142-x

Sabbadin F, Urresti S, Henrissat B, Avrova Anna O, Welsh Lydia RJ, Lindley Peter J, Csukai M, Squires Julie N, Walton Paul H, Davies Gideon J, Bruce Neil C, Whisson Stephen C, McQueen-Mason Simon J (2021) Secreted pectin monooxygenases drive plant infection by pathogenic oomycetes. Science 373(6556):774–779. https://doi.org/10.1126/science.abj1342

Slamova K, Bojarova P, Petraskova L, Kren V (2010) beta-N-acetylhexosaminidase: what's in a name...? Biotechnol Adv 28(6):682–693. https://doi.org/10.1016/j.biotechadv.2010.04.004

Su C, Tu G, Huang S, Yang Q, Shahzad MF, Li F (2016) Genome-wide analysis of chitinase genes and their varied functions in larval moult, pupation and eclosion in the rice striped stem borer, Chilo suppressalis. Insect Mol Biol 25(4):401–412. https://doi.org/10.1111/imb.12227

Tetreau G, Cao X, Chen YR, Muthukrishnan S, Jiang H, Blissard GW, Kanost MR, Wang P (2015) Overview of chitin metabolism enzymes in Manduca sexta: identification, domain organization, phylogenetic analysis and gene expression. Insect Biochem Mol Biol 62:114–126. https://doi.org/10.1016/j.ibmb.2015.01.006

Tews I, Perrakis A, Oppenheim A, Dauter Z, Wilson KS, Vorgias CE (1996) Bacterial chitobiase structure provides insight into catalytic mechanism and the basis of Tay-Sachs disease. Nat Struct Biol 3(7):638–648. https://doi.org/10.1038/nsb0796-638

Tomiya N, Narang S, Park J, Abdul-Rahman B, Choi O, Singh S, Hiratake J, Sakata K, Betenbaugh MJ, Palter KB, Lee YC (2006) Purification, characterization, and cloning of a Spodoptera frugiperda Sf9 beta-N-acetylhexosaminidase that hydrolyzes terminal N-acetylglucosamine on the N-glycan core. J Biol Chem 281(28):19545–19560. https://doi.org/10.1074/jbc.M603312200

Toprak U, Baldwin D, Erlandson M, Gillott C, Hou X, Coutu C, Hegedus DD (2008) A chitin deacetylase and putative insect intestinal lipases are components of the Mamestra configurata (Lepidoptera: Noctuidae) peritrophic matrix. Insect Mol Biol 17(5):573–585. https://doi.org/10.1111/j.1365-2583.2008.00829.x

Toprak U, Erlandson M, Baldwin D, Karcz S, Wan L, Coutu C, Gillott C, Hegedus DD (2016) Identification of the Mamestra configurata (Lepidoptera: Noctuidae) peritrophic matrix proteins and enzymes involved in peritrophic matrix chitin metabolism. Insect Sci 23(5):656–674. https://doi.org/10.1111/1744-7917.12225

Tsigos I, Martinou A, Kafetzopoulos D, Bouriotis V (2000) Chitin deacetylases: new, versatile tools in biotechnology. Trends Biotechnol 18(7):305–312. https://doi.org/10.1016/S0167-7799(00)01462-1

Uchiyama T, Katouno F, Nikaidou N, Nonaka T, Sugiyama J, Watanabe T (2001) Roles of the exposed aromatic residues in crystalline chitin hydrolysis by chitinase A from Serratia marcescens 2170. J Biol Chem 276(44):41343–41349. https://doi.org/10.1074/jbc.M103610200

Vaaje-Kolstad G, Horn SJ, van Aalten DMF, Synstad B, Eijsink VGH (2005) The Non-catalytic Chitin-binding Protein CBP21 from Serratia marcescens is essential for Chitin degradation. J Biol Chem 280(31):28492–28497. https://doi.org/10.1074/jbc.M504468200

Vaaje-Kolstad G, Houston DR, Riemen AH, Eijsink VG, van Aalten DM (2005) Crystal structure and binding properties of the Serratia marcescens chitin-binding protein CBP21. J Biol Chem 280(12):11313–11319. https://doi.org/10.1074/jbc.M407175200

Vaaje-Kolstad G, Westereng B, Horn SJ, Liu Z, Zhai H, Sorlie M, Eijsink VG (2010) An oxidative enzyme boosting the enzymatic conversion of recalcitrant polysaccharides. Science 330(6001):219–222. https://doi.org/10.1126/science.1192231

Vaaje-Kolstad G, Forsberg Z, Loose JS, Bissaro B, Eijsink VG (2017) Structural diversity of lytic polysaccharide monooxygenases. Curr Opin Struct Biol 44:67–76. https://doi.org/10.1016/j.sbi.2016.12.012

van Aalten DM, Komander D, Synstad B, Gaseidnes S, Peter MG, Eijsink VG (2001) Structural insights into the catalytic mechanism of a family 18 exo-chitinase. Proc Natl Acad Sci U S A 98(16):8979–8984. https://doi.org/10.1073/pnas.151103798

Wang S, Jayaram SA, Hemphala J, Senti KA, Tsarouhas V, Jin H, Samakovlis C (2006) Septate-junction-dependent luminal deposition of chitin deacetylases restricts tube elongation in the Drosophila trachea. Curr Biol 16(2):180–185. https://doi.org/10.1016/j.cub.2005.11.074

Wu Q, Liu T, Yang Q (2013) Cloning, expression and biocharacterization of OfCht5, the chitinase from the insect Ostrinia furnacalis. Insect Sci 20(2):147–157. https://doi.org/10.1111/j.1744-7917.2012.01512.x

Xi Y, Pan PL, Ye YX, Yu B, Zhang CX (2014) Chitin deacetylase family genes in the brown planthopper, Nilaparvata lugens (Hemiptera: Delphacidae). Insect Mol Biol 23(6):695–705. https://doi.org/10.1111/imb.12113

Xi Y, Pan PL, Ye YX, Yu B, Xu HJ, Zhang CX (2015) Chitinase-like gene family in the brown planthopper, Nilaparvata lugens. Insect Mol Biol 24(1):29–40. https://doi.org/10.1111/imb.12133

Yu R, Liu W, Li D, Zhao X, Ding G, Zhang M, Ma E, Zhu K, Li S, Moussian B, Zhang J (2016) Helicoidal organization of chitin in the cuticle of the migratory locust requires the function of the chitin deacetylase 2 enzyme (LmCDA2). J Biol Chem 291(47):24352–24363. https://doi.org/10.1074/jbc.M116.720581

Yu HZ, Li NY, Li B, Toufeeq S, Xie YX, Huang YL, Du YM, Zeng XD, Zhu B, Lu ZJ (2019) Immune functional analysis of chitin deacetylase 3 from the Asian citrus psyllid diaphorina citri. Int J Mol Sci 21(1). https://doi.org/10.3390/ijms21010064

Zhang D, Chen J, Yao Q, Pan Z, Chen J, Zhang W (2012) Functional analysis of two chitinase genes during the pupation and eclosion stages of the beet armyworm Spodoptera exigua by RNA interference. Arch Insect Biochem Physiol 79(4–5):220–234. https://doi.org/10.1002/arch.21018

Zheng Y, Krell PJ, Doucet D, Arif BM, Feng Q (2008) Cloning, expression, and localization of a molt-related β-*N*-acetylglucosaminidase in the spruce budworm Choristoneura fumiferana. Arch Insect Biochem Physiol 68(1):49–59. https://doi.org/10.1002/arch.20235

Zhong XW, Wang XH, Tan X, Xia QY, Xiang ZH, Zhao P (2014) Identification and molecular characterization of a chitin deacetylase from Bombyx mori peritrophic membrane. Int J Mol Sci 15(2):1946–1961. https://doi.org/10.3390/ijms15021946

Zhu Q, Arakane Y, Beeman RW, Kramer KJ, Muthukrishnan S (2008) Functional specialization among insect chitinase family genes revealed by RNA interference. Proc Natl Acad Sci U S A 105(18):6650–6655. https://doi.org/10.1073/pnas.0800739105

Zhu Q, Arakane Y, Beeman RW, Kramer KJ, Muthukrishnan S (2008) Characterization of recombinant chitinase-like proteins of Drosophila melanogaster and Tribolium castaneum. Insect Biochem Mol Biol 38(4):467–477. https://doi.org/10.1016/j.ibmb.2007.06.011

Zhu KY, Merzendorfer H, Zhang W, Zhang J, Muthukrishnan S (2016) Biosynthesis, turnover, and functions of chitin in insects. Annu Rev Entomol 61(1):177

Single Chitooligosaccharides and Their Roles in Various Bioactivities

Kecheng Li

1 Introduction

Chitin is a linear biopolymer consisting of β(1 → 4)-linked N-acetyl-D-glucosamine (GlcNAc, A) residues, which is considered as the second most abundant polysaccharide in nature, often existing as a structural component in the exoskeletons of nematodes, insects and crustaceans (e.g., crabs, lobsters and shrimps) and in the cell walls of fungi (Van Dyken and Locksley 2018). Chitosan is the deacetylation product of chitin, which is soluble in aqueous acidic media and behaves as a cationic polysaccharide due to amine groups. Chitosan can be further degraded into small oligosaccharides with high solubility in neutral aqueous solutions. In general, these degraded products with degree of polymerization (DP) lower than 20 are called chitooligosaccharides (COS), chitosan oligomers or chitooligomers. A large number of investigations have shown that COS possess diverse biological functions including antioxidant (dos Santos et al. 2008; Li et al. 2012), antimicrobial (Benhabiles et al. 2012; Oliveira et al. 2008), anti-inflammatory (Azuma et al. 2015; Chung et al. 2012; Jitprasertwong et al. 2021), antitumor (Fernandes et al. 2012; Liu et al. 2019), and other biomedical bioactivities (Chumjan et al. 2019; Jiang et al. 2014). COS could also elicit plant immune responses and activate defense system, which has huge application in agriculture (Hidangmayum et al. 2019; Mukhtar Ahmed et al. 2020; Xie et al. 2020).

Molecular structure determines biological function. Based on the original structure of heterogeneous chitosan, COS is correspondingly composed of β-1,4-linked GlcNAc and D-glucosamine (GlcN) residues. COS is described and classified

K. Li (✉)
CAS and Shandong Province Key Laboratory of Experimental Marine Biology, Center for Ocean Mega-Science, Institute of Oceanology, Chinese Academy of Sciences, Qingdao 266071, China
e-mail: lkc@qdio.ac.cn

Laboratory for Marine Drugs and Bioproducts, Pilot National Laboratory for Marine Science and Technology (Qingdao), Qingdao 266237, China

S.-K. Kim (ed.), *Chitooligosaccharides*,
https://doi.org/10.1007/978-3-030-92806-3_16

according to DP or average molecular weight (MW), degree of acetylation, pattern of acetylation (PA) or sequence (Hao et al. 2021a). Currently, the production techniques of COS generally lead to a complicated oligosaccharides mixture with various homogenous or heterogenous GlcNAc/GlcN oligomers. Consequently, many of these biological activities of COS have been unreliable and controversial. Sometimes, the above-mentioned bioactivities can be observed, and sometimes not. Even when the MW and degree of acetylation are provided, the oligomers sequences are often ambiguous. The reproducibility is an important issue. In this scenario, access to COS with accurate structure is unavoidable. It is definite that single COS with good characterization and high purity would greatly facilitate to make clear the structure–function relationships.

2 Access to Single Chitooligosaccharides

Based on its heterogeneity, COS describe a large and versatile family of oligosaccharides with different sequences. Figure 1 describes the possible fragments in a COS mixture. It mainly contains GlcN homomers and heterogeneous GlcN/GlcNAc sequences. Some minor homogeneous $(GlcNAc)_n$ are also expected to be present in a degradation product of chitosan. Currently, chemical synthesis and chromatographic separation are the two major ways to prepare single COS.

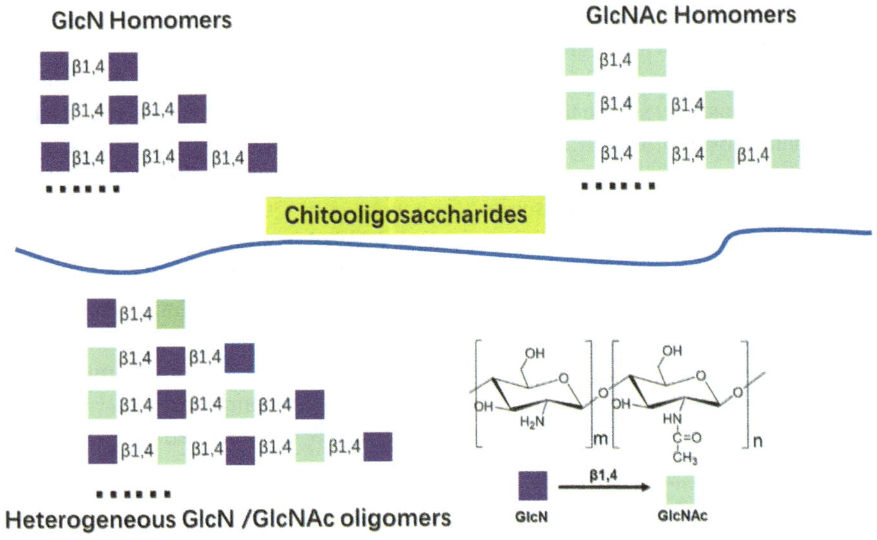

Fig. 1 Structure and possible sequences of chitooligosaccharides. ■ represents glucosamine, GlcN (D); ■ represents N-acetyl glucosamine, GlcNAc(A)

2.1 Chemical Synthesis

Single COS could be prepared by chemical synthesis using GlcN/GlcNAc as raw materials. During the synthesis process of COS, the glycosylation reaction to selectively form the β(1 → 4) glycosidic linkage is the key step. GlcN and GlcNAc have multiple hydroxyl groups, each of which may react with the hydroxyl group of another monosaccharide residue to form a new glycosidic bond. Therefore, before the glycosylation reaction, multiple steps of protection and deprotection are commonly needed.

As for the typical amino sugar, N-protection in the GlcN/GlcNAc donor play a vital role in the glycosylation reaction. In 1977, N-phthalyl group was first introduced to COS synthesis (a chitobiose derivative) by Lemieux et al. (1977). Subsequently, the synthesis of single COS has been extensively assayed via various GlcN donors/acceptors (Bedini et al. 2013; Berthelot et al. 2017; Huang et al. 2006; Yano et al. 2020). In principle, homo-or heterogenous single COS with random sequence could all be synthesized by carbohydrate chemists. However, the chemical synthesis of single COS is still a very challenging task, especially for those oligomers larger than trisaccharides that may need more than 20 reaction steps to be achieved. The final yield is still very low. Recently, automated solution-phase synthesizer provides an exciting new tool to prepare single COS, such as the chitotetraose (Nokami et al. 2015). The robotization would be an innovation technology to simplify the synthesis and shorten the reaction time. On the other hand, the chem-enzymatic synthesis of COS was also developed. Regioselective synthesis can be carried out by enzymatic methods under mild conditions and without protection and deprotection steps. Using the reverse mode of chitinase, $(GlcNAc)_2$ could be prepared by enzymatic glycosylation using a 1,2-oxazoline derivative of GlcNAc as glycosyl donor for chitinase (Kobayashi et al. 1997). By similar way, the single COS with higher DP, $(GlcNAc)_7$ could also be enzymatically synthesized using 1,2-oxazoline derivative of $(GlcNAc)_5$ and $(GlcNAc)_2$ as starting materials (Yoshida et al. 2012). However, the enzymatic transglycosylation was just used for those homogeneous GlcNAc oligomers that was subject to chitinase catalyzation.

2.2 Chromatographic Separation

In the past years, much effort has been made to develop the rapid and efficient separation technology of COS. The available single COS fragments are increasing. In comparison to chemical synthesis, chromatographic separation is relatively easy to be manipulated. Another great advantage of chromatographic separation is that multiple single COS molecules could be collected by a single round. Therefore, biological researchers preferred chromatographic method to prepare single COS for bioassays.

2.2.1 Charcoal Column Chromatography

Charcoal column chromatography was widely applied in the separation of various oligosaccharides. Activated charcoal absorbs polar molecules in polar solution and desorbs them in apolar solution (Fujimoto et al. 2009). COS contain plenty of hydroxyl and amino groups which could form hydrogen bond interaction with activated charcoal and are absorbed. The interaction between COS and activated charcoal is expected to become stronger with the chains increasing. In general, COS could be captured by activated charcoal column in water, and then each individual COS can be washed out by ethanol gradient (4, 8, 10, 13, 16, 30%). Correspondingly, di-N-acetylchitobiose, tri-N-acetylchitotriose, tetra-N-acetylchitotetraose and penta-N-acetylchitopentaose were separated with the purity of 70–85% (Semenuk et al. 2001).

2.2.2 Size Exclusion Chromatography

Size exclusion chromatography (SEC) is one of the most common techniques used for oligosaccharides separation. Single oligosaccharides could be separated according to their different sizes. The SEC separation of homogenous chitin oligosaccharides series, $(GlcNAc)_n$, is relatively easy. In the early 1990s, several GlcNAc homomers with single DP were isolated using a Toyopearl HW-40S column. Five fractions, including chitin hexamer, pentamer, tetramer, trimer, and dimer, were collected. The purity of each $(GlcNAc)_n$ fraction was higher than 92% after only one round of SEC (Aiba 1994). In addition, a preparative method of GlcNAc dimer to heptamer is reported using combination columns of Bio Gel P2 and P4, with a yield of $(GlcNAc)_n$ higher than 40 mg (Izume et al. 1992). Compared with $(GlcNAc)_n$, the homogenous deacetylated $(GlcN)_n$ is more difficult to be separated using SEC due to free amino groups that may lead to more complicated intermolecular force. Therefore, the resin with a larger pore size seems to be required for the SEC separation procedure of single $(GlcN)_n$. It is reported that five single $(GlcN)_n$ fractions could be collected using a Bio Gel P6 column combing with ÄKTA FPLC (Fast Protein Liquid Chromatogram) at a flow of 0.1 mol/L NH_4HCO_3 as eluent, including trimer (>95%) and tetramer to heptamer (>90%) (Gao et al. 2013).

2.2.3 Ion Exchange Chromatography

Chitooligosaccharides have cationic property considering the existence of free amino groups on the sugar chains. The free amino groups can be protonated in acidic buffer, and thus the charged oligomers can interact with the cation exchange resin and further be separated according to their different numbers of amino groups. Compared with SEC, ion-exchange chromatography takes advantage of large sample load, fast flow and high resolution, and become the most popular technique for the separation of single COS. Xiong et al. prepared several homogenous $(GlcN)_n$ by a strong acidic ion-exchange resin (Dowex 50WX8-200) and finally five purified $(GlcN)_n$ fractions, dimer to hexamer, were collected, with the purities of the dimer, trimer, and tetramer higher than 98%, and the purities of pentamer and hexamer above 90% and 80%, respectively (Xiong et al. 2009). Additionally, Wu et al. utilized a convenient ion-exchange chromatography (SP Histrap HP cation-exchange column) to separate COS and obtained five fractions primarily comprised chitobiose, chitotriose, chitotetraose, chitopentaose, and chitohexaose, respectively, with chromatographic purities over 90% (Wu et al. 2018). The commercial ion-exchange resin (CM Sephadex C25) is also very suitable for the separation of single COS. Li et al. developed a high-resolution chromatography for the preparative separation of a pure homogenous COS series using CM Sephadex C25. Six homogenous $(GlcN)_n$, chitobiose, chitotriose, chitotetraose, chitopentaose, chitohexamose and chitoheptaose, were separated, with purities over 98% for dimer to hexamer and a purity of 93% for heptamer. Furthermore, a chromatographic separation model for ion-exchange chromatography was established to separate homogenous $(GlcN)_n$. The capacity factor (k) of $(GlcN)_n$ and their DP exhibited a good correlation, which can be used to predict the retention time of COS with higher DP. Finally, a chitooctaose fraction with a purity of over 90% was successfully isolated in the eluent (Li et al. 2013a, 2016).

Additionally, the ion-exchange chromatography also produced some operational difficulties for subsequent desalination due to a mass of salt used in the eluted solution. The conventional desalination methods like gel chromatography and dialysis are time-consuming and have difficulty in desalting of small oligosaccharides. Activated charcoal extraction was an alternative to desalt the separated fractions. Activated charcoal could absorb COS but hardly absorbs inorganic salt. The separated COS fractions could be captured by activated charcoal while NaCl is still retained in water (Li et al. 2012).

Ion-exchange chromatography could also be used for the pre-separation of heterogenous GlcN/GlcNAc oligomers according to their numbers of amino group. For example, N-acetylchitotriose and N,N-diacetylchitotriose were separated from a chitotriose mixture by a CM Sepharose Fast Flow column (Li et al. 2013b). It is also reported that heterogenous GlcN/GlcNAc hexamers were successfully separated using a single round of ion-exchange chromatography, including N-acetylchitohexaose (D5A1), di-*N*-acetylchitohexaose (D4A2), tri-*N*-acetylchitohexaose (D3A3), tetra-*N*-acetylchitohexaose (D2A4), penta-*N*-acetylchitohexaose (D1A5), and hexa-*N*-acetylchitohexaose (A6) (Li et al. 2017). These N-acetylated chitohexaoses could be further used for the separation of some isomers with various sequences.

2.2.4 Hydrophilic Interaction Liquid Chromatography

Hydrophilic interaction liquid chromatography (HILIC) was proposed by Alpert in 1990 (Alpert 1990), where analytes are separated on a polar stationary phase and eluted by a binary mobile phase with the main component usually being 5–40% water in acetonitrile. In recent years, HILIC has been gaining intense interest as

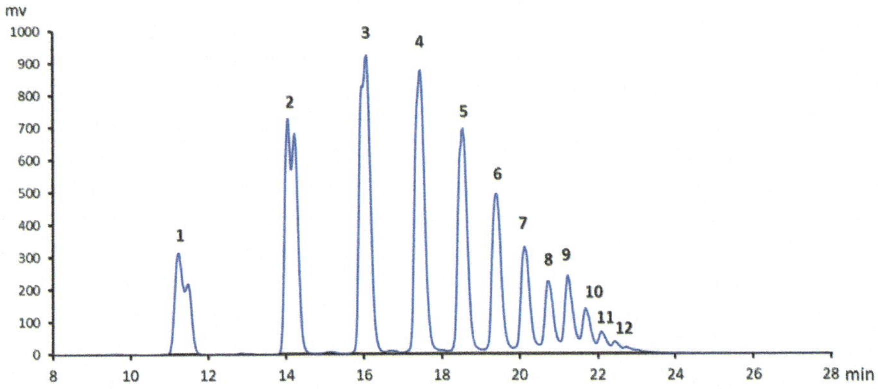

Fig. 2 HILIC separation of COS using a Click Maltose column

a powerful tool for polar compounds separation, especially for carbohydrates (Fu et al. 2010; Guo et al. 2009). HILIC is also a very effective technique for separating COS due to its polar structure. HILIC has high selectivity. The separation of more than ten single GlcN homomers could be achieved by HILIC using a Click Maltose column (Fig. 2). Furthermore, the advancing HILIC techniques have been applied to separate some specific GlcN/GlcNAc sequences. For example, all fourteen partially acetylated COS tetramers with specific sequences were successfully produced by a semi-preparative HILIC in combination with an enzymatic deacetylation using different chitin deacetylases (Hembach et al. 2017). However, due to the use of expensive acetonitrile, the cost of single COS separation using HILIC is very high and hard to achieve scale.

Overall, the separation approaches of single COS fragments were being improved in the past decades. Based on the above progress, some sequences according to the separation method are summarized as shown in Table 1. Anyway, more and more specific homo- or hetero-COS sequences are being available, which greatly contribute to the research on revealing the biological roles of chitin and chitosan in nature.

3 The Roles of Single Chitooligosaccharides in Bioactivities

3.1 Antioxidative Activity

Oxidative stress is one of the major risk factors for chronic diseases, such as cardiovascular diseases (Dhalla et al. 2000), atherogenesis (Surdu et al. 2021), and Parkinson's disease (Jenner 2003). Excess reactive oxygen species (ROS) generation contributes to proteins, lipids and DNA injury, leading to inflammation, tissue damage and subsequent cellular apoptosis (Valko et al. 2007). Antioxidants can scavenge free radicals and protect human body from highly toxic ROS, which to some extent retard the

Table 1 Separation of some single COS

Chromatography	Columns	Conditions	Components	Purity (%)	References
Charcoal column chromatography	1:1 charcoal and Cellite 545 (300 × 40 mm)	Ethanol gradient (8, 10, 13, 16, 30%)	$(GlcNAc)_n$, n = 2–5	$(GlcNAc)_{2–4}$ (>85%); $(GlcNAc)_5$ (70%)	Semenuk et al. (2001)
Size exclusion chromatography (SEC)	Toyopearl HW-40S column (5 × 68 cm)	Elutent: water Flow rate: 94 mL/h	$(GlcNAc)_{2–6}$	$(GlcNAc)_2$ (97%) $(GlcNAc)_3$ (98%) $(GlcNAc)_4$ (94%) $(GlcNAc)_5$ (92%) $(GlcNAc)_6$ (96%)	Aiba (1994)
	Bio Gel P2 (2.65 × 165 cm) and Bio Gel P4 (2.65 × 175 cm)	Elutent: Sodium formate buffer (pH 4.2)	$(GlcNAc)_{2–7}$	$(GlcNAc)_2$ (80%) $(GlcNAc)_{3–6}$ (95%) $(GlcNAc)_7$ (90%)	Izume et al. (1992)
	Bio Gel P6 (2.6 × 110 cm)	Elutent: 0.1 mol/L NH_4HCO_3	$(GlcN)_{3–7}$	$(GlcN)_3$ (95%) $(GlcN)_{4–7}$ (90%)	Gao et al. (2013)
Ion exchange chromatography	Dowex 50WX8-200 (250 × 4 mm)	HCl linear gradient elution Flow rate: 300 mL/h Column temperature: 30 °C	$(GlcN)_{2–6}$	$(GlcN)_{2–4}$ (98%) $(GlcN)_5$ (90%) $(GlcN)_6$ (80%)	Xiong et al. (2009)
	CM-Sephadex C-25 column (45 cm × 2.6 cm)	Gradient elution: NaCl (0–1.2 mol/L)-HAc buffer Flow rate: 3 mL/min	$(GlcN)_6$	93%	Wei et al. (2009)
	CM-Sephadex C-25 column (2.6 × 50 cm)	Gradient elution: NaCl (0–2 mol/L)-HAc buffer Flow rate: 3 mL/min	$(GlcN)_{2–8}$	$(GlcN)_{2–6}$ (>98%) $(GlcN)_7$ (93%) $(GlcN)_8$ (90%)	Li et al. (2013a, 2016)

(continued)

Table 1 (continued)

Chromatography	Columns	Conditions	Components	Purity (%)	References
	CM Sepharose Fast Flow (1.6 cm × 10 cm)	Gradient elution: NaCl–HAc buffer (0–0.8 mol/L) Flow rate: 2 mL/min	$(GlcN)_2(GlcNAc)_1$ $(GlcN)_1(GlcNAc)_2$	/	Li et al. (2013b)
	CM-Sephadex C-25 column (2.6 × 50 cm)	Gradient elution: NaCl (0–2 mol/L)-HAc buffer Flow rate: 3 mL/min	$(GlcN)_5(GlcNAc)_1$ $(GlcN)_4(GlcNAc)_2$ $(GlcN)_3(GlcNAc)_3$ $(GlcN)_2(GlcNAc)_4$ $(GlcN)_1(GlcNAc)_5$	/	Li et al. (2017)
Hydrophilic interaction liquid chromatography (HILIC)	Phenomenex Luna 5 μm CN column (100 Å, 250 mm × 10 mm)	Gradient elution: solvent A (acetonitrile: water, 8:2) and B (acetonitrile: water, 2:8) in formic buffer Flow rate: 2.0 ml/min	$(GlcN)_n(GlcNAc)_m$ (m + n = 4)	>95%	Hembach et al. (2017)
Metal affinity chromatography	Agarose CL-6B and silica functionalized with various chelating functions (150 mm × 10 mm)	Imidazole step gradient Flow rate of 1.0 mL/min	$(GlcN)_{2–4}$	$(GlcN)_{2–3}$ (95%) $(GlcN)_4$ (90%)	Ledevedec et al. (2008)

progress of many chronic diseases. COS has been reported to possess high antioxidant activity in vitro and vivo, such as radical scavenging and oxidative stress inhibition in cells (Je et al. 2004; Liu et al. 2009; Ngo et al. 2009; Yuan 2009). Spin-trapping electron spin resonance shows that COS could scavenge hydroxyl radicals, superoxide radicals, DPPH radicals, etc., and the scavenging activity closely depends on its MW (Park et al. 2003). Generally, those short chains COS were proved to have higher antioxidant activity (Li et al. 2012). For example, chitopentaose could improve cell viability and decrease the production of ROS induced by hydrogen peroxide (H_2O_2) in HaCaT cells (Zhang et al. 2020). Chitobiose and chitotriose were found to be more potent than three reference compounds (aminoguanidine, pyridoxamine and a water-soluble α-tocopherol analogue) in scavenging hydroxyl radical (Chen et al. 2003). On the other hand, in vivo assay suggested that chitobiose and chitotriose could be well absorbed from the gastrointestinal tract. When given orally, chitobiose and chitotriose are able to protect rats from acute hepatotoxicity induced by CCl_4-dependent lipid peroxidation (Chen et al. 2005a, b). Very recently, the effect of COS sequence on its antioxidant activity was revealed using all four chitosan dimers with different sequences. It is proved that the amino group at the reducing end plays a critical role in scavenging superoxide radicals and reducing power of the chitosan dimer. Furthermore, the fully deacetylated GlcN dimer was found to have the strongest DPPH scavenging activity. When the amino groups of chitosan dimer were acetylated, it showed better activity in scavenging hydroxyl radicals (Hao et al. 2021b).

3.2 *Anti-inflammatory Activity*

Numerous studies have reported the anti-inflammatory activity of COS. COS could decrease secretion of nitric oxide (NO) in lipopolysaccharide (LPS)-induced RAW 264.7 cells and BV-2 microglia (Vo et al. 2012; Yoon et al. 2007), mitigate the glycerol-induced inflammatory response in rat kidney (Yoon et al. 2008), and attenuate organ dysfunction in LPS-induced sepsis (Qiao et al. 2011). Besides murine macrophages and mouse models, the anti-inflammatory activity of COS was also studied using human THP-1 cells. COS exposure could reduce the production of multiple pro-inflammatory cytokines associated with LPS-stimulated inflammation without affecting cell viability (Jitprasertwong et al. 2021). In particular, chitohexaose could inhibit LPS-induced production of inflammatory molecules TNF-α, IL-1β and IL-6 by macrophages in vitro and in vivo in mice. It was found that chitohexaose activated macrophages to a non-inflammatory phenotype via a non-canonical TLR4 signaling pathway to produce IL-10 (Panda et al. 2012). A significant protection mediated by chitohexaose was also reported in mice suffering from acetaminophen-induced hepatotoxicity. Chitohexaose was shown to decrease mortality along with reduced hepatic damage, increase peritoneal migration of neutrophils, decrease mRNA expression of IL-1β as well as inhibition of inflammasome activation in liver (Barman et al. 2016). In addition, the anti-inflammatory activities of several single COS with different sizes (chitobiose to chitooctaose) were investigated in LPS-stimulated RAW264.7 cells as well as in a rat model with myocarditis. Among all single COS, the oligomers larger than tetraose showed good anti-inflammatory activities by reducing the levels of IL-1β, IL-17A and IFN- γ, and increasing the level of IL-10, in the order of chitopentaose < chitohexaose < chitoheptaose < chitooctaose. Interestingly, chitoheptaose showed better in vivo therapeutic effects on myocarditis by improving cardiac parameters, oxidative factors (ROS and RNS), inflammatory cytokines (IL-1β, IL-10, IL-17A, and IFN-γ) and apoptotic factors (Caspase 3, BAX, and BCL-2) compared with chitopentaose, chitohexaose, and chitooctaose (Zhao et al. 2020). However, the protective effects of different COSs on heart therapy are very complex. The exact functional molecular mechanism remains unclear.

3.3 Antitumor Activity

The antitumor activity of COS was reported in 1980s. Suzuki et al. first reported that $(GlcN)_6$ and $(GlcNAc)_6$ were capable of highly inhibiting the growth of sarcoma 180 and MM-46 solid tumors transplanted in mice (Suzuki et al. 1986). In contrast, the homomers with lower DP, $(GlcN)_5$ and $(GlcNAc)_5$ were unable to exhibit antitumor activity, and thus the molecular size was regarded as one of the essential elements for manifesting the antitumor effect (Tokoro et al. 1988). Similar results were reported by Xiong et al. (2009). COS has inhibitory effect on tumor angiogenesis, which is dependent on its molecular size. Xiong et al. investigated the antiangiogenic effect of five single COSs (chitobiose, chitotriose, chitotetraose, chitopentaose and chitohexaose), and found that chitohexaose showed the strongest inhibitory activity against the proliferation and migration of tumor-induced ECV304 cells. Further research demonstrated that $(GlcN)_6$ down-regulated the VEGF and *uPA* mRNA expressions in ECV304 cells. In addition, it is found that COS with different sequences may have different effects on angiogenesis. The N-acetylated chitobiose (DA) was speculated to be the primary structural basis for pro-angiogenic activity (Djordjevic et al. 2014). Chitohexaose was also reported to have the most potent inhibitory effect on A549 cell proliferation and significantly down-regulated *cyclin D1* and *bcl-xl* mRNA expression levels, mainly involving apoptosis of tumor cells (Li et al. 2011). Unlike previous study, Zhu et al. reported that $(GlcN)_5$ had the anti-proliferation effect on hepatocellular carcinoma. $(GlcN)_5$ was found to decrease mitochondrial membrane potential, activate the cleavage of Caspases9/3, thus inducing mitochondrial-mediated apoptosis in HepG2 cells (Zhu et al. 2021). Therefore, the primary structure of COS responsible for antitumor activity might be different caused by different tumor cells.

3.4 Biostimulant

The heavy use of agrochemicals has resulted in a series of serious problems, including soil fertility decrease, environmental damages and unsafe food (Vassilev et al. 2015). In such a scenario, there arises public awareness of the development of eco-friendly plant growth regulators for sustainable agriculture. The natural COS could promote plant growth and effectively induce plants against abiotic stress when applied in low quantities, which is defined as biostimulant, other than fertilizer (du Jardin 2015). Therefore, COS are currently regarded as one of the most potent plant growth regulators (Brown and Saa 2015). COS could promote plant growth in both crop and horticulture, such as increasing the yield of wheat (Wang et al. 2015), enhancing nutrient uptake of bean (Chatelain et al. 2014), and increasing the accumulative inflorescence number of orchid (Limpanavech et al. 2008). The key components of COS promoting plant growth gradually become clear as reported by several recent studies. Chitotetraose to chitooctaose and two COS mixtures with DP 8–10 and DP 10–12 were found to promote significantly the growth and photosynthesis of wheat seedlings. Among these GlcN homomers, chitoheptaose was the most efficient. However, in the same experiment condition, the small chitobiose and chitotriose were inactive (Zhang et al. 2016). It seems that the plant promotion effect of COS requires an essential structure with a DP of at least 4. Similarly, chitohexaose and chitoheptaose also exhibited the strongest improving effect on the beneficial secondary metabolites during soybean seeds germination (Jia et al. 2019). On the other hand, a comparative experiment was performed using chitin hexamer, chitosan hexamer and partially acetylated chitosan hexamer in order to make clear N-acetylation effect of COS on plant growth and photosynthesis (Li et al. 2020). Among these three chitin and chitosan hexamers, the homogenous GlcN hexamer exhibited the best activity.

Furthermore, in the past decades, there were numerous reports that COS could effectively induce plants against abiotic stress, including salinity (Turk 2019), chilling (Zhou et al. 2018), water deficit (Al-Ghamdi 2019; Muley et al. 2019) and heavy metal stress (Zong et al. 2017a). These stress conditions cause many negative effects on plant growth, mainly including the lipid peroxidation of membrane, inhibition of nutrients, water uptake decrease and toxicity due to ions accumulation. It is reported that spraying COS onto the leaves of edible rape could obviously alleviate the Cd stress (Zong et al. 2017a, b). Subsequently, it was proved that the chitohexaose, chitoheptaose and chitooctaose are the key components of COS inducing plants against abiotic stress (Zhang et al. 2017a). These three GlcN oligomers showed the most effective activity to increase the contents of chlorophyll and proline, activate antioxidant enzyme activities, and induce the expression of salt-related genes in wheat seedlings under salt stress. Similarly, chitohexaose and chitoheptaose also showed more effective activities alleviating the adverse effects in plants suffering from chilling stress than other single COSs (chitobiose to chitopentaose) (Zou et al. 2017).

Metabolite profiling revealed that $(GlcN)_7$ actually could act as a "metabolic enhancer" showing a pleiotropic modulation of carbon and nitrogen metabolism in plants (Zhang et al. 2017b). The photosynthetic carbon fixation was enhanced by GlcN heptamer with an increase in sucrose phosphate synthase and fructose 1, 6–2 phosphatase activities. Several metabolites in tricarboxylic acid cycle, including oxaloacetate and malate, were also increased by $(GlcN)_7$. Moreover, $(GlcN)_7$ could promote the nitrogen reduction and nitrogen assimilation. The activity of key enzymes of nitrogen reduction and glutamine synthetase/glutamate synthase cycle were enhanced. Four hundred differentially expressed genes responsive to $(GlcN)_7$ were found, including 268 up-regulated and 132 down-regulated mRNAs, many of which were correlated to photosynthesis, primary carbon and nitrogen metabolism as well as defense responses (Zhang et al. 2018).

3.5 Eliciting Activity

Chitin and chitosan are typical molecular signatures for whole classes of microbes. Plants could efficiently digest chitin and chitosan by chitinases into small COS fragments. These COS fragments are known as pathogen-associated molecular patterns (PMAPs) that could elicit plant immunity to control diseases. As for homogeneous $(GlcNAc)_n$, the oligomers larger than pentaose showed strong eliciting activity to induce the formation of phytoalexins in suspension-cultured rice cells, but the smaller trimer was inactive (Yamada et al. 1993), indicating the strict requirements for the size during the specific recognition and signal transduction between COS and plant cells. It is revealed that GlcNAc octamer but not tetramer or pentamer, can act as a bivalent ligand to induce the Arabidopsis cell surface receptor chitin elicitor receptor kinase 1 dimerization for its activation and transducing the signals (Liu et al. 2012).

The effectiveness of COS as an elicitor is also closely correlated with its sequences. Recently, it is found that COS required a minimum DP of 6 and at least four N-acetyl groups to trigger an oxidative burst inducing defense reactions in Arabidopsis seedlings (Gubaeva et al. 2018). In addition, a hypothesis was proposed that small COS are priming-active but elicitor-inactive (Basa et al. 2020). Besides elicitor activity, PAMPs can also enable plant cells to enter a state of elevated alertness that triggers plants to react more quickly and more strongly following the detection of a pathogen, known as priming. The priming activity of COS was recently revealed to be sequence-specific. A monoacetylated COS tetramer with the GlcNAc residue located at the non-reducing end (ADDD) showed the highest priming activity. In contrast, other partially-acetylated tetramers with an acetyl group at or close to the reducing end (DDDD, DDDA, DDAD, DAAA, ADAA, AADA) did not induce priming.

4 Conclusion and Prospects

Access to single COS with well-defined structure greatly facilitates to understand the role in its biological activities and would promote their application in the field of biomedicine and agriculture. The advancing separation would produce more and more single COS and enable their preparation easier. Regarding the preparation of single hetero-COS, it is a potential strategy that selective deacetylation by enzymic method combined with subsequent high-resolution separation techniques. The remarkable biological significance of single COS ensures a continued research interest globally. It is expected that a large library containing diverse COS with specific sequences would be established in the future. Furthermore, the structural complexity of COS brings great difficulty to understand its molecular mechanism although much pioneering work has been done. So far, whether the specific receptors of complex COS exist in plants and animals still remains controversial. The signal transduction network related to COS has not yet been elucidated clearly. It is still a very challenging task to clarify the action mode of complex COS.

References

Aiba S (1994) Preparation of N-acetylchitooligosaccharides by hydrolysis of chitosan with chitinase followed by N-acetylation. Carbohydr Res 265(2):323–328

Al-Ghamdi AA (2019) Marjoram physiological and molecular performance under water stress and chitosan treatment. Acta Physiol Plant 41(4)

Alpert AJ (1990) Hydrophilic-interaction chromatography for the separation of peptides, nucleic acids and other polar compounds. J Chromatogr A 499:177–196

Azuma K, Osaki T, Kurozumi S, Kiyose M, Tsuka T, Murahata Y, Okamoto Y (2015) Anti-inflammatory effects of orally administered glucosamine oligomer in an experimental model of inflammatory bowel disease. Carbohydr Polym 115:448–456

Barman PK, Mukherjee R, Prusty BK, Suklabaidya S, Senapati S, Ravindran B (2016) Chitohexaose protects against acetaminophen-induced hepatotoxicity in mice. Cell Death Disease 7

Basa S, Nampally M, Honorato T, Das SN, Podile AR, El Gueddari NE, Moerschbacher BM (2020) The pattern of acetylation defines the priming activity of chitosan tetramers. J Am Chem Soc 142(4):1975–1986

Bedini E, Cirillo L, Marchetti R, Basso S, Tufano D, Molinaro A, Parrilli M (2013) Synthesis of partially N-acetylated chitooligosaccharides and muropeptides. Synlett 25(03):365–370

Benhabiles MS, Salah R, Lounici H, Drouiche N, Goosen MFA, Mameri N (2012) Antibacterial activity of chitin, chitosan and its oligomers prepared from shrimp shell waste. Food Hydrocolloids 29(1):48–56

Berthelot N, Brossay A, Gasciolli V, Bono J-J, Baron A, Beau J-M, Vauzeilles B (2017) Synthesis of lipo-chitooligosaccharide analogues and their interaction with LYR3, a high affinity binding protein for Nod factors and Myc-LCOs. Org Biomol Chem 15(37):7802–7812

Brown P, Saa S (2015) Biostimulants in agriculture. Front Plant Sci 6

Chatelain PG, Pintado ME, Vasconcelos MW (2014) Evaluation of chitooligosaccharide application on mineral accumulation and plant growth in Phaseolus vulgaris. Plant Sci 215:134–140

Chen AS, Taguchi T, Sakai K, Kikuchi K, Wang MW, Miwa I (2003) Antioxidant activities of chitobiose and chitotriose. Biol Pharm Bull 26(9):1326–1330

Chen AS, Taguchi T, Okamoto H, Danjo K, Sakai K, Matahira Y, Miwa I (2005a) Pharmacokinetics of chitobiose and chitotriose administered intravenously or orally to rats. Biol Pharm Bull 28(3):545–548
Chen AS, Taguchi T, Sakai K, Matahira Y, Wang MW, Miwa I (2005b) Effect of chitobiose and chitotriose on carbon tetrachloride-induced acute hepatotoxicity in rats. Biol Pharm Bull 28(10):1971–1973
Chumjan W, Winterhalter M, Suginta W (2019) Effects of H-bonds on sugar binding to chitoporin from Vibrio harveyi. BBA-Biomembranes 1861(3):610–618
Chung MJ, Park JK, Park YI (2012) Anti-inflammatory effects of low-molecular weight chitosan oligosaccharides in IgE-antigen complex-stimulated RBL-2H3 cells and asthma model mice. Int Immunopharmacol 12(2):453–459
Dhalla NS, Temsah RM, Netticadan T (2000) Role of oxidative stress in cardiovascular diseases. J Hypertens 18(6)
Djordjevic MA, Bezos A, Susanti, Marmuse L, Driguez H, Samain E, Vauzeilles B, Beau J-M, Kordbacheh F, Rolfe BG, Schwörer R, Daines AM, Gresshoff PM, Christopher R (2014) Lipo-chitin oligosaccharides, plant symbiosis signalling molecules that modulate mammalian angiogenesis in vitro. PLoS One 9(12):e112635. https://doi.org/10.1371/journal.pone.0112635
dos Santos ALW, El Gueddari NE, Trombotto S, Moerschbacher BM (2008) Partially acetylated chitosan oligo- and polymers induce an oxidative burst in suspension cultured cells of the gymnosperm Araucaria angustifolia. Biomacromol 9(12):3411–3415
du Jardin P (2015) Plant biostimulants: definition, concept, main categories and regulation. Sci Hortic 196:3–14
Fernandes J, Sereno J, Garrido P, Parada B, Cunha M, Reis F, Santos-Silva A (2012) Inhibition of bladder tumor growth by chitooligosaccharides in an experimental carcinogenesis model. Mar Drugs 10(12):2661–2675
Fu Q, Liang T, Zhang X, Du Y, Guo Z, Liang X (2010) Carbohydrate separation by hydrophilic interaction liquid chromatography on a "click" maltose column. Carbohydr Res 345(18):2690–2697
Fujimoto Y, Hattori T, Uno S, Murata T, Usui T (2009) Enzymatic synthesis of gentiooligosaccharides by transglycosylation with β-glycosidases from Penicillium multicolor. Carbohydr Res 344(8):972–978
Gao L, Li C, Wang S, Zhao X, Guan H (2013) Preparation and analysis of chitooligosaccharide isomers with different degree. Chin J Mar Drugs 32(3):21–27
Gubaeva E, Gubaev A, Melcher RLJ, Cord-Landwehr S, Singh R, El Gueddari NE, Moerschbacher BM (2018) "Slipped sandwich" model for chitin and chitosan perception in arabidopsis. Mol Plant Microbe Interact 31(11):1145–1153
Guo Z, Jin Y, Liang T, Liu Y, Xu Q, Liang X, Lei A (2009) Synthesis, chromatographic evaluation and hydrophilic interaction/reversed-phase mixed-mode behavior of a "Click β-cyclodextrin" stationary phase. J Chromatogr A 1216(2):257–263
Hao W, Li K, Li P (2021a) Review: advances in preparation of chitooligosaccharides with heterogeneous sequences and their bioactivity. Carbohydr Polym 252:117206–117206
Hao W, Li K, Ma Y, Li R, Xing R, Yu H, Li P (2021b) Preparation and antioxidant activity of chitosan dimers with different sequences. Mar Drugs 19(7)
Hembach L, Cord-Landwehr S, Moerschbacher BM (2017) Enzymatic production of all fourteen partially acetylated chitosan tetramers using different chitin deacetylases acting in forward or reverse mode. Sci Rep 7
Hidangmayum A, Dwivedi P, Katiyar D, Hemantaranjan A (2019) Application of chitosan on plant responses with special reference to abiotic stress. Physiol Mol Biol Plants 25(2):313–326
Huang L, Wang Z, Li X, Ye XS, Huang X (2006) Iterative one-pot syntheses of chitotetroses. Carbohydr Res 341(10):1669–1679
Izume M, Nagae S, Kawagishi H, Ohtakara A (1992) Preparation of normal-acetylchitooligosaccharides from enzymatic hydrolyzates of chitosan. Biosci Biotechnol Biochem 56(8):1327–1328

Je J-Y, Park P-J, Kim S-K (2004) Radical scavenging activity of hetero-chitooligosaccharides. Eur Food Res Technol 219(1):60–65

Jenner P (2003) Oxidative stress in Parkinson's disease. Ann Neurol 53(S3):S26–S38

Jia Y, Ma Y, Zou P, Cheng G, Zhou J, Cai S (2019) Effects of different oligochitosans on isoflavone metabolites, antioxidant activity, and isoflavone biosynthetic genes in soybean (Glycine max) seeds during germination. J Agric Food Chem 67(16):4652–4661

Jiang M, Guo Z, Wang C, Yang Y, Liang X, Ding F (2014) Neural activity analysis of pure chito-oligomer components separated from a mixture of chitooligosaccharides. Neurosci Lett 581:32–36

Jitprasertwong P, Khamphio M, Petsrichuang P, Eijsink VGH, Poolsri W, Muanprasat C, Yamabhai M (2021) Anti-inflammatory activity of soluble chito-oligosaccharides (CHOS) on VitD3-induced human THP-1 monocytes. PLoS ONE 16(2):e0246381

Kobayashi S, Kiyosada T, Shoda S (1997) A novel method for synthesis of chitobiose via enzymatic glycosylation using a sugar oxazoline as glycosyl donor. Tetrahedron Lett 38(12):2111–2112

Ledevedec F, Bazinet L, Furtos A, Venne K, Brunet S, Mateescu M (2008) Separation of chitosan oligomers by immobilized metal affinity chromatography. J Chromatogr A 1194(2):165–171

Lemieux RU, Takeda T, Chung BY (1977) Synthesis of 2-amino-2-deoxy-β-d-glucopyranosides. Synth Methods Carbohydr 39:90–115

Li X, Wang J, Chen X, Tian J, Li L, Zhao M, Zhou C (2011) Effect of chitooligosaccharides on cyclin D1, bcl-xl and bcl-2 mRNA expression in A549 cells using quantitative PCR. Chin Sci Bull 56(15):1629–1632

Li K, Xing R, Liu S, Li R, Qin Y, Meng X, Li P (2012) Separation of chito-oligomers with several degrees of polymerization and study of their antioxidant activity. Carbohydr Polym 88(3):896–903

Li K, Liu S, Xing R, Yu H, Qin Y, Li R, Li P (2013a) High-resolution separation of homogeneous chitooligomers series from 2-mers to 7-mers by ion-exchange chromatography. J Sep Sci 36(7):1275–1282

Li K, Liu S, Xing R, Qin Y, Li P (2013b) Preparation, characterization and antioxidant activity of two partially N-acetylated chitotrioses. Carbohydr Polym 92(2):1730–1736

Li K, Xing R, Liu S, Li P (2016) Advances in preparation, analysis and biological activities of single chitooligosaccharides. Carbohydr Polym 139:178–190

Li K, Xing R, Liu S, Qin Y, Li P (2017) Access to N-acetylated chitohexaose with well-defined degrees of acetylation. BioMed Res Int 2017

Li KC, Zhang XQ, Yu Y, Xing RE, Liu S, Li PC (2020) Effect of chitin and chitosan hexamers on growth and photosynthetic characteristics of wheat seedlings. Photosynthetica 58(3):819–826

Limpanavech P, Chaiyasuta S, Vongpromek R, Pichyangkura R, Khunwasi C, Chadchawan S, Bangyeekhun T (2008) Chitosan effects on floral production, gene expression, and anatomical changes in the Dendrobium orchid. Sci Hortic 116(1):65–72

Liu H-T, Li W-M, Xu G, Li X-Y, Bai X-F, Wei P, Du Y-G (2009) Chitosan oligosaccharides attenuate hydrogen peroxide-induced stress injury in human umbilical vein endothelial cells. Pharmacol Res 59(3):167–175

Liu T, Liu Z, Song C, Hu Y, Han Z, She J, Chai J (2012) Chitin-induced dimerization activates a plant immune receptor. Science 336(6085):1160–1164

Liu X, Zhang Y, Liu Z, Xie X (2019) Anti-tumor effect of chitin oligosaccharide plus cisplatin in vitro and in vivo. Onco Targets Ther 12:7581–7590

Mukhtar Ahmed KB, Khan MMA, Siddiqui H, Jahan A (2020) Chitosan and its oligosaccharides, a promising option for sustainable crop production—a review. Carbohydr Polym 227:115331

Muley AB, Shingote PR, Patil AP, Dalvi SG, Suprasanna P (2019) Gamma radiation degradation of chitosan for application in growth promotion and induction of stress tolerance in potato (Solanum tuberosum L.). Carbohydr Polym 210:289–301

Ngo D-N, Lee S-H, Kim M-M, Kim S-K (2009) Production of chitin oligosaccharides with different molecular weights and their antioxidant effect in RAW 264.7 cells. J Funct Foods 1(2):188–198

Nokami T, Isoda Y, Sasaki N, Takaiso A, Hayase S, Itoh T, Yoshida J-I (2015) Automated electrochemical assembly of the protected potential TMG-chitotriomycin precursor based on rational optimization of the carbohydrate building block. Org Lett 17(6):1525–1528

Oliveira EN, El Gueddari NE, Moerschbacher BM, Peter MG, Franco TT (2008) Growth of phytopathogenic fungi in the presence of partially acetylated chitooligosaccharides. Mycopathologia 166(3):163–174

Panda SK, Kumar S, Tupperwar NC, Vaidya T, George A, Rath S, Ravindran B (2012) Chitohexaose activates macrophages by alternate pathway through TLR4 and blocks endotoxemia. PLoS Pathog 8(5):e1002717

Park P-J, Je J-Y, Kim S-K (2003) Free radical scavenging activity of chitooligosaccharides by electron spin resonance spectrometry. J Agric Food Chem 51(16):4624–4627

Qiao Y, Bai X-F, Du Y-G (2011) Chitosan oligosaccharides protect mice from LPS challenge by attenuation of inflammation and oxidative stress. Int Immunopharmacol 11(1):121–127

Semenuk T, Krist P, Pavlicek J, Bezouska K, Kuzma M, Novak P, Kren V (2001) Synthesis of chitooligomer-based glycoconjugates and their binding to the rat natural killer cell activation receptor NKR-P1. Glycoconj J 18(10):817–826

Surdu AM, Pinzariu O, Ciobanu D-M, Negru A-G, Cainap S-S, Lazea C, Cismaru G (2021) Vitamin D and its role in the lipid metabolism and the development of atherosclerosis. Biomedicines 9(2)

Suzuki K, Mikami T, Okawa Y, Tokoro A, Suzuki S, Suzuki M (1986) Antitumor effect of hexa-N-acetylchitohexaose and chitohexaose. Carbohydr Res 151:403–408

Tokoro A, Tatewaki N, Suzuki K, Mikami T, Suzuki S, Suzuki M (1988) Growth-inhibitory effect of hexa-N-acetylchitohexaose and chitohexaose against meth-a solid tumor. Chem Pharm Bull 36(2):784–790

Turk H (2019) Chitosan-induced enhanced expression and activation of alternative oxidase confer tolerance to salt stress in maize seedlings. Plant Physiol Biochem 141:415–422

Valko M, Leibfritz D, Moncol J, Cronin MTD, Mazur M, Telser J (2007) Free radicals and antioxidants in normal physiological functions and human disease. Int J Biochem Cell Biol 39(1):44–84

Van Dyken SJ, Locksley RM (2018) Chitins and chitinase activity in airway diseases. J Allergy Clin Immunol 142(2):364–369

Vassilev N, Vassileva M, Lopez A, Martos V, Reyes A, Maksimovic I, Malusa E (2015) Unexploited potential of some biotechnological techniques for biofertilizer production and formulation. Appl Microbiol Biotechnol 99(12):4983–4996

Vo TS, Ngo DH, Ta QV, Wijesekara I, Kong CS, Kim SK (2012) Protective effect of chitin oligosaccharides against lipopolysaccharide-induced inflammatory response in BV-2 microglia. Cell Immunol 277(1–2):14–21

Wang M, Chen Y, Zhang R, Wang W, Zhao X, Du Y, Yin H (2015) Effects of chitosan oligosaccharides on the yield components and production quality of different wheat cultivars (Triticum aestivum L.) in Northwest China. Field Crop Res 172:11–20

Wei X, Wang Y, Xiao J, Xia W (2009) Separation of chitooligosaccharides and the potent effects on gene expression of cell surface receptor CR3. Int J Biol Macromol 45(4):432–436

Wu Y, Lu W-P, Wang J, Gao Y, Guo Y (2018) Rapid and convenient separation of chitooligosaccharides by ion-exchange chromatography. In: Aleksandrova M, Szewczyk R (eds) 5th annual international conference on material science and engineering

Xie X, Yan Y, Liu T, Chen J, Huang M, Wang L, Li X (2020) Data-independent acquisition proteomic analysis of biochemical factors in rice seedlings following treatment with chitosan oligosaccharides. Pestic Biochem Physiol 170:104681

Xiong C, Wu H, Wei P, Pan M, Tuo Y, Kusakabe I, Du Y (2009) Potent angiogenic inhibition effects of deacetylated chitohexaose separated from chitooligosaccharides and its mechanism of action in vitro. Carbohydr Res 344(15):1975–1983

Yamada A, Shibuya N, Kodama O, Akatsuka T (1993) Induction of phytoalexin formation in suspension-cultured rice cells by N-acetylchitooligosaccharides. Biosci Biotechnol Biochem 57(3):405–409

Yano K, Itoh T, Nokami T (2020) Total synthesis of Myc-IV(C16:0, S) via automated electrochemical assembly. Carbohydr Res 492:108018

Yoon HJ, Moon ME, Park HS, Im SY, Kim YH (2007) Chitosan oligosaccharide (COS) inhibits LPS-induced inflammatory effects in RAW 264.7 macrophage cells. Biochem Biophys Res Commun 358(3):954–959

Yoon HJ, Moon ME, Park HS, Kim HW, Im SY, Lee JH, Kim YH (2008) Effects of chitosan oligosaccharide (COS) on the glycerol-induced acute renal failure in vitro and in vivo. Food Chem Toxicol 46(2):710–716

Yoshida N, Tanaka T, Noguchi M, Kobayashi A, Ishikura K, Ikenuma T, Shoda S-I (2012) One-pot chemoenzymatic route to chitoheptaose via specific transglycosylation of chitopentaose-oxazoline on chitinase-template. Chem Lett 41(7):689–690

Yuan W-P (2009) Antioxidant activity of chito-oligosaccharides on pancreatic islet cells in streptozotocin-induced diabetes in rats. World J Gastroenterol 15(11):1339

Zhang X, Li K, Liu S, Xing R, Yu H, Chen X, Li P (2016) Size effects of chitooligomers on the growth and photosynthetic characteristics of wheat seedlings. Carbohydr Polym 138:27–33

Zhang X, Li K, Liu S, Zou P, Xing R, Yu H, Li P (2017a) Relationship between the degree of polymerization of chitooligomers and their activity affecting the growth of wheat seedlings under salt stress. J Agric Food Chem 65(2):501–509

Zhang X, Li K, Xing R, Liu S, Li P (2017b) Metabolite profiling of wheat seedlings induced by chitosan: revelation of the enhanced carbon and nitrogen metabolism. Front Plant Sci 8

Zhang X, Li K, Xing R, Liu S, Chen X, Yang H, Li P (2018) miRNA and mRNA expression profiles reveal insight into chitosan-mediated regulation of plant growth. J Agric Food Chem 66(15):3810–3822

Zhang J, Wang W, Mao X (2020) Chitopentaose protects HaCaT cells against H2O2-induced oxidative damage through modulating MAPKs and Nrf2/ARE signaling pathways. J Funct Foods 72

Zhao Q, Yin L, Zhang L, Jiang D, Liu L, Ji H (2020) Chitoheptaose promotes heart rehabilitation in a rat myocarditis model by improving antioxidant, anti-inflammatory, and antiapoptotic properties. Oxid Med Cell Longev 2020

Zhou J, Chen Q, Zhang Y, Fan L, Qin Z, Chen Q, Zhao L (2018) Chitooligosaccharides enhance cold tolerance by repairing photodamaged PS II in rice. J Agric Sci 156(7):888–899

Zhu C, Zhao M, Fan L, Cao X, Xia Q, Zhou J, Zhao L (2021) Chitopentaose inhibits hepatocellular carcinoma by inducing mitochondrial mediated apoptosis and suppressing protective autophagy. Bioresour Bioprocess 8(1)

Zong H, Li K, Liu S, Song L, Xing R, Chen X, Li P (2017a) Improvement in cadmium tolerance of edible rape (Brassica rapa L.) with exogenous application of chitooligosaccharide. Chemosphere 181:92–100

Zong H, Liu S, Xing R, Chen X, Li P (2017b) Protective effect of chitosan on photosynthesis and antioxidative defense system in edible rape (Brassica rapa L.) in the presence of cadmium. Ecotoxicol Environ Saf 138:271–278

Zou P, Tian X, Dong B, Zhang C (2017) Size effects of chitooligomers with certain degrees of polymerization on the chilling tolerance of wheat seedlings. Carbohydr Polym 160:194–202

Isolation of Chitinolytic Enzymes and Development of Chitooligosaccharides in Indonesia

Yusro Nuri Fawzya and Ekowati Chasanah

1 Introduction

Chitooligosaccharides (COS) have become one of the great interests due to their excellent properties and beneficial biological effects including, controlling blood glucose level, having antihyperlipidemic and antioxidative activities (Katiyar et al. 2011), reducing high blood pressure, lowering blood cholesterol, enhancing calcium absorption, controlling arthritis, protective effects against infections, and improvement of antitumor properties (Lodhi et al. 2014). COS are produced by depolymerization of chitin or chitosan, a natural polymer generated from crustacean shells, fungi and algae cell walls, insects' exoskeletons, and endoskeleton of cephalopods, by chemical, physical, enzymatic methods, or a combination thereof. So far, crustacean shells are the main source of commercial chitin/chitosan traded in the world, due to the abundant availability of raw materials.

The market of chitosan in the world is predicted to reach USD 784.8 million in 2020, and will increase to USD 2,093.5 million in 2028, with growing at a CAGR of 12.9% during the forecast period. The market trends are "The increasing use of chitosan as one of the essential ingredients in dietary supplement" which is supported by its ability to restrict lipid absorption that lead to add chitosan into dietary supplement. The largest revenue share of the global chitosan market in 2020 was contributed by the Asia Pacific market which required chitosan in preparing pharmaceuticals to address the high prevalence of chronic diseases in the region. Based on the chitosan source, crustacean (particularly shrimp shells) accounted for the main revenue share in the world market in 2020 due to the high demand of shrimp chitosan for waste water treatment and also food application (Research and Market 2021).

Y. N. Fawzya (✉) · E. Chasanah
Research and Development Center for Marine and Fisheries Product Processing and Biotechnology, Jakarta, Indonesia
e-mail: nurifawzya@gmail.com

S.-K. Kim (ed.), *Chitooligosaccharides*,
https://doi.org/10.1007/978-3-030-92806-3_17

Crustacean shells as the main source of global chitin/chitosan are available in large quantities from shrimp or crab industries, reaching about 6–8 million tons per year, in which about 1.5 million is produced in Southeast Asia (Yan and Chen 2015). As the shrimp shell contains about 25–40% chitin and 15–20% chitin in crab shells, the approximate global chitin produced is about 1.5–2 million tons annually. Chitin can be prepared through deproteination and demineralization of crustacean shells, and the next process is deacetylation of chitin that will produce chitosan. Chitin and chitosan are high molecular weight polymers which are soluble in dilute aqueous acidic solution; they cannot be dissolved in water. This characteristic limits their uses in food, health, and other applications. Hydrolysis of chitin/chitosan into their low molecular weight oligomers can improve their solubility in water and their potential uses, including in the biomedical and food applications.

Chitooligosaccharides (COS), depolymerized products of chitosan, can be produced through hydrolysis process using physical, chemical, or enzymatic methods. The COS preparation by using physical methods including microwave, hydrothermal, ultrasonication, and gamma rays was reviewed by Liang et al. (2018), whereas the production of COS by using chemical methods which was reviewed by Liang et al. (2018) and Schmitz et al. (2019) can be applied by using acids, such as HNO_2, HF, HCl, H_3PO_4, or by oxidative degradation method using H_2O_2, ozone, sodium perborate, and potassium persulfate. Other acids studied for the COS preparation were lactic acid, trichloroacetic acid, formic acid, and acetic acid which was reported by Lodhi et al. (2014). The chemical degradation is the most commonly used method to produce COS, due to the simplicity and high efficiency process, even though the reaction is hard to control and may produce mixture of products with various degrees of polymerization (DP) which is difficult for purification. In addition, the harsh condition in acid-hydrolysis process causes the process environmentally unfriendly and hazardous (Kaczmarek et al. 2019).

Preparation of COS by enzymatic hydrolysis offers an alternative method which is worked under mild condition and more controllable. The process produces COS with a more uniform size of molecular weight (MW). However, this method is less developed for commercial scale production due to the higher cost and low availability of chitinolytic enzymes. Chitinolytic enzymes are hydrolytic enzymes which specifically break down glycosidic bonds in chitin/chitosan-producing COS. They include chitinase, chitosanase, glucanase, etc. Commercial hydrolytic enzymes such as protease, cellulase, lysozyme, amylase, etc. can be used as non-specific enzymes for an alternative of COS production (Lodhi et al. 2014).

Chitinolytic enzymes occur and distribute widely in various sources, including microorganisms (bacteria, archaea, fungi, and viruses), arthropod, higher plants, and mammals. They function depending upon the organism requirement, and have an activity in physiological and biotransformation processes (Dukariya and Kumar 2020). Their hydrolytic activity on chitin/chitosan that produces oligomers which have a considerable of beneficial application leads to bring global attention of the enzymes to be explored.

Indonesia, where 75% of the region is marine, is rich in various sources of chitinolytic enzymes and produces high volume of crustacean shells from shrimp-processing industries which can be used as chitin/chitosan raw material. This chapter presents the recent studies of microbial chitinolytic enzymes and the development of COS production using enzymatic methods, including the development of COS in Indonesia.

2 Chitinolytic Enzymes

Chitinolytic enzymes are a group of enzymes that hydrolyze chitin substrate, a β-1,4-linked polysaccharide of N-acetylglucosamine, to produce oligomer, a shorter polymer and or monomer glucosamine and or acetyl glucosamine. There are at least three enzymes in a chitinolytic enzymes, i.e., chitin deacetylase, chitinase, and chitosanase. The chitinolytic enzymes are produced by chitinolytic organisms including microorganism which produced these enzymes by secreting those chitinolytic enzymes, which can be applied for production of chitin-derived products or as fungal or pest control agents.

Chitin deacetylase (CDA) (E.C. 3.5.1.4.1) is an important, frontier enzymes that hydrolyze the acetamido group of chitin polymers producing chitosan, a deacetylated form of chitin polymer. This CDA enzyme is a key enzyme to replace the unfriendly processing of chitosan production, as nowdays chitosan produced, i.e., chemically processed. According to the CAZy classification system (CAZy 2021), CDAs belong to the family 4 of carbohydrate esterases (CE4) along with acetyl xylan and peptidoglycan. These enzymes may have important biological roles in fungi since this family of the CE4 enzymes are abundant in the genomes of chitin-containing fungi (CAZy 2021), indicating that these enzymes have important roles in fungi (Liu et al. 2017). Studies of CDA had been reported, such as CDAs from fungi (Hekmat et al. 2003; Kafetzopoulos et al. 1993; Naqvi et al. 2016), and chitin oligosaccharide deacetylase from *Vibrio parahaemolyticus* (Hirano et al. 2015) and *Vibrio cholera* (Pascual and Planas 2018).

While, chitinase (EC 3.2.1.14) and chitosanase (EC 3.2.1.132) hydrolyze chitin and chitosan polymer into its monomer form of chitin and chitosan, i.e., N-acetylglucosamine and N-glucosamine (Charles-Rodriguez et al. 2008; Patil et al. 2000), which can be applied in various industries due to their biological activities. These enzyme are in the group of glycosyl hydrolase (GH). Chitinases play important physiological roles in nutrition, morphogenesis, defense, and aggression. In fish and mollusca, the presence of chitinases is needed for digesting chitin present in their prey, while in insects and crustaceans, chitinases takes an important role in molting process catalyzing exoskeletal molting. Whereas in plant and seaweed chitinases act as self-defense proteins against fungi, and in bacteria, chitinolytic enzymes are used in producing simples molecules from chitin for their nutritional needs.

Related with COS production, of those three chitinolytic enzymes chitosanase is considered important. COS is essential ingredient in functional food and nutraceutical industries having beneficial bioactivities and are nontoxic, nonallergenic, and biodegradable functional biopolymers. Chitosanases are grouped into six glycoside hydrolase (GH) families based on overall structural similarities of their amino acid sequence. This classification is reported not closely correlated to sequence specificity since the products produced are not similar or different; therefore, Weikert et al. (2017) proposed the classification should be revisited. The bioactivity of COS as hydrolysis products is not only depending on DP (degree of polymerization) and FA (fractionation of acetylation) but also PA (pattern of acetylation). Chitosanases are grouped into four classes based on their cleavage specificity. All chitosanases can cleave D–D (D = deacetylated) and for class 1, it can cleave A–D, (A = acetylated) in addition to D–D, Class II chitosanase (GH 8) is only capable of cleaving the D–D, while class III chitosanases also cleave D–A in addition to cleave D–D, and class IV GH is able to cleave all bonds except A–A. The difference of chitosanase with other enzyme capable of hydrolyzing β 1–4 glycosidic linkage was chitosanases which are capable of hydrolyzing β 1–4 glycosidic linkage of commercial chitosan which have 30–60% acetyl groups, while chitinase, lysozyme, and cellulase enzymes are capable of hydrolyzing commercial chitosan with 20–35% acetyl groups (Tremblay et al. 2001).

Chitin/chitosan oligomer (COS) products are chitin and chitosan derivatives that can be more widely applied than their parent chitin and chitosan because by becoming shorter chain, the COS are more soluble. In addition to, the lower degree of polymeration (DP) of COS posesses better bioactivity. To produce COS with consistent bioactivity, the DP and FA (fraction of acetylation) as well as PA, i.e., the sequence of GlcN and GlcNAc units, are very important to considered (Weikert et al. 2017). Therefore, purification strategy to obtain pure enzyme is very important before COS production to asses more detail about the enzyme action, especially in the substrate spesificity and specific hydrolysis of the enzyme. Purification step is also increasing the specific activity of the enzyme after removing the impurites, including non-targetted enzyme presented in the crude enzyme.

2.1 Distribution of Chitinolytic Enzymes

Chitinolytic degrading microorganisms are ubiquitously present in nature where chitin–containing organism was found or in chitin-associated environment both terestrial and marine. Chitin has been reported as the second most abundant biopolymer in earth, chitin is present in fungal cell walls, crustacean exoskeletons, insect cuticles and nematode egg shells (Shinya and Fukamizo 2017; Souza et al. 2009), and plankton (Souza et al. 2009). Marine and aquatic ecosystems are reported to be the biggest sources of chitinolytic enzymes since marine and aquatic inhabitans are rich in chitin-containing organisms such as shrimp and other crustaceans.

Chitinases (EC 3.2.1.14) have been reported to be present in various organisms including viruses, bacteria, fungi, insects, higherplants, and animals with various functions as defense, nutrient digestion, morphogenesis, and pathogenesis in each organism (Aliabadi et al. 2016; Souza et al. 2009). Chitinase has been secreted by *Serratia marcescens* PT-6, a gram negative bacterium isolated from shrimp pond sediment (Agustiar et al. 2019) and *Cohnella* sp. A01 (Aliabadi et al. 2016), *Alcaligenes faecalis* AU02 from shrimp industrial waste such as seafood efluent (Annamalai et al. 2011), *Stenotrophomonas maltophilia* KPU 2123 from shrimp waste (head and shell of shrimp) (Fawzya et al. 2009a), and *Aeromonas media* KLU 11.16 (Chasanah et al. 2011). Chitinase from *Stenotrophomona maltophilia* was also found in shrimp rubbish (Salas-ovilla et al. 2019), while from fishery solid waste local fish market had been identified, *Achromobacter denitrificans* (Dai et al. 2015). Chitinase from *Providencia stuartii* isolated from rotten tiger shrimp (*Penaeus monodon*) shells has been reported by Hardoko et al. (2020).

The enzyme had also been isolated from the stomach of the greenling fish, *Hexagrammos otakii* (three chitinase isoforms), common mackerel, *Scomber japonica*, as well as an insect tobacco hornworm, *Manduca exta* (Matsumiya et al. 2006). The chitinase from those different sources was reported to have unique substrate specificities showing the enzyme's role in the host. Chitinases isolated from the stomach of the greenling fish were more active to α-chitin, where this fish lives in the bottom of the sea and feed on shrimp and crab or other crustaceans which have α-chitin in their exoskeletons, and Polychaeta which is rich β-chitin (Matsumiya et al. 2006). Chitinase-producing bacteria from the genus Pseudomonas has been isolated from the skin of blue swimmer crab (*Portunus pelagicus*) (Sulistijowati et al. 2021). *Achromobacter xylosoxidans* isolated from the shrimp waste disposal site using chitin minimal medium has been reported to produce chitinase (Subramanian et al. 2020).

Thermostable chitinase and chitosanase have been identified from *B. licheniformis* MB-2 which was isolated from the water of Tompaso geothermal springs, Manado, Indonesia. The purified chitinase enzyme was a monomer with an apparent molecular weight of 67 kDa (Toharisman et al. 2005). From the same bacteria, chitosanase enzyme had also been identified as monomer with molecular weight of 75 kDa (Chasanah et al. 2006). Both enzyme worked best at 70 °C and 6.0, for temperature and pH, respectively. The chitinase enzyme hydrolyzed colloidal chitin, glycol chitin, chitosan, and glycol chitosan and the first 13 N-terminal amino acids of the pure enzyme were determined as SGKNYKIIGYYPS, which is identical to those in chitinases from *B. licheniformis* and *B. circulan*. While the pure chitosanase worked best at 85% deacetylated (DDA) soluble chitosan, followed by 90% and 100% DDA chitosan. Traditional fermented products have become chtinolytic sources such as Ronto (Haryogya and Ustadi 2020) and rusip (Puspita et al. 2017). Ronto is a traditional fermented rebon shrimp (rebon shrimp, salt, rice) from South Kalimantan, while rusip is similar product from Bangka, Indonesia which was made of small shrimp, salt (25%), and palm sugar (10%). Some rusip producers sometimes add rice or starch as well. The chitinolytic bacteria identified from rusip were *Bacillus cereus*, *Bacillus thuringiensis*, *Stenotrophomonas maltophilia*, *Enterobacter cloacae*, and *Pseudomonas stutzeri*, while from ronto was *Bacillus* sp. (Haryogya and Ustadi 2020). Chitinolytic bacteria was also detected from traditional product, namely, terasi. The bacteria was identified as *Micromonospora* sp. T5a1 (Patantis et al. 2020).

Marine fungus are another important sources of chitinolytic enzymes. Chitinase from marine-derived *Aspergillus terreus* isolated from marine sediment, which has been able to inhibit the growth of *Aspergillus niger*, *Aspergillus oryzae*, *Penicillum oxysporium*, *Rhizocotonia solani*, *Candida albicans*, and *Fusarium solani*, while it could not inhibit the growth of *Rhizopus oryzae*, indicating that *A. terreus* could be a good candidate for biocontrol of pathogenic fungi. It was also able to inhibit pathogenic bacteria, i.e., *Staphylococcus aureus*, *Salmonella typhi*, and *Pseu domonas aeruginosa*, but not active against *Escherichia coli*, *Aeromonas hydrophila*, and *Photobacterium damsela* (Farag et al. 2016). *Clonostachys rosea* IG119 was one of the 28 marine fungus that have been isolated from the marine phanerogam *Posidonia oceanica* and its epiphytic algae *Dictyota dichotoma* (Pheophyta) and *Sphaerococcus coronopifolius* (Rhodophyta) (Pasqualetti et al. 2019).

Chtinolytic enzyme can be also found and extracted from plants. Plant chitinase is different from chitinase from fish stomach, and other sources in molecular weight, pI value, and substrate specifity as well. Chitinase for biocontrol of pathogenic fungi in cultured plants has been isolated and identified as *B. thuringiensis* and *B. subtilis* (Suryadi et al. 2013).

2.2 Production and Isolation of Chitinolytic Enzymes

In general, chitinolytic enzyme is secreted extracellularly by the microbe producers which are cultivated using minimal medium. For lab-scale production, cultivation and production of chitinase from mold as *Aspergillus terreus* have been reported using basal medium supplemented by shrimp-shell powder as a sole carbon source, i.e., (g/L) $(NH_4)_2SO_4$, 0.5; KCl, 0.5; 1.0; $MgSO_4.7H_2O$, $CaCl_2$, 0.02; $FeSO_4.7H_2O$, K_2HPO_4, 1.0; and traces, pH 5.0. Shrimp-shell powder was used at a concentration of 20 g/L. The production was conducted in Erlenmeyer containing 50 ml of the media. A two ml of a spore suspension (2×10^6 spore/ml) from slants was inoculated, and further the inoculated media was incubated at 30 °C in an incubator shaker (120 rpm). After incubation, the culture medium was filtered and centrifugated at 8000 g for 20 min. The supernatant was crude chitinase (Farag et al. 2016).

Response Surface Methodology (RSM) has been used to optimize the chitinolytic production from microorganism, both from bacteria and fungi. Subramanian et al. (2020) reported optimization of the potential isolate, identified as *A. xylosoxidans* for chitinase production, out of 12 isolates screened. The study was started with choosing variables using the Plackett–Burman design and the variables were used for RSM (responce surface method) study. The variables showing a positive impact on the enzyme production were colloidal chitin, peptone, KH_2PO_4, and $MgSO_4$ with the incubation of 37 °C for 3–5 days. Using central composite design (CCD), the optimal values of peptone, chitin, KH_2PO_4, and $MgSO_4$ which were estimated in actual units were 0.5545 g/L, 30.6061 g/L, 0.50 g/L, and 0.2818 g/L, respectively. The predicted chitinase activity was 464.2 U/mL, while the laboratory work was 467 U/mL confirming that the CCD model used was reliable. The same model also has been used for optimization of chitinase production by *Paenibacillus* sp. D1 (Singh et al. 2009), a marine microbial producing chitinolytic enzymes.

Cloning was another technique used to improve the enzyme production (Farag et al. 2016). Cloning and heterologous expression of the chitinase Chit42 protein of *Trichoderma harzianum* was used by Kidibule et al. (2018). The Chit42 protein was expressed in *Pichia pastoris* which was cultured using fed-batch fermentation to about 3 g/L. The cloned enzymes expressed had similar biochemical properties as the natural producer, i.e., having molecular weight of 42 kDa and working best at pH 5.5–6.5 and 30–40 °C. The enzyme was active 100% to colloidal chitin which had DD <5% but it is also able to degrade chitosan with DD >90% with the activity of 2–3%. It means that the enzymes were more efficiently hydrolyzed colloidal chitin than chitosan.

Production of chitinolytic enzyme from halophilic *Trichoderma lixii* IG127 and *Clonostachys rosea* IG119 has been reported using additional 38% NaCl in shaken cultures in media containing 1% colloidal chitin, 1% yeast nitrogen base (Pasqualetti et al. 2019). RSM was applied using 1% colloidal chitin and different concentrations of corn step liquor and yeast nitrogen base (0–0.5%). The maximum activity (394 U/L) was obtained after 120 h culture on medium containing Corn Steep Liquor 0.47% and Yeast Nitrogen Base 0.37%. Maximum of productivity (3.3 U/L.h.) was recorded at the same incubation time. RSM has also been used to improve production of chitnolytic enzyme from fungus *Penicillium janthinellum* P9. Production of chitinolytic enzyme in bench reactor has been conducted in 3L-bioreactor. The enzyme activity has increased about 65%, from 415 U/L to 686 U/L after optimization. It was reported that production of *Penicillium janthinellum* P9 chitinase apparently be more dependent on appropriate combination of different fermentation conditions (Fenice et al. 1998). However, study on the scale-up production using bigger fermentor is not available so far.

Some microorganisms secreted more than one chitinolytic enzyme that is very advantageous in producing various COS further according to the substrate and the products targetted. *B. licheniformis* isolated from hot spring as previously mentioned has extracellularly secreted thermostable chitinase and thermostable chitosanase as well. The chitosanase has been produced by cultivating the bacteria using 15% starter in liquid medium consisting of 0.24% (w/v) chitosan, 0.25% casiton, 1% $MgSO_4$, 1.4% K_2HPO_4, 0.02% $CaCl_2{\cdot}2H_2O$, 0.002% $FeSO_4{\cdot}7H_2O$, and incubated at 55 °C, 120-rpm shaker water bath for 60 h (Chasanah et al. 2006). The pure chitosanase was obtained by single step purification using hydrophobic interaction chromatography system (butyl Sepharose 4 FF) with apparent molecular mass of 75 kDa. On the other hand, the chitinase from *B. licheniformis* MB-2 had been produced using 10% seed culture in liquid medium consisting of 0.5% colloidal chitin, 0.7% $(NH_4)_2SO_4$, 0.1% K_2HPO_4, 0.1% NaCl, 0.01% $MgSO_4$. $7H_2O$, 0.05% yeast extract, and 1.5% agar (pH 7.0) and incubated at 55 °C on a rotary shaker at 180 rev min^{-1} (Toharisman et al. 2005). The pure chitinase enzyme, sizing 67 kDa, was obtained by using ion exchanger, namely, DEAE-Sephacel column followed by gel filtration chromatography, i.e., HiLoad Superdex 75. As *B. licheniformis* was isolated from hot spring, the strategy of heating the protein produced at 80 °C for 5 min was used as prepurification step for both enzymes. The chitinase activity was increased by 1.37-fold after heating treatment, while after the two-step purification conducted, the enzyme activity has increased 15.33-fold (Toharisman et al. 2005). Prepurification strategy using ammonium sulfate 65% (saturation) of crude chitinase *Bacillus licheniformis* B307 has improved their activity by 8.24-fold (Akeed et al. 2020). Application of ammonium sulfate to concentrate the chitosanase has also been used for isolating chitosanase from *Aeromonas media* KLU 11.16, before purification by ion exchange chromatography DEAE Sepharose Hi Trap FF matrix followed by gel filtration using Sephadex G-100 (Chasanah et al. 2011).

Isolation of chitinase from *Aspergillus terreus,* a marine fungi, was similar from that of from bacteria, i.e., prepurification using 65% ammonium sulfate precipitation, followed by gel filtration on Sephadex G-100 and DEAE-Sephadex A-50 ion exchange chromatography producing pure enzyme with 5.16-fold of purification and specific activity of 182.08 U/mg protein (Farag et al. 2016).

Prepurification aims to concentrate the crude enzyme before entering purification step which can also use ultrafiltration as reported by Patantis et al. (2020). The ultrafiltration using 10 kDa molecular cutt-off improved the activity by 1.76% and 1.17% for crude chitosanase produced by *Stenotrophomonas maltophilia* KPU 2123 and *Micromonospora sp. T5a1*, respectively. Further purification using DEAE Sepharose ion exchange and Sephacryl S-300 gel filtration chromatography improving the specific activity by 2.35-fold, yields 30.90% and 3.32-fold, yields 22.79%, respectively.

3 Chitooligosaccharides

3.1 Production of Chitooligosaccharides (COS) by Enzymatic Method

Chitooligosaccharides (COS) are depolymerization products of chitin and/or chitosan (usually with degree of polymerization up to 20), composed of homo- or hetero-oligomers of GlcN and/or GlcNAc connected with β(1 → 4) bonds (Miguez et al. 2021). Both chitinolytic enzymes and commercial hydrolytic enzymes which do not specifically digest chitin/chitosan can be employed for producing chitooligosaccharides. Unlike non-specific enzymes, so far, commercial chitinolytic enzymes, especially chitosanases, are of limited availability. Many researches on COS preparation have been conducted with less variation in the production process, commonly using commercial chitin/chitosan that is chemically produced from crustacean shell. Firstly, crustacean shells are processed into chitin through demineralization and deproteination. The chitin is then deacetylated using sodium hydroxide at high concentration and high temperatures to produce chitosan. Then, COS are prepared by hydrolyzing chitosan enzymatically at optimum condition of the enzyme.

Chitin can be extracted biologically from shrimp waste using lactic acid bacteria *Lactobacillus acidophilus* FNCC 116 to demineralize shrimp shells and deproteinize using *Bacillus licheniformis* F11.1. It was reported that a better chitin yield was obtained when demineralization process conducted first followed by the deproteination process resulting in reduction of 79.61% protein and 88.65% ash (Wahyuntari et al. 2011). Chitin produced by biologically process has been claimed to have higher viscocity than that of chemically proccesed.

Chitinolytic enzymes hydrolyzed chitin/chitosan molecules at the glycoside bonds between the sugars. After hydrolysis process, the next step is enzyme inactivation which is usually conducted by heating at 90–100 °C for about 5–10 min. COS is a supernatant produced by separation of the hydrolysates.

Crude enzyme is commonly used for COS production, usually in the form of concentrated liquid or solid powders. Due to the hard access for enzyme to break down the strong crystalline chitin/chitosan, pretreatment is required to convert them into modified chitin/chitosan (such as colloidal chitin, swollen chitin, or soluble chitosan) to increase the accessibility of enzyme to the substrate. However, these methods are neither economically nor environmentally feasible for scaling up toward industrial purposes. Physical, mechanical, chemical, or biotechnological pathways have been done to overcome this problem (Arnold et al. 2020).

Chitinase from *Vibrio anguillarum* E-383a isolated from seawater was employed to hydrolyzes colloidal chitin, usually prepared by addition of concentrated HCl. The reaction produced $(GlcNAc)_2$ with the maximum yield reaching 40.3%, and the small amounts of *N*-acetylglucosamine and *N,N',N*-triacetylchitotriose. The chitinase might work as an exo-type chitinase which hydrolyzes the glycosidic bonds of chitin into N,N'-diacetylchitobiose units (Takiguchi and Shimahara 1988). Hydrolysis of swollen chitin and chitosan which are prepared by using 85% phosphoric acid was performed by a thermostable chitinase Chi 1 isolated from a filamentous fungus *Myceliophthora thermophila* C1 producing GlcNAc, $(GlcNAc)_2$, and $(GlcNAc)_3$ mixtures with the $(GlcNAc)_2$ as a major product. Meanwhile, swollen chitosan from chitosan 90 DDA/100 which was hydrolyzed with the same enzyme at various hydrolysis time produced hetero-oligosaccharides containing GlcNAc and GlcN units with a polymerization degree (DP) of 2–12 (Krolicka et al. 2018). Soluble chitosan, the most familiar substrates for COS preparation, is commonly served by dissolving chitosan in acetic acid-sodium acetate buffer. The use of soluble chitosan for production of COS using chitosanolytic crude enzyme from *Paenibacillus ehimensis* resulted in hydrolysates containing $(GlcN)_2$ and $(GlcN)_3$ (Pagnoncelli et al. 2010). A potential thermo-tolerant and cold-adapted chitosanase from marine bacterium *Pseudoalteromonas* sp. SY39 explored its hydrolytic pattern and reaction products. It was investigated that the enzyme acted as endo-chitosanase with chitobiose as the main product, with the yield of 69.9% of total products, after incubation of chitosan with the enzyme at 30 °C for 24 h (Zhou et al. 2019). Generally, hydrolysis of chitosan by chitinolytic enzymes produces COS with DP of 1–7 depending on the reaction condition. Prolonged hydrolysis time commonly produces chitosan monomer as the major product (Jung and Park 2014). We summarize some of the COS production technology using chitinolytic enzymes which have been published or recently studied, as presented in Table 1.

Commercial non-specific enzymes can also be applied for COS preparation, such as cellulase, lipase, amylase, etc. Water-soluble chitosan with low molecular weight ranging from 1000 to 4000 Da can be obtained by hydrolyzing chitosan using complex enzymes composed of commercial pectinase, α-amylase, and cellulase. The reaction lasted for 2 h and the products had no change in the structure of chitosan glycosidic ring as well as deacetylation degree (Xie et al. 2011). Degradation of chitosan by a commercial lipase showed that the enzyme worked on chitosan in both exo- and endo-hydrolytic types. The product yield was 93.8% produced by the 24 h hydrolysis at 37 °C, and was composed of glucosamine and chitooligosaccharides with DP of 2–6 and above (Lee et al. 2008). Other researches on COS preparation using non-specific enzymes are presented in Table 2.

A scaling-up production of COS was initiated by Jeon and Kim (2000) using a dual reactor system consisting of an ultrafiltration (UF) membrane reactor and a column reactor packed with an immobilized enzyme. The application of immobilized chitosanase from *Bacillus pumilus* BM-262 and the use of a membrane reactor were based on their experience and previous researches in the enzymatic process for protein and polysaccharide hydrolysis using membrane reactor system. Two steps of COS production consisted of preparation of partially hydrolyzed chitosan (PHC) using the column reactor packed with immobilized enzyme and the COS production from PHC in the UF membrane reactor. Based on the observation on the effect of outflow rates of PHC on the membrane fouling, the PHC with a 5 mL/min overflow rate was considered as the most acceptable substrate without producing fouling and resulting in higher productivity. The COS produced could be fractioned by suitable membranes with molecular weight desired. The use of 10-, 5-, and 1-kDa-MWCO membranes produced COSs with a significant molecular weight difference in the gel permeation chromatograph patterns. COS as permeate from 1-kDa-MWCO membrane composed of pentamer, hexamer, and heptamer. The use of this method could produce COS with higher molecular weights (Jeon and Kim 2000).

Similar research was done for production of COS in a dual reactor system using an immobilized enzyme. The first reactor was a batch reactor for hydrolysis of chitosan, and the second reactor was a packed-bed reactor (PBR) for passing the COS produced which preventing clogging of the column. Research was carried out in a small lab scale using a commercial α-amylase from *Bacillus amyloliquefaciens* (BAN) which had a chitosanolytic activity. The immobilized enzyme was used both in the first reactor, a continuous stirred tank reactor, and in the second reactor, a packed-bed reactor with the immobilized enzyme. The hydrolysis of chitosan by this immobilized enzyme in the first reactor produced partially hydrolyzed chitosan (PHC) with an obvious reduction of viscosity that supported the substrate flow through the PBR, producing enhanced productivity (Paloma Santos-Moriano et al. 2016).

Development of COS production was studied by producing a fully defined structure of chitooligosaccharides which was produced by degradation of chitin using chitin deacetylases (CDA) (Liang et al. 2018). Hembach et al. (2017) used pure bacterial, fungal, and viral chitin deacetylases (CDAs) to produce partially acetylated chitosan tetramers. The three bacterial CDAs were NodB from *Rhizobium* sp. GRH2, *Vc*CDA from *Vibrio cholerae*, and *Bc*CDA5 from *Bacillus licheniformis*. The fungal CDAs used were *Cn*CDA2 and *Cn*CDA4 from *Cryptococcus neoformans*, *Pes*CDA from *Pestalotiopsis* spec., *Pgt*CDA from *Puccinia graminis* f. sp. tritici, and *Pa*CDA from *Podospora anserine*. Whereas the viral CDAs applied was *Cv*CDA from *Chlorovirus* CVK2. All CDAs were expressed in either *E. coli* or *Hansenula polymorpha*. It was observed that CDAs from various origins can *N*-acetylate chitosan oligomers efficiently, showing the same region specificity as their conventional deacetylation. The CDAs also worked on partially acetylated chitooligosaccharides, in forward and reverse modes as both deacetylation and *N*-acetylation. There were 14 possible partially acetylated chitosan tetramers resulting from these treatments which were analyzed using UHPLC-ESI-MS. The products might be further purified regarding the degree of acetylation, degree of polymerization, and pattern of acetylation. These methods might contribute for the research on correlation of chitooligosaccharides structure and function. Other study on preparation of defined chitosan oligomers with specific pattern of acetylation (PA) was reported by Hamer et al. (2015). Two CDAs were used, NodB from *Rhizobium* sp. GRH2 and COD from *Vibrio cholera*. The combined CDAs worked to deacetylate chitin oligomers at the first two units starting from the non-reducing end producing doubly deacetylated chitosan oligomers. The NodB CDA deacetylated the first unit starting from the non-reducing end, while the COD CDA deacetylated the second unit starting from the non-reducing end. So far, the defined chitosan oligomers have not been widely reported due to the limited available CDAs.

Recently, studies on COS productions have been developed by involving the lytic polysaccharide monooxygenases (LPMOs) that is used in synergy with chitinolytic enzymes to depolymerize recalcitrant crystalline chitin/chitosan. The LPMOs are copper-dependent enzymes which catalyze an oxidation reaction either C_1 or C_4 of the glucopyranose ring of polysaccharides, including chitin/chitosan and cellulose molecules. The reaction produces oxidized chain ends of chitin/chitosan molecules which then facilitate chitinases/chitosanases in depolymerization of chitin/chitosan. The activity of LPMOs needs an oxygen co-substrate, which was at the beginning thought to be O_2, or may also be H_2O_2 (Arnold et al. 2020; Bissaro et al. 2017; Forsberg et al. 2019). These enzymes may be produced by some microorganisms, such as *Serratia marcescens* (Vaaje-Kolstad et al. 2010), *Bacillus cereus* (Mutahir et al. 2018), and *Vibrio cholerae* (Loose et al. 2014). The LPMO from *B. cereus* or *S. marcescens* and GH18 chitinases enzymes acted synergistically in the degradation of shrimp-shell chitin. The use of both enzymes simultaneously produced yield of 26% higher compared to the sum of products yield prepared by individual enzyme (LPMO and chitinase). It was also demonstrated that *N*-acetyl glucosamine produced correlated with the oxidized products yields (Mutahir et al. 2018).

Other development procedure for COS preparation was recently reported by Kumar et al. (2021). A two-step chemo-enzymatic method for production of COS has been studied for eco-friendly and cost-effective handling of coastal waste resources. Chitin was treated by sulfuric acid, sodium hydroxide, and ionic liquids (1-ethyl-3-methylimidazolium bromide and 2,4,4-trimethylpentyl phosphinate) prior to hydrolyzation with purified chitinase from *Thermomyces lanuginosus* ITCC 8895. The chemical treatment changed the crystalline chitin into amorphous form that supports enzyme accessibility. COS produced contained $(GlcNAc)_2$ and $(GlcNAc)_3$ which increased to 1.5- and 1.3-fold yield by optimization (Kumar et al. 2021).

Further important step of COS production is separation and purification of COS. Enzymatic hydrolysis of chitin or chitosan commonly produces a mixture of oligomers which needs separation and purification to obtain each component. A number of techniques for COS separation and purification have been reviewed by Jung and Park (2014), including ultrafiltration, gel filtration, and ion exchange and metal affinity chromatography. Separation of COS based on their size can be done using preparative separation through size exclusion chromatography, while COS with identical DP can be separated based on the amount of deacetylated units using cation-exchange chromatography (Jung and Park 2014).

3.2 Development of Chitooligosaccharides in Indonesia

Indonesia, as an archipelagic country, has high fishery resources, including shrimp and other crustacean which are the main contributors of raw material for chitin/chitosan production. Shrimp is Indonesia's mainstay fishery export product; the total shrimp export reached 239,000 MT in 2020, up 15% from 2019 (Dao 2021), despite a pandemic situation. The by-product of shrimp shells produced was around 119,000 tons, a large amount of raw material for chitin/chitosan manufactures. However, the crustacean shells processed into chitosan are still limited in quantities, because most of the materials are distributed for feed purposes.

In Indonesia, researches on chitin/chitosan have been carried out since the 1990s; however, the modification of chitin/chitosan, including water-soluble chitosan and chitosan oligomers, was just studied about 10 years later. Studies on production of water-soluble carboxymethyl chitosan through etherification of chitosan using monochloroacetic acid were reported by Basmal et al. (2007), Fawzya et al. (2008), Oktavia et al. (2005), Putra et al. (2016). Other water-soluble chitosan prepared by combination of physical and chemical methods has been carried out through hydrolysis of chitosan using HCl under pressure condition at 0.8–0.9 atm, 110 °C. The optimum condition was determined using Box-Behnken Response Surface Methodology (RSM-Box-Behnken) analyses, and the recommended optimum formula for water-soluble chitosan production was the concentration of HCl 1.75%, ratio of chitosan to HCl 1:10 with the hydrolysis time of 45 min. However, the verification of the optimum formula produced the hydrolysis product which has not completely dissolved in water, but 66.38%, with the yield of 80% (Sudianto et al. 2020). Development of water-soluble chitosan production was conducted by depolymerization of chitosan using ultraviolet radiation preceded by HCl hydrolysis. The use of 2% HCl

Table 1 Production of chitin- and chitooligosaccharides by enzymatic methods using specific enzymes

Source of chitin/chitosan	Source of enzyme	Enzyme	Condition	Products and yield	Analysis	Reference	Remarks
Colloidal chitin	*Vibrio anguillarum* E-383a	Chitinase	30 °C, 6 days	$(GlcNAc)_2$ 40.3%	HPLC	Takiguchi and Shimahara (1988)	
Chitosan (85; 90;100% DD)	*Bacillus licheniformis MB-2*	Chitosanase	70 °C, 24 h	$(GlcN)_5$ and $(GlcN)_6$	HPLC	Chasanah et al. (2006)	
Soluble chitosan	*Paenibacillus ehimensis*	Chitosanolytic (crude enzyme)	55 °C, 9 h	(GlcN)2 and (GlcN)3	TLC	Pagnoncelli et al. (2010)	
Soluble chitosan	*Streptomyces griseus*	Immobilized chitosanase	37 ◦C, 4 h	$(GlcN)_3$ to $(GlcN)_6$	TLC	Song et al. (2014)	
Rhizopus oligosporus NRRL2710 cells	*Streptomyces* sp. N174	Chitosanase from (CsnN174)	40 °C, 24 h	GlcN–GlcNAc 1.1%, $(GlcN)_2$–GlcNAc 0.4%, (GlcN)2 1.3%	NMR and MALDI-TOF–MS	Mahata et al. (2014)	Hydrolysis 67%
Rhizopus oligosporus NRRL2710 cells	*Pyrococcus furiosus*	Chitinase (PfChit)	40 and 70 °C, 24 h	–	–	Mahata et al. (2014)	Hydrolysis 36%
Rhizopus oligosporus NRRL2710 cells	*Trichoderma viride*	Chitinase (TvChit)	40 °C, 24 h	–	–	Mahata et al. (2014)	Hydrolysis 35%
Soluble chitosan	*Stenotrophomonas maltophilia KPU 2123*	Chitosanase	30 °C, 16 h	(GlcN), $(GlcN)_2$, $(GlcNAc)_2$, $(GlcN)_3$, $(GlcNAc)_6$, $(GlcN)_7$, $(GlcN)_8$, $(GlcNAc)_8$, $(GlcN)_{10}$, $(GlcN)_{14}$, $(GlcNAc)_{16}$	MALDI-TOF MS	Martosuyono et al. (2014)	COS was desalted by Gel Filtration Chromatography
Colloidal chitin	*Trichoderma harzianum*	Chitinase Chit 42 expressed in *Pichia pastoris*	35 °C, 24 h	0.13 g/L GlcNAc, 0.99 g/L $GlcNAc)_2$, 0.10 g/L $(GlcNAc)_3$	HPAEC-PAD	Kidibule et al. (2018)	– Only 15% of the substrate was transformed into small – Size acetylated COS – Lack of reference markers

(continued)

Table 1 (continued)

Source of chitin/chitosan	Source of enzyme	Enzyme	Condition	Products and yield	Analysis	Reference	Remarks
Chitosan QS 1 (81% DD)	*Trichoderma harzianum*	Chitinase Chit 42 expressed in *Pichia pastoris*	35 °C, 24 h	0.25 GlcNAc 0.14 g/L (GlcNAc)2	HPAEC-PAD	Kidibule et al. (2018)	– Only ~ 5% of the substrate was transformed into small – Size acetylated COS – Lack of reference markers
Colloidal chitin	*Trichoderma harzianum* in *Pichia pastoris*	Chitinase Chit42	35 °C, pH 6, 24 h	GlcNAc, $(GlcNAc)_2$, and $(GlcNAc)_3$	HPAEC-PAD, MALDI-TOF MS	Santos-Moriano et al. (2019)	
Chitosan QS 1 (81% DD)	*Trichoderma harzianum* in *Pichia pastoris*	Chitinase Chit42	35 °C, pH 6, 24 h	at least 11 COS with different proportions of GlcNAc and GlcN	HPAEC-PAD, MALDI-TOF MS	Santos-Moriano et al. (2019)	
Colloidal chitin	*Streptomyces diastaticus* strain CS1801	Chitinase	5 d	$(GlcN)_2$, $(GlcN)_3$, $(GlcN)_4$,$(GlcN)_5$	UPLC-MS	Xu et al. (2020)	Total conversion rate = 1.085%
Soluble chitosan	*Stenotrophomonas maltophilia KPU 2123*	Chitosanase	30 °C, 16 h	(GlcNAc) to $(GlcNAc)_5$	HPLC	Patantis et al. (2020)	
Soluble chitosan	*Micromonospora* sp. T5a1	Chitosanase	37 °C, 16 h	(GlcNAc)	HPLC	Patantis et al. (2020)	
Partially acetylated chitosan	*Paenibacillus pabuli* (chitinase) *Fusarium fujikuroi* (lytic polysaccharide monooxygenases /LPMO)	Recombinant chitinase, *Pp*Chi, LPMO	40 °C for 24 h	$GlcN_2GlcNAc_1$ (DP3), $GlcN_3GlcNAc_1$ (DP4), and $GlcN_4GlcNAc_1$ (DP5),	HPAEC-PAD and MALDI-TOF MS;	Li et al. (2021)	Sixfold increase of yield when was applied in two steps (LPMO → chitinase) 1.4-fold increase of yield when was combined in one-pot reaction
Colloidal chitosan	*Bacillus atrophaeus*	Rekombinan Chitosanase In *Escherichia coli* BL21	pH 5.0 and 45 °C, 24 h	$(GlcN)_2$, $(GlcN)_3$, and $(GlcN)_4$	TLC; ESI–MS	Cui et al. (2021)	

Table 2 Production of chitin- and chitooligosaccharides by enzymatic methods using non-specific enzymes

Source of chitin/chitosan	Source of enzyme	Enzyme	Condition	Products and yield	Analysis	References	Remarks
β-Chitin	*Aspergillus niger*	Cellulase and lipase (9:1)	37 °C, 4 days	GlcNAc 61%	HPLC NH2P-50	Sukwattanasinitt et al. (2002)	
Colloidal chitosan	*Trichoderma viride* (Commercial enzyme, Sigma)	Cellulase	55 °C, pH 5.2, 9 h	DP 1–6	HPLC	Wu and Tsa (2004)	
Chitosan (82.8% DD)	Commercial enzyme (Novozyme)	Lipase	37 °C, 24 h 37 °C, 6 h	DP 1 93.8% DP 1–6 58.2%	TLC	Lee et al. (2008)	
Soluble chitosan	*Trichoderma reesei*	Cellulase	60 °C 3 h	$(GlcNAc)_2$ to $(GlcNAc)_5$	TLC	Fawzya et al. (2009b)	
Soluble chitosan	*Carica papaya*	Papain	40 °C, 16 h	na	–	Laokuldilok et al. (2017)	MW 4.3 kDa
Soluble chitosan	Chicken egg white	Lysozyme	30 °C, 16 h	na	–	Laokuldilok et al. (2017)	MW 11.7 kDa
Soluble chitosan	*Aspergillus niger*	Cellulase	37 °C, 16 h	na	–	Laokuldilok et al. (2017)	MW 11.7 kDa
Chitosan CHIT 600 (Chitosan DD >90%)	*Bacillus amyloliquefaciens*	Neutrase 1.8L	50 °C, pH 5, 24 h	GlcN, $(GlcN)_2$, $(GlcN)_3$, $(GlcN)_4$, and $(GlcN)_5$ [$(GlcN)_2$ and $(GlcN)_3$] were major products	HPAEC-PAD, MALDI-TOF MS	Santos-Moriano et al. (2019)	

followed by UV radiation for 75 min resulted in yield of 70.07% with good water solubility (99.80%), and molecular weight of 169.46 kDa (Nuryadin et al. 2020).

Production of glucosamine, a chitosan monomer, has been investigated by Cahyono et al. (2014), Rosmiati et al. (2016), Suptijah et al. (2014), Tanasale et al. (2019) using chemical method with mostly using HCl under certain pressure condition. Hydrolysis of chitosan with 5–8% HCl under atmospheric pressure (0.5–1 atm) for an hour, produced water-soluble glucosamine-HCl with yield of 65–70% and melting point of 190–192 °C. Both sample and standard of glucosamine had similar pattern of FTIR absorption bands (Cahyono et al. 2014; Suptijah et al. 2014). Similar result was reported by Rosmiati et al. (2016) who worked on silkworm (*Bombyx mori* L.) pupae shell chitosan for glucosamine-HCl preparation with yield of 89.25% and solubility of 93.89%. Slight process modification of glucosamine production was applied by the use of hydrogen peroxide to replace HCl. The optimal condition hydrolysis process was achieved at the H_2O_2 concentration of 4%, at 30 °C for 4 h. The yield of glucosamine reached 97.5% with the molecular weight of 4.2×10^3 (Tanasale et al. 2019). A study on chitosan degradation by the combination of sonication and hydrothermal process in the presence of supercritical CO_2 demonstrated that the hydrolysis product from the sonication at 60 °C for 120 min before the hydrothermal process at 200 °C and pressure of 23 MPa for 3–5 min composed of glucose and 5-hydroxy methyl furfural (5-HMF) with the yield of 0.17–0.61% and 1.91–14.39%, respectively. The structure of chitosan did not affect by the hydrothermal process except for the inter- and intramolecular of chitosan changes and the crystallinity degree of the chitosan residue varying from 19.2 to 28.9% (Savitri and Rosyadi 2015). Preparation of N-acetil glucosamine by enzymatic method was reported by Hardoko et al. (2019). They used crude intracellular chitinase from *Providencia stuartii* to hydrolyze chitin through chitin fermentation at various concentrations of chitin and fermentation time. They demonstrated that the optimum process occurred at 1.0% chitin, pH 5.0, temperature of 40 °C for 6 h fermentation, with the N-acetyl glucosamine produced which was 1,680.06 ppm. A longer fermentation time was required for producing N-acetyl glucosamine from shrimp-shell chitin hydrolyzed with chitinolytic fungus *Beauveria bassiana*. N-acetyl glucosamine yield of 14.3 ppm was produced from the chitin fermentation at temperature of 25 °C, pH 7 for 7 days (Handayani 2018).

Researches on preparation of chitooligosaccharides by enzymatic hydrolysis was becoming in great interest since around the 2000s. It was preceded by the exploration of chitinolytic enzymes which was mostly generated from microbial sources. Unfortunately, none of these studies have been developed to a pilot scale. Early studies on microbial chitinolytic enzymes were explored from thermophilic sources as reported by Yuli et al. (2004) who investigated the characteristics of a thermostable chitinase from *Bacillus* sp. 13.26 and Toharisman et al. (2005) isolated chitinase from *B. licheniformis*. Both isolates were originated from Tompaso hotspring in North Sulawesi, Indonesia. Exploration of chitinases and chitin deacetylases from thermophilic bacteria was also conducted from *Bacillus* K29-14 isolated from geothermal area in West Jawa (Rahayu et al. 2004) and *Bacillus thermoleovorans* LW-4-11 isolated from Langoan hotspring water in North Sulawesi (Toharisman and Suhartono

2007). Since then, a lot of studies have been carried out to explore the chitinolytic enzymes from various sources, including chitinolytic enzymes produced by sponges-associated bacteria (Chasanah et al. 2009; Ginting et al. 2021), bacteria isolated from shrimp and crab waste (Chasanah et al. 2012; Dliyauddin et al. 2020; Masri et al. 2021; Sulistijowati et al. 2021), bacteria isolated from traditional fermented shrimp products (Noviendri et al. 2006; Puspita et al. 2017), bacteria isolated from soil or shrimp pond sediment (Agustiar et al. 2019; Wibowo et al. 2020), bacteria isolated from digestive tract (Pardosi et al. 2018), rhizosphere's bacteria (Mubarik et al. 2010), and fungi (Elawati et al. 2018). However, studies on the COS production are not as much the exploration of chitinolytic enzymes.

Hydrolysis chitosan at 70 °C overnight by using chitosanase *Bacillus licheniformis* MB-2 isolated from hot spring water in Manado produced chitooligosaccharide mixture containing of $(GlcN)_5$ and $(GlcN)_6$, with the 85% deacetylated soluble chitosan which was the most susceptible to the enzyme and showed inhibitory effect on *Pseudomonas aeruginosa, Salmonella typhimurium, Listeria monocytogenes, Bacillus cereus, Escherichia coli*, and *Staphylococcus aureus* (Chasanah et al. 2006). A cellulase *Trichoderma reesei* could hydrolyze soluble chitosan (DD 85%) non-specifically producing COS mixture with the viscosity of 2.15 cPs, and composed of $(GlcNAc)_2$ to $(GlcNAc)_5$ (Fawzya et al. 2009b). Degradation of chitosan by chitosanase *Klebsiella* sp. (the activity of 5,235 U/mg) at 40 °C for 3 h was observed by Sarni et al. (2016). The hydrolysis reaction reduced the molecular weight of chitosan from 25,410 to 1,065 g/mol which was determined by intrinsic viscosities approach, and produced COS containing a mixture of chitosan monomer to octamer.

Study on the effect of microwave irradiation on the cellulase hydrolysis of chitosan compared to conventional method using shaker incubator exhibited that the reaction time required for the same yield of reducing sugar was much faster for the treatment of microwave irradiation. However, the rate of relative viscosity decrease was faster in the treatment of shaker incubator. The *Vmax*. and *Km* of Michaelis–Menten kinetics was found to be 1.98 mg/mL h and 4.228 mg/mL for microwave irradiation treatment, and 0.0814 mg/mL h and 6.516 mg/mL for conventional method or shaker incubator (Rokhati et al. 2018).

Chitosanase from *Stenotrophomonas maltophilia* KPU 2123 isolated from shrimp shell has been produced, characterized, and applied for COS preparation, both from crude and purified enzymes. The three-step purification of this enzyme using ultrafiltration, DEAE Sepharose ion exchange, and Sepacryl S-300 gel filtration chromatography increased the specific activity by 2.35-fold, and resulted in yield of 30.90%. Identification of COS analyzed by HPLC showed that the COS composed at least monomer to hexamer, indicating that the enzyme categorized as endo-chitosanase (Patantis et al. 2020). In the previous study, crude enzyme was employed for COS preparation and analyzed the product's cytotoxicity against He La cell which resulted positively with the LC_{50} of 120 ppm (Fawzya et al. 2009a). However, when the crude enzyme used for scaling-up production of COS by 25 times, the product quality was not as good as the previous product quality, due to the high ash/mineral content of the product (Fawzya et al. 2018). This problem was related to the crude enzyme activity which decreased during storage, and needed more volumes of enzyme to hydrolyze chitosan. It has been known that the enzyme stability of wild-type bacteria is one of the major problems in using strains, in which protein engineering is usually applied to improve the stability, activity, or specificity of enzymes. Desalination of COS was applied by dialysis and gel filtration chromatography. The later method successfully removed most of salt which reached up to 100%, and based on the COS identification using MALDI-TOF, this method produced COS with more complete various degrees of polymerization of COS compared to dialysis (Martosuyono et al. 2014). Limited COS standard is a major problem in the identification of COS. A standard composed of chitosan monomer to hexamer was widely used in many COS studies.

4 Conclusion

Chitinolytic enzymes have been extensively explored for the bioconversion of chitin/chitosan into chitin/chitooligosaccharides as the alternative method to the destructive chemical method with no additional product modification. The bioactivity of COS is affected by degree of polymerization, fractionation of acetylation as well as pattern of acetylation. As the sequence of GlcN and GlcNAc units determines the COS bioactivity, the use of pure enzyme is considered to obtain COS with consistent bioactivities.

Bioproduction of COS has developed through various technical modifications to improve yield as well as the desired molecular size of COS, for example, the use of CDA for producing a fully defined structure of COS, the application of LPMOs in synergy with chitinolytic enzymes for depolymerization of crystalline chitin/chitosan, combination of physical or chemical treatment with enzyme application for COS preparation, and production of COS by using a dual reactor system. The limited availability and high cost of chitinases and chitosanases pose a challenge to scaling up COS bioproduction. In addition, the main distribution of industrial shrimp processing waste for feed comes to be an obstacle for the development of the COS industry in Indonesia.

Authors' Contributions Yusro Nuri Fawzya (ORCID No. 0000-0001-8581-7674) and Ekowati Chasanah (ORCID No. 000-0002-5974-0690) contribute equally as the main contributors in preparing the draft of manuscript. All authors read and approved the final manuscript.

Conflict of Interests The authors declare that there is no conflict of interests regarding the publication of this article.

References

Agustiar AA, Faturrohmah I, Sari BW, Isnaini NB, Puspita ID, Triyanto HA, Ustadi (2019) Increasing chitinase activity of Serratia marcescens PT-6 through optimization of medium composition. Squalen Bull Mar Fish Postharvest Biotechnol 14(3):113–120

Akeed Y, Atrash F, Naffaa W (2020) Partial purification and characterization of chitinase produced by Bacillus licheniformis B307. Heliyon 6(5):e03858. https://doi.org/10.1016/j.heliyon.2020.e03858

Aliabadi N, Aminzadeh S, Karkhane AA, Haghbeen K (2016) Thermostable chitinase from Cohnella sp. A01: isolation and product optimization. Brazilian J Microbiol 47(4):931–940. https://doi.org/10.1016/j.bjm.2016.07.009

Annamalai N, Rajeswari MV, Vijayalaksmi S, Balasubramania T (2011) Purification and characterization of chitinase from Alcaligenes faecalis AU02 by utilizing marine wastes and its antioxidant activity. Ann Microbiol 61:801–807. https://doi.org/10.1007/s13213-011-0198-5

Arnold ND, Brück WM, Garbe D, Brück TB (2020) Enzymatic modification of native chitin and conversion to specialty chemical products. Mar Drugs 18(2):1–27. https://doi.org/10.3390/md18020093

Basmal J, Prasetyo P, Farida F (2007) Pengaruh suhu eterifikasi terhadap kualitas dan kuantitas kitosan larut air ayang dibuat dari cangkang rajungan. Jurnal Pascapanen Dan Bioteknologi Kelautan Dan Perikanan 2(2):99–106

Bissaro B, Røhr ÅK, Müller G, Chylenski P, Skaugen M, Forsberg Z, Horn SJ, Vaaje-Kolstad G, Eijsink VGH (2017) Oxidative cleavage of polysaccharides by monocopper enzymes depends on H_2O_2. Nat Chem Biol 13(10):1123–1128. https://doi.org/10.1038/nchembio.2470

Cahyono E, Suptijah P, Wientarsih I (2014) Development of a pressurized hydrolysis method for producing glucosamine. Asian J Agric Food Sci 02(05):390–396

CAZy (2021) Welcome to the carbohydrate-active enZYmes database. http://www.cazy.org/

Charles-Rodriguez AV, Mauricio-benavides JE, Garca-Garcia Y, Aguilar CN, Rodriguez-Herrera R, Rodriguez J, Contreras-Esquivel JC (2008) Chitosanase production by a new bacterial sources. Res J Biol Sci 3(8):957–963

Chasanah E, Purwiyatno H, Witarto AB, Hwang JK, Suhartono MT (2006) Biochemical characteristics of chitosanase from the Indonesian Bacillus licheniformis MB-2. Mol Biotechnol 33(2):93–102. https://doi.org/10.1385/MB:33:2:93

Chasanah E, Zilda DS, Uria AR (2009) Screening and characterization of bacterial chitosanase from marine environment. J Coast Dev 12(2):64–72

Chasanah E, Patantis G, Zilda DS, Ali M, Risjani Y (2011) Purification and characterization of Aeromonas media KLU 11.16 chitosanase isolated from shrimp waste. J Coast Dev 15(1):104–113

Chasanah E, Ali M, Ilmi M (2012) Identification and partial characterization of crude extracellular enzymes from bacteria isolated from shrimp waste processing. Squalen Bull Mar Fish Postharvest Biotechnol 7(1):11. https://doi.org/10.15578/squalen.v7i1.11

Cui D, Yang J, Lu B, Shen H (2021) Efficient preparation of chitooligosaccharide with a potential chitosanase Csn-SH and its application for fungi disease protection. Front Microbiol 12(June):1–12. https://doi.org/10.3389/fmicb.2021.682829

Dai BA, Bhushan K, Rameez S, Gaur A, Malarvili G (2015) Novel chitinolytic potential of Achromobacter denitrificans isolated from fishery waste. IOSR J Pharm Biol Sci Ver III 10(4):2319–7676. https://doi.org/10.9790/3008-10433136

Dao T (2021) Shrimp Insights_ India, Indonesia, and Vietnam all poised for continued growth. https://www.seafoodsource.com/news/supply-trade/shrimp-insights-india-indonesia-and-vietnam-all-poised-for-continued-growth

Dliyauddin MOH, Ardyati TRI, Suharjono (2020) Evaluation of enzymatic activity and identification of potent proteolytic and chitinolytic bacteria isolated from crab shell waste. Biodiversitas 21(1):211–218. https://doi.org/10.13057/biodiv/d210127

Dukariya G, Kumar A (2020) Distribution and biotechnological applications of chitinase: a review. Int J Biochem Biophys 8(2):17–29. https://doi.org/10.13189/ijbb.2020.080201

Elawati NE, Pujiyanto S, Kusdiyantini E (2018) Karakteristik dan sifat kinetika enzim kitinase asal jamur entomopatogen Beauveria bassiana. Jurnal Bioteknologi Biosains Indonesia (JBBI) 5(1):1. https://doi.org/10.29122/jbbi.v5i1.2587

Farag AM, Abd-Elnabey HM, Ibrahim HAH, El-Shenawy M (2016) Purification, characterization and antimicrobial activity of chitinase from marine-derived *Aspergillus terreus*. Egypt J Aquat Res 42(2):185–192. https://doi.org/10.1016/j.ejar.2016.04.004

Fawzya YN, Novianty R, Wibowo S (2008) Effect of deacetylation and alkalination on the characteristics of carboxymethyl chitosan. J Fish Sci 1(1):64–75. https://doi.org/10.33369/pendipa.v1i1.2982

Fawzya YN, Pratitis A, Chasanah E (2009a) Karakterisasi enzim kitosanase dari isolat bakteri KPU 2123 dan aplikasinya untuk produksi oligomer kitosan. Jurnal Pasacapanen Dan Bioteknologi Kelautan Dan Perikanan 4(1):69–78

Fawzya YN, Sihotang MY, Syarmalina, Pratitis A (2009b) Produksi Kitooligosakarida menggunakan selulase dari *Trichoderma reesei* dan bioaktivitasnya sebagai antibakteri. Jurnal Pascapanen dan Bioteknologi Kelautan dan Perikanan 4(2):105. https://doi.org/10.15578/jpbkp.v4i2.442

Fawzya Y, Rahmawati A, Patantis G (2018) Physicochemical properties of chitooligosaccharide prepared by using chitosanase from *Stenotrophomonas maltophilia* KPU 2123. IOP Conf Ser: Earth Environ Sci 139(1). https://doi.org/10.1088/1755-1315/139/1/012045

Fenice M, Leuba J-L, Federici F (1998) Chitinolytic enzyme activity of Penicillium janthinellum P9 in bench-top reactor. J Ferment Bioeng 86(5):620–623

Forsberg Z, Sørlie M, Petrović D, Courtade G, Aachmann FL, Vaaje-Kolstad G, Bissaro B, Røhr ÅK, Eijsink VG (2019) Polysaccharide degradation by lytic polysaccharide monooxygenases. Curr Opin Struct Biol 59:54–64. https://doi.org/10.1016/j.sbi.2019.02.015

Ginting L, Poluan GG, Warouw V, Wullur S (2021) Screening and identification of sponge-associated chitinolytic bacteria by forming chitosan from Manado Bay, Indonesia. Pakistan J Biol Sci: PJBS 24(2):227–234

Hamer SN, Cord-Landwehr S, Biarnés X, Planas A, Waegeman H, Moerschbacher BM, Kolkenbrock S (2015) Enzymatic production of defined chitosan oligomers with a specific pattern of acetylation using a combination of chitin oligosaccharide deacetylases. Sci Rep 5:1–9. https://doi.org/10.1038/srep08716

Handayani R, Fabiola F (2018) Pembuatan N-Asetilglukosamin dari Limbah Udang (Penaeus monodon) dengan Bantuan Kapang Beauveria bassiana. Prosiding Simposium Nasional Kelautan Dan Perikanan V Universitas Hasanuddin, Makassar, pp 227–232

Hardoko, Mastuti TS, Puspasari D, Halim Y (2019) Utilization of crude intracellular chitinase enzyme from Providencia stuartii for glucosamine production from shrimp shells. 19(2):62–67

Hardoko, Josephine C, Handayani R, Halim Y (2020) Isolation, identification and chitinolytic index of bacteria from rotten tiger shrimp (Penaeus monodon) shells. AACL Bioflux 13(1):360–371

Haryogya AM, Ustadi (2020) Isolation and molecular identification of chitinolytic bacteria from ronto. E3S Web Conf 147:0–7. https://doi.org/10.1051/e3sconf/202014703030

Hekmat O, Tokuyasu K, Withers SG (2003) Subsite structure of the endo-type chitin deacetylase from a Deuteromycete, Colletotrichum lindemuthianum: an investigation using steady-state kinetic analysis and MS. Biochem J 374:369–380

Hembach L, Cord-Landwehr S, Moerschbacher BM (2017) Enzymatic production of all fourteen partially acetylated chitosan tetramers using different chitin deacetylases acting in forward or reverse mode. Sci Rep 7(1):1–11. https://doi.org/10.1038/s41598-017-17950-6

Hirano T, Sugiyama K, Sakaki Y, Hakamata W, Park S, Nishio T (2015) Structure-based analysis of domain function of chitin oligosaccharide deacetylase from Vibrio parahaemolyticus. FEBS Lett 589(1):145–151. https://doi.org/10.1016/j.febslet.2014.11.039

Jeon YJ, Kim SK (2000) Continuous production of chitooligosaccharides using a dual reactor system. Process Biochem 35(6):623–632. https://doi.org/10.1016/S0032-9592(99)00118-1

Jung W, Park R (2014) Bioproduction of chitooligosaccharides: present and perspective. Mar Drugs 5328–5356. https://doi.org/10.3390/md12105328

Kaczmarek MB, Struszczyk-Swita K, Li X, Szczęsna-Antczak M, Daroch M (2019) Enzymatic modifications of chitin, chitosan, and chitooligosaccharides. Front Bioeng Biotechnol 7(SEP). https://doi.org/10.3389/fbioe.2019.00243

Kafetzopoulos D, Martinou A, Bouriotis V (1993) Bioconversion of chitin to chitosan : Purification and characterization of chitin deacetylase from Mucor rouxii. Proc Natl Acad Sci 90(April 1993):2564–2568

Katiyar D, Singh B, Lall AM, Haldar C (2011) Efficacy of chitooligosaccharides for the management of diabetes in alloxan induced mice: a correlative study with antihyperlipidemic and antioxidative activity. Eur J Pharm Sci 44(4):534–543. https://doi.org/10.1016/j.ejps.2011.09.015

Kidibule PE, Santos-Moriano P, Jiménez-Ortega E, Ramírez-Escudero M, Limón MC, Remacha M, Plou FJ, Sanz-Aparicio J, Fernández-Lobato M (2018) Use of chitin and chitosan to produce new chitooligosaccharides by chitinase Chit42: enzymatic activity and structural basis of protein specificity. Microb Cell Fact 17(1):1–13. https://doi.org/10.1186/s12934-018-0895-x

Krolicka M, Hinz SWA, Koetsier MJ, Joosten R, Eggink G, Van Den Broek LAM, Boeriu CG (2018) Chitinase Chi1 from Myceliophthora thermophila C1, a thermostable enzyme for chitin and chitosan depolymerization. J Agric Food Chem 66(7):1658–1669. https://doi.org/10.1021/acs.jafc.7b04032

Kumar M, Madhuprakash J, Balan V, Singh AK, Vivekanand V, Conceptulization NP (2021) Chemoenzymatic production of chitooligosaccharides employing ionic liquids and Thermomyces lanuginosus chitinase. Bioresour Technol 125399

Laokuldilok T, Potivas T, Kanha N, Surawang S, Seesuriyachan P, Wangtueai S, Phimolsiripol Y, Regenstein JM (2017) Physicochemical, antioxidant, and antimicrobial properties of chitooligosaccharides produced using three different enzyme treatments. Food Biosci 18:28–33. https://doi.org/10.1016/j.fbio.2017.03.004

Lee DX, Xia WS, Zhang JL (2008) Enzymatic preparation of chitooligosaccharides by commercial lipase. Food Chem 111(2):291–295. https://doi.org/10.1016/j.foodchem.2008.03.054

Li J, Wang D, Chang SC, Liang PH, Srivastava V, Guu SY, Shie JJ, Khoo KH, Bulone V, Hsieh YSY (2021) Production of structurally defined chito-oligosaccharides with a single N-acetylation at their reducing end using a newly discovered chitinase from Paenibacillus pabuli. J Agric Food Chem 69(11):3371–3379. https://doi.org/10.1021/acs.jafc.0c06804

Liang S, Sun Y, Dai X (2018) A review of the preparation, analysis and biological functions of chitooligosaccharide. Int J Mol Sci 19(8). https://doi.org/10.3390/ijms19082197

Liu Z, Gay LM, Tuveng TR, Agger JW, Westereng B, Mathiesen G, Horn SJ, Vaaje-Kolstad G, Van Aalten DMF, Eijsink VGH (2017) Structure and function of a broad-specificity chitin deacetylase from Aspergillus nidulans FGSC A4. Sci Rep 7:1–12. https://doi.org/10.1038/s41598-017-020 43-1

Lodhi G, Kim YS, Hwang JW, Kim SK, Jeon YJ, Je JY, Ahn CB, Moon SH, Jeon BT, Park PJ (2014) Chitooligosaccharide and its derivatives: preparation and biological applications. Biomed Res Int 2014. https://doi.org/10.1155/2014/654913

Loose JSM, Forsberg Z, Fraaije MW, Eijsink VGH, Vaaje-Kolstad G (2014) A rapid quantitative activity assay shows that the Vibrio cholerae colonization factor GbpA is an active lytic polysaccharide monooxygenase. FEBS Lett 588(18):3435–3440. https://doi.org/10.1016/j.febslet.2014.07.036

Mahata M, Shinya S, Masaki E, Yamamoto T, Ohnuma T, Brzezinski R, Mazumder TK, Yamashita K, Narihiro K, Fukamizo T (2014) Production of chitooligosaccharides from Rhizopus oligosporus NRRL2710 cells by chitosanase digestion. Carbohydr Res 383:27–33. https://doi.org/10.1016/j.carres.2013.06.002

Martosuyono P, Pratitis A, Prasetya A, Prabawati EK (2014) Desalination of chitooligosaccharides using gel filtration and utrafiltration. Squalen Bull Mar Fish Postharvest Biotechnol 9(3):127–136. https://doi.org/10.15578/squalen.v9i3.110

Masri M, Sukmawaty E, Aditia L (2021) Novel chitinolytic bacteria from the shrimp shell processing waste. Biodiversitas 22(5):2672–2681. https://doi.org/10.13057/biodiv/d220527

Matsumiya M, Arakane Y, Haga A, Muthukrishnan S, Kramer KJ (2006) Substrate specificity of chitinases from two species of fish, greenling, Hexagrammos otakii, and common Mackerel, Scomber japonicus, and the insect, tobacco hornworm, Manduca sexta. Biosci Biotechnol Biochem 70(4):971–979. https://doi.org/10.1271/bbb.70.971

Miguez N, Kidibule P, Santos-Moriano P, Ballesteros AO, Fernandez-Lobato M, Plou FJ (2021) Enzymatic synthesis and characterization of different families of chitooligosaccharides and their bioactive properties. Appl Sci (Switzerland) 11(7). https://doi.org/10.3390/app11073212

Mubarik NR, Mahagiani I, Anindyaputri A, Agatis J (2010) Chitinolytic bacteria isolated from chili rhizosphere: chitinase characterization and its application as biocontrol for whitefly (Bemisia tabaci Genn.). Am J Agric Biol Sci 5(4):430–435

Mutahir Z, Mekasha S, Loose JSM, Abbas F, Vaaje-Kolstad G, Eijsink VGH, Forsberg Z (2018) Characterization and synergistic action of a tetra-modular lytic polysaccharide monooxygenase from Bacillus cereus. FEBS Lett 592(15):2562–2571. https://doi.org/10.1002/1873-3468.13189

Naqvi S, Cord-landwehr S, Singh R, Bernard F, Kolkenbrock S, Gueddari NEEN, Moerschbacher BM (2016) A recombinant fungal chitin deacetylase produces fully defined chitosan oligomers with novel patterns of acetylation. Appl Environ Microbiol 82(22):6645–6655. https://doi.org/10.1128/AEM.01961-16.Editor

Noviendri D, Chasanah E, Fawzya YN (2006) Karakterisasi enzim kitinase yang diproduksi oleh isolat bakteri JB4 dari terasi. Jurnal Pascapanen Dan Bioteknologi Kelautan Dan Perikanan 1(2):85–92

Nuryadin DF, Setyaningsih I, Hardiningtyas SD (2020) Depolimerisasi kitosan menggunakan sinar ultra violet dan katalis asam khlorida (Depolymerization of chitosan using ultraviolet light and chloride acid as catalyst). Jurnal Pengolahan Hasil Perikanan Indonesia. 23:412–422

Oktavia DA, Wibowo S, Fawzya YN (2005) Pengaruh jumlah monokhloroasetat terhadap karakteristik karboksimetil kitosan dari kitosan cangkang dan kaki rajungan. Jurnal Penelitian Perikanan Indonesia 11(4):79–88

Pagnoncelli MGB, de Araújo NK, da Silva NMP, de Assis CF, Rodrigues S, de Macedo GR (2010) Chitosanase production by Paenibacillus ehimensis and its application for chitosan hydrolysis. Braz Arch Biol Technol 53(6):1461–1468. https://doi.org/10.1590/S1516-89132010000600023

Pardosi L, Suryanto D, Siregar AZ (2018) Isolation of chitinolytic bacteria from two lizard digestive tract and characterization of their crude chitinase. Int J ChemTech Res 11(07):17–28. https://doi.org/10.20902/ijctr.2018.110703

Pascual S, Planas A (2018) Screening assay for directed evolution of chitin deacetylases: application to vibrio cholerae deacetylase mutant libraries for engineered specificity. Anal Chem 90:10654–10658. https://doi.org/10.1021/acs.analchem.8b02729

Pasqualetti M, Barghini P, Giovannini V, Fenice M (2019) High production of chitinolytic activity in halophilic conditions by a new marine strain of Clonostachys rosea. Molecules 24:1–13

Patantis G, Zilda D, Fawzya Y, Chasanah E (2020) Purification of chitosanase from Stenotrophomonas maltophilia KPU 2123 and Micromonospora sp . T5a1 for chitooligosacharide production. IOP Conf Ser: Earth Environ Sci 404 012078:1–8. https://doi.org/10.1088/1755-1315/404/1/012078

Patil RS, Ghormade V, Deshpande MV (2000) Chitinolytic enzymes: an exploration. Enzyme Microb Technol 26(7):473–483. https://doi.org/10.1016/S0141-0229(00)00134-4

Puspita ID, Wardani A, Puspitasari ROA, Nugraheni PS, Putra MMP, Pudjiraharti S, Ustadi (2017) Occurrence of chitinolytic bacteria in shrimp rusip and measurement of their chitin-degrading enzyme activities. Biodiversitas 18(3):1275–1281. https://doi.org/10.13057/biodiv/d180354

Putra P, Husni A, Puspita ID (2016) Characterization and application of N, O-carboxy methyl chitosan produced at different tempereratureof etherification. Int J Pharm Clin Res 8(11):1493–1498

Rahayu S, Tanuwidjaja F, Rukayadi Y, Suwanto A, Thenawidjaja Suhartono M, Hwang JK, Pyun YR (2004) Study of thermostable chitinase enzymes from Indonesian Bacillus K29–14. J Microbiol Biotechnol 14(4):647–652

Research and Market R (2021) Chitosan market size, trends & growth, by grade, by source, by application, and by region, forecast to 2028. https://www.researchandmarkets.com/reports/5312383/chitosan-market-size-trends-and-growth-by. Accessed 26 Aug 2021

Rokhati N, Pramudono B, Istirokhatun T, Susanto H (2018) Microwave irradiation-assisted chitosan hydrolysis using cellulase enzyme. Bull Chem React Eng Catal 13(3):466–474. https://doi.org/10.9767/bcrec.13.3.2378.466-474

Rosmiati R, Kusharto CM, Anwar F, Suptijah P (2016) Physicochemical properties of silkworm pupae shell (Bombyx mori L.) glucosamine hydrochloride. 53–65

Salas-ovilla R, Gálvez-López D, Vázquez-Ovando A, Salvador-Figueroa M, Rosas-Quijano R (2019) Isolation and identification of marine strains of Stenotrophomona maltophilia with high chitinolytic activity. Peer J 1–12. https://doi.org/10.7717/peerj.6102

Santos-Moriano P, Woodley JM, Plou FJ (2016) Continuous production of chitooligosaccharides by an immobilized enzyme in a dual-reactor system. J Mol Catal B Enzym 133:211–217. https://doi.org/10.1016/j.molcatb.2016.09.001

Santos-Moriano P, Kidibule P, Míguez N, Fernández-Arrojo L, Ballesteros AO, Fernández-Lobato M, Plou FJ (2019) Tailored enzymatic synthesis of chitooligosaccharides with different deacetylation degrees and their anti-inflammatory activity. Catalysts 9(5). https://doi.org/10.3390/catal9050405

Sarni, Natsir H, Dali S (2016) Chitosan oligomer production from waste tiger shrimp (Penaeus monodon) using enzymes chitosanase of bacterial isolates Klebsiella sp. Ind J Chem Res 3:283–289

Savitri E, Rosyadi A (2015) Degradation of chitosan by hydrothermal process in the presence of sonication pre-treatment with supercritical CO_2 as pressurized fluid. 212–219. https://doi.org/10.1002/masy.201550329

Schmitz C, Auza LG, Koberidze D, Rasche S, Fischer R, Bortesi L (2019) Conversion of chitin to defined chitosan oligomers: current status and future prospects. Mar Drugs 17(8):1–22. https://doi.org/10.3390/md17080452

Shinya S, Fukamizo T (2017) Interaction between chitosan and its related enzymes: a review. Int J Biol Macromol 104:1422–1435

Singh AK, Mehta G, Chhatpar HS (2009) Optimization of medium constituents for improved chitinase production by Paenibacillus sp. D1 using statistical approach. Lett Appl Microbiol 49:708–714. https://doi.org/10.1111/j.1472-765X.2009.02731.x

Song JY, Alnaeeli M, Park JK (2014) Efficient digestion of chitosan using chitosanase immobilized on silica-gel for the production of multisize chitooligosaccharides. Process Biochem 49(12):2107–2113. https://doi.org/10.1016/j.procbio.2014.09.003

Souza CP, Burbano-Rosero EM, Almeida BC, Martins GG, Albertini LS, Rivera ING (2009) Culture medium for isolating chitinolytic bacteria from seawater and plankton. World J Microbiol Biotechnol 25(11):2079–2082. https://doi.org/10.1007/s11274-009-0098-z

Subramanian K, Sadaiappan B, Aruni W (2020) Bioconversion of chitin and concomitant production of chitinase and N—acetylglucosamine by novel Achromobacter xylosoxidans isolated from shrimp waste disposal area. Sci Rep 1–14. https://doi.org/10.1038/s41598-020-68772-y

Sudianto, Suseno S, Suptijah P (2020) Optimasi Produksi Kitosan Larut Air Menggunakan Metode Hidrolisis Bertekanan (Optimatization of Production Water Soluble Chitosan Using Pressurized Hidrolysis Method). Jurnal Pengolahan Hasil Perikanan Indonesia 23:441–446

Sukwattanasinitt M, Zhu H, Sashiwa H, Aiba SI (2002) Utilization of commercial non-chitinase enzymes from fungi for preparation of 2-acetamido-2-deoxy-D-glucose from β-chitin. Carbohydr Res 337(2):133–137. https://doi.org/10.1016/S0008-6215(01)00298-1

Sulistijowati R, Sudin, Harmain RM (2021) Chitinase activity potential and identification of chitinolytic bacteria isolated of swimmer crab's cell. Int J Agric Biol Eng 14(3):228–231. https://doi.org/10.25165/j.ijabe.20211403.5273

Suptijah P, Ibrahim B, Ernawati E (2014) Pemanfaatan Limbah Krustasea Dalam Pembuatan Glukosamin Hidroklorida Dengan Metode Autoklaf. Jurnal Teknologi Perikanan Dan Kelautan 5(2):173–181. https://doi.org/10.24319/jtpk.5.171-179

Suryadi Y, Priyatno TP, Susilowati DN, Samudra IM, Yudhistira N, Purwakusumah E (2013) Isolasi dan Karakterisasi Kitinase asal Bacillus cereus 11 UJ (Isolation and chitinase characterization of Bacillus cereus 11 UJ). Jurnal Biologi Indonesia 9(1):51–62

Takiguchi Y, Shimahara K (1988) N, N-Diacetylchitobiose production from chitin by Vibrio anguillarum strain E-383a. Lett Appl Microbiol 6(6):129–131. https://doi.org/10.1111/j.1472-765X.1988.tb01231.x

Tanasale MFJDP, Bijang CM, Rumpakwara E (2019) Preparation of chitosan with various molecular weight and its effect on depolymerization of chitosan with hydrogen peroxide using conventional technique. 12(1):112–120

Toharisman A, Suhartono MT (2007) Partial purification and characterization of chitin deacetylase produced by Bacillus thermoleovorans LW-4-11. Jurnal Biologi Indonesia 4(4):215–223

Toharisman A, Suhartono MT, Spindler-Barth M, Hwang JK, Pyun YR (2005) Purification and characterization of a thermostable chitinase from Bacillus licheniformis Mb-2. World J Microbiol Biotechnol 21(5):733–738. https://doi.org/10.1007/s11274-004-4797-1

Tremblay H, Yamaguchi T, Fukamizo T, Brzezinski R (2001) Mechanism of chitosanase-oligosaccharide interaction: subsite structure of Streptomyces sp. N174 chitosanase and the role of Asp57 carboxylate. J Biochem 130(5):679–686. https://doi.org/10.1093/oxfordjournals.jbchem.a003034

Vaaje-Kolstad G, Westereng B, Horn SJ, Liu Z, Zhai H, Sørlie M, Eijsink VGH (2010) An oxidative enzyme boosting the enzymatic conversion of recalcitrant polysaccharides. Science 330(6001):219–222. https://doi.org/10.1126/science.1192231

Wahyuntari B, Junianto, Setyahadi S (2011) Process design of microbiological chitin extraction process design of microbiological chitin extraction. Microbiology 5(1):39–45. https://doi.org/10.5454/mi.5.1.7

Weikert T, Niehues A, Moerschbacher BM, Cord-landwehr S, Hellmann MJ (2017) Reassessment of chitosanase substrate specificities and classification. Nat Commun 1–11. https://doi.org/10.1038/s41467-017-01667-1

Wibowo RH, Sipriyadi S, Mubarik NR, Rusmana I (2020) Soil chitinolytic bacteria from Jambi Province to produce antifungal of plant pathogens. Mangifera Edu 5(1):26–37. https://doi.org/10.31943/mangiferaedu.v5i1.95

Wu GJ, Tsai GJ (2004) Cellulase degradation of shrimp chitosan for the preparation of a water-soluble hydrolysate with immunoactivity. Fish Sci 70(6):1113–1120. https://doi.org/10.1111/j.1444-2906.2004.00912.x

Xie H, Jia Z, Huang J, Zhang C (2011) Preparation of low molecular weight chitosan by complex enzymes hydrolysis. Int J Chem 3(2):47–53. https://doi.org/10.5539/ijc.v3n2p180

Xu T, Qi M, Liu H, Cao D, Xu C, Wang L, Qi B (2020) Chitin degradation potential and whole-genome sequence of Streptomyces diastaticus strain CS1801. AMB Express 10(1). https://doi.org/10.1186/s13568-020-0963-6

Yan N, Chen X (2015) Don't waste seafood waste. Nature 524:155–157

Yuli PE, Suhartono MT, Rukayadi Y, Hwang JK, Pyun YR (2004) Characteristics of thermostable chitinase enzymes from the indonesian Bacillus sp. 13.26. Enzyme Microbial Technol 35(2–3):147–153

Zhou Y, Chen X, Li X, Han Y, Wang Y, Yao R, Li S (2019) Purification and characterization of a new cold-adapted and thermo-tolerant chitosanase from marine bacterium Pseudoalteromonas sp. SY39. Molecules 24(1). https://doi.org/10.3390/molecules24010183

Antimicrobial Activity of Chitooligosaccharides

Krishika Sambyal, Palak Sharma, and Rahul Vikram Singh

1 Introduction

Chitin (Fig. 1) is a natural biopolymer polysaccharide distributed among the shells of sea arthropods. Every year approximately six to eight million tons of shell waste are produced by seafood industry due to its improper discard in environment. However, chitin is extracted at a commercial scale due to its vast applications aiding in reduction of marine waste (Yan and Chen 2015; Singh et al. 2021). The partial deacetylation of chitin generates an N-acetyl-D-glucosamine (GlcNAc) and D-glucosamine (GlcN) copolymer known as chitosan (Šimat et al. 2020) with low solubility in universal solution which makes its application difficult in the food and biomedical sectors. Moreover, high molecular mass, the pattern of acetylation and deacetylation degree are the few variables that acts as barriers for its extensive use in the biomedical field (Li and Zhuang 2020). Few mechanisms such as perturbation of ion transport on the microorganisms have been suggested for the antimicrobial property of chitosan (Yin et al. 2020).

But to overcome these boundaries, hydrolysis of chitosan has been performed yielding small, soluble chitooligosaccharides (COSs) with higher solubility and low viscosity as compared to chitosan along with several biological properties such as antimicrobial activity against a broad spectrum of food-borne microbes acknowledging it as a natural food preservative (Devlieghere et al. 2004; Silva et al. 2021). COSs (Fig. 2) are the chitosans with a degree of polymerization less than 20 and an average molecular weight less than 3.9 kDa. They can also be called as chitosan oligomers or chitooligomers (Lodhi et al. 2014). Due to their hydrophilic nature

K. Sambyal
University Institute of Biotechnology, Chandigarh University, Gharuan, Punjab, India

P. Sharma
Department of Biotechnology, Shri Mata Vaishno Devi University, Jammu & Kashmir, India

R. V. Singh (✉)
Academy of Scientific and Innovative Research (AcSIR), Ghaziabad 201002, India

S.-K. Kim (ed.), *Chitooligosaccharides*,
https://doi.org/10.1007/978-3-030-92806-3_18

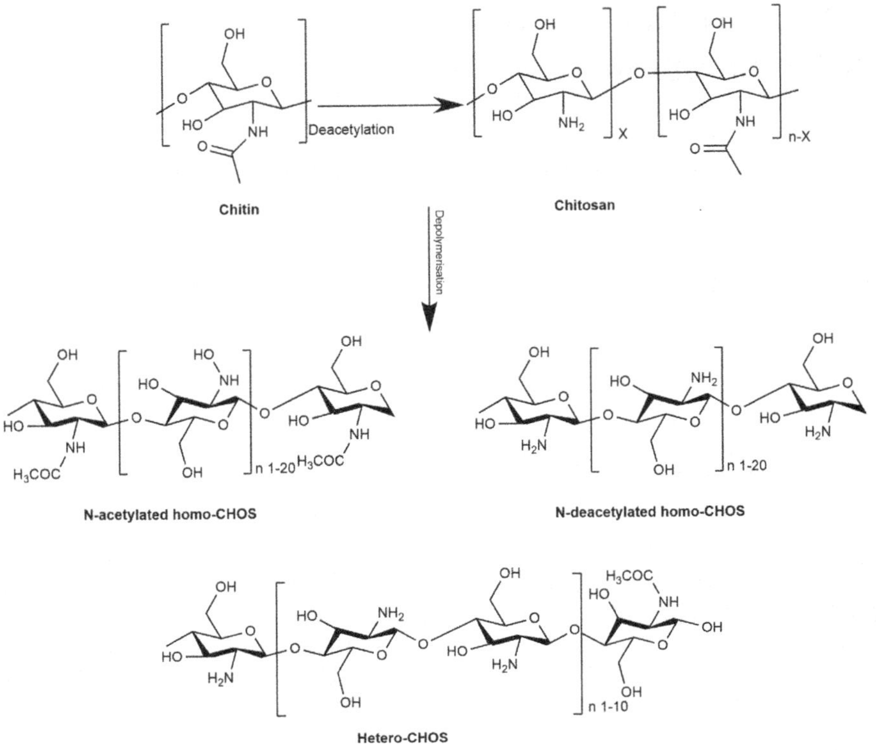

Fig. 1 Chemical structure of chitin and its derivatives (Jafari et al. 2020)

and low molecular weight, they are of greater demand than their precursor molecule chitosan (Liaqat and Eltem 2018). The shorter chain length and free amino groups in D-glucosamine units in COSs can be considered as a reason for their water solubility and lower viscosity (Bahrke 2008). In contrast to chitin and chitosan, COS has properties like cell membrane penetrability, easy absorption other than water solubility which makes it an extremely valuable product (Liaqat and Eltem 2018). Therefore, the enzyme-mediated production of chitooligosaccharides has been done from chitosan. The glucosamine and acetylated glucosamine rings in COS are connected with each other by β 1,4-glycosidic bond (El-Sayed et al. 2017). It is known that the principle characteristic factors directly affecting the physicochemical and biological activity of COS are the degree of polymerization and molecular weight. COS with a degree of polymerization six or higher and molecular weight low are more biologically active than the COS with low degree of polymerization and high molecular weight (Prashanth and Tharanathan 2007).

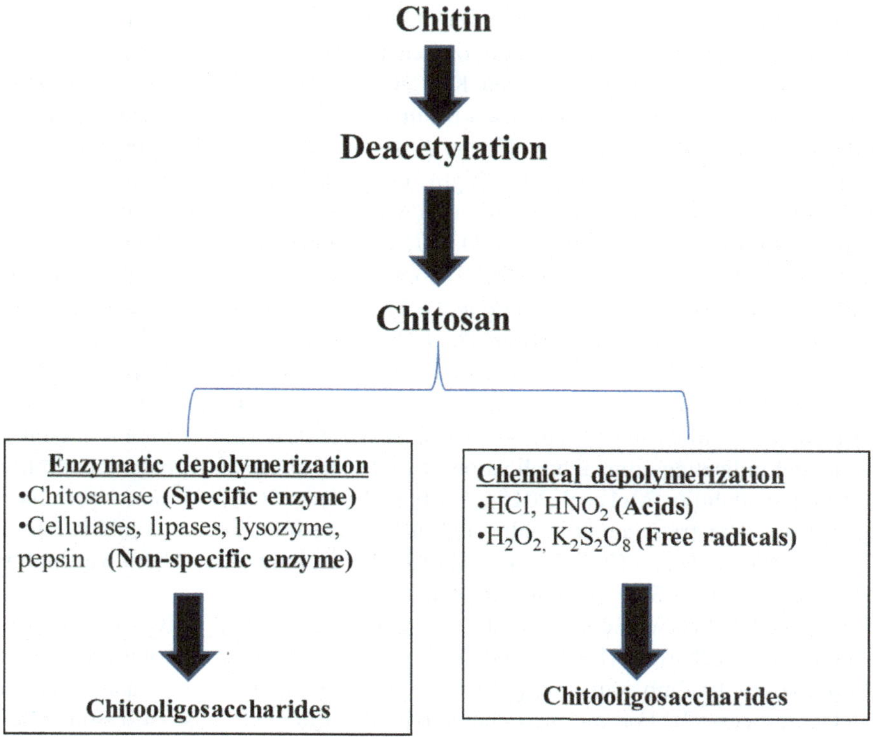

Fig. 2 Production of chitooligosaccharides by depolymerization of chitin

2 Antimicrobial Activity of Chitooligosaccharides

In 1979, the broad-spectrum antimicrobial activity of chitosan and its derivatives was reported by Allan and Hadwiger for the first-time (1979). Since then, chitosan and COS have been used by industries for their antimicrobial properties. To define the antimicrobial potential of COS and its utilization in agriculture, medical and food sectors, the use of well-characterized COS is necessary for in vitro as well as in vivo studies (Liaqat and Eltem 2018). Due to the antimicrobial resistance in various microbes, there is a need for new compounds with enhanced activity presenting options for the treatment of several bacterial, yeast and parasitic diseases (El-Sayed et al. 2017; Bloom and Cadarette 2019). The degree of polymerization, degree of N-acetylation and physicochemical characteristics, microorganism type are few factors responsible for the antimicrobial activities of COS (Chung et al. 2004). Moreover, the existence of the primary amino groups in the structure of COS is also supposed to be the reason for its antimicrobial property for now, since the molecular reason is not known yet. The alteration in cell membrane permeability by COS causes the death of the microbial cell due to the release of cell constituents and uncontrolled entry of materials into the cell from outer environment (Choi et al.

2001; Mei et al. 2015). Through the use of UF membrane in association with enzymatic reactor, three different chitooligosaccharides fractions were prepared based on their molecular weight (Jeon and Kim 2000). These were used to determine the antimicrobial activity against Gram-positive, negative and lactic acid bacteria in order to explore the effects of molecular mass on the microbial population (Silva et al. 2021). COS was observed to be highly effective for *S. typhi* that was inhibited 90% (approximately) at 0.1% concentration with 0.12% MIC (Minimum Inhibitory Concentration). For *E. coli* and *E. coli* O-157, high molecular weight COS possessed the highest antibacterial effect with MIC 0.06 and 0.12%, respectively. All COSs indicated a suppressed growth for *Pseudomonas aeruginosa* with antimicrobial activity 50% or lower at 0.1% concentration. The COS with higher molecular weight (>10 KDa) showed effective activity against pathogens rather than non-pathogens (Jeon et al. 2001). An enzymatically produced, well-characterized COS has been reported for its in vivo anti-dermatophytic activity against *Trichophyton rubrum* in a guinea pig model indicating clinical application of COS due to its safety, biocompatibility and biodegradability (Mei et al. 2015). Antifungal effect of COS was also determined against the *Botrytris cinerea* in vitro and in vivo condition with a combination of synthetic fungicide and COS. Results rebelled that combination of both was highly effective against the *Botrytris cinerea* (Rahman et al. 2015).

El-Sayed and co-workers reported the in vitro antimicrobial activity of four sets of chitooligosaccharides (I, II, III and IV) were prepared using immobilized pepper chitosanase with a molecular weight greater than 100, 100–10, 10–1 and less than 1 KDa, respectively. For antimicrobial activity determination, disc diffusion plates method (50 μg/ml) was used to detect in vitro antimicrobial activity against the pathogens namely *B. cereus, B. subtilis, Staphylococcus aureus, Pseudomonas aeruginosa, Candida albicans* and *Saccharomyces chevalieri* in comparison to chitosan. Results showed that immobilized pepper chitosanase can be helpful for the industrial production of antimicrobial chitooligosaccharides from chitosan (El-Sayed et al. 2017). The combination of COS in chitosan films was used to improve the antimicrobial activity against the pathogens such as *Lactobacillus plantarum, S. aureus, B. cereus Serratia liquefaciens* and *E. coli* (Fernández-de Castro et al. 2016). This serves as a proof of the fruitful use of COS in food packaging films due to their antimicrobial activities. The white-leg shrimp shells have also been used for the extraction of chitosan and COS through chemical treatment to determine their antibacterial and antifungal activities against several bacterial and fungal species (spoilage and pathogenic microbes). COS was highly soluble in water than chitosan due to its low molecular weight and high degree of deacetylation (13 kDa and 54.83%). The marked inhibitory activity was shown by both COS and chitosan against food-borne pathogenic, spoilage bacterial and fungal strains during antimicrobial assays. Higher inhibitory effect or antimicrobial effect have been observed against *Bacillus cereus* and *Rhizopus* sp. in bread upon substitution of flour by COS (1/100 g total weight basis) as compared to use of chitosan; along with delay in the fruity odor generally present in bread (Rakkhumkaew and Pengsuk 2018). A study was conducted to determine the antimicrobial activity of COS in compression of broad-spectrum antibiotic flomox and kluacid. Four groups of COS (1, 2, 3 and 4) were fractionated on the basis of their molecular weights by ultrafiltration. MIC was determined against *B. cereus*, *P. aeruginosa* and *C. albicans*. The inhibitory activities of COS 1 and 3 were highly effective like flomox and kluacid against *B. Cereus*. COS 4 (0.11 mg/ml concentration, less than 1.0 KDa) completely inhibited the precise growth rate of *C. albicans* and was stronger than kluacid (0.42 mg/ml). COS 3 (1–10 KDa) had 1.67 mg/ml MIC against *P. aeruginosa*. These outcomes suggest that COS can play a key role in antimicrobial activity (El-Sayed et al. 2019).

The squid processing industry generally discards squid (*Loligo formosana*) pen that is a great source of β-chitin. This β-chitin can be easily deacetylated to chitosan due to its higher reactivity toward various solvents and loose structure, further generating COS upon hydrolysis using non-specific enzymes (Singh et al. 2019a). In presence of reactive groups like amino, hydroxyl groups and smaller size, COS has more antioxidant and antimicrobial activity than chitosan (Lodhi et al. 2014). In another study, COS–EGCG (epigallocatechin-3-gallate) conjugates were prepared to determine the antimicrobial potential against *E. coli* and *Listeria monocytogenes*. The MIC values for *L. monocytogenes* (Gram-positive) were observed to be higher for both COS and C1-E0.5-conjugate as compared to *E. coli* due to its thick cell wall, therefore more tolerant against antimicrobial agents (Singh et al. 2020). The antibacterial potential of substances can be determined by MBC/MIC ratio (Minimum Bactericidal Concentration/Minimum Inhibitory Concentration) (Singh et al. 2019b). Antimicrobial agent with MBC/MIC ratio =/<2 is bacteriostatic and MBC/MIC ratio >2 is bactericidal. For COS, the MBC/MIC ratios against *L. monocytogenes* and *E. coli* were 1 and 2, respectively. For C1-E0.5-conjugate, MBC/MIC ratios were 1 and 5 toward *L. monocytogenes* and *E. coli,* respectively, suggesting COS activity as a bactericidal agent for both bacteria, whereas C1-E0.5-conjugate as bactericidal and bacteriostatic agent for *E. coil* and *L. monocytogenes*, respectively (Singh et al. 2020). COSs antimicrobial activity against several bacteria and yeasts such as *E. coli, P. aeruginosa, K. pneumonia, S. aureus, S. epidermidis, E. faecalis* and *C. albicans* was determined using microdilution method in Mueller Hinton broth. A higher antimicrobial effect upon Gram-positive bacteria and *Candida* yeast was exhibited by COS. The MIC for *S. epidermidis* and *C. albicans* were less than 0.25 and 0.5 mg/ml, respectively (Silva et al. 2021). Therefore, the broader biological activity of COSs represents them as active agents for use in various industrial sectors.

3 Future Prospects

Presently, there is a need for novel compounds in the pharmaceutical industry showing potent biological activities. Antimicrobial resistance has created an emergency of novel antibiotics; therefore, COS-mediated derivatives can be an alternate of novel antimicrobial agents against various microbial-mediated diseases. The research on biological activities of COS will provide a new emerging area for emerging biotechnological industries. Moreover, entire knowledge regarding the molecular study of biological activities for COS is not available. Therefore, there is a great need for prospective research centered on the molecular functioning so that complete biochemical functioning of COS could be known.

4 Conclusion

COS has a vast range of biological properties and significant potential for numerous industrial applications. Upon citing the previous literature on the antimicrobial activity of chitooligosaccharide, it's utilization as a potential antimicrobial agent against several pathogens like bacteria, yeast and fungi have been summarized. There is a great need for well characterization of COS for utilizing it as antimicrobial agent in food, medical and agriculture sectors.

References

Allan CR, Hadwiger LA (1979) The fungicidal effect of chitosan on fungi of varying cell wall composition. Exp Mycol 3(3):285–287

Bahrke S (2008) Mass spectrometric analysis of chitooligosaccharides and their interaction with proteins. Doctoral dissertation, PhD thesis, Universität Potsdam, Potsdam, Germany

Bloom DE, Cadarette D (2019) Infectious disease threats in the twenty-first century: strengthening the global response. Front Immunol 10:549

Choi BK, Kim KY, Yoo YJ, Oh SJ, Choi JH, Kim CY (2001) In vitro antimicrobial activity of a chitooligosaccharide mixture against Actinobacillus actinomycetemcomitans and Streptococcus mutans. Int J Antimicrob Agents 18(6):553–557

Chung YC, Su YP, Chen CC, Jia G, Wang HL, Wu JG, Lin JG (2004) Relationship between antibacterial activity of chitosan and surface characteristics of cell wall. Acta Pharmacol Sin 25(7):932–936

Devlieghere F, Vermeulen A, Debevere J (2004) Chitosan: antimicrobial activity, interactions with food components and applicability as a coating on fruit and vegetables. Food Microbiol 21(6):703–714

El-Sayed ST, Ali AM, El-Sayed EM, Shousha WG, Omar NI (2017) Characterization and potential antimicrobial effect of novel chitooligosaccharides against pathogenic microorganisms. J Appl Pharm Sci 7:6–12

El-Sayed ST, Ali AM, Omar NI (2019) A comparative evaluation of antimicrobial activity of chitooligosaccharides with broad spectrum antibiotics on growth of some pathogenic microorganisms. Biocatal Agric Biotechnol 22:101382

Fernández-de Castro L, Mengíbar M, Sánchez Á, Arroyo L, Villarán MC, de Apodaca ED, Heras Á (2016) Films of chitosan and chitosan-oligosaccharide neutralized and thermally treated: effects on its antibacterial and other activities. LWT 73:368–374

Jafari H, Bernaerts KV, Dodi G, Shavandi A (2020) Chitooligosaccharides for wound healing biomaterials engineering. Mater Sci Eng C 111266

Jeon YJ, Kim SK (2000) Continuous production of chitooligosaccharides using a dual reactor system. Process Biochem 35(6):623–632

Jeon YJ, Park PJ, Kim SK (2001) Antimicrobial effect of chitooligosaccharides produced by bioreactor. Carbohydr Polym 44(1):71–76

Li J, Zhuang S (2020) Antibacterial activity of chitosan and its derivatives and their interaction mechanism with bacteria: current state and perspectives. Eur Polym J 138:109984

Liaqat F, Eltem R (2018) Chitooligosaccharides and their biological activities: a comprehensive review. Carbohydr Polym 184:243–259

Lodhi G, Kim YS, Hwang JW, Kim SK, Jeon YJ, Je JY, Ahn CB, Moon SH, Jeon BT, Park PJ (2014) Chitooligosaccharide and its derivatives: preparation and biological applications. Biomed Res Int

Mei YX, Dai XY, Yang W, Xu XW, Liang YX (2015) Antifungal activity of chitooligosaccharides against the dermatophyte Trichophyton rubrum. Int J Biol Macromol 77:330–335

Prashanth KH, Tharanathan RN (2007) Chitin/chitosan: modifications and their unlimited application potential—an overview. Trends Food Sci Technol 18(3):117–131

Rahman MH, Hjeljord LG, Aam BB, Sørlie M, Tronsmo A (2015) Antifungal effect of chito-oligosaccharides with different degrees of polymerization. Eur J Plant Pathol 141(1):147–158

Rakkhumkaew N, Pengsuk C (2018) Chitosan and chitooligosaccharides from shrimp shell waste: characterization, antimicrobial and shelf life extension in bread. Food Sci Biotechnol 27(4):1201–1208

Silva NS, Araújo NK, Daniele-Silva A, Oliveira JW, Medeiros JM, Araújo RM, Ferreira LD, Rocha HA, Silva-Junior AA, Silva MS, Fernandes-Pedrosa MD (2021) Antimicrobial activity of chitosan oligosaccharides with special attention to antiparasitic potential. Mar Drugs 19(2):110

Šimat V, Elabed N, Kulawik P, Ceylan Z, Jamroz E, Yazgan H, Čagalj M, Regenstein JM, Özogul F (2020) Recent advances in marine-based nutraceuticals and their health benefits. Mar Drugs 18(12):627

Singh A, Benjakul S, Prodpran T (2019a) Chitooligosaccharides from squid pen prepared using different enzymes: characteristics and the effect on quality of surimi gel during refrigerated storage. Food Prod Process Nutr 1(1):1

Singh A, Benjakul S, Prodpran T (2019b) Ultrasound-assisted extraction of chitosan from squid pen: molecular characterization and fat binding capacity. J Food Sci 84(2):224–234

Singh A, Benjakul S, Huda N, Xu C, Wu P (2020) Preparation and characterization of squid pen chitooligosaccharide–epigallocatechin gallate conjugates and their antioxidant and antimicrobial activities. RSC Adv 10(55):33196–33204

Singh RV, Sambyal K, Negi A, Sonwani S, Mahajan R (2021) Chitinases production: a robust enzyme and its industrial applications. Biocatal Biotransform 4;39(3):161–89

Yan N, Chen X (2015) Sustainability: don't waste seafood waste. Nat News 524(7564):155

Yin M, Wang Y, Zhang Y, Ren X, Qiu Y, Huang TS (2020) Novel quaternarized N-halamine chitosan and polyvinyl alcohol nanofibrous membranes as hemostatic materials with excellent antibacterial properties. Carbohydr Polym 232:115823

Chitooligosaccharides for Drug Delivery

Nishat Tabassum, Shoeb Ahmed, and M. Azam Ali

Abbreviations

COS	Chitosan oligomer/chitooligosaccharide
DDS	Drug delivery systems
DP	Degree of polymerization
DD	Degree of deacetylation
MW	Molecular weight
TNF-α	Tumor necrosis factor α
IL-6	Interleukin-6
iNOS	Inducible nitric oxide synthase
PI3K/Akt	Phosphatidylinositol 3-kinase-Akt
VEGF	Vascular endothelial growth factor
uPA	Urokinase plasminogen activator
COX-2	Cyclooxygenase
MCP-1	Monocyte Chemoattractant Protein-1
TMCO	Trimethylated chitosan oligomer
EPR	Enhanced permeability and retention
GP	Glycerolphosphate
LCST	Lower critical solution temperature

N. Tabassum · S. Ahmed
Department of Chemical Engineering, Bangladesh University of Engineering and Technology (BUET), Dhaka 1000, Bangladesh
e-mail: ntabassum14@che.buet.ac.bd

S. Ahmed
e-mail: shoebahmed@che.buet.ac.bd

M. Azam Ali (✉)
Faculty of Dentistry, Division of Health Sciences, Center for Bioengineering and Nanomedicine, University of Otago, PO Box 56, Dunedin 9054, New Zealand
e-mail: azam.ali@otago.ac.nz

S.-K. Kim (ed.), *Chitooligosaccharides*,
https://doi.org/10.1007/978-3-030-92806-3_19

siRNA	Small interfering ribonucleic acids
miRNA	Micro RNA mimics
ASO	Antisense oligonucleotides
mRNA	Messenger RNA
NP	Nanoparticles
PBA	Phenylboronic acid
PTX	Paclitaxel
ATRA	All-trans-retinoic acid
PDLLA	Poly(DL-lactic) acid
PARP	Poly (ADP-ribose) polymerase
Bak1	BCL2 Antagonist/Killer 1
Bax	Bcl-2-associated X protein
Bcl-2	B-cell lymphoma 2
PD-L1	Programmed death-ligand 1
AmB	Amphotericin B

1 Introduction

Chitosan's high molecular weight and viscosity often limit its application in biochemical and biomedical settings (Cheung et al. 2015). One way of overcoming the limitation is to depolymerize the long chain, producing oligomers of chitosan, also known as chitooligosaccharides (COS) that have average molecular weight (MW) <10 kDa and/or degree of polymerization (DP) <10, and often, degrees of deacetylation (DD) different from the original chain (Trzciński 2006; Muanprasat and Chatsudthipong 2017; Naqvi and Moerschbacher 2017; Naveed et al. 2019). These changes impart not just aqueous solubility, but enhance the bioactive properties, which include the ability to absorb into epithelial cells (Chae et al. 2005; Aam et al. 2010), angio-inhibitory and antitumor response (Harish Prashanth and Tharanathan 2007), anti-inflammatory characteristics (Salah et al. 2013; Muanprasat and Chatsudthipong 2017), biodegradability and biocompatibility (Onishi and Machida 1999; Shao et al. 2015). These properties are also important pharmacological considerations in the choice of suitable polymeric carriers for drugs, suggesting the superiority that COS may have over the native chitosan in this role (Bernkop-Schnürch and Dünnhaupt 2012).

Active pharmaceutical agents also tend to have limited bioavailability and cellular uptake and need to be protected against enzymatic and acidic degradation in vivo (Svenson 2004). Furthermore, it is crucial that the drug only targets diseased tissues in order to minimize off-target effects (Manzari et al. 2021). Carriers not only protect the cargo, they are highly modifiable scaffolds. Their morphologies can be tuned and they can be functionalized with polymeric carriers being the most versatile (Ageitos et al. 2016). Carriers directly affect the potency, pharmacokinetics, distribution, absorption, clearance and cytotoxicity of the drugs. Other considerations include sufficient encapsulation of the drug while maintaining reasonable release profiles (Fenton et al. 2018). MW and DD are powerful parameters to modulate the actions of COS, offering a strategy to improve its efficacy in drug delivery systems (DDS). These findings indicate the strong potential of COS in the delivery of therapeutics. In a recent review, the authors highlighted the applications of COS in hydrogels, with particular focus on the impact of MW and DD (Tabassum et al. 2021). This chapter expands on the previous work, providing an overview of the therapeutic potential of COS and the relationship between COS' physical properties (MW and DD) and the pharmacology of COS-based DDS. To put these ideas into context, successful applications of COS-based DDS have been discussed.

2 Therapeutic Response of COS

COS has been reported to suppress the expression of the pro-inflammatory NF-κB gene, which has been implicated in various chronic diseases (Mendis et al. 2007). As immune cells generate excessive levels of reactive oxygen species and reactive nitrogen species as an inflammatory response, treatment with anti-oxidative drugs can protect against the damage, and COS having molecular weights <5 kDa have been effective in scavenging these radicals by increasing the activity of anti-oxidative enzymes (Park et al. 2003; Mengíbar et al. 2013; Zheng et al. 2020). There is evidence that reducing the MW improves the anti-oxidative response, for example, COS of 3 kDa performed better against iNOS and TNF-α, compared to COS of 50 kDa (Wu et al. 2015). This proves that the application of COS, especially a DDS based on COS can be beneficial in treating inflammatory bowel and pulmonary diseases (Yousef et al. 2012; Zhao et al. 2018). Vo et al. also discovered that COS prevented the activation of mast cells by the allergen calcium ionophore A23187. Release of histamine and β-hexosaminidase was curbed, and production of inflammatory cytokines such as tumor necrosis factor (TNF)-α, interleukin (IL)-1β, IL-4 and IL-6 was suppressed. COS in the lowest MW (1–3 kDa) category exhibited the strongest response (Vo et al. 2011). The ability to downregulate the expression of cytokines was the key to COS' ability to alleviate allergic asthma (Chung et al. 2012).

Therapeutic efficacy of COS against tumors has also been validated in several reports, where lower MW was more inhibitory to tumor proliferation (Wu et al. 2012; Salah et al. 2013). COS targets the phosphatidylinositol 3-kinase-Akt (PI3K/Akt) signaling network, whose dysfunction prevents cell apoptosis and promotes uncontrolled cell proliferation (Amirani et al. 2020). In vitro tests showed that MW <2 kDa was effective in inducing apoptosis in human leukemia cells by dephosphorylating the Akt kinase, thus inactivating it (Kim et al. 2013). Phosphorylation of Akt leads to the production of angiogenic cytokines, among which the vascular endothelial growth factor (VEGF) plays the major role in promoting tumor angiogenesis (Liu et al. 2011). Changes in the degree of polymerization (DP) of COS also impact therapeutic efficacy; COS having a DP of 6 suppresses the expression of VEGF and the protease, urokinase-type plasminogen activator (uPA) (Xiong et al. 2009), which has also been implicated in tumor metastasis (Duffy et al. 1999; Dublin et al. 2000). This capability of COS was attributed to its inactivation of the PI3K/Akt pathways. COS conjugated with the chemotherapeutic agent selenium (Se) blocked the proliferation of human gastric cancer cell in vitro, likely by decreasing levels of CD34 and VEGF (Jiang et al. 2021). Mallakuntla et al. studied COS' ability to promote apoptosis in human breast cancer cells BT-474 (Mallakuntla et al. 2021). Impact of COS increased as the dose rate was raised, inhibiting the phosphorylation of the epidermal growth factor receptor that is responsible for tumor proliferation (Parachoniak and Park 2012).

COS having DP between 4 and 11 exhibited strong immuno-stimulatory activity (Mei et al. 2013). Zheng et al. reported that 3 kDa was effective in the treatment of immunosuppressive diseases, as it reduced the levels of the inflammatory mediators COX-2, IL-10 and MCP-1 (Zheng et al. 2016). DP 6 was more effective at influencing delayed-type hypersensitivity among DP 2 to 6 (Xing et al. 2017). It is proposed that COS can be beneficial for obesity-induced insulin resistance since it can upregulate lipid and cholesterol metabolism (Liu et al. 2021a). COS can be more effective than chitosan due to its ability to cross the intestinal epithelia. In type-2 diabetes models, COS lowered the fasting glucose levels by increasing the expression of GLUT2, which regulates sugar levels by signaling insulin secretion, and inhibiting intestinal α-glucosidase (a glucose hydrolyzing enzyme) (Yu et al. 2017). Coupled with its ROS-scavenging and anti-inflammatory behaviors, COS can prevent the onset of diabetic neuropathy (Sutthasupha and Lungkaphin 2020). In addition, since chitosan has been reported to induce hemostasis by acting in synergy with platelets and erythrocytes, it follows therefore that COS can also exhibit hemostatic activity (Zhang et al. 2010). The $-NH_2$ groups in the chains were shown to be responsible for platelet aggregation, while (Okamoto et al. 2003) the coagulation of erythrocytes depended on the DD and MW (Klokkevold et al. 1999; Wu et al. 2008). Because of the stronger anti-inflammatory and hemostatic effects, COS can enhance the healing of wounds (Jafari et al. 2020).

3 Impact of COS Properties on Drug Delivery

3.1 Drug Absorption Enhancement

Polar, hydrophilic drugs are absorbed to a lower extent through the mucosal lipid membranes of tissues, which prevents effective mucosal delivery of these drugs (Bowe et al. 1997). By adhering to the negative regions of the membrane and promoting reorganization of the junction proteins, chitosan facilitates contact of the drugs with the membrane and carries them through the channels that form as the junction proteins restructure (Schipper et al. 1996; Sandri et al. 2012). However, the polymer chain must be positively charged in vivo under pH conditions above 6, where chitosan is not appreciably protonated, contrary to COS. Being more capable of binding in these conditions, COS can be better absorbed through negatively charged mucous membranes of tissues (Thanou et al. 2001). Gaserod et al. showed that decreasing the molecular weight leads to an increase in the binding strength of chitosan to polyanionic complexes (e.g., DNA) and mucous membranes. This may improve the uptake rate of COS-based DDS through targeted tissues because of prolonged retention at the site (Shaikh et al. 2011). Sandri et al. offered an explanation for this behavior; low MW leads to higher chain mobility that allows greater interpenetration into the mucin layers (Sandri et al. 2012). Since MW reduction aids in the permeability of the system through the membranes (Gåserød et al. 1998), Dong et al. discovered that reduction of MW resulted in increased uptake of chitosan derivatives from the mucous abdominal lumen to peripheral tissues (Dong et al. 2012). The impermeability of the mucosal interface also precludes effective pulmonary delivery (Morales et al. 2017). COS can aid in such cases, as was determined by Yamade et al. where the use of COS of DP 6 increased the pulmonary absorption of interferon-alpha by 2.5 times, whereas chitosan of molecular weights 22–96 kDa were not effective at all (Yamada et al. 2005).

The DD also governs binding; the impact of DD on charge density is greater at low MW (Maganti et al. 2011; Sacco et al. 2018). This may positively impact interaction of COS with oppositely charged copolymers or active drugs; however, it can also have the side-effect of reducing cell viability because of the much stronger interactions (Schipper et al. 1996; Chae et al. 2005). As such, when using COS of high DD in drug delivery, there is a need to control the dose rate, optimizing it to limit cytotoxic effects (Schipper et al. 1997; Tabassum et al. 2021).

To enhance the cationic character, there is often a need to modify chitosan/COS. Trimethylation of the primary amino group has proven to be an optimum strategy to increase permeation of the chitosan derivatives through the mucosal epithelia of the intestine (Thanou et al. 2001). Studies involving trimethylated chitosan oligomer (TMCO) found similar improvement in permeation rates of plasmid DNA complexed with the modified-COS derivative. Thanou et al. found that TMCO complexed with luciferase plasmid DNA had the highest transfection efficiency in Caco-2 cells (Thanou et al. 2002), while Zheng et al. determined the increased expression of GFP in the gastric and upper intestinal mucosa with TMCO nanoparticles compared to when high MW chitosan (45 kDa and 230 kDa) was used (Zheng et al. 2007). These results were an additional validation for COS' superiority over chitosan in gene delivery. Earlier studies had confirmed that plasmid DNA released at a faster rate when complexed with COS (Köping-Höggård et al. 2001; Issa et al. 2006).

3.2 *Drug Loading and Release*

Experiments with chitosan revealed that drug loading capacity increased when molecular weight was reduced (Genta et al. 1998; Yang and Hon 2009); however, the detailed mechanism of this behavior is not known and the capacity is also dependent on the drug type and drug-to-polymer weight ratio (Genta et al. 1998; Kim 1998).

Another important consideration in the design of drug delivery depots is controlling the degradation kinetics to achieve either sustained or controlled release. The primary mechanisms of release include diffusion across a semi-permeable membrane, bulk diffusion along the pores of a polymer matrix, release due to surface erosion of the polymer layers, diffusion engendered by solvent-mediated swelling of the matrix and bulk erosion of the polymeric matrix (often due to external stimuli, e.g., hydrolysis, thermal or irradiation) (Lee and Yeo 2015; Xing and Zhao 2018; Parhi 2020). The mechanism governs release kinetics (Fig. 1). Solute diffusion models for hydrogels provide insight into how increase in swelling of the matrix leads to higher diffusion rates of the entrapped solute (Peppas and Reinhart 1983). A study by Gupta and Jabrail determined that reducing the MW of chitosan improved the swelling ratios of glutaraldehyde cross-linked chitosan microspheres containing the drug Centchroman (Gupta and Jabrail 2007). MW, therefore, can be used to modulate the release rates, implying that COS would be more suitable if faster release is required (Genta et al. 1998). Furthermore, since COS is water-soluble, COS matrices can erode more quickly. Rapid dissociation can be important in the delivery of therapeutic plasmids, for faster onset of action (Köping-Höggård et al. 2001). However, release rates are also a function of the size of the payload, with larger solutes such as proteins experiencing greater hindrance to diffusion and consequently, slower release (Rehmann et al. 2017).

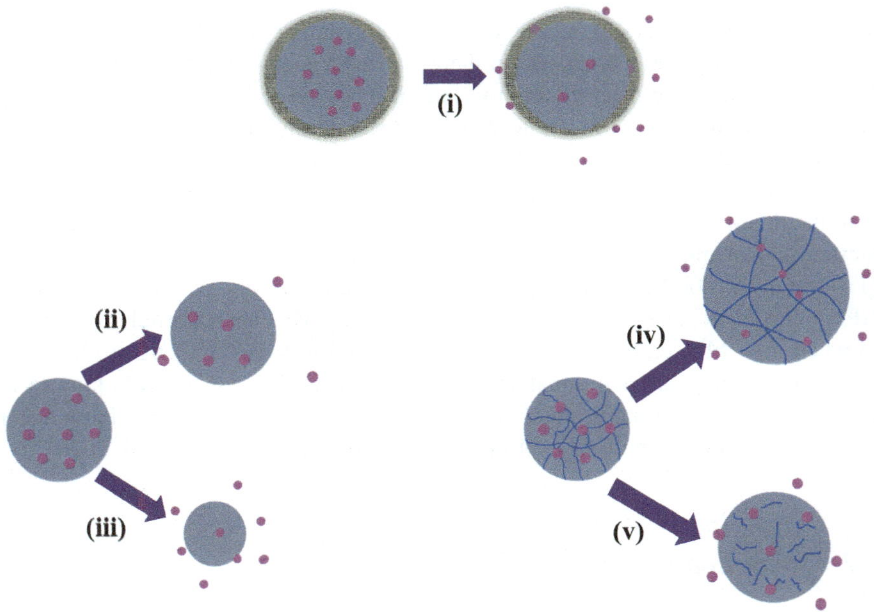

Fig. 1 Drug release mechanism in DDS (i) diffusion across a semi-permeable membrane from a drug reservoir, (ii) diffusion from a porous matrix, (iii) surface erosion-controlled, (iv) solvent-induced transport due to swelling of matrix and (v) bulk erosion of matrix

In case of targeted delivery, it is also necessary that the drug does not degrade while traveling to the tissue of interest. The delivery system must be designed to be stimuli-responsive, releasing its contents under biophysical or biochemical cues (e.g., reactive oxygen species) (Lopes et al. 2013). Drugs destined for the lower gastrointestinal tract should only degrade at the region's higher pH levels, while remaining intact in the stomach. This behavior was observed in COS-based DDS (Crouse et al. 2015; Stealey et al. 2020), whereas chitosan tended to precipitate out in the lower intestines (Zeng et al. 2008), rendering them unsuitable for drug delivery to these areas.

DD's effect on drug loading is not as simple. Gupta and Jabrail found that upon increasing DD, the degree of swelling of chitosan nanoparticles decreased, likely due to more cross-linking between the GlcN resides. This explained the lower diffusion coefficients and consequently, the reduction in loading and release rates for the encapsulated solute. However, the highest loading capacity was seen in chitosan microspheres where DD was 62%. The authors hypothesized that loading capacity is a strong function of the mesh size between the cross-links, and for 62% DD, the size was optimum for the drug molecules (Gupta and Jabrail 2006). Chitosan having 62% DD also released a higher percentage of drug in a controlled manner than either 48% or 75% DD (Gupta and Jabrail 2007). Presence of other overriding factors can alter this trend, for example, if the drug is negatively charged, encapsulation efficiency increases and drug release rates fall if DD is increased, because of stronger electrostatic interactions (Yang and Hon 2010).

3.3 Biodegradability and Bioaccumulation

Hepatic enzymes can metabolize chitosan. Therefore, DDS developed from chitosan or its derivatives are biodegradable, producing substances that are natural metabolites of the body (Onishi and Machida 1999; Shao et al. 2015). This means the risk of bioaccumulation is negligible, becoming lower as the MW decreases, since the chains are shorter and easy to be cleared (Richardson et al. 1999; Lee et al. 2003). In a study by Zeng et al., COS of MW 1 kDa existed at levels below 1 mg/g in several major organs, 25 min after oral administration (Zeng et al. 2008). Degradation rates need to be adjusted depending on the required retention time in the organs to which the DDS is directed. DD can be a useful parameter to achieve this. COS having 95% DD enhanced the stability of pectin-based DDS intended for the lower GI tract, maintaining its integrity in the intestines *en route* to the colon (Zhang et al. 2016). In fact, COS is critical to stability and carriers degrade negligibly in both gastric and intestinal fluids (Stealey et al. 2020). Yuan et al. concluded that for low MW chitosan, increasing the DD would lead to lower degradation in the kidneys, important if the COS-based DDS is intended to treat renal disorders (Yuan et al. 2011). Lysozyme-mediated degradation of COS is reduced if the DD is high because the enzyme is less active toward D-D linkages (Yomota et al. 1990; Kristiansen et al. 1998); in fact, greater charge densities correlated with decreased lysozyme hydrolysis (Yue et al.

2011). This suggests that when the drugs are targeting the lower organs, namely the colon and kidneys, high DD values should be chosen.

3.4 *Targeting Precision*

Chemical and physical modifications to the DDS impart selective targeting capabilities when administered systemically. Physical alterations achieve passive targeting, where the size, shape and surface charge govern which tissues the system binds to and absorbs into (Inggrid Setyawati et al. 2015). For instance, nanoparticles less than 200 nm in size preferentially extravasate into fenestrated tumor vasculature where the residence time is also longer due to poor lymphatic clearance in the tumors, an effect appropriately known as enhanced permeability and retention (EPR) (Yhee et al. 2013). COS already has an advantage because of its size, and targeting precision can be adjusted using MW and DD (Herdiana et al. 2021). Nevertheless, strict control of passive targeting is difficult and the endosomal escape of the DDS depends on the route via which the DDS is internalized, which is governed by the DDS-cell interactions. Influencing the fate of the DDS is thus very challenging (Inggrid Setyawati et al. 2015).

Active targeting is more suitable and effective, relying on surface moieties that act as targeting ligands (in the form of aptamers, antibody, peptides, metal complexes, etc.) (Srinivasarao and Low 2017). Chitosan conjugated with these ligands has been extensively studied as a delivery system, and COS, having stronger biological activities and ability to improve drug absorption, is likely to be a better candidate (Yue et al. 2011; Seong et al. 2018). Active ligands bind exclusively to overexpressed receptors in diseased tissues (e.g., HER2, folate receptor, CD44, etc.) (Cheng et al. 2012), and the DDS gets readily absorbed into the tissues because of COS (Song et al. 2019; Elbehairi et al. 2020). Thus, a COS-based active-targeting DDS can produce a more efficient therapeutic response, limiting off-target side effects. However, tumor heterogeneity limits the success of active-targeting strategies, as the tumors that do not sufficiently express the target receptor may end up surviving (Rosenblum et al. 2018).

3.5 *Responsiveness to External Stimuli*

A goal in targeted drug delivery is to prevent drug release in off-target regions while ensuring sufficient release occurs at the tissue of interest. Utilizing environmental stimuli as actuators for drug release is a well-established control strategy (Lopes et al. 2013). COS has been used in DDS to achieve pH-responsive delivery, releasing the drug inside the cells of diseased tissues after uptake because of the moderate acidic environment, while maintaining integrity in neutral pH (7.4) (Moorthy et al.

2019). Tang et al. obtained similar results with PEG-COS metal coordination self-healing hydrogels (Tang et al. 2019). COS also exhibited temperature-mediated gelation; aqueous solution of COS, when cross-linked with glycerolphosphate (GP), formed a soluble complex. On heating above the lower critical solution temperature (LCST), the complex spontaneously aggregated to stable nanoparticles. The dielectric constant of the aqueous solution governed the LCST, with increase in dielectric constant leading to a drop in LCST and destabilization of the aggregated complex (Burova et al. 2020). Analysis of the impact of DD on the dielectric constant of chitosan suspensions revealed that up to a DD of 87%, the dielectric constant increased. With further increase in DD, dielectric constant dropped (Ko et al. 2011). Modulation of the DD can therefore be used to alter the LCST, a control parameter for achieving temperature-responsive drug release. Reduction-sensitive disulfide bonds have been used to confer responsiveness to reducing agents present in tumor cells. Liposomes delivering drugs to tumors often have a polymeric coating that limits the drug release in off-target regions. Yin et al. used COS for the purpose, attaching it to the liposomes via disulfide bonds to achieve additional control over the drug release, which was triggered by the reduction of the bonds in the tumors with high concentrations of glutathione (Yin et al. 2018).

4 Applications of COS-Based DDS

4.1 Gene Delivery

Gene therapy can be more efficacious in managing chronic illnesses, and RNA-based therapeutics, namely small-interfering RNAs (siRNA), micro RNA mimics (miRNA), antisense oligonucleotides (ASO), in addition to synthetic mRNAs (mRNA) have shown promise (Fig. 2) (Berger et al. 2020). However, when administered systemically, the immune system, specifically the toll-like receptor, recognizes it as a pathogen (Whitehead et al. 2011). Therapeutic nucleic acids also suffer from limited cellular uptake, as they have to pass through several physiological barriers before reaching the nucleus of the target cells, the region of interest (Kanasty et al. 2013). Finding suitable vectors and delivery systems to deliver genes to target organs is thus imperative. Garaivo et al. determined that chitosan oligomers were superior to chitosan for intracellular delivery of genes, because they are not taken up by endocytic vesicles coated with the clathrin protein, which transfers the polyplexes to the lysosome for degradation—a fate that linear chain chitosan-DNA nanoplexes tend to face (Garaiova et al. 2012). COS governs the fate of the delivery systems. For transport proteins such as bovine serum albumin, which rapidly undergo clathrin-mediated endocytosis (Nishikawa et al. 1992; Fujita et al. 1994), a 70% reduction in lysosome degradation was observed when complexed with COS (Kumari et al. 2020). In rat corneas injected with COS-DNA nanoparticles (MW 5.7 kDa), gene expression was 5.4-fold higher than when polyethylenimine-DNA (current treatment

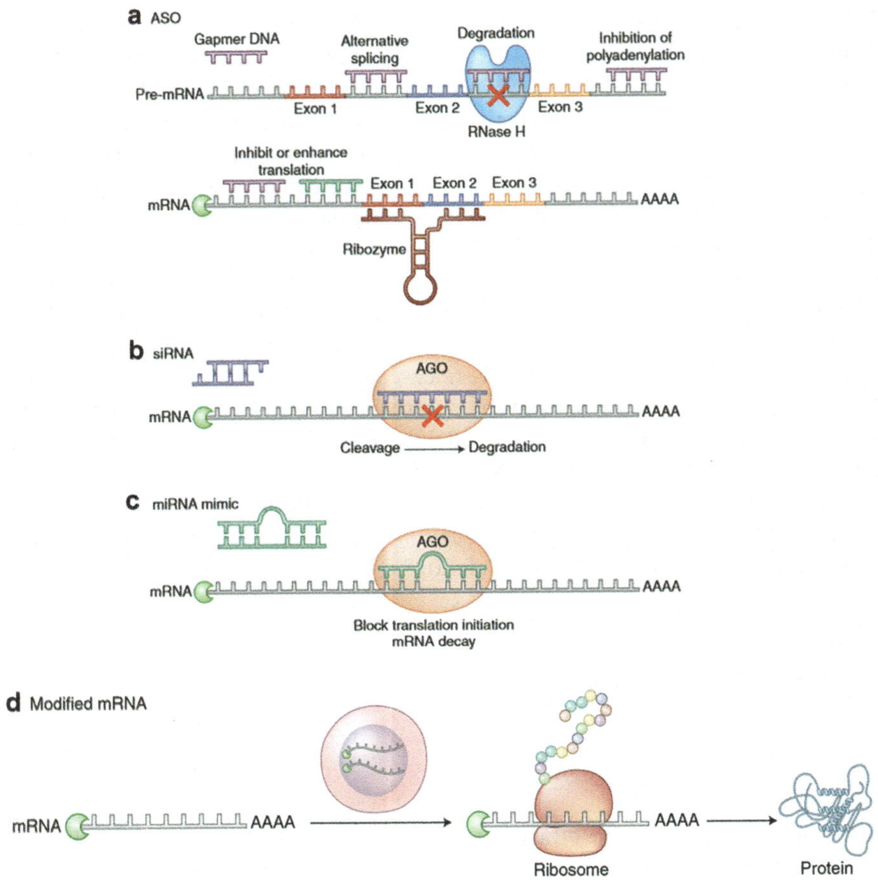

Fig. 2 RNA-based therapeutics. **a** ASOs are single-stranded nucleic acid chains acting on precursor mRNAs affecting splicing and initiating RNA decay and/or degradation, **b** siRNAs and **c** miRNAs interfere with translation of disease-specific genes, **d** transcription of modified mRNA replaces mutated proteins. Adapted with permission (Lieberman 2018)

standard) was administered and most of the nanoparticles targeted the corneal fibroblasts (Klausner et al. 2010). The study proved the suitability of COS as a novel gene carrier for the treatment of inherited and acquired corneal diseases. When conjugated with hyaluronic acid, COS NPs delivered plasmid DNA to corneal and conjunctival cells. The presence of hyaluronic acid facilitated uptake by hyaluronan receptors in the cell membrane (Contreras-Ruiz et al. 2011). Wong et al. grafted polyethylenimine with COS (3.4 kDa), injecting the copolymer into the common bile duct in rat livers. Transfection efficiency was superior to when either of the polymers was administered alone. COS also improved the biocompatibility of the polyethylenimine (Wong et al. 2006).

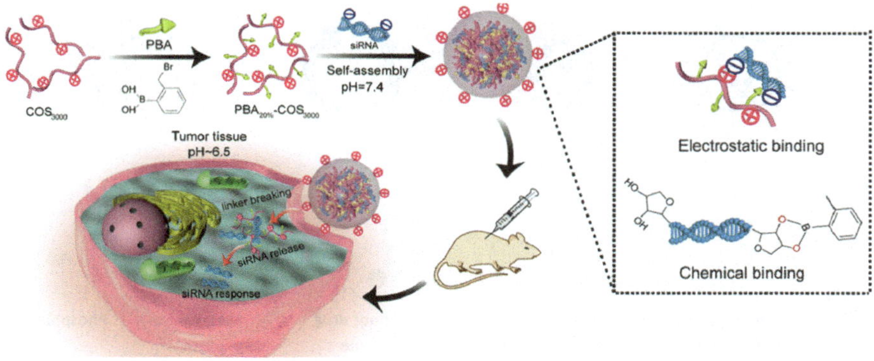

Fig. 3 COS-PBA nanoparticles encapsulating siRNA strands. Reprinted with permission (Liu et al. 2021b)

Small-interfering RNA is a powerful therapeutic for degenerative diseases as it can downregulate expressions of genes associated with a disorder by modifying the mRNA transcripts. However, suitable carriers preserving their stability in vivo are limited, which reduces preclinical utility of siRNA-based treatment regimens (Koutsilieri et al. 2007; Mahmoodi Chalbatani et al. 2019). Carriers developed from COS hold promise, as it does not agglomerate in physiological fluids, thus the integrity of COS-based complexes remain intact. Liu et al. determined that phenylboronic acid (PBA)-modified-COS nanoparticles loaded with survivin-targeted siRNA effectively silenced the genes that promote the metastasis of melanoma in vivo (Liu et al. 2021b) (Fig. 3). It is also possible to further improve the stability by conjugation with natural macromers such as silk fibroin, which was shown to enhance the loading capacity, uptake and biocompatibility of the siRNA–COS polyplexes (Shahbazi et al. 2015).

4.2 Tumor-Targeting Drug Delivery

COS has been incorporated into several types of delivery systems for tumor-targeting drugs (Fig. 4). Paclitaxel (PTX), a poorly soluble chemotherapy drug, has limited oral bioavailability (30%) (Sparreboom et al. 1997). This necessitates intravenous injection, an inconvenient strategy that induces side effects, because the formulation contains the clinically active Cremophor EL as excipient (Gelderblom et al. 2001). There is a need to use hydrophilic, low toxicity carriers to improve the bioavailability of the drug. Chitosan can be a more biocompatible alternative to deliver PTX, having stronger tumor specificity and better permeation capability (Saravanakumar et al. 2009; Koo et al. 2013). It follows therefore that COS should be equally, if not more, suitable for the purpose. In fact, studies by Zhang et al. and Jiang et al. have proven COS' viability to deliver a larger payload of PTX to targeted tumors. PTX

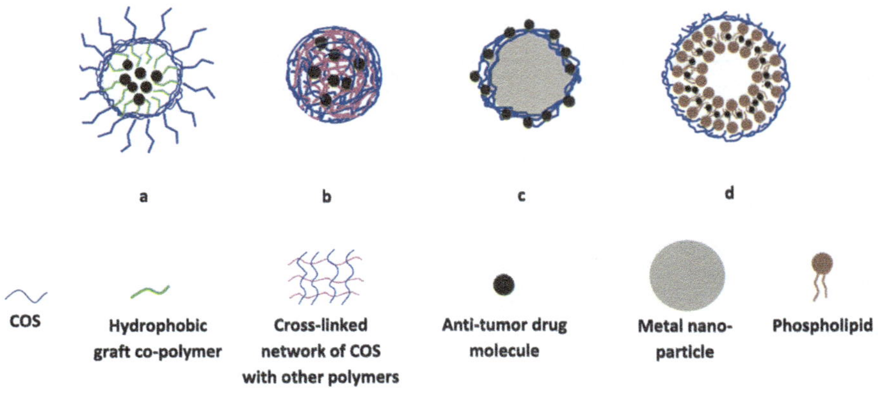

Fig. 4 Common systems for antitumor drug delivery using COS: **a** micelles, **b** polymeric nanoparticles, **c** COS-coated metal nanoparticles and **d** liposomes

with all-trans-retinoic acid (ATRA) is a potent combination chemotherapy formulation that requires a carrier where both drugs can be stable (Karmakar et al. 2008; Hou et al. 2012). COS nanoparticles having MW of 3.0 kDa and conjugated with ATRA was used to encapsulate PTX. In vitro testing revealed effective action against Hep G2 cancer cells (Zhang et al. 2015). Jiang et al. used deoxycholic acid-modified chitooligosaccharide/mPEG-PDLLA mixed micelles as carriers, where COS was specifically intended to improve the permeation extent. The encapsulation efficiency of PTX was over 97% and tumor cell uptake was high (Jiang et al. 2014). Raloxifene, an estrogen receptor modulator that combats postmenopausal osteoporosis and estrogen-dependent breast cancer, is amphiphobic, and almost 98% of the drug is eliminated from circulation when delivered orally (Kr et al. 2000). To improve the bioavailability, raloxifene was loaded into cyclodextrin and COS nanoparticles and the maximal plasma concentration (C_{max}) measured following oral administration. C_{max} increased by around 300 times relative to raloxifene suspensions. Bioavailability also increased by 361%. Cellular uptake, intestinal permeability and transepithelial absorption were other improvements attributed to COS and cyclodextrin (Wang and Li 2018). Another antitumor drug, docetaxel was also loaded into COS-coated lecithin system. More than 80% of human breast adenocarcinoma cells and 100% of human lung carcinoma cells had absorbed the drug within two hours (Lozano et al. 2008). Addition of COS to poly (lactic-co-glycolic acid) (PLGA) nanoparticles designed to encapsulate Osimertinib, a recently approved medication for lung carcinomas, led to better uptake rates in H1975 lung cancer cells (Hu et al. 2020). Proteins such as epidermal growth factor receptor (EGFR), Poly (ADP-ribose) polymerase (PARP), BCL2 Antagonist/Killer 1 (Bak1), caspase-9, Bcl-2-associated X protein (Bax) and B-cell lymphoma 2 (Bcl-2) were downregulated, and COS specifically reduced the expression of Programmed death-ligand 1 (PD-L1) that would otherwise bind to tumor cells, preventing their recognition by the immune system (Azuma et al. 2014). The antitumor drug doxorubicin, commonly administered as an injectable, has a high toxicological profile (Park et al. 2012). To minimize off-target effects of the drug, liposome-encapsulated forms have been developed from synthetic polymers such as PEG (Barenholz 2012). Chitosan and COS have stronger mucoadhesive properties compared to PEG, and facilitate longer retention at the epithelium surface, which gives COS an advantage over PEG as carriers (De Campos et al. 2003; Dyawanapelly et al. 2016). Stearic acid-grafted COS micelles were used to encapsulate the antitumor drug doxorubicin (Hu et al. 2008, 2009). The encapsulation lowered the drug's toxicity to healthy cells, in addition to making the release pH-responsive (Du et al. 2010). Liposome-based DDS, when conjugated with COS via disulfide linkers, became responsive to reducing agents such as glutathione or dithiothreitol, which allowed drug release to be triggered via reduction. Such systems have been effective against osteosarcomas (Yin et al. 2017, 2018). In an effort to make the DDS pH-responsive, mesoporous silica nanoparticles' surface was coated with COS. Drug release was triggered under acidic conditions (Moorthy et al. 2019). Although all these studies demonstrated the in vitro feasibility of COS to carry hydrophobic antitumor drugs, it remains to be tested how well it performs in vivo.

4.3 Wound Healing Drugs

COS has been observed to stimulate the growth of types I and III collagen in wounds, repair damaged nerves, accelerate hemostasis and protect against infections, which make it a potent wound healing agent in its own right (Chandika et al. 2015; Hou et al. 2016; Sandri et al. 2017; Park et al. 2018; Oh et al. 2021). It indicates that incorporation of wound care drugs with COS would likely have a stronger healing activity. Such agents (e.g. curcumin) may also suffer from rapid degradation and have poor water solubility, which would limit their in vivo administration. Use of COS can overcome these challenges. Nguyen et al. have extensively studied liposome-based delivery systems for curcumin complexed with COS (Nguyen et al. 2019a, b), concluding that when COS was used, the complex did not agglomerate and remained stable for over 72 h (whereas with chitosan, the stability was less than 24 h). Within seven days, the wound had closed by 90% compared to when curcumin had been administered by itself (Nguyen et al. 2019a). Studies comparing COS and *N*-acetyl-L-cystein-modified chitosan revealed that liposome carriers containing the latter were better suited to deliver curcumin in the cornea, achieving controlled release for over 72 h and a 6.4-fold higher permeability efficiency. Retention of curcumin in the cornea was also greater (Li et al. 2016). This reveals an important insight; although chemical modification of chitosan may improve its activity, the biocompatibility of the chemical groups has to be ensured to avoid secondary effects. COS can therefore be a superior choice, as it exhibits a stronger biological activity than chitosan, without requiring any modification.

4.4 Other Drugs

Nano-complexes developed from COS have been used to target a variety of conditions, and have been delivered through different routes. Baicalin, an antiviral drug for the treatment of atopic dermatitis, viral hepatitis and HIV infection, has short circulation times after intravenous injection and 46% of the drug is metabolized after oral administration (Lai et al. 2003; Qiu et al. 2003). When co-delivered with COS of various molecular weights (1–5 kDa) via the transdermal route, the in vivo bioavailability after 24 h and the permeation rate were the highest for the complex containing COS of MW 1 kDa (Zhou et al. 2010). Another treatment for atopic dermatitis, gallic acid, was encapsulated in an injectable pluronic F-127 and COS-hyaluronic acid conjugate-based hydrogel. The system was designed to be pH and temperature-responsive, and showed negligible toxicity to keratinocyte cells, making it viable for transdermal delivery (Chatterjee et al. 2020). Medications for treating systemic fungal infection include amphotericin B (AmB), which is usually administered via injection. However, rapid agglomeration not only limits its efficacy but it can also induce toxic side effects such as nephrotoxicity, cardiotoxicity, hepatoxicity, leukopenia and thrombocytopenia (Bolard et al. 1993). This can be avoided by using stable carriers (Wang et al. 2016). Linoleic acid-modified-COS (MW 1 kDa) was grafted with methoxy-poly(ethylene glycol) that had been modified with linoleic acid. Hemolytic toxicity dropped dramatically from 90% when only AmB was used to 20% at the highest dose level. Further evaluation revealed the synergistic effect of COS and linoleic acid on *Candida albicans* (Song et al. 2019).

COS can effectively convey drugs across the blood–brain barrier due to its size. Periventricular leukomalacia (PVL), a brain injury associated with cerebral palsy, commonly occurring in premature infants due to hypoxia or ischemia (Zaghloul et al. 2017). Erythropoietin (EPO), known to reverse chronic anemia, can be beneficial for counteracting such conditions (Mizuno et al. 2008). However, protease degradation limits its availability. COS nanoparticles loaded with EPO were injected intraperitoneally for the treatment of PVL in five-day old rats and EPO levels in the brain were maintained at around 100 IU/ml 12 h after injection. Furthermore, only 50 IU/kg EPO sufficiently protected against white matter loss, compared to the standard 5,000 IU/kg when EPO is administered alone (Wang et al. 2012). Table 1 summarizes a few other recent in vivo studies on COS-based DDS delivering a variety of therapeutic and medical agents.

5 Challenges and Future Prospects

The studies attest to the versatility and efficacy of COS as not just a carrier for a wide range of drugs, but as a biologically active agent itself for biomedical applications. COS therefore can act in synergy with other therapeutic agents to provide an effective, targeted treatment strategy for chronic and acute conditions. Various

Table 1 Recent COS-based DDS for delivery of unconventional therapeutic agents

Medication	Application	Carrier formulation	Results	References
Gadolinium (III) complexed with diethylenetri-aminepentaacetic acid	Angiographic imaging agent	COS-stearic acid nanomicelles consisting of redox-responsive disulfide bonds as linking groups	Sustained release after 12 h of injection; tumor boundaries remained visible for imaging for up to two hours after administration	Xin et al. (2018)
Honokiol	Anti-fungal plant derivative	COS (1 kDa)-pluronic conjugate	88% encapsulation efficiency; 13% drug loading capacity; minimum inhibition capacity (MIC_{90}) same as that for pure honokiol	Song et al. (2018)
Nanoceria	Cerium ions for the treatment of age-related macular degeneration	COS-coated nanoceria loaded within alginate hydrogels	IL-6 decreased by 40% in ARPE-19 cells; hydrogen peroxide induced apoptosis prevented; expression of pro-angiogenic factor VEGF reduced	Wang et al. (2018)
Immunoglobulin Y (IgY)	Antibody	Alginate microcapsules coated with COS	IgY remained stable in acidic gastric fluid after four hours; >84% activity was retained	Zhang et al. (2020)

routes of administration have been successfully tested, using drugs that are either poorly soluble or have limited bioavailability and permeation efficiency across tissue barriers. Nevertheless, most of the research has focused on in vitro analyses or has narrow scope, where the absorption/distribution and toxicity studies are confined to a single target organ. Furthermore, there is a need to decipher how the modulation of COS' properties (MW, DD and chemical modifications) affects cellular response in addition to the toxicological profile of the DDS. By answering these fundamental questions, especially the impact on several organs, it is possible to progress from these proof-of-concept analyses to clinical evaluation. Unlike well-characterized synthetic polymers, natural materials have yet to find commercial use as carriers (Fenton et al. 2018). Future challenges also include: (1) choosing suitable forms of delivery (e.g. hydrogels, nanoparticles), (2) finding better design strategies that allow more control over the functionality of the system, (3) studying local delivery and not just being confined to systemic delivery and (4) comparison studies on routes of administration. These studies can offer important insight into providing a definitive evaluation of COS in DDS and guidelines for its translation to clinical settings.

References

Aam BB et al (2010) Production of chitooligosaccharides and their potential applications in medicine. Mar Drugs 8(5):1482–1517. https://doi.org/10.3390/md8051482

Ageitos JM, Chuah J-A, Numata K (2016) Chapter 1 Design considerations for properties of nanocarriers on disposition and efficiency of drug and gene delivery, pp 1–22. https://doi.org/10.1039/9781782622536-00001

Amirani E et al (2020) Effects of chitosan and oligochitosans on the phosphatidylinositol 3-kinase-AKT pathway in cancer therapy. Int J Biol Macromol 164:456–467. https://doi.org/10.1016/j.ijbiomac.2020.07.137

Azuma K et al (2014) Association of PD-L1 overexpression with activating EGFR mutations in surgically resected nonsmall-cell lung cancer. Ann Oncol 25(10):1935–1940. https://doi.org/10.1093/annonc/mdu242

Barenholz Y (2012) Doxil®—the first FDA-approved nano-drug: lessons learned. J Control Release: off J Control Release Soc 160(2):117–134. https://doi.org/10.1016/j.jconrel.2012.03.020

Berger AG, Chou JJ, Hammond PT (2020) Approaches to modulate the chronic wound environment using localized nucleic acid delivery. Adv Wound Care [Preprint]. https://doi.org/10.1089/wound.2020.1167

Bernkop-Schnürch A, Dünnhaupt S (2012) Chitosan-based drug delivery systems. Eur J Pharm Biopharm 81(3):463–469. https://doi.org/10.1016/j.ejpb.2012.04.007

Bolard J, Joly V, Yeni P (1993) Mechanism of action of amphotericin B at the cellular level. Its modulation by delivery systems. J Liposome Res 3(3):409–427. https://doi.org/10.3109/08982109309150728

Bowe CL et al (1997) Design of compounds that increase the absorption of polar molecules. Proc Natl Acad Sci USA 94(22):12218–12223

Burova TV et al (2020) Biodegradable thermoresponsive oligochitosan nanoparticles: mechanisms of phase transition and drug binding-release. Int J Biol Macromol 164:1451–1460. https://doi.org/10.1016/j.ijbiomac.2020.07.203

Chae SY, Jang M-K, Nah J-W (2005) Influence of molecular weight on oral absorption of water soluble chitosans. J Control Release 102(2):383–394. https://doi.org/10.1016/j.jconrel.2004.10.012

Chandika P et al (2015) Fish collagen/alginate/chitooligosaccharides integrated scaffold for skin tissue regeneration application. Int J Biol Macromol 81:504–513. https://doi.org/10.1016/j.ijbiomac.2015.08.038

Chatterjee S et al (2020) Drug delivery system of dual-responsive PF127 hydrogel with polysaccharide-based nano-conjugate for textile-based transdermal therapy. Carbohydr Polym 236:116074. https://doi.org/10.1016/j.carbpol.2020.116074

Cheng Z et al (2012) Multifunctional nanoparticles: cost versus benefit of adding targeting and imaging capabilities. Science (New York, N.Y.) 338(6109):903–910. https://doi.org/10.1126/science.1226338

Cheung RCF et al (2015) Chitosan: an update on potential biomedical and pharmaceutical applications. Mar Drugs 13(8):5156–5186. https://doi.org/10.3390/md13085156

Chung MJ, Park JK, Park YI (2012) Anti-inflammatory effects of low-molecular weight chitosan oligosaccharides in IgE–antigen complex-stimulated RBL-2H3 cells and asthma model mice. Int Immunopharmacol 12(2):453–459. https://doi.org/10.1016/j.intimp.2011.12.027

Contreras-Ruiz L et al (2011) Intracellular trafficking of hyaluronic acid-chitosan oligomer-based nanoparticles in cultured human ocular surface cells. Mol vis 17:279–290

Crouse JZ et al (2015) Development of a microscale red blood cell-shaped pectin-oligochitosan hydrogel system using an electrospray-vibration method: preparation and characterization. J Appl Biomater Funct Mater 13(4):e326-331. https://doi.org/10.5301/jabfm.5000250

De Campos AM et al (2003) The effect of a PEG versus a chitosan coating on the interaction of drug colloidal carriers with the ocular mucosa. Eur J Pharm Sci off J Eur Feder Pharm Sci 20(1):73–81. https://doi.org/10.1016/s0928-0987(03)00178-7

Dong W et al (2012) Effects of molecular weights on the absorption, distribution and urinary excretion of intraperitoneally administrated carboxymethyl chitosan in rats. J Mater Sci Mater Med 23(12):2945–2952. https://doi.org/10.1007/s10856-012-4747-7

Du Y-Z et al (2010) Stearic acid grafted chitosan oligosaccharide micelle as a promising vector for gene delivery system: Factors affecting the complexation. Int J Pharm 391(1):260–266. https://doi.org/10.1016/j.ijpharm.2010.02.017

Dublin E et al (2000) Immunohistochemical expression of uPA, uPAR, and PAI-1 in breast carcinoma. Fibroblastic expression has strong associations with tumor pathology. Am J Pathol 157(4):1219–1227. https://doi.org/10.1016/S0002-9440(10)64637-8

Duffy MJ et al (1999) Urokinase plasminogen activator: a prognostic marker in multiple types of cancer. J Surg Oncol 71(2):130–135. https://doi.org/10.1002/(sici)1096-9098(199906)71:2%3c130::aid-jso14%3e3.0.co;2-9

Dyawanapelly S et al (2016) Improved mucoadhesion and cell uptake of chitosan and chitosan oligosaccharide surface-modified polymer nanoparticles for mucosal delivery of proteins. Drug Deliv Transl Res 6(4):365–379. https://doi.org/10.1007/s13346-016-0295-x

Elbehairi SEI et al (2020) Role of Pd(II)–chitooligosaccharides–Gboxin analog in oxidative phosphorylation inhibition and energy depletion: targeting mitochondrial dynamics. Chem Biol Drug Des 96(4):1148–1161. https://doi.org/10.1111/cbdd.13703

Fenton OS et al (2018) Advances in biomaterials for drug delivery. Adv Mater 30(29):1705328. https://doi.org/10.1002/adma.201705328

Fujita T et al (1994) Control of in vivo fate of albumin derivatives utilizing combined chemical modification. J Drug Target 2(2):157–165. https://doi.org/10.3109/10611869409015905

Garaiova Z et al (2012) Cellular uptake of DNA-chitosan nanoparticles: the role of clathrin- and caveolae-mediated pathways. Int J Biol Macromol 51(5):1043–1051. https://doi.org/10.1016/j.ijbiomac.2012.08.016

Gåserød O, Smidsrød O, Skjåk-Bræk G (1998) Microcapsules of alginate-chitosan—I: A quantitative study of the interaction between alginate and chitosan. Biomaterials 19(20):1815–1825. https://doi.org/10.1016/S0142-9612(98)00073-8

Gelderblom H et al (2001) Cremophor EL: the drawbacks and advantages of vehicle selection for drug formulation. Eur J Cancer (Oxford, England: 1990) 37(13):1590–1598. https://doi.org/10.1016/s0959-8049(01)00171-x

Genta I, Perugini P, Pavanetto F (1998) Different molecular weight chitosan microspheres: influence on drug loading and drug release. Drug Dev Ind Pharm 24(8):779–784. https://doi.org/10.3109/03639049809082726

Gupta K, Jabrail F (2006) Effects of degree of deacetylation and cross-linking on physical characteristics, swelling and release behavior of chitosan microspheres. Carbohydr Polym 66:43–54. https://doi.org/10.1016/j.carbpol.2006.02.019

Gupta KC, Jabrail F (2007) Glutaraldehyde cross-linked chitosan microspheres for controlled release of centchroman. Carbohydr Res 342:2244–2252. https://doi.org/10.1016/j.carres.2007.06.009

Harish Prashanth KV, Tharanathan RN (2007) Chitin/chitosan: modifications and their unlimited application potential—an overview. Trends Food Sci Technol 18(3):117–131. https://doi.org/10.1016/j.tifs.2006.10.022

Herdiana Y et al (2021) Chitosan-based nanoparticles of targeted drug delivery system in breast cancer treatment. Polymers 13(11):1717. https://doi.org/10.3390/polym13111717

Hou L et al (2012) Pharmacokinetics of a paclitaxel-loaded low molecular weight heparin-all-trans-retinoid acid conjugate ternary nanoparticulate drug delivery system. Biomaterials 33(21):5431–5440. https://doi.org/10.1016/j.biomaterials.2012.03.070

Hou H et al (2016) Chitooligosaccharide inhibits scar formation and enhances functional recovery in a mouse model of sciatic nerve injury. Mol Neurobiol 53(4):2249–2257. https://doi.org/10.1007/s12035-015-9196-0

Hu F-Q et al (2008) Cellular uptake and cytotoxicity of shell crosslinked stearic acid-grafted chitosan oligosaccharide micelles encapsulating doxorubicin. Eur J Pharm Biopharm 69(1):117–125. https://doi.org/10.1016/j.ejpb.2007.09.018

Hu F-Q et al (2009) Synthesis and antitumor activity of doxorubicin conjugated stearic acid-g-chitosan oligosaccharide polymeric micelles. Biomaterials 30(36):6955–6963. https://doi.org/10.1016/j.biomaterials.2009.09.008

Hu X et al (2020) Chitooligosaccharides-modified PLGA nanoparticles enhance the antitumor efficacy of AZD9291 (Osimertinib) by promoting apoptosis. Int J Biol Macromol 162:262–272. https://doi.org/10.1016/j.ijbiomac.2020.06.154

Inggrid Setyawati M et al (2015) Understanding and exploiting nanoparticles' intimacy with the blood vessel and blood. Chem Soc Rev 44(22):8174–8199. https://doi.org/10.1039/C5CS00499C

Issa MM et al (2006) Targeted gene delivery with trisaccharide-substituted chitosan oligomers in vitro and after lung administration in vivo. J Control Release 115(1):103–112. https://doi.org/10.1016/j.jconrel.2006.06.029

Jafari H et al (2020) Chitooligosaccharides for wound healing biomaterials engineering. Mater Sci Eng C Mater Biol Appl 117:111266. https://doi.org/10.1016/j.msec.2020.111266

Jiang C et al (2014) Deoxycholic acid-modified chitooligosaccharide/mPEG-PDLLA mixed micelles loaded with paclitaxel for enhanced antitumor efficacy. Int J Pharm 475(1–2):60–68. https://doi.org/10.1016/j.ijpharm.2014.08.037

Jiang Z et al (2021) Effect of chitosan oligosaccharide-conjugated selenium on improving immune function and blocking gastric cancer growth. Eur J Pharmacol 891:173673. https://doi.org/10.1016/j.ejphar.2020.173673

Kanasty R et al (2013) Delivery materials for siRNA therapeutics. Nat Mater 12(11):967–977. https://doi.org/10.1038/nmat3765

Karmakar S, Banik NL, Ray SK (2008) Combination of all-trans retinoic acid and paclitaxel-induced differentiation and apoptosis in human glioblastoma U87MG xenografts in nude mice. Cancer 112(3):596–607. https://doi.org/10.1002/cncr.23223

Kim CJ (1998) Effects of drug solubility, drug loading, and polymer molecular weight on drug release from Polyox tablets. Drug Dev Ind Pharm 24(7):645–651. https://doi.org/10.3109/03639049809082366

Kim M-O et al (2013) Water-soluble chitosan sensitizes apoptosis in human leukemia cells via the downregulation of Bcl-2 and dephosphorylation of akt. J Food Biochem 37(3):270–277. https://doi.org/10.1111/j.1745-4514.2011.00628.x

Klausner EA et al (2010) Ultrapure chitosan oligomers as carriers for corneal gene transfer. Biomaterials 31(7):1814–1820. https://doi.org/10.1016/j.biomaterials.2009.10.031

Klokkevold PR et al (1999) The effect of chitosan (poly-N-acetyl glucosamine) on lingual hemostasis in heparinized rabbits. J Oral Maxillofac Surg 57(1):49–52. https://doi.org/10.1016/S0278-2391(99)90632-8

Ko YG et al (2011) Gelation of chitin and chitosan dispersed suspensions under electric field: effect of degree of deacetylation. ACS Appl Mater Interfaces 3(4):1289–1298. https://doi.org/10.1021/am200091r

Koo H et al (2013) Enhanced drug-loading and therapeutic efficacy of hydrotropic oligomer-conjugated glycol chitosan nanoparticles for tumor-targeted paclitaxel delivery. J Control Release: off J Control Release Soc 172(3):823–831. https://doi.org/10.1016/j.jconrel.2013.08.297

Köping-Höggård M et al (2001) Chitosan as a nonviral gene delivery system. Structure–property relationships and characteristics compared with polyethylenimine in vitro and after lung administration in vivo. Gene Ther 8(14):1108–1121. https://doi.org/10.1038/sj.gt.3301492

Koutsilieri E, Rethwilm A, Scheller C (2007) The therapeutic potential of siRNA in gene therapy of neurodegenerative disorders. J Neural Transm Suppl 72:43–49. https://doi.org/10.1007/978-3-211-73574-9_7

Kr S, N S, Jm M (2000) Raloxifene hydrochloride. Am J Health-Syst Pharm: AJHP: Off J Am Soc Health-Syst Pharm 57(18). https://pubmed.ncbi.nlm.nih.gov/11006795/. Accessed 25 July 2021

Kristiansen A, Vårum KM, Grasdalen H (1998) The interactions between highly de-N-acetylated chitosans and lysozyme from chicken egg white studied by 1H-NMR spectroscopy. Eur J Biochem 251(1–2):335–342. https://doi.org/10.1046/j.1432-1327.1998.2510335.x

Kumari M, Liu C-H, Wu W-C (2020) Oligochitosan modified albumin as plasmid DNA delivery vector: endocytic trafficking, polyplex fate, in vivo compatibility. Int J Biol Macromol 142:492–502. https://doi.org/10.1016/j.ijbiomac.2019.09.121

Lai M-Y et al (2003) Comparison of metabolic pharmacokinetics of baicalin and baicalein in rats. J Pharm Pharmacol 55(2):205–209. https://doi.org/10.1211/002235702522

Lee DW et al (2003) Effect of gamma-irradiation on degradation of alginate. J Agric Food Chem 51(16):4819–4823. https://doi.org/10.1021/jf021053y

Lee JH, Yeo Y (2015) Controlled drug release from pharmaceutical nanocarriers. Chem Eng Sci 125:75–84. https://doi.org/10.1016/j.ces.2014.08.046

Li J et al (2016) A comparative study on the efficiency of chitosan-N-acetylcysteine, chitosan oligosaccharides or carboxymethyl chitosan surface modified nanostructured lipid carrier for ophthalmic delivery of curcumin. Carbohydr Polym 146:435–444. https://doi.org/10.1016/j.carbpol.2016.03.079

Lieberman J (2018) Tapping the RNA world for therapeutics. Nat Struct Mol Biol 25(5):357–364. https://doi.org/10.1038/s41594-018-0054-4

Liu W et al (2011) Tumor-derived vascular endothelial growth factor (VEGF)-a facilitates tumor metastasis through the VEGF-VEGFR1 signaling pathway. Int J Oncol 39(5):1213–1220. https://doi.org/10.3892/ijo.2011.1138

Liu S-H, Chen F-W, Chiang M-T (2021a) Chitosan oligosaccharide alleviates abnormal glucose metabolism without inhibition of hepatic lipid accumulation in a high-fat diet/streptozotocin-induced diabetic rat model. Mar Drugs 19(7):360. https://doi.org/10.3390/md19070360

Liu X et al (2021b) Enhancing anti-melanoma outcomes in mice using novel chitooligosaccharide nanoparticles loaded with therapeutic survivin-targeted siRNA. Eur J Pharm Sci 158:105641. https://doi.org/10.1016/j.ejps.2020.105641

Lopes JR et al (2013) Physical and chemical stimuli-responsive drug delivery systems: targeted delivery and main routes of administration. Curr Pharm Des 19(41):7169–7184. https://doi.org/10.2174/13816128113199990698

Lozano MV et al (2008) Highly efficient system to deliver taxanes into tumor cells: docetaxel-loaded chitosan oligomer colloidal carriers. Biomacromol 9(8):2186–2193. https://doi.org/10.1021/bm800298u

Maganti N et al (2011) Structure-process-property relationship of biomimetic chitosan-based nanocomposite scaffolds for tissue engineering: biological, physico-chemical, and mechanical functions. Adv Eng Mater 13(3):B108–B122. https://doi.org/10.1002/adem.201080094

Mahmoodi Chalbatani G et al (2019) Small interfering RNAs (siRNAs) in cancer therapy: a nano-based approach. Int J Nanomed 14:3111–3128. https://doi.org/10.2147/IJN.S200253

Mallakuntla MK et al (2021) Chitooligosaccharides induce apoptosis in human breast cancer cells. Carbohydr Polym Technol Appl 2:100077. https://doi.org/10.1016/j.carpta.2021.100077

Manzari MT et al (2021) Targeted drug delivery strategies for precision medicines. Nat Rev Mater 6(4):351–370. https://doi.org/10.1038/s41578-020-00269-6

Mei Y et al (2013) Protective effect of chitooligosaccharides against cyclophosphamide-induced immunosuppression in mice. Int J Biol Macromol 62:330–335. https://doi.org/10.1016/j.ijbiomac.2013.09.038

Mendis E et al (2007) An in vitro cellular analysis of the radical scavenging efficacy of chitooligosaccharides. Life Sci 80(23):2118–2127. https://doi.org/10.1016/j.lfs.2007.03.016

Mengíbar M et al (2013) Influence of the physico-chemical characteristics of chito-oligosaccharides (COS) on antioxidant activity. Carbohydr Polym 97(2):776–782. https://doi.org/10.1016/j.carbpol.2013.05.035

Mizuno K et al (2008) Pretreatment with low doses of erythropoietin ameliorates brain damage in periventricular leukomalacia by targeting late oligodendrocyte progenitors: a rat model. Neonatology 94(4):255–266. https://doi.org/10.1159/000151644

Moorthy MS et al (2019) Chitosan oligosaccharide coated mesoporous silica nanoparticles for pH-stimuli responsive drug delivery applications. J Porous Mater 26(1):217–226. https://doi.org/10.1007/s10934-018-0646-8

Morales JO et al (2017) Challenges and future prospects for the delivery of biologics: oral mucosal, pulmonary, and transdermal routes. AAPS J 19(3):652–668. https://doi.org/10.1208/s12248-017-0054-z

Muanprasat C, Chatsudthipong V (2017) Chitosan oligosaccharide: biological activities and potential therapeutic applications. Pharmacol Ther 170:80–97. https://doi.org/10.1016/j.pharmthera.2016.10.013

Naqvi S, Moerschbacher BM (2017) The cell factory approach toward biotechnological production of high-value chitosan oligomers and their derivatives: an update. Crit Rev Biotechnol 37(1):11–25. https://doi.org/10.3109/07388551.2015.1104289

Naveed M et al (2019) Chitosan oligosaccharide (COS): an overview. Int J Biol Macromol 129:827–843. https://doi.org/10.1016/j.ijbiomac.2019.01.192

Nguyen M-H, Lee SE et al (2019a) A simple strategy to enhance the in vivo wound-healing activity of curcumin in the form of self-assembled nanoparticle complex of curcumin and oligochitosan. Mater Sci Eng C Mater Biol Appl 98:54–64. https://doi.org/10.1016/j.msec.2018.12.091

Nguyen M-H, Vu N-B-D et al (2019b) In vivo comparison of wound healing and scar treatment effect between curcumin-oligochitosan nanoparticle complex and oligochitosan-coated curcumin-loaded-liposome. J Microencapsul 36(2):156–168. https://doi.org/10.1080/02652048.2019.1612476

Nishikawa M et al (1992) Pharmacokinetics of receptor-mediated hepatic uptake of glycosylated albumin in mice. Int J Pharm 85(1):75–85. https://doi.org/10.1016/0378-5173(92)90136-P

Oh G-W et al (2021) Characterization of an oxidized alginate-gelatin hydrogel incorporating a COS-salicylic acid conjugate for wound healing. Carbohydr Polym 252:117145. https://doi.org/10.1016/j.carbpol.2020.117145

Okamoto Y et al (2003) Effects of chitin and chitosan on blood coagulation. Carbohyd Polym 53(3):337–342. https://doi.org/10.1016/S0144-8617(03)00076-6

Onishi H, Machida Y (1999) Biodegradation and distribution of water-soluble chitosan in mice. Biomaterials 20(2):175–182. https://doi.org/10.1016/S0142-9612(98)00159-8

Parachoniak CA, Park M (2012) Dynamics of receptor trafficking in tumorigenicity. Trends Cell Biol 22(5):231–240. https://doi.org/10.1016/j.tcb.2012.02.002
Parhi R (2020) Drug delivery applications of chitin and chitosan: a review. Environ Chem Lett 18(3):577–594. https://doi.org/10.1007/s10311-020-00963-5
Park E-J et al (2012) Doxorubicin induces cytotoxicity through upregulation of pERK–dependent ATF3. PLoS ONE 7(9):e44990. https://doi.org/10.1371/journal.pone.0044990
Park H-H et al (2018) Characterization and biological activity of PVA hydrogel containing chitooligosaccharides conjugated with gallic acid. Carbohydr Polym 198:197–205. https://doi.org/10.1016/j.carbpol.2018.06.070
Park P-J, Je J-Y, Kim S-K (2003) Free radical scavenging activity of chitooligosaccharides by electron spin resonance spectrometry. J Agric Food Chem 51(16):4624–4627. https://doi.org/10.1021/jf034039+
Peppas NA, Reinhart CT (1983) Solute diffusion in swollen membranes. Part I. A new theory. J Membr Sci 15(3):275–287. https://doi.org/10.1016/S0376-7388(00)82304-2
Qiu F, He Z-G, Li H-Z (2003) HPLC analyses and pharmacokinetic studies of baicalin and oxymatrine in rabbits. Pharmazie 58(9):616–619
Rehmann MS et al (2017) Tuning and predicting mesh size and protein release from step growth hydrogels. Biomacromol 18(10):3131–3142. https://doi.org/10.1021/acs.biomac.7b00781
Richardson SC, Kolbe HV, Duncan R (1999) Potential of low molecular mass chitosan as a DNA delivery system: biocompatibility, body distribution and ability to complex and protect DNA. Int J Pharm 178(2):231–243. https://doi.org/10.1016/s0378-5173(98)00378-0
Rosenblum D et al (2018) Progress and challenges towards targeted delivery of cancer therapeutics. Nat Commun 9(1):1410. https://doi.org/10.1038/s41467-018-03705-y
Sacco P et al (2018) The role played by the molecular weight and acetylation degree in modulating the stiffness and elasticity of chitosan gels. Carbohyd Polym 196:405–413. https://doi.org/10.1016/j.carbpol.2018.05.060
Salah R et al (2013) Anticancer activity of chemically prepared shrimp low molecular weight chitin evaluation with the human monocyte leukaemia cell line, THP-1. Int J Biol Macromol 52:333–339. https://doi.org/10.1016/j.ijbiomac.2012.10.009
Sandri G et al (2012) The role of chitosan as a mucoadhesive agent in mucosal drug delivery. J Drug Deliv Sci Technol 22(4):275–284. https://doi.org/10.1016/S1773-2247(12)50046-8
Sandri G et al (2017) Halloysite and chitosan oligosaccharide nanocomposite for wound healing. Acta Biomater 57:216–224. https://doi.org/10.1016/j.actbio.2017.05.032
Saravanakumar G et al (2009) Hydrotropic oligomer-conjugated glycol chitosan as a carrier of paclitaxel: synthesis, characterization, and in vivo biodistribution. J Control Release: off J Control Release Soc 140(3):210–217. https://doi.org/10.1016/j.jconrel.2009.06.015
Schipper NG et al (1997) Chitosans as absorption enhancers for poorly absorbable drugs 2: mechanism of absorption enhancement. Pharm Res 14(7):923–929. https://doi.org/10.1023/a:1012160102740
Schipper NG, Vårum KM, Artursson P (1996) Chitosans as absorption enhancers for poorly absorbable drugs. 1: influence of molecular weight and degree of acetylation on drug transport across human intestinal epithelial (Caco-2) cells. Pharm Res 13(11):1686–1692. https://doi.org/10.1023/a:1016444808000
Seong JS, Yun ME, Park SN (2018) Surfactant-stable and pH-sensitive liposomes coated with N-succinyl-chitosan and chitooligosaccharide for delivery of quercetin. Carbohydr Polym 181:659–667. https://doi.org/10.1016/j.carbpol.2017.11.098
Shahbazi B et al (2015) Preparation and characterization of silk fibroin/oligochitosan nanoparticles for siRNA delivery. Colloids Surf B Biointerfaces 136:867–877. https://doi.org/10.1016/j.colsurfb.2015.10.044
Shaikh R et al (2011) Mucoadhesive drug delivery systems. J Pharm Bioallied Sci 3(1):89–100. https://doi.org/10.4103/0975-7406.76478
Shao K et al (2015) Pharmacokinetics and biodegradation performance of a hydroxypropyl chitosan derivative. J Ocean Univ China 14(5):888–896. https://doi.org/10.1007/s11802-015-2600-6

Song Z et al (2018) Oligochitosan-pluronic 127 conjugate for delivery of honokiol. Artif Cells Nanomed Biotechnol 46(sup1):740–750. https://doi.org/10.1080/21691401.2018.1434785
Song Z et al (2019) Linolenic acid-modified methoxy poly (ethylene glycol)-oligochitosan conjugate micelles for encapsulation of amphotericin B. Carbohydr Polym 205:571–580. https://doi.org/10.1016/j.carbpol.2018.10.086
Sparreboom A et al (1997) Limited oral bioavailability and active epithelial excretion of paclitaxel (Taxol) caused by P-glycoprotein in the intestine. Proc Natl Acad Sci 94(5):2031–2035. https://doi.org/10.1073/pnas.94.5.2031
Srinivasarao M, Low PS (2017) Ligand-targeted drug delivery. Chem Rev 117(19):12133–12164. https://doi.org/10.1021/acs.chemrev.7b00013
Stealey S et al (2020) Calcium-oligochitosan-pectin microcarrier for colonic drug delivery. Pharm Dev Technol 25(2):260–265. https://doi.org/10.1080/10837450.2019.1691591
Sutthasupha P, Lungkaphin A (2020) The potential roles of chitosan oligosaccharide in prevention of kidney injury in obese and diabetic conditions. Food Funct 11(9):7371–7388. https://doi.org/10.1039/d0fo00302f
Svenson S (2004) Carrier-based drug delivery. In: Carrier-based drug delivery. American Chemical Society (ACS Symposium Series, vol 879), pp 2–23. https://doi.org/10.1021/bk-2004-0879.ch001.
Tabassum N, Ahmed S, Ali MA (2021) Chitooligosaccharides and their structural-functional effect on hydrogels: a review. Carbohydr Polym 261:117882. https://doi.org/10.1016/j.carbpol.2021.117882
Tang Q et al (2019) A pH-responsive self-healing hydrogel based on multivalent coordination of Ni2+ with polyhistidine-terminated PEG and IDA-modified oligochitosan. J Mater Chem B 7(1):30–42. https://doi.org/10.1039/c8tb02360c
Thanou M, Verhoef JC, Junginger HE (2001) Chitosan and its derivatives as intestinal absorption enhancers. Adv Drug Deliv Rev 50(Suppl 1):S91-101. https://doi.org/10.1016/s0169-409x(01)00180-6
Thanou M et al (2002) Quaternized chitosan oligomers as novel gene delivery vectors in epithelial cell lines. Biomaterials 23(1):153–159. https://doi.org/10.1016/s0142-9612(01)00090-4
Trzciński S (2006) 13. Combined degradation of chitosans, p 9
Vo T-S, Kong C-S, Kim S-K (2011) Inhibitory effects of chitooligosaccharides on degranulation and cytokine generation in rat basophilic leukemia RBL-2H3 cells. Carbohydr Polym 84(1):649–655. https://doi.org/10.1016/j.carbpol.2010.12.046
Wang Z, Li Y (2018) Raloxifene/SBE-β-CD inclusion complexes formulated into nanoparticles with chitosan to overcome the absorption barrier for bioavailability enhancement. Pharmaceutics 10(3):E76. https://doi.org/10.3390/pharmaceutics10030076
Wang T et al (2012) Erythropoietin-loaded oligochitosan nanoparticles for treatment of periventricular leukomalacia. Int J Pharm 422(1–2):462–471. https://doi.org/10.1016/j.ijpharm.2011.10.058
Wang Y et al (2016) Biodegradable functional polycarbonate micelles for controlled release of amphotericin B. Acta Biomater 46:211–220. https://doi.org/10.1016/j.actbio.2016.09.036
Wang K et al (2018) Nanoceria-loaded injectable hydrogels for potential age-related macular degeneration treatment. J Biomed Mater Res Part A 106(11):2795–2804. https://doi.org/10.1002/jbm.a.36450
Whitehead KA et al (2011) Silencing or stimulation? siRNA delivery and the immune system. Annu Rev Chem Biomol Eng 2:77–96. https://doi.org/10.1146/annurev-chembioeng-061010-114133
Wong K et al (2006) PEI-g-chitosan, a novel gene delivery system with transfection efficiency comparable to polyethylenimine in vitro and after liver administration in vivo. Bioconjug Chem 17(1):152–158. https://doi.org/10.1021/bc0501597
Wu Y et al (2008) Coagulation property of hyaluronic acid–collagen/chitosan complex film. J Mater Sci Mater Med 19(12):3621–3629. https://doi.org/10.1007/s10856-008-3477-3

Wu H et al (2012) Inhibition of angiogenesis by chitooligosaccharides with specific degrees of acetylation and polymerization. Carbohydr Polym 89(2):511–518. https://doi.org/10.1016/j.carbpol.2012.03.037

Wu N et al (2015) Immunostimulative activity of low molecular weight chitosans in RAW264.7 macrophages. Mar Drugs 13(10):6210–6225. https://doi.org/10.3390/md13106210

Xin C et al (2018) Stearic acid-grafted chitooligosaccharide nanomicelle system with biocleavable gadolinium chelates as a multifunctional agent for tumor imaging and drug delivery. Pharm Res 36(1):10. https://doi.org/10.1007/s11095-018-2530-2

Xing R et al (2017) Monomer composition of chitooligosaccharides obtained by different degradation methods and their effects on immunomodulatory activities. Carbohydr Polym 157:1288–1297. https://doi.org/10.1016/j.carbpol.2016.11.001

Xing P, Zhao Y (2018) Supramolecular vesicles for stimulus-responsive drug delivery. Small Methods 2(4):1700364. https://doi.org/10.1002/smtd.201700364

Xiong C et al (2009) Potent angiogenic inhibition effects of deacetylated chitohexaose separated from chitooligosaccharides and its mechanism of action in vitro. Carbohydr Res 344(15):1975–1983. https://doi.org/10.1016/j.carres.2009.06.036

Yamada K et al (2005) Chitosan oligomers as potential and safe absorption enhancers for improving the pulmonary absorption of interferon-alpha in rats. J Pharm Sci 94(11):2432–2440. https://doi.org/10.1002/jps.20454

Yang H-C, Hon M-H (2009) The effect of the molecular weight of chitosan nanoparticles and its application on drug delivery. Microchem J 92(1):87–91. https://doi.org/10.1016/j.microc.2009.02.001

Yang H-C, Hon M-H (2010) The effect of the degree of deacetylation of chitosan nanoparticles and its characterization and encapsulation efficiency on drug delivery. Polym-Plast Technol Eng 49(12):1292–1296. https://doi.org/10.1080/03602559.2010.482076

Yhee JY et al (2013) The EPR effect in cancer therapy. In: Bae YH, Mrsny RJ, Park K (eds) Cancer targeted drug delivery: an elusive dream. Springer, New York, NY, pp 621–632. https://doi.org/10.1007/978-1-4614-7876-8_23

Yin X et al (2017) Chitooligosaccharides modified reduction-sensitive liposomes: enhanced cytoplasmic drug delivery and osteosarcomas-tumor inhibition in animal models. Pharm Res 34(10):2172–2184. https://doi.org/10.1007/s11095-017-2225-0

Yin X et al (2018) Estrogen-functionalized liposomes grafted with glutathione-responsive sheddable chotooligosaccharides for the therapy of osteosarcoma. Drug Deliv 25(1):900–908. https://doi.org/10.1080/10717544.2018.1458920

Yomota C, Komuro T, Kimura T (1990) Studies on the degradation of chitosan films by lysozyme and release of loaded chemicals. Yakugaku Zasshi: J Pharm Soc Jpn 110(6):442–448. https://doi.org/10.1248/yakushi1947.110.6_442

Yousef M et al (2012) Chitosan oligosaccharide as potential therapy of inflammatory bowel disease: therapeutic efficacy and possible mechanisms of action. Pharmacol Res 66(1):66–79. https://doi.org/10.1016/j.phrs.2012.03.013

Yu S-Y et al (2017) Antidiabetic effect of chitosan oligosaccharide (GO2KA1) is mediated via inhibition of intestinal alpha-glucosidase and glucose transporters and PPARγ expression. BioFactors 43(1):90–99. https://doi.org/10.1002/biof.1311

Yuan Z et al (2011) Enhanced accumulation of low-molecular-weight chitosan in kidneys: a study on the influence of N-acetylation of chitosan on the renal targeting. J Drug Target 19(7):540–551. https://doi.org/10.3109/1061186X.2010.521158

Yue Z-G et al (2011) Surface charge affects cellular uptake and intracellular trafficking of chitosan-based nanoparticles. Biomacromol 12(7):2440–2446. https://doi.org/10.1021/bm101482r

Zaghloul N, Patel H, Ahmed MN (2017) A model of Periventricular Leukomalacia (PVL) in neonate mice with histopathological and neurodevelopmental outcomes mimicking human PVL in neonates. PLoS ONE 12(4):e0175438. https://doi.org/10.1371/journal.pone.0175438

Zeng L et al (2008) Absorption and distribution of chitosan in mice after oral administration. Carbohydr Polym 71(3):435–440. https://doi.org/10.1016/j.carbpol.2007.06.016

Zhang J et al (2010) Chitosan modification and pharmaceutical/biomedical applications. Mar Drugs 8(7):1962–1987. https://doi.org/10.3390/md8071962

Zhang J et al (2015) Polymeric nanoparticles based on chitooligosaccharide as drug carriers for co-delivery of all-trans-retinoic acid and paclitaxel. Carbohydr Polym 129:25–34. https://doi.org/10.1016/j.carbpol.2015.04.036

Zhang W et al (2016) Novel pectin-based carriers for colonic drug delivery. Pharm Dev Technol 21(1):127–130. https://doi.org/10.3109/10837450.2014.965327

Zhang J et al (2020) Microencapsulation of immunoglobulin Y: optimization with response surface morphology and controlled release during simulated gastrointestinal digestion. J Zhejiang Univ Sci B 21(8):611–627. https://doi.org/10.1631/jzus.B2000172

Zhao Y et al (2018) Chitosan oligosaccharides alleviate PM2.5-induced lung inflammation in rats. Environ Sci Pollut Res 25. https://doi.org/10.1007/s11356-018-3365-4

Zheng F et al (2007) Chitosan nanoparticle as gene therapy vector via gastrointestinal mucosa administration: results of an in vitro and in vivo study. Life Sci 80(4):388–396. https://doi.org/10.1016/j.lfs.2006.09.040

Zheng B et al (2016) Molecular weight-dependent immunostimulative activity of low molecular weight chitosan via regulating NF-κB and AP-1 signaling pathways in RAW264.7 macrophages. Mar Drugs 14(9):169. https://doi.org/10.3390/md14090169

Zheng Y et al (2020) Protective effect of chitosan oligosaccharide against oxidative damage of peripheral blood mononuclear cells in dairy cows induced by diethylenetriamine/nitric oxide via NF-κB signalling pathway. Ital J Anim Sci 19(1):602–609. https://doi.org/10.1080/1828051X.2020.1772131

Zhou X et al (2010) Effect of low molecular weight chitosans on drug permeation through mouse skin: 1. Transdermal delivery of baicalin. J Pharm Sci 99(7):2991–2998. https://doi.org/10.1002/jps.22063

Influence of Isolation Conditions on the Physicochemical and Biological Properties of Chitosan and Chitosan Oligosaccharides from Marine Crustacean Shell Wastes

S. Sabu, A. Sasidharan, and V. Venugopal

1 Introduction

Natural biodegradable polymers are slowly replacing synthetic polymers because of the advantages of the latter due to their significant capability in applications such as biomedicine and pharmaceuticals (Naveed et al. 2019). The advantages of natural biopolymers include biodegradability, biocompatibility, and non-toxicity. Natural macromolecules, especially polysaccharides, possess significant bioactive properties and are generally isolated from crustacean shell wastes and agricultural feedstocks (Ramesh and Tharanathan 2003).

The shellfish processing industry generates vast quantities of shell wastes, which are mostly discarded, leading to serious environmental hazards. An estimated 3.4 million tons of shrimp are globally harvested; many countries in the Asian continent are grossly engaged in shrimp production (Tolesa et al. 2019). Shrimp shell wastes generally have 20–30% chitin, 30–50% minerals, and 30–40% proteins, besides a small amount of pigments and lipids (Younes et al. 2016; Liu et al. 2019). The major minerals found in the exoskeleton of shrimp are calcium as calcium carbonate ($CaCO_3$), followed by magnesium (Percot et al. 2003). The calcium and protein, as well as protein derivatives such as hydrolysates and peptides could be suitable food additives, food products development, and ingredients for environment-friendly plant fertilizers (El Knidri et al. 2018; Hu et al. 2020).

S. Sabu (✉)
School of Industrial Fisheries, Cochin University of Science and Technology, Kochi, Kerala 682016, India
e-mail: sabuif@cusat.ac.in

A. Sasidharan · V. Venugopal
Department of Fish Processing Technology, Kerala University of Fisheries and Ocean Studies, Kochi, Kerala 682506, India

S.-K. Kim (ed.), *Chitooligosaccharides*,
https://doi.org/10.1007/978-3-030-92806-3_20

Chitin and its deacetylated form, chitosan, is reported to have great interest among researchers and industrialists as a natural biopolymer with versatile, functional properties (Alishahi and Aïder 2012; Venugopal 2009, 2011; Elwakeel 2010). Chitooligoosaccharides or chitosan oligosaccharides (COSs) are the degraded products of chitin or chitosan, which are obtained by acid hydrolysis, enzymatic degradation, or both. COSs have shorter chain lengths and free amino groups attached to D-glucosamine units. They are less viscous and are soluble in water (Naveed et al. 2019; Lodhi et al. 2014; Du and Bui 2019). COSs can be defined as chitosans with degree of polymerization (DP) less than 20 and an average molecular weight less than 3,900 (Mourya et al. 2011). Their positive physiological effects, non-toxicity, and higher solubility attracted the scientific community to use these compounds in pharmaceutical and medicinal applications as well as nutraceuticals and cosmeceuticals (Areco et al. 2015; Swiatkiewicz et al. 2015; Lodhi et al. 2014).

2 Raw Materials for COS Preparation

Chitin is a polymer of β (1 → 4) N-acetyl D-glucosamine units and is predominantly found in the outer shells of crustaceans. It is the second most ubiquitous biopolymer in nature after cellulose (Abdel-Rahman et al. 2015; Trache 2018; Trache et al. 2020). Chitin is regarded as a derivative of cellulose, in which the C-2 hydroxyl group (–OH) is substituted by an acetamido group (Abdou et al. 2008; Abdel-Rahman et al. 2015; Gbenebor et al. 2017; Trache et al. 2020). Both are polymers of monosaccharide made up of β-(1–4)-2-acetamido-2-deoxy-β-D-glucose and β-(1–4)-2-deoxy-β-D-glucopyranose units, respectively. Thus, chitin is poly (β-(1–4)-N-acetyl-D-glucosamine) (Pillai et al. 2009). Chitin varies in crystal structure and includes α, β, and γ allomorphic forms. Alpha-chitin is aligned in an anti-parallel manner with strong intra- and inter-sheets hydrogen bonds and is generally isolated from the exoskeletons of crustaceans. Beta-chitin has a parallel arrangement with weak intra-sheets hydrogen bonds and is usually isolated from squid pens. Gamma-chitin is a mixture of anti-parallel and parallel chains and is obtained from fungi and yeast (Cuong et al. 2016; Kaya et al. 2017). Shells, particularly those of crabs and shrimps, are commercial sources of chitin. The polysaccharide remains bound to proteins and minerals in the shell (Kaur and Dhillon 2015; El Knidri et al. 2018).

3 Conversion of Chitin and Chitosan into Oligosaccharides

Chitin is the principal raw material for the manufacture of the commercially more pertinent polymer, chitosan, and chitosan oligomers (COSs). Chitosan, like chitin, is a copolymer of GlcN and GlcNAc units, GlcN being the predominant constituent (Rinaudo 2006). Chitosan is thus considered a deacetylated form of chitin, frequently defined by the degree of acetylation (DA). Even though chitosan transforms to chitin when the DA reaches 50%, for practical reasons, the forename chitosan is used for molecules with a lower DA, and chitin is normally reticent about portraying polymers where the DA is ~100%. Other than DA, the properties of chitin and chitosan, including the degree of polymerization (DP), indicate the chain length and associate with the molecular weight (MW) in kDa (No et al. 2002). For bigger oligomers and polymers, the chain length has a propensity to vary within a certain range, and therefore the DP may be given an estimated or average value, or the average MW in kDa may be used in its place. Schmitz et al. (2019) while considering these facts, observed that the consistent production of COS with definite properties is essential to guarantee unswerving biological and physiological activity. They also observed that COSs are more soluble and physiologically active than chitosan. They stated that given the never-ending variety of molecules that could be created by methodically scheming the DA, DP, and PA, the construction of COSs could permit a new mechanism of action and potentially new biological and physiological applications.

Depolymerization of chitin or chitosan to COSs can be achieved by (i) chemical, (ii) physical, or (iii) enzymatic method (Mourya et al. 2011; Aam et al. 2010). Controlled acid or enzymatic hydrolysis of chitin or chitosan gives COSs. Hydrochloric acid, nitrous acid, phosphoric acid, hydrofluoric acid, and acids with electrolytes are used in acid hydrolysis. Chitinase enzyme and other non-specific enzymes are used in the enzymatic hydrolysis methods. Methods employing oxidative reductase with hydrogen peroxide or persulfate are also used (Il'ina and Varlamov 2007). Other acids such as formic acid, trichloroacetic acid, lactic acid, and acetic acids are also tried for degrading chitin and chitosan (Mourya et al. 2011). A pictorial representation of conversion of shell waste into COS is depicted in Fig. 1.

3.1 Chemical Methods

The chemical deprivation method is uncomplicated to execute, but the products can have wide variations in MWs and as a result, it is rather cumbersome to separate and purify the products. The major methods involved in the chemical degradation of COS include acid and oxidative degradation. Jia and Shen (2002) prepared low-molecular-weight chitosan with 85% phosphoric acid, which could be attained by increasing hydrolysis time. Trombotto et al. (2008) examined a different acid hydrolysis method to prepare COSs. Initially, fully *N*-deacetylated chitosan was depolymerized by concentrated HCl, which resulted in the production of glucosamine (GlcN) oligomers having a reduced degree of polymerization. Toward the end of the process, activated charcoal was used to bleach; the surplus HCl was removed by co-vaporization. This was followed by hyper-filtration to isolate COSs having the lowest degree of polymerization. The hydrolysate hyper filtrated through a cellulose acetate membrane having a 500 g/mol cutoff was subjected to precipitation by absolute ethanol. Oligomers having the highest degree of polymerization were obtained by enhancing its pH to nine by the addition of NaOH. The fraction having

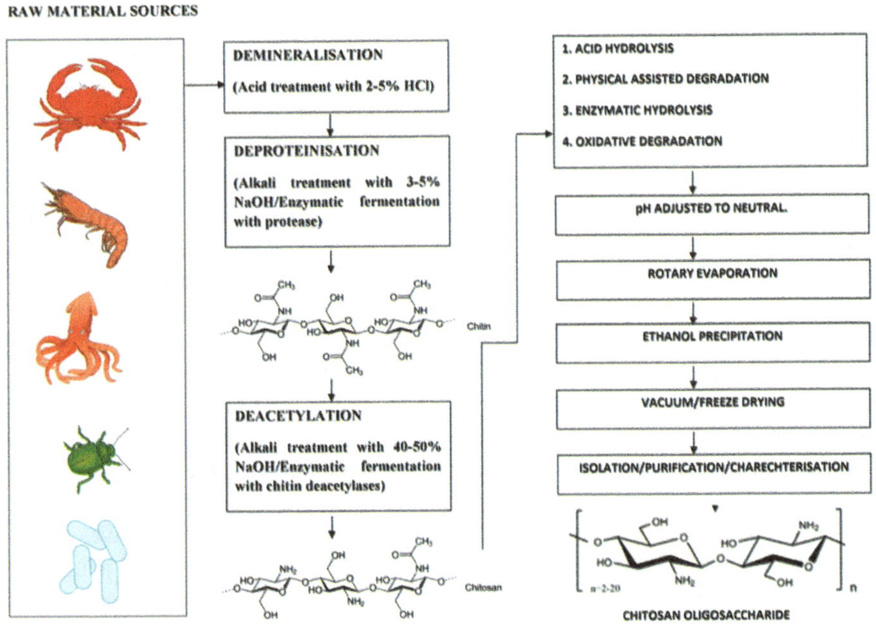

Fig. 1 Process of conversion of raw materials into COSs

a low degree of polymerization GlcN oligomers, making up 7.5% of the raw material weight, could be isolated by centrifugation or freeze-drying. Nitrous acid could also be utilized to prepare COS (Tømmeraas et al. 2002). Alternately, an oxidative degradation method could be used to prepare COSs where oxidants such as hydrogen peroxide, ozone, sodium perborate, and potassium persulfate were used (Tishchenko et al. 2011). Xia et al. (2013a, b) used oxidation by two percent (v/v) hydrogen peroxide and 0.1% phosphotungstic acid at 65 °C for a duration of 40 min. The final product obtained was white with a degree of polymerization of about seven.

3.2 *Physical Methods*

Degradation by physical means is an easier method to obtain COS having an enhanced purity. Liang et al. (2018) pointed out that ultrasonic, microwave, and gamma rays could be used to prepare low-molecular-weight chitosan with diminutive contamination. However, the productivity was low. Baxter et al. (2005) utilized acetic acid to treat chitosan with ultrasonic assistance. It was observed that ultrasonic energy in the low to medium power range can modify the molecular weight of chitosan. Xing et al. (2005) reported that hydrolysis could be speeded up by the addition of inorganic salts in a microwave irradiation field. Zainol et al. (2009) found that Co^{60} γ-ray irradiation could degrade chitosan with mild browning effect on the polysaccharide, while functional groups are not significantly affected. Hydrodynamic cavitation is another potential method to degrade organic polymers (Amin et al. 2010). Swirling cavitation, a kind of hydrodynamic cavitation, was utilized by Wu et al. (2014) for degrading chitosan. The extent of degradation was influenced by treatment time, pressure, temperature, and solution concentration. The optimum conditions were solution concentration of 1 g L^{-1}, 0.4 MPa, 60 °C, and a treatment time of three hours. Treatment under these conditions lowered the original viscosity of the sample by 83.65%.

3.3 Enzymatic Methods

Enzymatic method is easier to operate and monitor than chemical hydrolysis. Further, the products can be obtained without supplementary modifications. The enzymes used may include specific hydrolases such as chitinase, chitosanase, glucanase, etc., and non-specific enzymes such as lysozyme, protease, lipase, amylase, cellulose, among others (Cheng et al. 2006). Chitosanases (EC 3.2.1.132) are glycosyl hydrolases that catalyze the endohydrolysis of β-1,4-glycosidic bonds of partially acetylated chitosan to release chitosan oligosaccharides (COS). Chitosanases are isolated, purified, and characterized from different sources mainly from bacteria and fungi (Thadathil and Velappan 2014). Chitosanases have received much attention due to their wide range of applications. The cost of the process may, however, be influenced by the availability and specificity of the enzyme used (Kumar et al. 2004). *Bacillus alvei* produces an extracellular enzyme; the enzyme could reduce the molecular weight of chitosan from 88,000 to 5,000 Da, associated with a significant reduction in the viscosity during a treatment time of two hours at a pH 5.5 and 37 °C, two hours, at a chitosan concentration of one percent w/v (Hernandez et al. 2017). A commercial α-amylase can also be utilized for the production of COS from *Clanis bilineata*. The larvae skin was treated with the α-amylase which resulted in demineralization and deproteination and deacetylation. The prepared COS had a degree of polymerization in the range of two to eight and antibacterial activity (Wu 2012). LIChi18A is a recombinant chitinase extracted from *Lactococcus lactis* and over-expressed in *Escherichia coli* BL21 (DE3). The enzyme was used to extract COS from fungal waste mycelia (Lv et al. 2016). Mallakuntla et al. (2017) extracted a single domain hyper-transglycosylating chitinase from *Enterobacter cloacae* subsp. *cloacae* 13047 (*EcChi1*), the treatment with which could generate longer chitin oligosaccharides under optimal conditions of 40 °C and pH 5.0. Sánchez et al. (2017) prepared COS with one-step (P1) and two-step (P2) processes. Chitosan was diluted in 0.2 M acetic/acetate buffer in P1 and depolymerized by chitosanases from *Streptomyces griseus*. In P2, chitosan was initially dissolved in 0.1 M HCl and chemically hydrolyzed by KNO_2. The solution was then precipitated by NH_4OH and washed repeatedly with ethanol/distilled water. The resulting low-molecular-weight chitosan was depolymerized with the same chitosanase in P1. P1 gave a COS preparation, which was composed of 27% of fully deacetylated sequences while the preparation from P2 had 63% fully deacetylated sequences. Utilizing complex enzymes composed of commercial α-amylase, cellulose, and pectinase for chitosan hydrolysis, resulted in COS with a molecular weight below 4,000 Da after two hours, and more than 90% COS had good water solubility (Xie et al. 2011). Chitin obtained by fermentation by successive two-step fermentation maintains excellent physicochemical and structural properties of commercial chitin (Liu et al. 2020).

4 Influence of Isolation Conditions on the Physicochemical Properties of Chitosan and COS

The quality of isolated chitin depends on its source, temperature, duration of extraction treatment, and the type of extractants used (Kaya et al. 2015). The order of removals of protein (deproteinization) and minerals (demineralization) could be interchanged without serious adverse effects on the yield and quality of chitin (No and Meyers 1995; Kjartansson et al. 2006; Kaur and Dhillon 2015; Tolesa et al. 2019; Oyatogun et al. 2020). Degree of deacetylation (DA) and molecular weight (Mw) are considered critical quality parameters of chitin and chitosan. Gbenebor et al. (2017) reported the influence of diverse acid and alkaline treatments on DA, Mw, solubility, and crystallinity of chitin. It was observed that the chemical process could have a pronounced effect on the DA of chitin (Kaur and Dhillon 2015). Rasweefali et al. (2021) reported that the process of isolation had a strong impact on the DD and Mw of chitosan.

Characterization of the chitosan is crucial according to the structure–properties relationship to define their possible industrial applications (de Alvarenga et al. 2010). Baskar and Kumar (2009) observed that variations in alkali concentration/temperature/reaction time could enhance the deacetylation process, resulting in a range of chitosan molecules having different characteristics and applications. Chemical extraction involving acids and alkali is considered the best method for chitosan industrial production because of its shorter reaction time and capacity to remove complete organic salts and proteins. Several studies have focused on optimizing the factors influencing extraction of chitosan from various crustacean sources (Baskar and Kumar 2009; Al Sagheer et al. 2009; Bajaj et al. 2011; Abdel-Rahman et al. 2015). Table 1 represents a few recent studies following three of the major methods of CAS isolation such as acid hydrolysis, oxidation degradation, and enzymatic hydrolysis and the effect of experimental conditions on the final product quality.

Table 1 Major COS preparation methods and their highlights

Method	Raw material	Experimental conditions	Product characteristics	Reference
Chemical methods				
	Chitosan (Cuban lobster) DA 12%, MW 82.9 kDa	Concentrated HCl, 72 °C, 30 min	DPs up to 16	Cabrera and Cutsem (2005)
	α-Chitosan (shrimp shells) DD 82%, 658 kDa	Concentrated HCl, 70 °C	DP 6, DP 11	Li et al. (2012a, b)
	Chitosan DD 97%, 300 kDa	AcOH (6, 9, 12%) 65, 75, 85 °C; 0–7,500 min	32 kDa to 2 kDa	Tsao et al. (2011)
	Chitosan, DA 11.6 ± 4%	Dissolve in two percent AcOH and add equal volume of nitrite	DP 1–11	Han et al. (2015)
	Chitosan (shrimp shells) DA 0.85, MW 560 kDa	HCl (2%), add inorganic salt, react for 25 min with the assistance of 480–800 W microwave	MW 3 × 104 Da, structures of the main chain of chitosan and product were the same	Ibrahim et al. (2016)
	Chitosan DD 64.9%, MW 123 kDa	H_2 O_2-AcOH molar ratio 1.7, 50 °C, eight hours	Yield 86%, COS H7 DD 69.3%, MW 9.03 kDa	Tishchenko et al. (2011)
	Chitosan (shells) DD 95.1%, MW 247 ± 31.2 kg mol^{-1}	5.5% H_2 O_2, 37 °C, four hours	DD 95%, MW 2.7 ± 0.40 kg mol^{-1}	Avelelas et al. (2019)
	Chitosan (shrimp shells) DD 93.5%, MW 41 × 104 Da	Two percent H_2O_2, 65 °C, 40 min in 0.1% phosphotungstic acid	Yield 92.3%, average DP 7, structures of the main chain of chitosan and product were the same	Xia et al. (2013a, b)
	Chitosan (Metapenaeus dobsoni shrimp) DD 91.5%	Processing in plasma treatment 60 mM H_2 O_2, one hour	Yield 23%, DD 87.5%, DP 2–9	Chokradjaroen et al. (2017)
	Chitosan (crab shells) DD 82%, MW 658 × 103 Da	1% H_2O_2, with microwave irradiation at 800 W at 80 °C for 25 min	Yield 70%, DP 2–6	Xing et al. (2017)
Enzymatic methods				
	Chitosan DD 93%	Chitosanase (*Pseudoalteromonas* sp. SY39), enzyme dosage 18 U g^{-1}, pH 6.0, 40 °C, 10 h	Yield 47.8%, DP 2–5	Zhou et al. (2020)
	Chitosan DD 98%	Chitosanase (*Paenibacillus* sp. TKU047), pH 7.0, 37 °C, 24 h	Yield 68.44%, DP 2–9, DD 100%	Doan et al. (2020)

(continued)

Table 1 (continued)

Method	Raw material	Experimental conditions	Product characteristics	Reference
	4-MU-(GlcNAc) 3 (fungal chitin)	Chitinase (LlChi18A) pH 3.8, enzyme dosage 20 nM, 40 °C, 5 days	Yield 24%, purity 78%, N,N′-diacetylchitobiose	Lv et al. (2016)
	Chitosan (shrimp) DD 51.44%	Chitosanase (Aeromonas) 11.16), enzyme dosage 8 U g^{-1}, 30 °C, 4 h	Yield 96.14%, DP 1–6	Fawzya et al. (2019)
	Chitosan (shrimp shells) DD 91.5%, MW 995 ± 5 kDa	Enzyme mixture including cellulase and chitosanase, at pH 5.3, 45 °C, 6 h	Yield 79.8%, DP 6–8	Dong et al. (2015)

4.1 Molecular Weight (MW)

Molecular weight (MW) is one of the significant parameters to ensure the quality of chitosan (Rasweefali et al. 2021). In general, the Mw of chitosan determines its applications. DD and Mw of chitosan are greatly affected by reaction conditions like temperature, reagents concentrations, repetition of alkaline steps, time, and atmospheric conditions of the deacetylation (Tolaimate et al. 2003). Shrimp chitosan with 78% deacetylation (DD) exhibited a molecular weight of 6.27 kDa during treatment with 40% KOH for six hours at a temperature of 90 °C (Kumari et al. 2017). Factors influencing the average molecular weight of chitosan are DD, raw material, the concentration of alkali, extraction process, insoluble in the solution (Younes et al. 2014), chitin treatment, shear stress, dissolved oxygen concentration, and particle size (Kucukgulmez et al. 2011).

Based on MW, chitosan is classified into three different types, namely low-molecular-weight chitosan (LMWC; <50 kDa), medium-molecular-weight chitosan (MMWC; 50–250 kDa), and high-molecular-weight chitosan (HMWC) (>250 kDa). Chitosan with high molecular weight are suitable for removing biochemical oxygen demand, chemical oxygen demand, and turbidity, while low molecular weight and higher DD are suitable for NH_3 and PO_4^{3-} from wastewater since chitosan (pKa ~ 6) carry a large positive charge below pH 6 (Chung 2006; Chung et al. 2005). Chitosan with medium-molecular-weight possesses anti-cancer activities, and chitosan with low molecular weight has antitumor, antioxidant, and antibacterial properties (Anraku et al. 2011; Kara and Stevens 2002).

According to a study on chitin nanocrystal by Tian et al. (2017), the crystalline structure, thermal stability, and water solubility were considerably influenced by the MW, and with the decrease in MW, and change in the crystalline structure. The thermal stability was found to be decreasing with the reduction of the MW. It was observed by Sahariah et al. (2019) that the antimicrobial activity of trimethyl chitosan against *Staphylococcus aureus* increased significantly with an increase in molecular weight until a critical Mws of 20 and 50 kDa were reached for trimethyl chitosan and chitosan, respectively. Additional increase in Mw did not affect the antimicrobial activity. Bui (2019) observed that water-soluble chitosan with lower molecular weight exhibited considerably higher antioxidant activity than the one with higher molecular weight.

4.2 *Viscosity*

Chitosan, when dissolved in an acidic medium, gives a viscous solution. The viscosity of chitosan solution is related to the Mw, DD, concentration, pH, and temperature of chitosan solution. The viscosity of chitosan solution, at the molecular level, is a direct measure of the volume of the polymer molecules, which in turn is governed by the molecular size or chain length (Chattopadhyay and Inamdar 2010). Jeon et al. (2002) reported that viscosities of crab chitosan (having DD 86.40–91.30%) decreased from 369, 57, to 14 cP with increasing deacetylation times from four, ten, to 20 h, respectively. Similar findings were reported by Rasweefali et al. (2021) who observed correlation with DD and viscosity. Chitosan extracted at 100 ± 2 °C for a period of six hours recorded its highest DD and lowest viscosity (Rasweefali et al. 2021). The antimicrobial property of chitosan depends on its viscosity (Shirvan et al. 2018). The biological properties of chitosan are observed to be largely dependent on its polymerization degree (PD), deacetylation degree (DA), and viscosity (Basa et al. 2020; Ahmed et al. 2020). Cho et al. (1998) reported that chitosan's antimicrobial activity increased with decreasing viscosity from 1,000 to ten cP. Intrinsic viscosity (η) is a measurement of the hydrodynamic volume taken by macromolecules in solution and is consequently an indication of their size. It is highly dependent on the concentration of the polymer in the solution. This property is significant for understanding the performance of chitosan in solution, and its rheology since films, membranes, fibers, and hydrogels are prepared from solutions of chitosan polysaccharide (Dima et al. 2013). Khalil-Bagy et al. (2020) studied chitosan of different molecular weights and different viscosity ranges (from 4.12 to 12.19) and found that the nematicidal activity of chitosan augmented as the molecular weight and viscosity decreased. They also indicated that as the molecular weight and viscosity increased, the nematicidal effect decreased significantly.

4.3 *Degree of Deacetylation (DD)*

The deacetylation process refers to removing acetyl groups from the molecular chain of chitin and the substitution of amino groups (NH_2). Applications of chitosan mainly depend on reactive amino groups contained in the structure (Shirvan et al. 2018). The term chitosan is applicable only when the degree of deacetylation is above 70%. DD influences physicochemical and biological properties, viz. ability to interact with metal ions, biodegradability, acid–base and electrostatic properties, sorption properties, and self-aggregation (Kumari et al. 2017). Factors affecting the DD of chitosan are the isolation method of chitin, sources, species, alkaline concentration, reaction time, and temperature (El Knidri et al. 2018; Shirvan et al. 2018). NaOH concentration up to 60% was suggested for obtaining chitosan having significantly higher DD (Hargono and Djaeni 2010). The DD of chitosan has also been observed to affect the physical properties of chitosan-based papers (Bhardwaj et al. 2020). The impact of DD on ionic conductivity of chitosan was studied by Senthilkumar et al. (2019), who observed that when compared to chitosan with high degree of deacetylation (HDD), the ionic conductivity for the chitosan with low degree of deacetylation (LDD) was high. The conductivity was also found to be influenced by increases in concentration and temperature. The observations made by Mathaba and Daramola (2020) demonstrated that the degree of deacetylation influenced performance and the efficiency of chitosan-infused polyether sulphone membrane.

5 Biological Activities of Chitosan and COS

Biological properties of chitosan and COS are greatly influenced by the degree of deacetylation (DD), Mw, solubility, crystallinity, and the distribution of GlcNAc and GlcN units along the polymeric chain (Younes et al. 2014). Effects of Mw on antimicrobial properties of the chitosan and its derivatives have been examined (Lim and Hudson 2004; Younes et al. 2014; Rabea et al. 2003). Chitosan can disrupt the physiological activities of the microorganisms (Zheng and Zhu 2003). Sudarshan et al. (1992) postulated that low Mw chitosan break/penetrate the cell wall of bacteria more easily and interact with DNA. The influence of the polycationic nature of chitosan and its effect on permeability and antibacterial activity was reported by Hudson (Lim and Hudson 2004). Leakage of intracellular components is the result of the interactions between chitosan and cell membrane leading to the death of microorganisms (Sudarshan; et al. 1992; Shahidi and Abuzaytoun 2005). DD and the associated positive charge densities of chitosan may influence the antimicrobial activity (Kong et al. 2010). Younes and Rinaudo (2015) reported a similar kind of property of chitosan related to the pH of the chitosan solution. This indicates that penetration of chitosan to the bacterial cell wall occurs at lower pH due to the increase in positive ionic charge of chitosan. Low DA, low Mw, and low pH offered higher antibacterial efficiency (Younes and Rinaudo 2015).

Many authors have recently researched the antioxidant activities of chitosan and COSs influenced by their varying Mws (Anraku et al. 2011; Chang et al. 2018). Significantly higher antioxidant property for lower Mw chitosan was reported by Anraku et al. (2011) and Chang et al., (2018). Remarkable scavenging activity against 2,2-diphenyl-i-picrylhydrazyl (DPPH) radicals and better ferrous ion chelating activity were observed for low Mw chitosan (12 kDa) than medium or high Mw chitosan (95 and 318 kDa, respectively) (Chang et al. 2018). Low Mw chitosans capability to scavenge hydroxyl radicals and reduce the cupric acid ions and its resultant superior antioxidant properties were reported by Tomida et al. (2009).

6 Influence of Isolation Conditions on the Properties of COSs

COSs have attracted huge interest among the scientific community in recent years due to their attractive functional properties (Kerch 2015). Major applications of COSs and their derivatives include their uses in pharmaceutical (Zhang et al. 2010), nutraceutical (Swiatkiewicz et al. 2015), biomedical (Naveed et al. 2019), and cosmaceutical fields (Lodhi et al. 2014). Their physiological properties include antioxidant (Yang et al. 2017), antifungal (Rahman et al. 2015), antibacterial (Liu et al. 2019; Sun et al. 2018), anti-inflammatory (Azuma et al. 2015), anti-diabetic (Lee et al. 2003), anti-cholesterolemia (Swiatkiewicz et al. 2015), antitumor, anti-obesity, anti-Alzheimer's disease (Naveed et al. 2019), calcium absorption (Areco et al. 2015), and immune stimulating activities (Lee et al. 2003). Others encompass their drug/gene delivery (Manivasagan et al. 2017), tissue engineering (Nitta et al. 2003), and plant immune-boosting functions. They can also be used as flocculants for water treatment and recovery of biomolecules such as proteins from their colloidal suspensions (Wan Ngah et al. 2011; Naveed et al. 2019).

7 Influence of Enzymatic Hydrolysis on COS

Wang et al. (2012) reported the successful production of chitooligosaccharides by the enzymatic hydrolysis of chitin or chitosan by chitinase. The enzyme could be produced crude enzyme-containing chitinases by submerged fermentation of *Lecanicillium fungicola* in presence of α-Chitin (90% of solubility, 55% DD, and MW of 343.5 kDa) or ß-Chitin (5 mg/ml). The authors added 0.02% NaN_3 as an antimicrobial agent and 5 mg/ml of crude chitinase in 50 mM of citrate–phosphate buffer into the substrate used (Ramirez et al. 2006). Wang et al. (2012) prepared crude extracellular chitinases from *Bacillus cereus* TKU027 (I and II) using chitosan having 60% deacetylation (0.5%, w/v in 50 mM phosphate buffer) as substrate using optimum reaction conditions, namely pH 5 and 40 °C. The authors separated the COSs by filtration of the reaction medium. The rate of reaction was significantly higher with α-chitin compared to ß form. The use of partially deacetylated chitin yielded up to 16-fold COSs compared with that obtained with crystalline α-chitin. The preparation had a DP in the 4–9 range (Wang et al. 2012).

Mengíbar et al. (2013) examined the relationship between the physicochemical characteristics, average Mw, DA, DP, and especially sequence composition of COSs on antioxidant properties. These oligosaccharides were obtained by enzymatic depolymerization with chitosanase and lysozyme using a specific chitosan and its reacetylated product. The COS fraction below 5 kDa obtained from chitosanase depolymerization showed the highest capacity to scavenge DPPH radicals and to reduce Fe^{3+}. A correlation was found between the relative amount of molecules with a given ratio of acetylated versus deacetylated units within the COS and their antioxidant activity,

A digestive chitosanase from blue crab (*Portunus segnis*) having optimum activity at pH 4.0 and 60 °C was applied to produce COSs having varying M_w and DA. The hydrolysis process at 40 °C for 24 h allowed preparation of a low Mw water-soluble chitosan depolymerization product (CDP) ($H_{24\,h}$, <4.4 kDa) with DP up to 6 and a high Mw CDP having 142.19 kDa. Interestingly, as compared to the native chitosan, CDP, especially low M_w derivatives exhibited potent antioxidant activities, while high M_w derivatives, especially $C_{24\,h}$, markedly inhibited the growth of all tested bacteria and fungi.

Since Chitin is an organic polymer with modified glucose monomers, cellulase, which cleaves ß-1,4 glycosidic linkages, can also be employed to hydrolyze chitin (Laokuldilok et al. 2017). Li et al. (2020) treated chitosan (DD 94.3%) (one kilogram) suspended in glacial acetic acid (0.315 L), deionized water (9.45 L), and hydrogen peroxide (30% w/w, 0.420 L) at 60 °C. The pH of the mixture was adjusted with NaOH solution to ~5.6. Adding cellulase solution (500 mL, 378.4 g/L), the enzymatic hydrolysis was performed at 50 ± 10 °C for ~10 h, 2.3. After 30 min, catalase (21 g) was added to terminate the reaction. Membrane separation, spray drying, centrifugation, and ultrafiltration gave chitosan hydrolysate. The preparation had COSs ranging in Mw from 1.06 to 41.2. Laokuldilok et al. (2017) examined lysozyme, cellulose, and papain separately on two chitosans with DD of 80 and 90 at pH 4, for a period up to 16 h. Chitosan solutions (1%, w/v) were dissolved in 0.1 mol/l sodium acetate buffer, pH 4.0. Enzymatic hydrolysis was done at 40, 30, and 37 °C for papain, lysozyme, and cellulase, respectively. After hydrolysis, the reaction was stopped by heating in boiling water for ten minutes. The resulting materials were centrifuged at 6,240 *g* for 15 min. Papain showed the highest Mw reduction of chitosan DD. After 16 h of hydrolysis with papain, the average Mw was 4.3 kDa. Mengíbar et al. (2013) tried lysozyme treatment of chitosan for six days at pH 4.5 and 37 °C. The enzyme, however, was comparatively less efficient than chitosanase.

Chemical, physical, and enzymatic methods along with modern genetic manipulation techniques have been widely explored for the production of COSs having required degree of polymerization for specific functional activities. However, no method has been found sufficiently effective in the production of defined COS. Enzymatic processes supplemented with green production methods are emerging for the production of COS, some of which have been discussed in this article. These processes are known to cause minimum environmental damage.

8 Conclusion

Chitin is the principal raw material for the manufacture of the commercially important polymer chitosan and its hydrolyzed products, chitosan oligomers (COSs). The extraction conditions have a profound effect on the physicochemical and functional properties of chitosan. Properties of chitosan including molecular weight (Mw), degree of deacetylation (DD), and their structural characteristics significantly influence its biological activities. Degradation of the linear chitosan significantly affects the Mw and other physical properties such as viscosity, and solubility and also the biological properties of the COSs. Chemical extraction methods for COSs have limitations. Therefore, replacing these methods with enzymatic hydrolysis, preferably by chitosanases, have scope for industrial production of COSs. Refining the enzymatic methods through innovations including combination of green methods are the likely future directions for the production of biologically active COS for varied applications.

References

Aam BB, Heggset EB, Norberg AL, Sørlie M, Vårum KM, Eijsink VG (2010) Production of chitooligosaccharides and their potential applications in medicine. Mar Drugs 8(5):1482–517. https://doi.org/10.3390/md8051482

Abdel-Rahman RM, Hrdina R, Abdel-Mohsen AM, Fouda MM, Soliman AY, Mohamed FK, Mohsin K, Pinto TD (2015) Chitin and chitosan from Brazilian Atlantic Coast: isolation, characterization and antibacterial activity. Int J Biol Macromol 80:107–120. https://doi.org/10.1016/j.ijbiomac.2015.06.027

Abdou ES, Nagy KSA, Elsabee MZ (2008) Extraction and characterization of chitin and chitosan from local sources. Biores Technol 99(5):1359–1367. https://doi.org/10.1016/j.biortech.2007.01.051

Ahmed F, Soliman FM, Adly MA, Soliman HAM, El-Matbouli M, Saleh M (2020) In vitro assessment of the antimicrobial efficacy of chitosan nanoparticles against major fish pathogens and their cytotoxicity to fish cell lines. J Fish Dis 43:1049–1063. https://doi.org/10.1111/jfd.13212

Al Sagheer FA, Al-Sughayer MA, Muslim S, Elsabee MZ (2009) Extraction and characterization of chitin and chitosan from marine sources in Arabian Gulf. Carbohydr Polym 77(2):410–419. https://doi.org/10.1016/j.carbpol.2009.01.032

Alishahi A, Aïder M (2012) Applications of chitosan in the seafood industry and aquaculture: a review. Food Bioprocess Technol 5:817–830. https://doi.org/10.1007/s11947-011-0664-x

Amin LP, Gogate PR, Burgess AE, Bremner DH (2010) Optimization of a hydrodynamic cavitation reactor using salicylic acid dosimetry. Chem Eng J 156:165–169. https://doi.org/10.1016/j.cej.2009.09.043

Anraku M, Fujii T, Kondo Y, Kojima E, Hata T, Tabuchi N, Tsuchiya D, Goromaru T, Tsutsumi H, Kadowaki D, Maruyama T, Otagiri M, Tomida H (2011) Antioxidant properties of high molecular weight dietary chitosan in vitro and in vivo. Carbohyd Polym 83:501–505. https://doi.org/10.1016/j.carbpol.2010.08.009

Areco V, Rivoira M, Rodriguez V, Marchionatti A, Carpentieri A, Tolosa de Talamoni N (2015) Dietary and pharmacological compounds altering intestinal calcium absorption in humans and animals. Nutr Res Rev 28(2):83–99. https://doi.org/10.1017/S0954422415000050

Avelelas F, Horta A, Pinto LFV, Cotrim Marques S, Marques Nunes P, Pedrosa R, Leandro SM (2019) Antifungal and antioxidant properties of chitosan polymers obtained from nontraditional *Polybius henslowii* sources. Mar Drugs 17(4):239. https://doi.org/10.3390/md17040239
Azuma K, Izumi R, Osaki T, Ifuku S, Morimoto M, Saimoto H, Minami S, Okamoto Y (2015) Chitin, chitosan, and its derivatives for wound healing: old and new materials. J Funct Biomater 6(1):104–142. https://doi.org/10.3390/jfb6010104
Bajaj M, Winter J, Gallert C (2011) Effect of deproteination and deacetylation conditions on viscosity of chitin and chitosan extracted from Crangon crangon shrimp waste. Biochem Eng J 56(1–2):51–62. https://doi.org/10.1016/j.bej.2011.05.006
Basa S, Nampally M, Honorato T, Das SN, Podile AR, El Gueddari NE, Moerschbacher BM (2020) The pattern of acetylation defines the priming activity of chitosan tetramers. J Am Chem Soc 142:1975–1986. https://doi.org/10.1021/jacs.9b11466
Baskar D, Sampath Kumar TS (2009) Effect of deacetylation time on the preparation, properties and swelling behavior of chitosan films. Carbohydr Polym 78(4):767–772. https://doi.org/10.1016/j.carbpol.2009.06.013
Baxter S, Zivanovic S, Weiss J (2005) Molecular weight and degree of acetylation of high-intensity ultrasonicated chitosan. Food Hydrocolloids 19:821–830. https://doi.org/10.1016/j.foodhyd.2004.11.002
Bhardwaj S, Bhardwaj NK, Negi YS (2020) Effect of degree of deacetylation of chitosan on its performance as surface application chemical for paper-based packaging. Cellulose 27:5337–5352. https://doi.org/10.1007/s10570-020-03134-5
Cabrera JC, Cutsem PV (2005) Preparation of chitooligosaccharides with degree of polymerization higher than 6 by acid or enzymatic degradation of chitosan. Biochem Eng J 25(2):165–172. https://doi.org/10.1016/j.bej.2005.04.025
Chang SH, Wu CH, Tsai GJ (2018) Effects of chitosan molecular weight on its antioxidant and antimutagenic properties. Carbohydr Polym 1(181):1026–1032. https://doi.org/10.1016/j.carbpol.2017.11.047
Chattopadhyay D, Inamdar M (2010) Aqueous behaviour of chitosan. Int J Polym Sci. https://doi.org/10.1155/2010/939536
Cheng CY, Chang CH, Wu YJ, Li YK (2006) Exploration of glycosyl hydrolase family 75, a chitosanase from Aspergillus fumigatus. J Biol Chem 281:3137–3144. https://doi.org/10.1074/jbc.M512506200
Cho HR, Chang DS, Lee WD, Jeong ET, Lee EW (1998) Utilization of chitosan hydrolysate as a natural food preservative for fish meat paste products. Korean J Food Sci Technol 30:817–822 JAKO199803042077379.pdf (koreascience.or.kr)
Chokradjaroen C, Rujiravanit R, Watthanaphanit A, Theeramunkong S, Saito N, Yamashita K, Arakawa R (2017) Enhanced degradation of chitosan by applying plasma treatment in combination with oxidizing agents for potential use as an anticancer agent. Carbohydr Polym 167:1–11. https://doi.org/10.1016/j.carbpol.2017.03.006
Chung YC (2006) Improvement of aquaculture wastewater using chitosan of different degrees of deacetylation. Environ Technol 27(11):1199–1208. https://doi.org/10.1080/09593332708618734
Chung YC, Li YH, Chen CC (2005) Pollutant removal from aquaculture wastewater using the biopolymer chitosan at different molecular weights. J Environ Sci Health Part A Toxic/Hazard Substan Environ Eng 40(9):1775–1790. https://doi.org/10.1081/ESE-200068058
Cuong HN, Minh NC, Van Hoa N, Trung TS (2016) Preparation and characterization of high purity β-chitin from squid pens (Loligo chenisis). Int J Biol Macromol 93(A):442–447. https://doi.org/10.1016/j.ijbiomac.2016.08.085
de Alvarenga ES, de Oliveira CP, Bellato CR (2010) An approach to understanding the deacetylation degree of chitosan. Carbohydr Polym 80(4):1155–1160. https://doi.org/10.1016/j.carbpol.2010.01.037
Dima JB, Sequeiros C, Zaritky NE (2013) Optimización de la obtención de quitosano de crustáceos patagónicos (Puerto Madryn, Chubut): Desarrollo de micropartículas y evaluación de su acción

bactericida en patógenos de usual frecuencia en maricultura. Rev Asociación Argent Ing Quím 57:1–19
Doan CT, Tran TN, Nguyen VB, Tran TD, Nguyen AD, Wang S-L (2020) Bioprocessing of squid pens waste into chitosanase by paenibacillus sp. tku047 and its application in low-molecular weight chitosan oligosaccharides production. Polymers 12(5):1163. https://doi.org/10.3390/polym12051163
Dong H, Wang Y, Zhao L, Zhou J, Xia Q, Qiu Y (2015) Enzymatic preparation for DP 6–8 chitooligosaccharides. J Food Process Eng 38:336–344. https://doi.org/10.1111/jfpe.12159
Du D, Bui X (2019) Study on preparation of water-soluble chitosan with varying molecular weights and its antioxidant activity. Adv Mater Sci Eng 1–8. https://doi.org/10.1155/2019/8781013
El Knidri H, Belaabed R, Addaou A, Laajeb A, Lahsini A (2018) Extraction, chemical modification, and characterization of chitin and chitosan. Int J Biol Macromol 120:1181–1189. https://doi.org/10.1016/j.ijbiomac.2018.08.139
Elwakeel KZ (2010) Environmental application of chitosan resins for the treatment of water and wastewater: a review. J Dispersion Sci Technol 31(3):273–288. https://doi.org/10.1080/01932690903167178
Fawzya YN, Wibowo TS, Noriko N (2019) Antifungal activity of chitosan oligomer prepared using chitosanase of Aeromonas media KLU 11.16. IOP Conf Ser: Earth Environ Sci 278:012026
Gbenebor OP, Akpan EI, Adeosun SO (2017) Thermal, structural and acetylation behavior of snail and periwinkle shells chitin. Prog Biomater 6:97–111. https://doi.org/10.1007/s40204-017-0070-1
Han Z, Zeng Y, Lu H, Zhang L (2015) Determination of the degree of acetylation and the distribution of acetyl groups in chitosan by HPLC analysis of nitrous acid degraded and PMP labeled products. Carbohydr Res 2(413):75–84. https://doi.org/10.1016/j.carres.2015.03.002
Hargono H, Djaeni M (2010) Utilization of chitosan prepared from shrimp shell as fat diluents. J Coast Dev 7(1):31–37
Hu X, Tian Z, Li X, Wang S, Pei H, Sun H, Zhang Z (2020) Green, simple, and effective process for the comprehensive utilization of shrimp shell waste. ACS Omega 5(30):19227–19235. https://doi.org/10.1021/acsomega.0c02705
Ibrahim KA, El-Eswed BI, Abu-Sbeih KA, Arafat TA, Al Omari MM, Darras FH, Badwan AA (2016) Preparation of chito-oligomers by hydrolysis of chitosan in the presence of zeolite as adsorbent. Mar Drugs 14(8):43. https://doi.org/10.3390/md14080043
Il'ina AV, Varlamov VP (2007) Enzymatic depolymerization of N-succinylchitosan. Russ J Bioorg Chem 33:146–149. https://doi.org/10.1134/S1068162007010165
Jeon YJ, Kamil JY, Shahidi F (2002) Chitosan as an edible invisible film for quality preservation of herring and Atlantic cod. J Agric Food Chem 50:167–5178. https://doi.org/10.1021/jf011693l
Jia Z, Shen D (2002) Effect of reaction temperature and reaction time on the preparation of low-molecular-weight chitosan using phosphoric acid. Carbohydr Polym 49:393–396. https://doi.org/10.1016/S0144-8617(02)00026-7
Kara A, Stevens R (2002) Characterization of biscuit fired bone china body microstructure. Part I: XRD and SEM of crystalline phases. J Eur Ceram Soc 22(5):731–736. https://doi.org/10.1016/S0955-2219(01)00371-5
Kaur S, Dhillon GS (2015) Recent trends in biological extraction of chitin from marine shell wastes: a review. Crit Rev Biotechnol 35(1):44–61. https://doi.org/10.3109/07388551.2013.798256
Kaya M, Mujtaba M, Bulut E, Akyuz B, Zelencova L, Sofi K (2015) Fluctuation in physicochemical properties of chitins extracted from different body parts of honeybee. Carbohyd Polym 132:9–16. https://doi.org/10.1016/j.carbpol.2015.06.008
Kaya M, Mujtaba M, Ehrlich H, Salaberria AM, Baran T, Amemiya CT, Galli R, Akyuz L, Sargin I, Labidi J (2017) On chemistry of γ-chitin. Carbohydr Polym 15(176):177–186. https://doi.org/10.1016/j.carbpol.2017.08.076
Kerch G (2015) The potential of chitosan and its derivatives in prevention and treatment of age-related diseases. Mar Drugs 13(4):2158–2182. https://doi.org/10.3390/md13042158

Khalil-Bagy HMM, Ibtesam BFM, Abou-Zaid EAA, Sabah BM, Nashwa SMA (2020) Control of green mold disease using chitosan and its effect on orange properties during cold storage. Arch Phytopathol Plant Prot 54(11–1):570–585. https://doi.org/10.1080/03235408.2020.1847568

Kjartansson GT, Zivanovic S, Kristbergsson K, Weiss J (2006) Sonication-assisted extraction of chitin from shells of fresh water prawns (*Macrobrachium rosenbergii*). J Agric Food Chem 54:3317–3323. https://doi.org/10.1021/jf052184c

Kong M, Chen XG, Ke, Xing, Park HJ (2010) Antimicrobial properties of chitosan and mode of action: a state of the art review. Int J Food Microbiol 144(1):51–63. https://doi.org/10.1016/j.ijfoodmicro.2010.09.012

Kucukgulmez A, Celik M, Yanar Y, Sen D, Polat H, Kadak AE (2011) Physicochemical characterization of chitosan extracted from Metapenaeus stebbingi shells. Food Chem 126(3):1144–1148. https://doi.org/10.1016/j.foodchem.2010.11.148

Kumar ABV, Varadaraj MC, Lalitha RG, Tharanathan RN (2004) Low molecular weight chitosans: preparation with the aid of papain and characterization. Biophys Acta General Subjects 1670:137–146. https://doi.org/10.1016/j.bbagen.2003.11.004

Kumari S, Kumar Annamareddy SH, Abanti S, Kumar Rath P (2017) Physicochemical properties and characterization of chitosan synthesized from fish scales, crab, and shrimp shells. Int J Biol Macromol 104:1697–1705. https://doi.org/10.1016/j.ijbiomac.2017.04.119

Laokuldilok T, Potivas T, Kanha N, Surawang S, Seesuriyachan P, Wangtueai S, Phimolsiripol Y, Regenstein JM (2017) Physicochemical, antioxidant, and antimicrobial properties of chitooligosaccharides produced using three different enzyme treatments. Food Biosci 18:28–33. https://doi.org/10.1016/j.fbio.2017.03.004

Lee HW, Park YS, Choi JW, Yi SY, Shin WS (2003) Antidiabetic effects of chitosan oligosaccharides in neonatal streptozotocin-induced noninsulin-dependent diabetes mellitus in rats. Biol Pharm Bull 26:1100–1103

Li K, Xing R, Liu S, Qin Y, Li B, Wang X, Li P (2012a) Separation and scavenging superoxide radical activity of chitooligomers with degree of polymerization 6–16. Int J Biol Macromol 51(5):826–830. https://doi.org/10.1016/j.ijbiomac.2012.07.031

Li K, Xing R, Liu S, Li R, Qin Y, Meng X, Li P (2012b) Separation of chito-oligomers with several degrees of polymerization and study of their antioxidant activity. Carbohydr Polym 88(3):896–903. https://doi.org/10.1016/j.carbpol.2012.01.033

Li M, Xie R, Liu J, Gan L, Long M (2020) Preparation of chitooligosaccharide acetate salts with narrow molecular size distribution and the antioxidative activity. Polym Degrad Stab 182:109393. https://doi.org/10.1016/j.polymdegradstab.2020.109393

Liang S, Sun Y, Dai X (2018) A review of the preparation, analysis and biological functions of chitooligosaccharide. Int J Mol Sci 19:2197. https://doi.org/10.3390/ijms19082197

Lim SH, Hudson SM (2004) Synthesis and antimicrobial activity of a water-soluble chitosan derivative with a fiber-reactive group. Carbohydr Res 22;339(2):313–9. https://doi.org/10.1016/j.carres.2003.10.024

Liu F, Chang W, Chen M, Xu F, Ma J, Zhong F (2019) Tailoring physicochemical properties of chitosan films and their protective effects on meat by varying drying temperature. Carbohydr Polym 212:150–159. https://doi.org/10.1016/j.carbpol.2019.02.019

Liu Y, Xing R, Yang H, Liu S, Qin Y, Li K, Yu H, Li P (2020) Chitin extraction from shrimp (*Litopenaeus vannamei*) shells by successive two-step fermentation with Lactobacillus rhamnoides and Bacillus amyloliquefaciens. Int J Biol Macromol 148:424–433. https://doi.org/10.1016/j.ijbiomac.2020.01.124

Lodhi G, Kim Y-S, Hwang J-W, Kim S-K, Jeon Y-J, Je J-Y, Ahn C-B, Moon S-H, Jeon B-T, Park P-J (2014) Chitooligosaccharide and its derivatives: preparation and biological applications. Biomed Res Int. https://doi.org/10.1155/2014/654913

Lv M, Hu Y, Gänzle MG, Lin J, Wang C, Cai J (2016) Preparation of chitooligosaccharides from fungal waste mycelium by recombinant chitinase. Carbohydr Res 22(430):1–7. https://doi.org/10.1016/j.carres.2016.04.019

Mallakuntla MK, Vaikuntapu PR, Bhuvanachandra B, Das SN, Podile AR (2017) Transglycosylation by a chitinase from Enterobacter cloacae subsp. cloacae generates longer chitin oligosaccharides. Sci Rep 7:5113–5125. https://doi.org/10.1038/s41598-017-05140-3

Manivasagan P, Bui NQ, Bharathiraja S, Moorthy MS, Oh, Yun-Ok, Song K, Seo H, Yoon M, Oh, Junghwan (2017) Multifunctional biocompatible chitosan-polypyrrole nanocomposites as novel agents for photoacoustic imaging-guided photothermal ablation of cancer. Sci Rep 7:43593. https://doi.org/10.1038/srep43593

Mathaba M, Daramola MO (2020) Effect of chitosan's degree of deacetylation on the performance of PES membrane infused with chitosan during AMD treatment. Membranes 10(3):52. https://doi.org/10.3390/membranes10030052

Mengíbar M, Mateos-Aparicio I, Miralles B, Heras Á (2013) Influence of the physico-chemical characteristics of chito-oligosaccharides (COS) on antioxidant activity. Carbohyd Polym 97(2):776–782. https://doi.org/10.1016/j.carbpol.2013.05.035

Mourya VK, Inamdar NN, Choudhari YM (2011) Chitooligosaccharides: synthesis, characterization and applications. Polym Sci, Ser A 53:583–612. https://doi.org/10.1134/S0965545X11070066

Naveed M, Phil L, Sohail M, Hasnat M, Baig MMFA, Ihsan AU, Shumzaid M, Kakar MU, Khan TM, Akabar MD, Hussain MI, Zhou Q-G (2019) Chitosan oligosaccharide (COS): an overview. Int J Biol Macromol 129:827–843. https://doi.org/10.1016/j.ijbiomac.2019.01.192

Nitta T, Hata M, Gotoh S, Seo Y, Sasaki H, Hashimoto N, Furuse M, Tsukita S (2003) Size-selective loosening of the blood-brain barrier in claudin-5-deficient mice. J Cell Biol 161(3):653–660. https://doi.org/10.1083/jcb.200302070

No HK, Meyers SP (1995) Preparation and characterization of chitin and chitosan—a Review. J Aquat Food Prod Technol 4(2):27–52. https://doi.org/10.1300/J030v04n02_03

No HK, Park NY, Lee SH, Meyers SP (2002) Antibacterial activity of chitosans and chitosan oligomers with different molecular weights. Int J Food Microbiol 25;74(1–2):65–72. https://doi.org/10.1016/s0168-1605(01)00717-6

Oyatogun GM, Esan TA, Akpan EI, Adeosun SO, Popoola API, Imasogie BI, Soboyejo WO, Afonja AA, Ibitoye SA, Abere VD, Oyatogun AO, Oluwasegun KM, Akinwole IE, Akinluwade KJ (2020) Chitin, chitosan, marine to market. In: Gopi S, Thomas S, Pius A (eds) Handbook of chitin and chitosan. Elsevier, pp 341–381. ISBN 9780128179666. https://doi.org/10.1016/B978-0-12-817966-6.00011-X

Percot A, Viton C, Domard A (2003) Optimization of chitin extraction from shrimp shells. Biomacromol 4(1):12–18. https://doi.org/10.1021/bm025602k PMID: 12523840

Pillai CKS, Paul W, Sharma CP (2009) Chitin and chitosan polymers: Chemistry, solubility and fiber formation. Prog Polym Sci 34(7):641–678. https://doi.org/10.1016/j.progpolymsci.2009.04.001

Rabea E, Badawy M, Stevens C, Smagghe G, Steurbaut W (2003) Chitosan as antimicrobial agent: applications and mode of action. Biomacromol 4:1457–1465. https://doi.org/10.1021/bm034130m

Ramesh HP, Tharanathan RN (2003) Carbohydrates—the renewable raw materials of high biotechnological value. Crit Rev Biotechnol 23:149–173. https://doi.org/10.1080/713609312

Ramirez CL, Marın C, Huerta S, Revah S, Shirai K (2006) Enzymatic hydrolysis of chitin in the production of oligosaccharides using *Lecanicillium fungicola* chitinases. Process Biochem 41:1106–1110. https://doi.org/10.1016/j.procbio.2005.11.021

Rasweefali MK, Sabu S, Sunooj KV, Abhilash S, Xavier KAM (2021) Consequences of chemical deacetylation on physicochemical, structural and functional characteristics of chitosan extracted from deep-sea mud shrimp. Carbohydr Polym Technol Appl 2. https://doi.org/10.1016/j.carpta.2020.100032

Rinaudo M (2006) Chitin and chitosan: properties and applications. Prog Polym Sci 31(7):603–632. https://doi.org/10.1016/j.progpolymsci.2006.06.001

Sahariah P, Cibor D, Zielińska D, Hjálmarsdóttir MÁ, Stawski D, Másson M (2019) The effect of molecular weight on the antibacterial activity of N, N, N-trimethyl chitosan (TMC). Int J Mol Sci 20(7):1743. https://doi.org/10.3390/ijms20071743

Sánchez Á, Mengíbar M, Rivera-Rodríguez G, Moerchbacher B, Acosta N, Heras A (2017) The effect of preparation processes on the physicochemical characteristics and antibacterial activity of chitooligosaccharides. Carbohydr polym 157:251–257. https://doi.org/10.1016/j.carbpol.2016.09.055

Schmitz C, Auza LG, Koberidze D, Rasche S, Fischer R, Bortesi L (2019) Conversion of chitin to defined chitosan oligomers: current status and future prospects. Mar Drugs 17(8):452. https://doi.org/10.3390/md17080452

Senthilkumar P, Yaswant G, Kavitha S, Chandramohan E, Kowsalya G, Vijay R, Babu S, Kumar PS (2019) Preparation and characterization of hybrid chitosan-silver nanoparticles (Chi-Ag NPs); a potential antibacterial agent. Int J Biol Macromol 141. https://doi.org/10.1016/j.ijbiomac.2019.08.234

Shahidi F, Abuzaytoun R (2005) Chitin, chitosan, and co-products: chemistry, production, applications, and health effects. Adv Food Nutr Res 49:93–135. https://doi.org/10.1016/S1043-4526(05)49003-8

Shirvan AR, Shakeri M, Bashari A (2018) Recent advances in application of chitosan and its derivatives in functional finishing of textiles. In: The impact and prospects of green chemistry for textile technology. Elsevier, pp 107–133. https://doi.org/10.1016/B978-0-08-102491-1.00005-8.

Sudarshan NR, Hoover DG, Knorr D (1992) Antibacterial action of chitosan. Food Biotechnol 6(3):257–272. https://doi.org/10.1080/08905439209549838

Sun L, Sun J, Liu D, Fu, Minghai, Xi, Yang, Guo Y (2018) The preservative effects of chitosan film incorporated with thinned young apple polyphenols on the quality of grass carp (*Ctenopharyngodon idellus*) fillets during cold storage: correlation between the preservative effects and the active properties of the film. Food Packag Shelf Life 17:1–10. https://doi.org/10.1016/j.fpsl.2018.04.006

Swiatkiewicz S, Swiatkiewicz M, Arczewska-Wlosek A, Jozefiak D (2015) Chitosan and its oligosaccharide derivatives (chito-oligosaccharides) as feed supplements in poultry and swine nutrition. J Anim Physiol Anim Nutr 99:1–12. https://doi.org/10.1111/jpn.12222

Thadathil N, Velappan SP (2014) Recent developments in chitosanase research and its biotechnological applications: a review. Food Chem 1(150):392–399. https://doi.org/10.1016/j.foodchem.2013.10.083

Tian Y, Liang K, Wang X, Ji Y (2017) Fabrication of nanocomposite bioelastomer porous scaffold based on chitin nanocrystal supported emulsion-freeze-casting. ACS Sustain Chem Eng 5:3305–3313. https://doi.org/10.1021/acssuschemeng.6b03146

Tishchenko G, Šimůnek J, Brus J, Netopilík M, Pekárek M, Walterová Z, Koppová I, Lenfeld J (2011) Low-molecular-weight chitosans: preparation and characterization. Carbohydr Polym 86(2):1077–1081. https://doi.org/10.1016/j.carbpol.2011.04.073

Tolaimate A, Desbrieres J, Rhazi M, Alagui A (2003) Contribution to the preparation of chitins and chitosans with controlled physico-chemical properties. Polymer 44(26):7939–7952. https://doi.org/10.1016/j.polymer.2003.10.025

Tolesa LD, Gupta B, Lee M (2019) Chitin and chitosan production from shrimp shells using ammonium-based ionic liquids. Int J Biol Macromol 130:818–826. https://doi.org/10.1016/j.ijbiomac.2019.03.018

Tomida H, Fujii T, Furutani N, Michihara A, Yasufuku T, Akasaki K, Maruyama T, Otagiri M, Gebicki JM, Anraku M (2009) Antioxidant properties of some different molecular weight chitosans. Carbohydr Res 8;344(13):1690–1696. https://doi.org/10.1016/j.carres.2009.05.006.

Tømmeraas K, Köping-Höggård M, Vårum KM, Christensen BE, Artursson P, Smidsrød O (2002) Preparation and characterisation of chitosans with oligosaccharide branches. Carbohydr Res 337:2455–2462. https://doi.org/10.1016/S0008-6215(02)00334-8

Trache D (2018) Nanocellulose as a promising sustainable material for biomedical applications. AIMS Mater Sci 5(2):201–205. https://doi.org/10.3934/matersci.2018.2.201

Trache D, Tarchoun AF, Derradji M, Hamidon TS, Masruchin N, Brosse N, Hussin MH (2020) Nanocellulose: from fundamentals to advanced applications. Front Chem 8:392. https://doi.org/10.3389/fchem.2020.00392

Trombotto S, Ladavière C, Delolme F, Domard A (2008) Chemical preparation and structural characterization of a homogeneous series of chitin/chitosan oligomers. Biomacromol 9(7):1731–1738. https://doi.org/10.1021/bm800157x

Tsao CT, Chang CH, Lin YY, Wu, Ming Fung, Han JL, Hsieh KH (2011) Kinetic study of acid depolymerization of chitosan and effects of low molecular weight chitosan on erythrocyte rouleaux formation. Carbohydr Res 346(1):94–102. https://doi.org/10.1016/j.carres.2010.10.010

Venugopal V (2009) Marine products for healthcare: functional and bioactive nutraceutical compounds from the ocean. Taylor & Francis, Fl, p 514

Venugopal V (2011) Marine polysaccharides: food applications. Taylor & Francis, Fl, p 396

Wan Ngah WS, Teong LC, Hanafiah MAKM (2011) Adsorption of dyes and heavy metal ions by chitosan composites: a review. Carbohydr Polym 83(4):1446–1456. https://doi.org/10.1016/j.carbpol.2010.11.004

Wang SL, Liu CP, Liang TW (2012) Fermented and enzymatic production of chitin/chitosan oligosaccharides by extracellular chitinases from Bacillus cereus TKU027. Carbohydr Polym 90(3):1305–1313. https://doi.org/10.1016/j.carbpol.2012.06.077

Wu S (2012) Preparation of chitooligosaccharides from Clanis bilineata larvae skin and their antibacterial activity. Int J Biol Macromol 51(5):1147–1150. https://doi.org/10.1016/j.ijbiomac.2012.08.035

Wu Y, Huang Y, Zhou Y, Ren X, Yang F (2014) Degradation of chitosan by swirling cavitation. Innov Food Sci Emerg Technol 23:188–193. https://doi.org/10.1016/j.ifset.2014.02.001

Xia Z, Chen J, Wu S (2013a) Hypolipidemic activity of the chitooligosaccharides from Clanis bilineata (Lepidoptera), an edible insect. Int J Biol Macromol 59:96–98. https://doi.org/10.1016/S0141-0229(02)00138-2

Xia Z, Wu S, Chen J (2013b) Preparation of water soluble chitosan by hydrolysis using hydrogen peroxide. Int J Biol Macromol 59:242–245. https://doi.org/10.1016/j.ijbiomac.2013.04.034

Xie H, Jia Z, Huang J, Zhang C (2011) Preparation of low molecular weight chitosan by complex enzymes hydrolysis. Int J Chem 3(2). https://doi.org/10.5539/ijc.v3n2p180

Xing R, Liu S, Yu, Huahua, Guo Z, Wang P, Li C, Li Z, Li P (2005) Salt-assisted acid hydrolysis of chitosan to oligomers under microwave irradiation. Carbohydr Res 340(13):2150–2153. https://doi.org/10.1016/j.carres.2005.06.028

Xing R, Liu Y, Li K, Yu, Huahua, Liu S, Yang Y, Chen X, Li P (2017) Monomer composition of chitooligosaccharides obtained by different degradation methods and their effects on immunomodulatory activities. Carbohydr Polym 157:1288–1297. https://doi.org/10.1016/j.carbpol.2016.11.001

Yang F, Luan B, Sun Z, Yang C, Yu Z, Li X (2017) Application of chitooligosaccharides as antioxidants in beer to improve the flavour stability by protecting against beer staling during storage. Biotechnol Lett 39:305–310. https://doi.org/10.1007/s10529-016-2248-3

Younes I, Rinaudo M (2015) Chitin and chitosan preparation from marine sources. Structure, properties and applications. Mar Drugs 13(3):1133–1174. https://doi.org/10.3390/md13031133

Younes I, Sellimi S, Rinaudo M, Jellouli K, Nasri M (2014) Influence of acetylation degree and molecular weight of homogeneous chitosans on antibacterial and antifungal activities. Int J Food Microbiol 18(185):57–63. https://doi.org/10.1016/j.ijfoodmicro.2014.04.029

Younes I, Hajji S, Rinaudo M, Chaabouni M, Jellouli K, Nasri M (2016) Optimization of proteins and minerals removal from shrimp shells to produce highly acetylated chitin. Int J Biol Macromol 84:246–253. https://doi.org/10.1016/j.ijbiomac.2015.08.034

Zainol I, Akil HM, Mastor A (2009) Effect of γ-irradiation on the physical and mechanical properties of chitosan powder. Mater Sci Eng C 29:292–297. https://doi.org/10.1016/j.msec.2008.06.026

Zhang X, Zhang J, Zhu KY (2010) Chitosan/double-stranded RNA nanoparticle-mediated RNA interference to silence chitin synthase genes through larval feeding in the African malaria mosquito (*Anopheles gambiae*). Insect Mol Biol 19(5):683–693. https://doi.org/10.1111/j.1365-2583.2010.01029.x

Zheng L-Y, Zhu J-F (2003) Study on antimicrobial activity of chitosan with different molecular weights. Carbohydr Polym 54(4):527–530. https://doi.org/10.1016/j.carbpol.2003.07.009

Zhou Y, Li S, Li D, Wang S, Zhao W, Lv Z, Li X, Li H, Han Y (2020) Enzymatic preparation of chitooligosaccharides and their anti-obesity application. Biosci Biotechnol Biochem 84(7):1460–1466. https://doi.org/10.1080/09168451.2020.1744110

The manufacturer's authorised representative in the EU is Springer Nature Customer Service Centre GmbH, Europaplatz 3, 69115 Heidelberg, Germany. If you have any concerns regarding our products, please contact ProductSafety@springernature.com

Printed and bound by CPI Group (UK) Ltd, Croydon, CR0 4YY
30/09/2026
02225228-0013